ICE Themes Self-Compacting Concrete

The *ICE Themes* series showcases cutting-edge research and practical guidance in all branches of civil engineering. Each title focuses on a key issue or challenge in civil engineering and includes research from the industry's finest thinkers and influencers published through the ICE Publishing programme. Themes in the series include climate change resilience, advances in construction management, developments in renewable energy, and innovations in construction materials plus many more.

Also available in the *ICE Themes* series:

ICE Themes Flood Resilience (2018)
Edited by Manuela Escarameia and Andrew Tagg. ISBN 978-0-7277-6393-8
ICE Themes Low Carbon Concrete (2019)
Edited by Ravindra K Dhir OBE and Kevin Paine. ISBN 978-0-7277-6459-1
ICE Themes Recycled Aggregates: Use in Concrete (2019)
Edited by Ravindra K Dhir OBE and Chao Qun Lye. ISBN 978-0-7277-6463-8
ICE Themes Renewable Energy: Geothermal Energy, Heat Exchange Systems and Energy Piles (2018)
Edited by William Craig and Kevin Gavin. ISBN 978-0-7277-6398-3
ICE Themes Renewable Energy: Wind Turbine Foundations (2018)
Edited by Kenneth Gavin and William Craig. ISBN 978-0-7277-6396-9
ICE Themes Smart Concrete (2019)
Edited by Ravindra K Dhir OBE and Kevin Paine. ISBN 978-0-7277-6457-7

ice
Institution of Civil Engineers
Publishing

ICE Themes
Self-Compacting Concrete

Edited by Ravindra K Dhir OBE and Chao Qun Lye

Published by ICE Publishing, One Great George Street, Westminster, London SW1P 3AA.

Full details of ICE Publishing representatives and distributors can be found at: www.icebookshop.com/bookshop_contact.asp

Other titles by ICE Publishing:

Steel-Concrete Composite Bridges, Second edition.
David Collings. ISBN 978-0-7277-5810-1
Deterioration and Maintenance of Pavements
Derek Pearson. ISBN 978-0-7277-4114-1
Finite-Element Design of Concrete Structures, Second edition
Guenter A. Rombach. ISBN 978-0-7277-4189-9

www.icebookshop.com

A catalogue record for this book is available from the British Library

ISBN 978-0-7277-6461-4

ICE Publishing is a division of Thomas Telford Ltd, a wholly-owned subsidiary of the Institution of Civil Engineers (ICE).

Cover photo: Morpheus hotel, luxury accommodation in Macau. tuiimages.com/ Alamy stock photo

Commissioning Editor: James Hobbs
Production Editor: Madhubanti Bhattacharyya
Marketing Specialist: April Asta Brodie

MIX
Paper from responsible sources
FSC
www.fsc.org
FSC® C013056

Typeset by The Manila Typesetting Company
Index created by The Manila Typesetting Company
Printed and bound in Great Britain by TJ International, Padstow

Contents

Foreword

Self-compacting concrete (SCC), or self-consolidating concrete as it is known in the United States of America, is a remarkable development in the field of concrete technology. However, though its use is growing, it is still not in the mainstream of the concrete construction industry. As the name implies, SCC is a concrete mixture that can fill the spaces by its own and achieve full compaction without needing further work to expel the entrapped air. The common features of SCC include highly flowable, total filling ability without voids, flowing through tight openings without blocking and remaining homogeneous in the fresh state.

Self-compacting concrete was first developed properly in Japan by **H Okamura** in the mid-1980s, though concrete of close to similar nature has been around in Europe earlier. Thereafter, it attracted a broadly based research, covering from devising various test methods for assessment of its fresh properties to producing a reliable and consistent product for use in construction. Today, standards and specifications relevant to SCC have been established in various countries, such as EN 206:2013+A1 (2016) in Europe, which provides guidelines for the requirements of SCC in the fresh state, and its fresh properties can be evaluated in accordance with the EN 12350 series. Similar test methods are also available in the United States as ASTM standards.

The constituent materials of this concrete are similar to those normally used for conventional concrete but differ greatly in terms of the mix proportions. Additionally, polycarboxylate-based (PCE based) high range superplasticisers, together with viscosity modifying admixtures (VMAs), are normally used in SCC in order to achieve the desired fresh properties. It is widely known that SCC offers many benefits, including perfect solution for congested reinforcement casting, improvement in productivity, less labour dependent and safer working environment. When designed properly, it can be a sustainable solution for long-life durable concrete. Despite that, the actual use of SCC in the construction industry is not well-received in general, being more popular with precast industry than the ready-mix concrete producers. Indeed, this poses an important question, with a challenge of how to promote a greater use of SCC and establish its position in the construction industry, by overcoming the barriers that are holding back the wider acceptance of this concrete - sooner.

The first task undertaken in developing this ICE Themes book was to search the ICE virtual library for publications on SCC, covering its full spectrum, from laboratory research to actual site practice. The sourced papers then went through a series of data sorting and parking, using a newly developed *analytical systemisation method* (**Dhir *et al.***, 2019), where (i) the background information of the papers such as researchers involved, their respective countries and institutions, type of publications, as well as (ii) the content of the papers such as the types of materials used and the research subjects, applications, case studies and standards and specifications, were sorted in a prescribed manner.

In total, the ICE Virtual Library had a collection of 210 papers on self-compacting concrete, published over a period of 54 years, 1966–2019, though it was not until 1999 when the publications on the subject started to flow. The first paper published in 1966 was by **D J Palmer and G R Holland**, presented at the symposium on Large Bored Piles organised by Institute of Civil Engineers and Reinforced Concrete Association, which was held in London. The paper was not fully dedicated to SCC, but it briefly mentioned its use in in-situ piling work. Thereafter, no relevant SCC work was published until 1999, when 10 papers appearing in the proceedings of the international congress entitled 'Creating with Concrete' organised by the Concrete Technology Unit at the University of Dundee, UK. After that the rate of published papers on SCC within ICE publishing has been around 10 per annum, though showing a sign of slight increase since 2010.

Based on the first authors, 40 countries have actively been publishing their work on SCC with the ICE publishing, with 19 from Europe (UK ranking top), followed by 16 in Asia (India ranking top), North America and South America with two each and one from Oceania. Overall, the countries that have been actively publishing in this area in the ICE publications, with at least 10 papers, in descending order, are: India (31 papers), the United Kingdom (29 papers), China (16 papers), the United States of America (13 papers), Iran (12 papers), Italy (12 papers), Canada (12 papers) and Australia (10 papers).

Amongst the researchers involved in the area, **F Aslani** from the University of Western Australia, with a total of seven, has the highest number of papers published. This is followed by **A A A Hassan** from Memorial University of Newfoundland, Canada, with six papers. The researchers with five SCC publications include **B L Karihaloo** from Cardiff University, UK, and **M Sri Rama Chand and P Rathish Kumar** both from National Institute of Technology Warangal, India. In terms of the institutions, the top five contributors have been National Institute of Technology Warangal, India (8 papers), Memorial University of Newfoundland, Canada (6 papers), Cardiff University, UK (5 papers), Islamic Azad University, Iran (4 papers) and University of Technology Sydney, Australia (4 papers).

The SCC work published by ICE Publishing has been diverse and inspiring, involving a wide range of materials used in designing SCC mixes, covering from the common cements as specified in EN 197-1 (2011) to ashes and nanoparticles, various types of aggregate specified in EN 12620: 2002 + A1 (2008) as natural aggregates, lightweight aggregates, recycled and secondary aggregates, a variety of fibres, and an array of admixtures with different performance and enhancement. The performance of SCC mixes, especially the fresh properties in terms of flowability, passing ability, filling ability and segregation resistance have normally been investigated, followed by the strength properties. However, little work has been done on the creep and abrasion properties of SCC, and these hardened properties should be given more attention due to the high cement paste content in these mixes.

This book, consisting of 20 papers, selected from a total of 221 papers found in the ICE virtual library, has been divided into six sections, covering a wide spectrum of topics, including the history and development, designing methods, constituent materials, fresh and hardened properties, structural properties and practical use of SCC.

The book begins with its first section consisting of two papers discussing briefly the history and introduction to SCC, as well as some glimpses of its development in Canada and some countries in Asia and Europe. An overview is also provided for the use of (i) various test methods in the fresh state such, as V-funnel and box tests, and (ii) the rational mix design methods in working with SCC.

The following section provides some useful concepts and guidelines in designing SCC, based on different design elements. These include the design methods (i) incorporating fly ash up to 70% as a replacement for Portland cement, for various strength grades up to 90 MPa, and (ii) designing normal and high strength SCC mixes for strength up to 160 MPa, made with or without steel fibres. Apart from these, the fine/total aggregate ratio of a SCC mix is shown to affect its fresh properties and the optimum dosage of superplasticiser is found to be dependent on both the water/cement and fine/total aggregate ratios of the mix. The paste/voids volume ratio of SCC has a significant effect on its slump flow and strength development, but not so on the shrinkage and surface resistivity properties.

The third section focuses on the SCC made with special and unconventional constituent materials. The first paper is a review paper on the subject of nanoparticles which may be considered as new material in the area of concrete technology. The review covers the effect of various types of nanoparticles on the fresh and hardened properties of SCC such as mixing procedure, mix proportions, heat of hydration, workability, setting times, strength and durability. The next two papers are related to research on SCC made with lightweight aggregates, investigating the fresh properties and hardened properties of concrete at elevated temperatures respectively. The last paper compares the performance of SCC made with and without steel and polypropylene fibres and their combinations, in terms of fresh properties, as well as their strength and deformation properties.

The next section covers a selection of important properties of SCC in both the fresh and hardened states. In terms of fresh concrete, the selected property is the stability for which a novel image processing method has been proposed to assess the segregation of concrete. In terms of hardened concrete, the deformation properties are selected, presenting models for estimating the creep and shrinkage of normal strength and high strength SCC. These models were developed from the work undertaken by the Japanese Society of Civil Engineers. The microstructure, permeation and durability in the form of chloride ingress of SCC made with metakaolin are presented. The addition of self-curing compound to SCC has shown to result in a denser microstructure, thereby reducing the transportation of water and chloride ions into the system.

Section 5 deals with the study for the use of SCC in structural applications. The first paper presents a study on the bond between reinforcing bars and concrete made with recycled concrete aggregates, in which models to estimate the bond strength and stress-slip relationship are proposed. The study on the reinforced SCC frames has drawn a conclusion that the lower coarse aggregate content in SCC may reduce the aggregate interlocking effect to the total shear resistance, though further research is needed to substantiate such findings. The next two studies investigate the structural behaviour of SCC beams made with or without steel fibres and crumb rubber in terms of its shear capacity and first cracking moment, as well as its fatigue lives.

The book ends with a section consisting of two papers dealing for the practical issues relating to the use of SCC on construction sites. The first research studies the hydrostatic pressure induced by SCC on formwork, which is an important factor to consider for contractors in designing an appropriate formwork when using such a concrete. The next research is about an extensive interview programme undertaken with UK contractors for their view on the use of SCC, which has generated some interesting viewpoints that need to be addressed.

This book aims to provide readers, from different backgrounds, with both academically rigorous and practically relevant insights into valuable research findings from the work published in the 20 papers selected from 210 sourced from the ICE virtual library. It is hoped that the work presented in this book will help the readers to recognise the size and scale of progress made in developing SCC and its use in practice. It is believed that research and development work in the field of SCC will continue to grow in further refining the product and its role in practice, with the aim to bring SCC into the category of a viable material and preferred engineering solution in the construction industry.

Finally, the editors are grateful to the contributing authors for developing their publications in the first place and hope that the readers will find this book a useful reference source in their efforts to further the frontiers of knowledge in the field of self-compacting concrete.

Ravindra K Dhir OBE and Chao Qun Lye

REFERENCES

1. Dhir R K, de Brito J, Silva R V and Lye C Q, 2019. Sustainable Construction Materials: Recycled Aggregates. Woodhead Publishing, Elsevier, Cambridge, UK, 635 pp.
2. EN 197-1, 2011. Cement. Composition, specifications and conformity criteria for common cements. Comité Européen de Normalisation (CEN), Brussels, Belgium.
3. EN 12350 (series), 2009. Testing of fresh concrete. Comité Européen de Normalisation (CEN), Brussels, Belgium.
4. EN 206:2013+A1, 2016. Concrete. Specification, performance, production and conformity. Comité Européen de Normalisation (CEN), Brussels, Belgium.

About the editors

Professor Ravindra K Dhir OBE, an Emeritus Professor of Concrete Technology at the University of Dundee in the UK, also holds the positions of adjunct professor at Trinity College Dublin Ireland and honorary professor at the University of Birmingham UK. He remains actively engaged with his teaching, research and publishing work, as well his consultancy practice, *Applying Concrete Knowledge*. As a founding Director of Concrete Technology Unit, at Dundee (1988–2008), his hallmark was the development of a world renowned Centre of Excellence with extensive research facilities, as well as working very closely with industry in research and its dissemination into practice. He won numerous recognitions, awards and honours including the Order of the British Empire (OBE) from the Queen, and Honorary Fellowships of the Institute of Concrete Technology, UK and Indian Concrete Institute. He was Chairman of Concrete Society, Scotland (1986–87) and President of Concrete Society, UK (2009–2010). He serves on the Editorial Board of the *Magazine of Concrete Research* and is the Chairman of UK-India Education Research Initiative Concrete Congress.

Dr Chao Qun Lye is a Senior Technical Manager with G&W Group, Singapore, where he is responsible for R&D and Quality Control of concrete products. Having previously graduated from the National University of Singapore with a bachelors degree in civil engineering, he recently obtained his degree of Doctor of Philosophy from the University of Birmingham UK, for his research into the use of recycled and secondary aggregates in concrete and their effects on the load-dependent and load-independent deformation properties of concrete, such as elastic modulus, creep and shrinkage. He holds a strong interest in sustainability and innovation, applied to the use of cement additions such as fly ash, ground granulated blast-furnace slag and limestone as well as the use of recycled and secondary materials in concrete, geotechnics and road pavement applications.

Section 1
History, development and overview

ICE Publishing

Dhir and Lye
ISBN 978-0-7277-6461-4
https://doi.org/10.1680/icetscc.64614.003

Chapter 1
Self-compacting concrete

H. Okamura
Kochi University of Technology, Kochi, Japan

K. Ozawa
University of Tokyo, Tokyo, Japan

M. Ouchi
Kochi University of Technology, Kochi, Japan

Self-compacting concrete was first developed 1988 in order to achieve durable concrete structures. Since then, various investigations have been carried out, and the concrete has been used in practical structures in Japan, mainly by large construction companies. Investigations for establishing a rational mix design method and self-compactability testing methods have been carried out from the viewpoint of making it a standard concrete. In addition to Japan, investigations have been started in many countries, and it has been applied to practical structures especially in Canada, Sweden, The Netherlands, Thailand and Taiwan. Recommendations and manuals for self-compacting concrete have also been published in Japan. In this article, the current condition of self-compacting concrete is summarized based on reports given at the International Workshop on Self-compacting Concrete, Kochi, Japan, in 1998.

Introduction

Purpose of development

For several years, beginning in 1983, the problem of the durability of concrete structures has been a major topic of interest in Japan. To make durable concrete structures, sufficient compaction by skilled workers is required. However, the gradual reduction in the number of skilled workers in Japan's construction industry has led to a similar reduction in the quality of construction work. One solution for the achievement of durable concrete structures independent of the quality of construction work is the employment of self-compacting concrete, which can be compacted into every corner of a formwork, purely by means of its own weight and without the need for vibrating compaction (Figure 1). The necessity of this type of concrete was proposed by Okamura in 1986. Studies to develop self-compacting concrete, including a fundamental study on the workability of concrete, have been carried out by Ozawa and Maekawa at the University of Tokyo.[1]

The prototype of self-compacting concrete was first made in 1988 using materials already on the market (Figure 2). The prototype performed satisfactorily with regard to drying and hardening shrinkage, heat of hydration, density after hardening, and other properties.[1,2]

Fig. 1 Reasons for the necessity of self-compacting concrete

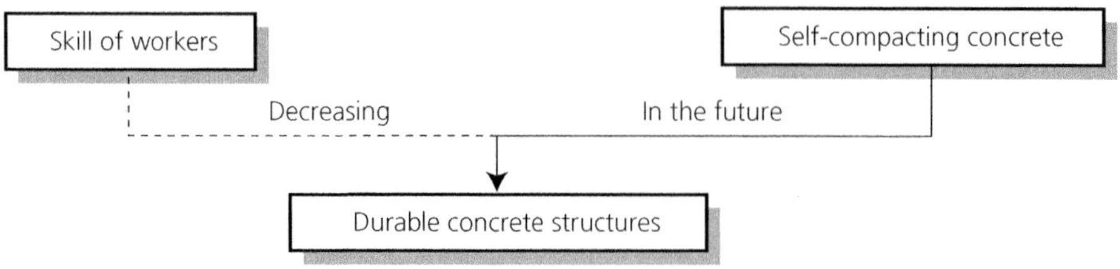

Fig. 2 (a) Prototype self-compacting concrete (1988, superplasticizer and viscosity agent used) compared with (b) conventional concrete

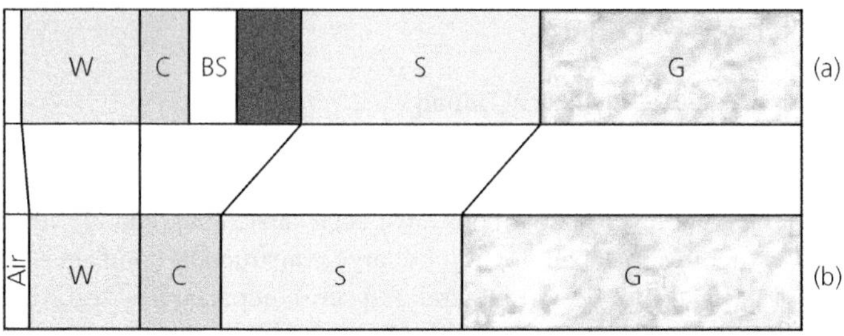

This concrete was named 'high-performance concrete', and was defined as follows at the three stages of concrete:

1. fresh – self-compactable
2. early age – avoidance of initial defects
3. hardened – protection against external factors.

At almost the same time, 'high-performance concrete' was also defined as a concrete with high durability due to a low water/cement ratio by Aitcin and coworkers.[3] Since then, the term 'highperformance concrete' has been used around the world to refer to high durability concrete. Therefore, we have changed the term for our proposed concrete to 'self-compacting high-performance concrete'.

Methods for achieving self-compactability

The method for achieving self-compactability involves not only high deformability of the paste or mortar but also resistance to segregation between coarse aggregate and mortar when the concrete flows through the confined zone of reinforcing bars. Okamura and Ozawa have employed the following methods to achieve self-compactability (Figure 3):[4]

1. limited aggregate content
2. low water/powder ratio
3. use of superplasticizer.

The frequency of collision and contact between aggregate particles can increase as the relative distance between the particles decreases and then internal stress can increase when concrete is

Fig. 3 Methods for achieving self-compactability (SP, superplasticizer; w/c, water/cement ratio)

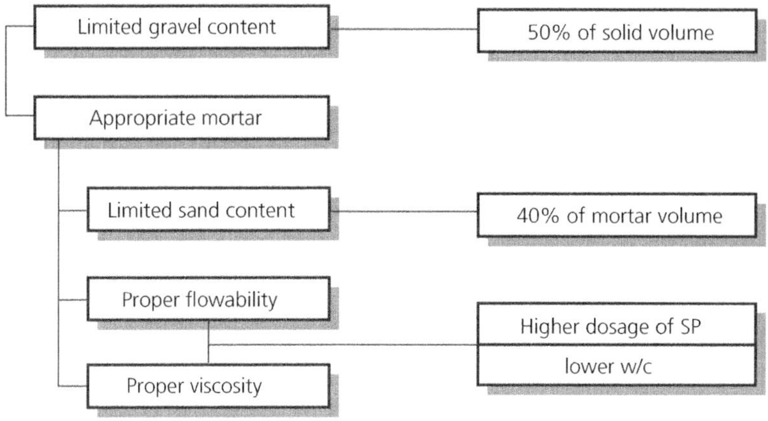

deformed, particularly near obstacles. It has been found that the energy required for flowing is consumed by the increased internal stress, resulting in blockage of aggregate particles. Limiting the coarse aggregate content, whose energy consumption is particularly intense, to a level lower than normal is effective in avoiding this kind of blockage.

A highly viscous paste is also required to avoid the blockage of coarse aggregate when concrete flows through obstacles. When concrete is deformed, paste with a high viscosity also prevents localized increases in the internal stress due to the approach of coarse aggregate particles. High deformability can be achieved only by the employment of a superplasticizer, keeping the water/powder ratio very low value.

Spread of concept of self-compacting concrete

The first paper on self-compacting concrete was presented by Ozawa at the Second East-Asia and Pacific Conference on Structural Engineering and Construction (EASEC-2) in January 1989.[1] The presentation by Ozawa at the CANMET and ACI International Conference, Istanbul, in May 1992 accelerated the spread of the concept to the world.[5]

After the ACI workshop on high-performance concrete hosted by Professor Paul Zia in Bangkok in November 1994, self-compacting concrete became a common subject of interest for researchers and engineers interested in the durability of concrete and in rational construction systems around the world.[6] In addition, the 1996 Ferguson Lecture by Okamura at the ACI Fall Convention in New Orleans stirred up interest in self-compacting concrete among researchers and engineers in North America.[7] As a result, research on self-compacting concrete started worldwide. These activities included investigations in Canada currently being carried out by Professor Aictin and coworkers. In January 1997 the RILEM committee on self-compacting concrete was formed, and a symposium was held in Stockholm in September 1999.

To put these various activities into perspective, the first international workshop on self-compacting concrete was held in Kochi, Japan, in August, 1998.[8] At the workshop, the foundation of an information exchange network on self-compacting concrete using the Internet

Web site 'International Network for Self-Compacting Concrete (SCC-Net)' was agreed. This network started in February 1999 (http://www.infra.kochi-tech.ac.jp/sccnet/).

Applications of self-compacting concrete

Applications in Japan

Current condition on application of self-compacting concrete in Japan

After the development of the prototype self-compacting concrete at the University of Tokyo, intensive research began in many places, especially in the research institutes of large construction companies. As a result, self-compacting concrete has now been used in many practical structures.[9–11] The first application of self-compacting concrete was in a building in June 1990. Self-compacting concrete was then used in the towers of a pre-stressed concrete cable-stayed bridge in 1991 (Figure 4).[12] Lightweight self-compacting concrete was used in the main girder of a cable-stayed bridge in 1992.[13] Since then, the use of self-compacting concrete in actual structures has gradually increased. Currently, the main reasons for the employment of self-compacting concrete can be summarized as follows:

1. to shorten construction period
2. to assure compaction in the structure: especially in confined zones where compaction by vibrator is difficult
3. to eliminate noise due to vibration.

The volume of self-compacting concrete used in Japan since 1990 is shown in Figure 5. The production of self-compacting concrete is currently only 0.1% of the total Japanese ready-mixed concrete production.[14] The current status of self-compacting concrete is as 'special concrete' rather than 'standard concrete'.

Fig. 4 Towers of a cable-stayed bridge using self-compacting concrete

Fig. 5 Annual production of self-compacting concrete in Japan (the production of ready-mixed concrete in Japan in 1997 was 167 200 000 m[30])

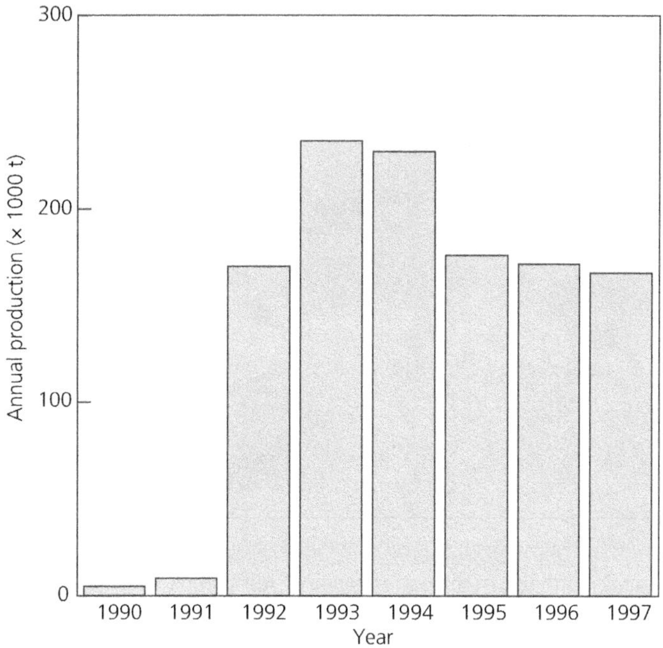

Examples of the application of self-compacting concrete are summarized below:

- bridges (anchorages, arches, beams, towers, joints)
- box culvert
- building
- concrete-filled steel column
- tunnel (lining, immersed tunnel, filling of survey tunnel)
- dam (concrete around structure)
- concrete products (blocks, culverts, walls, water tanks, slabs and segments)
- diaphragm wall
- tank (side wall, joint between side wall and slab)
- pipe roof.

Large-scale construction

Self-compacting concrete is currently being used to shorten the construction period of large-scale constructions.

The anchorages of Akashi-Kaikyo (Straits) Bridge opened in April 1998, a suspension bridge with the longest span in the world (1991 m), is a typical example (Figure 6).[15] Self-compacting concrete was used in the construction of the two anchorages of the bridge. A new construction

Fig. 6 Anchorage 4A of the Akashi-Kaikyo Bridge

system, which makes full use of the performance of self-compacting concrete, was introduced for this. The concrete was mixed at a batching plant beside the site and was pumped 200 m through pipes to the casting site, where the pipes were arranged in rows 3–5 m apart. The concrete was cast from gate valves located at 5 m intervals along the pipes.

These valves were automatically controlled so that a surface level of the cast concrete could be maintained. The maximum size of the coarse aggregate used at this site was 40 mm. The concrete fell as much as 3 m without segregation, despite the large size of coarse aggregate. In the final analysis, the use of self-compacting concrete shortened the anchorage construction period by 20%, from 2.5 to 2 years.

Self-compacting concrete was used for the wall of a large LNG tank belonging to the Osaka Gas Company.[16] The adoption of self-compacting concrete meant that:

1. the number of pours decreased from 14 to 10, as the height of each pour was increased
2. the number of concrete workers was reduced from 150 to 50
3. the construction period for the structure decreased from 22 to 18 months.

In addition, a rational acceptance test for self-compactability at the site was introduced (Figure 7). The concreting was completed in June 1998.

Concrete products

Self-compacting concrete is now often employed in concrete products to eliminate the noise of vibration (Figure 8).[17] This improves the working environment at plants and makes it possible for the plants to be located in urban areas. In addition, use of self-compacting

Fig. 7 Acceptance testing apparatus on site, installed between an agitator truck and pump

Fig. 8 Casting of self-compacting concrete for a tunnel segment

Fig. 9 Annual production of concrete products using self-compacting concrete in Japan

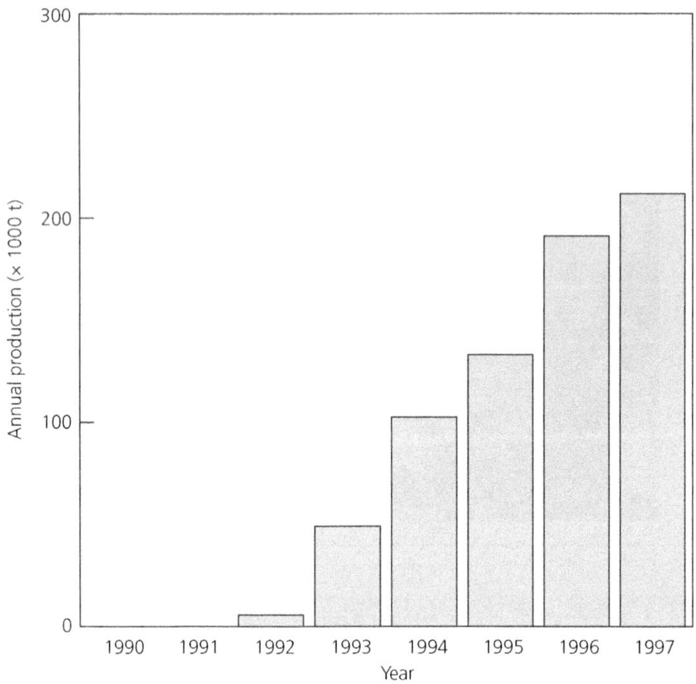

concrete increases the lifetime of the moulds. The production of concrete products using self-compacting concrete has been gradually increasing (Figure 9).[14]

Applications around the world

With the spread of the concept around the world, investigations on self-compacting concrete and applications in practical structures have been reported. The following information on current conditions in each country was summarized at the International Workshop on Self-Compacting Concrete, Kochi, Japan, in 1998.[8]

Canada

Investigations on self-compacting concrete in Canada started in 1992 under the leadership of the University of Sherbrooke, relating to high-performance concrete. Self-compacting concrete was used in the repair of a severely damaged beam in an external parking garage; rehabilitation of restricted slab and wall elements in a hydroelectric power plant, construction of two experimental basement walls and construction of a strong reaction wall for structural testing. The most promising markets for self-compacting concrete in Canada are:

1. basements – 25% of market for concrete in Canada
2. rehabilitation
3. 'silent concrete' in urban area.

Sweden

Self-compacting concrete technology for bridge applications has been developed in a joint project between a contractor and a research institute. During early 1998 three bridges were cast with encouraging results, and further bridge applications are planned. In an EU project on industrialization of concrete construction, self-compacting concrete for house building is being developed, and the first applications have been carried out, using self-compacting concrete with fibre. In addition, training activities on mix design, material production and testing methods were also carried out, and all major mixed concrete plants in Sweden are already producing self-compacting concrete.

The Netherlands

In the period 1995–1996 early experience was gained with self-compacting concrete in The Netherlands. After the Japanese mix design method was introduced in 1997, it was shown that self-compacting concrete with segregation resistance could be made using Dutch materials. Self-compacting concrete was applied to a building in 1997. In addition, in 1998, the Dutch Precast Concrete Industry formed a research group to study the application of self-compacting concrete in concrete products.

Thailand

Self-compacting concrete has been used in practical structures in Thailand since 1992. Since only 10% of the available fly ash is currently utilized in Thailand, the use of self-compacting concrete is seen as one way to use this resource. Some examples of the application of self-compacting concrete in Thailand are:

1. the water supply structure for a cooling tower in a coal generating plant – 4000 m^3
2. an overpass in an expressway project – 432 m^3
3. steel composite long columns in an office building – 429 m^3.

Taiwan

Investigation on self-compacting concrete started in Taiwan in 1994 and, since then, there have been many applications in practical structures due to the increasing demand for construction and the shortage of skilled workers.[18] Some examples are:

1. concrete-filled steel columns for high-rise buildings
2. an overpass in a highway interchange
3. a bus-privileged driveway
4. buildings.

In addition, self-compacting concrete will be employed in the Taiwan high-speed railway project.

Other countries

Investigations into self-compacting concrete have also been reported from the USA, Austria, the UK, France, Korea and Iceland.

State of the art of self-compacting concrete

Current status of self-compacting concrete

Self-compacting concrete has been used as a 'special concrete' only by large general construction companies in Japan. In order for self-compacting concrete to be used as a standard concrete rather than a special one, new systems for the design, manufacture and construction of self-compacting concrete need to be established, and a number of organizations, both within Japan and internationally, have set up committees to develop guidance. Among them, a system by which the ready mixed concrete industry can produce self-compacting concrete as a normal concrete would seem the most effective since, in Japan, as much as 70% of the total concrete produced is from the ready mixed concrete industry. Assuming general supply from ready mixed concrete plants, investigations to establish the following items have been carried out mainly at the University of Tokyo:

1. self-compactability testing
2. mix design
3. acceptance testing on site.

The first book on self-compacting concrete, *High Performance Concrete*, written by Okamura *et al.*, was published in 1993.[19]

Self-compactability evaluation

There are three requirements for self-compactability tests, as shown in Figure 10:

- test 1 – to check self-compactibility
- test 2 – to adjust the mix proportion when self-compactability is not sufficient
- test 3 – to evaluate materials.

To check self-compactibility

As a test for mix proportioning in the laboratory (i.e. test 1), the so-called U test or box test (Figures 11, 12 and 13) is recommended.[20] The U test was developed by Taisei's group. In this test, the degree of compactability can be indicated by the height that the concrete reaches after

Fig. 10 Types of self-compactability evaluation for mix proportioning and inspection

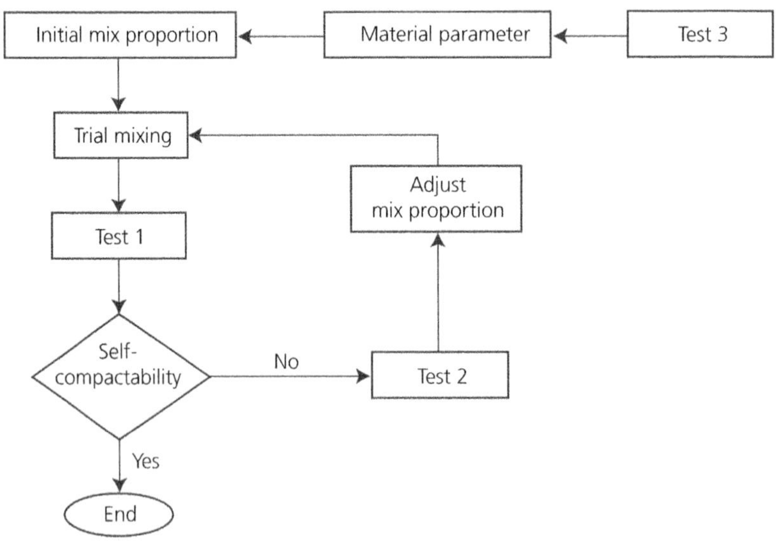

Fig. 11 U test (dimensions in mm)

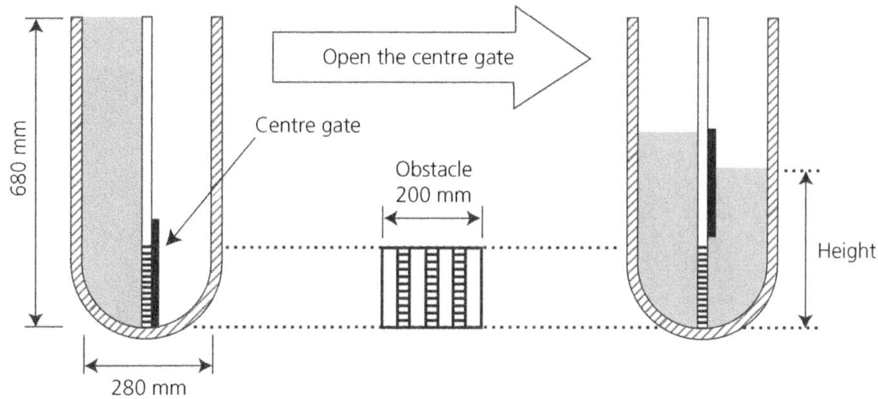

Fig. 12 Box test

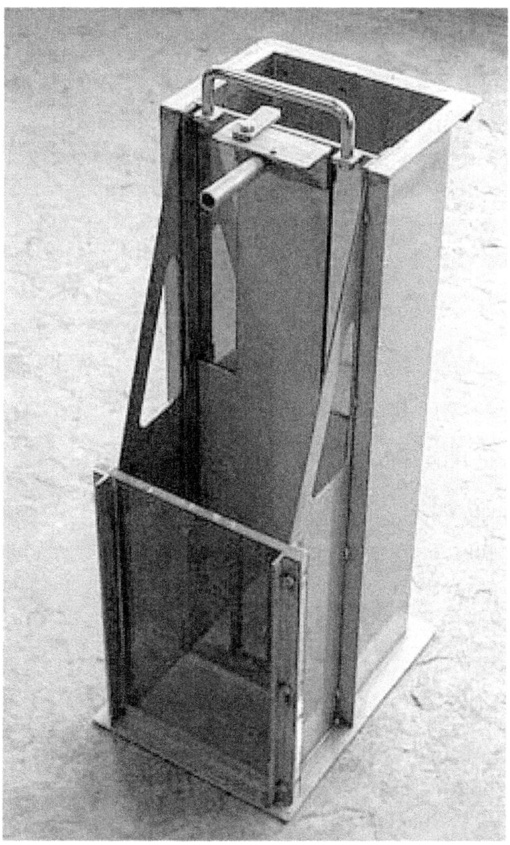

Fig. 13 Obstacles for the U or box test (left, R2; right, R1)

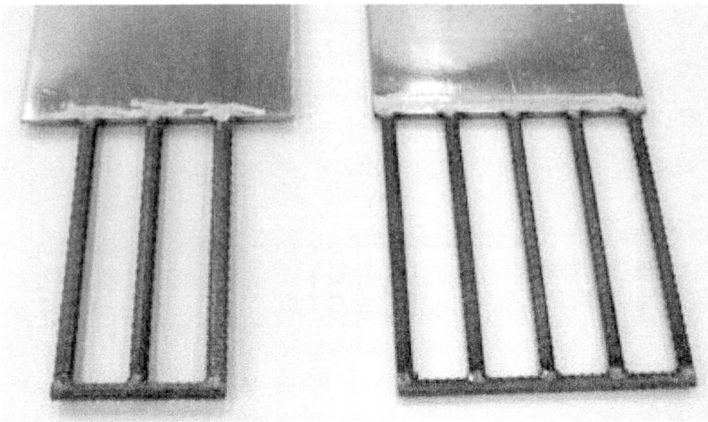

flowing through an obstacle. The obstacle with the highest requirements should be chosen. Concrete with a filling height of over 300 mm can be judged to be self-compacting. The box test is more suitable for detecting concrete with a higher possibility of segregation between coarse aggregate and mortar.

To adjust mix proportions

If the concrete is judged to have insufficient self-compactability with test 1, the cause has to be established quantitatively so that the mix proportions can be adjusted. Slump flow and funnel tests (Figure 14) have been proposed for testing the deformability and viscosity. These give the indices G_c and R_c, defined as follows:

$$G_c = (S_{fl1}S_{fl2} - S_{fl0}^2)/S_{fl0}^2$$

where S_{fl1} and S_{fl2} are measured flow diameters and S_{fl0} is the slump cone diameter, and

$$R_c = 10/t$$

where t is the measured time (in seconds) for concrete to flow through the funnel.

To evaluate materials

Flow and funnel tests for mortar or paste have been proposed for evaluating materials used in self-compacting concrete, e.g. powder material, sand and superplasticizer. Testing methods for mortar properties have been proposed which give the indices G_m and R_m for deformability and viscosity, respectively (Figures 15 and 16):[21]

$$G_m = (d_1 d_2 - d_0^2)/d_0^2$$

where d_1 and d_2 are measured flow diameters and d_0 is the flow cone diameter, and

$$R_m = 10/t$$

where t is the measured time (in seconds) for mortar to flow through the funnel.

Fig. 14 V funnel

Fig. 15 Mortar flow test and the index G_m

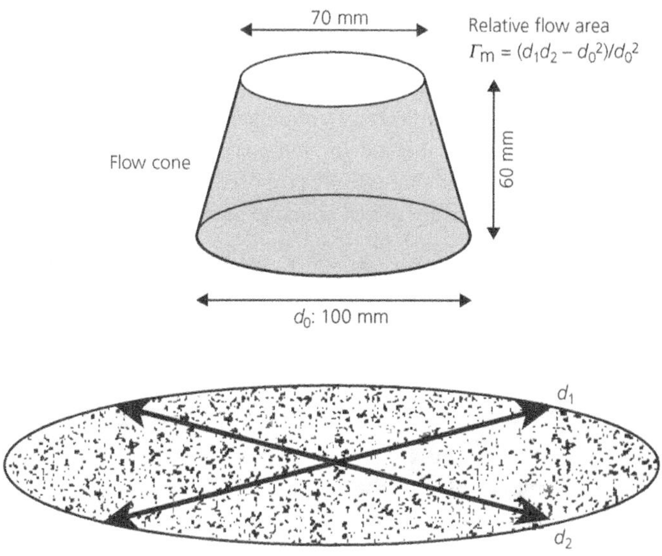

Fig. 16 Mortar funnel test and the index R_m

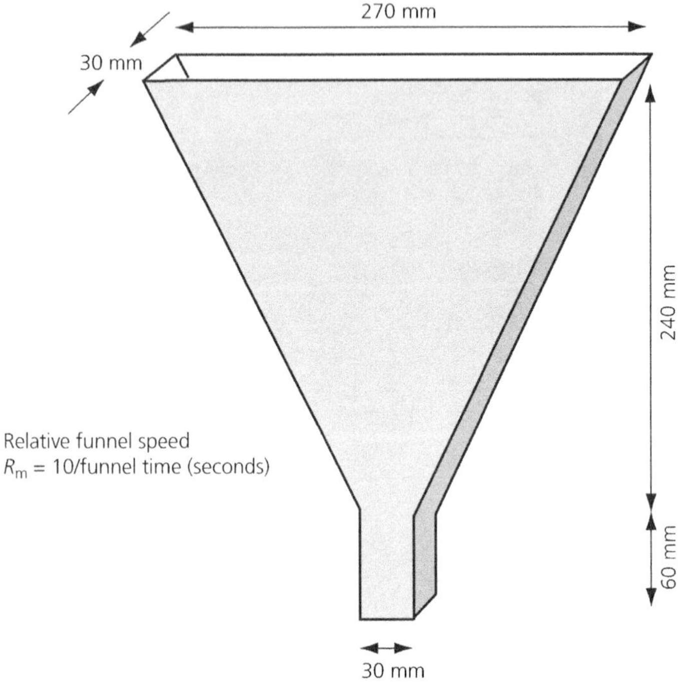

Larger G_m values indicate higher deformability and smaller R_m values indicate higher viscosity. Evaluation criteria for materials were proposed in terms of G_m and R_m.[3,4]

Mix design method

Rational mix design method

Self-compactability can be largely affected by the characteristics of materials and the mix proportion. A rational mix design method for self-compacting concrete using a variety of materials is necessary. Okamura and Ozawa have proposed a simple mix-proportioning system assuming a general supply from ready mixed concrete plants.[4] The coarse and fine aggregate contents are fixed so that self-compactability can be achieved easily by adjusting the water/powder ratio and superplasticizer dosage only:

1. the coarse aggregate content in concrete is fixed at 50% of the solid volume.
2. the fine aggregate content is fixed at 40% of the mortar volume.
3. the water/powder ratio is assumed to be 0.9–1.0 by volume, depending on the properties of the powder.
4. the superplasticizer dosage and the final water/powder ratio are determined so as to ensure self-compactability.

In the mix proportioning of conventional concrete, the water/cement ratio is fixed at first from the viewpoint of obtaining the required strength. With self-compacting concrete, however, the water/powder ratio has to be decided taking self-compactability into account because self-compactability is very sensitive to this ratio. In most cases, the required strength does not govern the water/cement ratio because the water/powder ratio is small enough to give the required strength for ordinary structures unless most of the powder materials in use are not reactive.

The mortar or paste in self-compacting concrete requires high viscosity as well as high deformability. This can be achieved by the employment of a superplasticizer, which results in a low water/powder ratio for high deformability.

Adjustment of the water/powder ratio and superplasticizer dosage

The characteristics of the powder and superplasticizer largely affect the mortar property, and so an appropriate water/powder ratio and superplasticizer dosage cannot be fixed without trial mixing at this stage. Therefore, once the mix proportion has been decided on, self-compactability has to be tested by the U-type, slump–flow and funnel tests. Methods of judging whether the water/powder ratio or superplasticizer dosage is larger or smaller than the required value by using the test results, and of estimating the required values, are necessary. The relationships between the properties of the mortar in self-compacting concrete and the mix proportion have been investigated and formulated.

These formulae can be used for establishing a rational method for adjusting the water/powder ratio and superplasticizer dosage for achieving appropriate deformability and viscosity.

Evaluation of materials with mortar or paste tests

Evaluation of materials is very important for self-compacting concrete since the characteristics of the materials greatly affect the self-compactability of fresh concrete. It is also desirable that the appropriate mix proportion can be estimated with a minimum number of trial mixings.

Dispersing effect of the superplasticizer

It was found that the ratio of G_m to R_m is almost constant with variation of V_w/V_p (volume ratio of water to powder) provided that Sp/P (the weight ratio of superplasticizer to powder) is constant (Figure 17).[21] The gradient of the G_m–R_m line corresponding to each Sp/P indicates the effect of superplasticizer, which is independent of V_w/V_p. This is effective in evaluating Sp/P by using only one pair of experimental results (G_m, R_m). In this study, G_m/R_m is proposed as the index for the effect of superplasticizer from the viewpoint of achieving both deformability and viscosity at the combination of powder material used. Larger values of G_m/R_m indicate a greater effect of superplasticizer, that is, higher deformability (G_m) without decrease of viscosity ($1/R_m$) (Figure 18).

Fig. 17 Relationship between G_m and R_m of mortar (sand content 40% by volume, moderate heat cement, *Sp/P* fixed on the line)

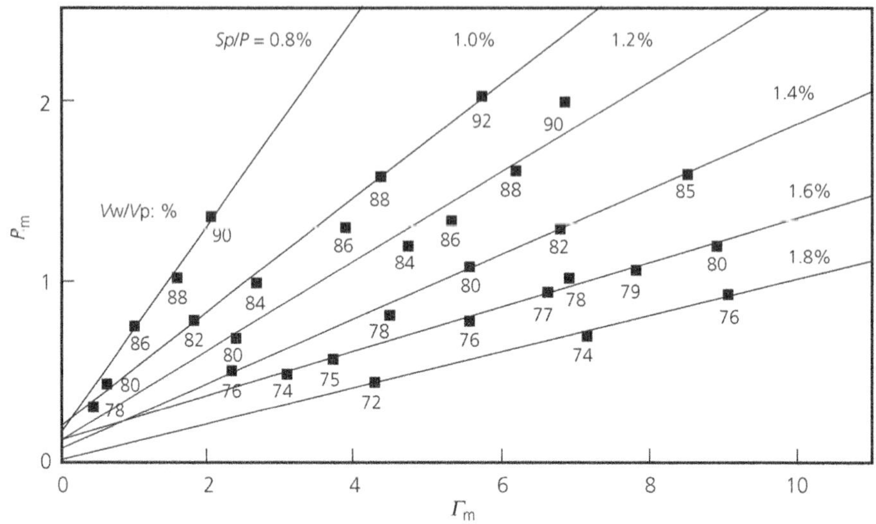

Fig. 18 Index for the effect of superplasticizer: G_m/R_m

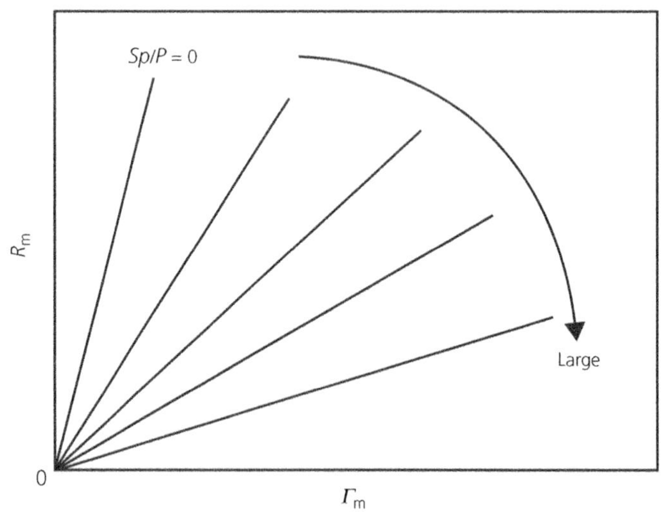

Fig. 19 Relationship between Sp/P and G_m/R_m (sand content 40% in mortar; MC, moderate heat cement (BS 4000); blast furnace slag with a Blaine value of 4000; FA, fly ash; SP, superplasticizer)

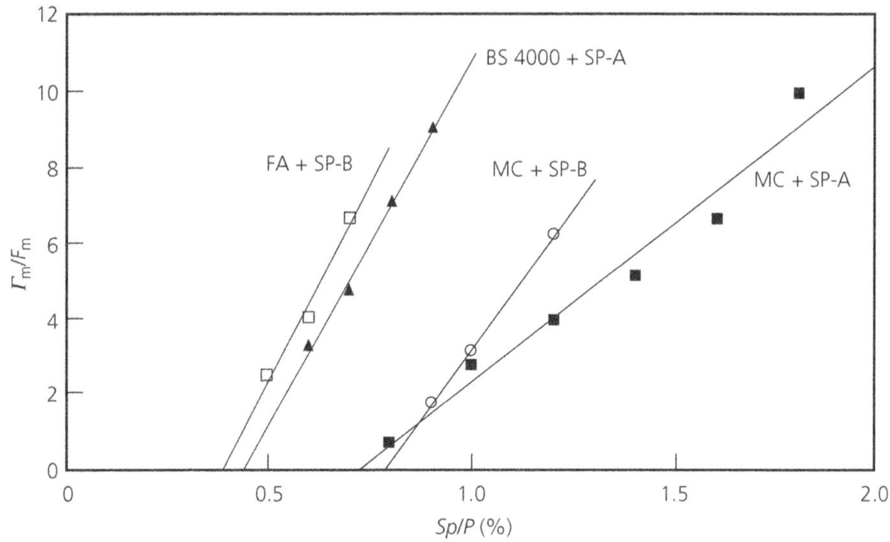

The relationship between Sp/P and G_m/R_m is significantly affected by the combination of superplasticizer and powder material used (Figure 19). At this stage, the relationship cannot be estimated without tests due to the chemical effect of superplasticizer depending on combination with the particular powder material used.

Particles in mortar: flowability

The relationship between G_m and R_m with variation of Sp/P for constant V_w/V_p has already been proposed (Figures 20 and 21):[21]

$$R_m = A\,G_m^{0.4}$$

where $A = K(V_w/V_p)$, i.e. a linear relationship exists between A and V_w/V_p

The relationships between V_w/V_p and A for mortar with variation of the powder materials are shown in Figure 22. It was found that the gradient depends on the property of the solid particles in the mortar. The gradient for fly ash mortar is the largest, and the gradients for mortar with ordinary powder having Blaine values of 3000–4000 cm^2/g are almost the same: around 4. It also found that the gradient is independent of the type of superplasticizer used.

Fig. 20 Relationship between G_m and R_m (sand content 40% by volume, moderate heat cement, V_w/V_p fixed on the curve)

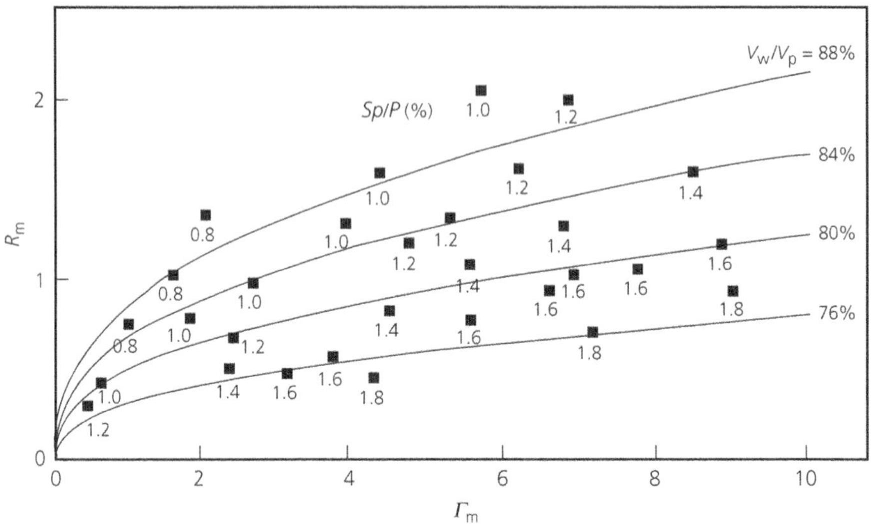

Fig. 21 Relationship between V_p/V_w and $R_m/G_m^{0.4}$

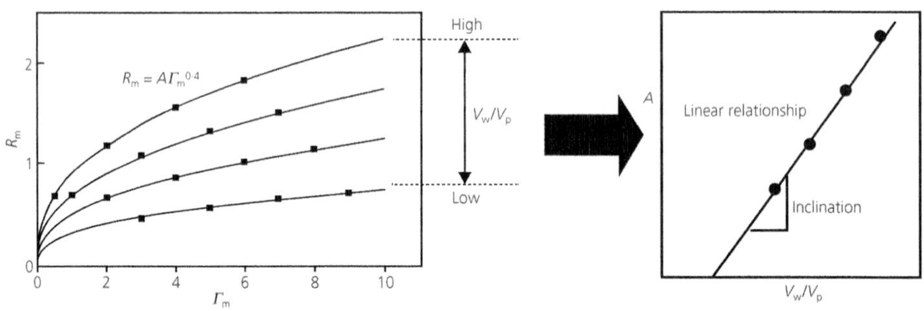

Particles in mortar: interaction with coarse aggregate

A testing method for the interaction between coarse aggregate and mortar particles in self-compacting concrete was proposed in which the ratio of funnel speed of the mortar with model coarse aggregate and that without model coarse aggregate is compared.[22] The conventional funnel for mortar in self-compacting concrete was adopted as the testing apparatus. Glass beads with a diameter of 10 mm were adopted as the model coarse aggregate, and the content of the model coarse aggregate was chosen as 20% of the total mortar volume so that blocking cannot occur easily. The index for the interaction was proposed as the ratio of the funnel speed of mortar with glass beads (R_{mb}) to that without glass beads (R_m) (Figure 23). It was found out

Fig. 22 Relationship between V_p/V_w and $R_m/G_m^{0.4}$ for variation in powder materials (sand content 40% by volume; SP, superplasticizer)

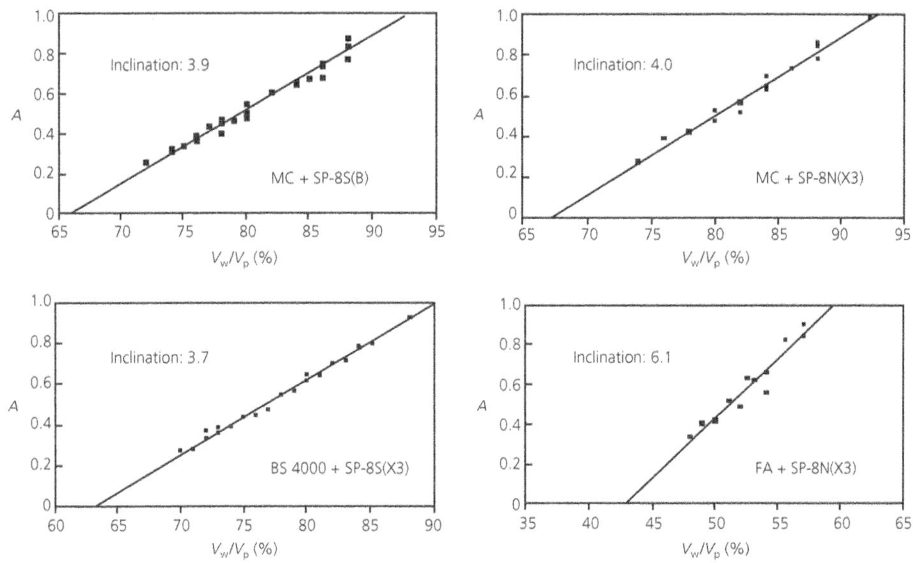

Fig. 23 Index R_{mb}/R_m for the degree of interaction of mortar with coarse aggregate

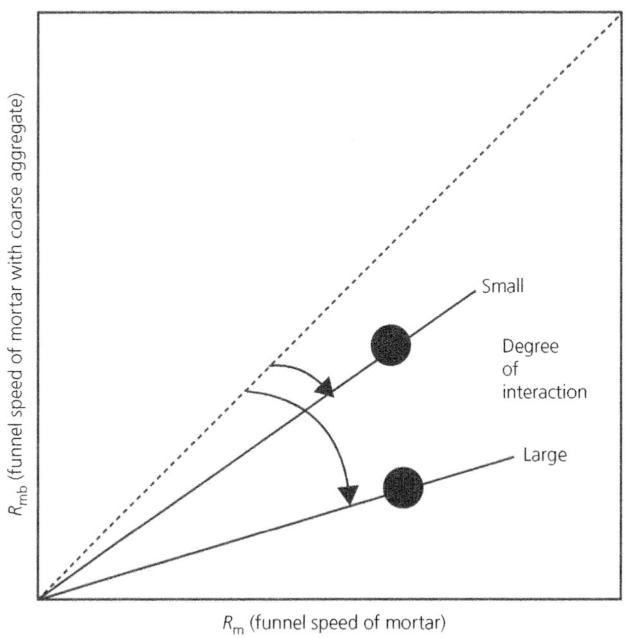

that the ratio R_{mb}/R_m was constant provided that the deformability or viscosity of mortar itself is within the range necessary for achieving self-compactability. It is possible that this test can be substituted for the self-compactability test for examining the sand content in the mortar (Figures 24 and 25).

Fig. 24 Relationships between sand content in mortar (v/v) and R_{mb}/R_m: 1, OPC (ordinary Portland cement) + CS (crushed sand); 2, FA (fly ash) + CS; 3, FA + RS (river sand); 4, OPC + RS; 5, OPC + MS (mountain sand)

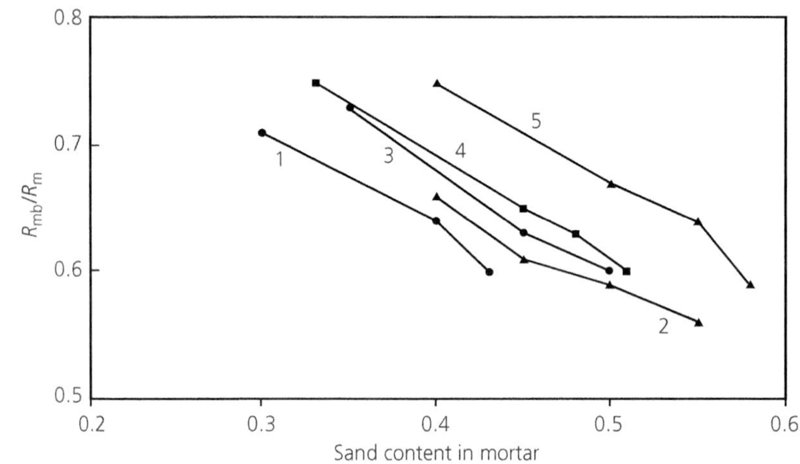

Fig. 25 Unique relationship between R_{mb}/R_m and filling height of the box test (obstacle R1) independent of the characteristics of the powder or sand (see Figure 24 for definitions of abbreviations)

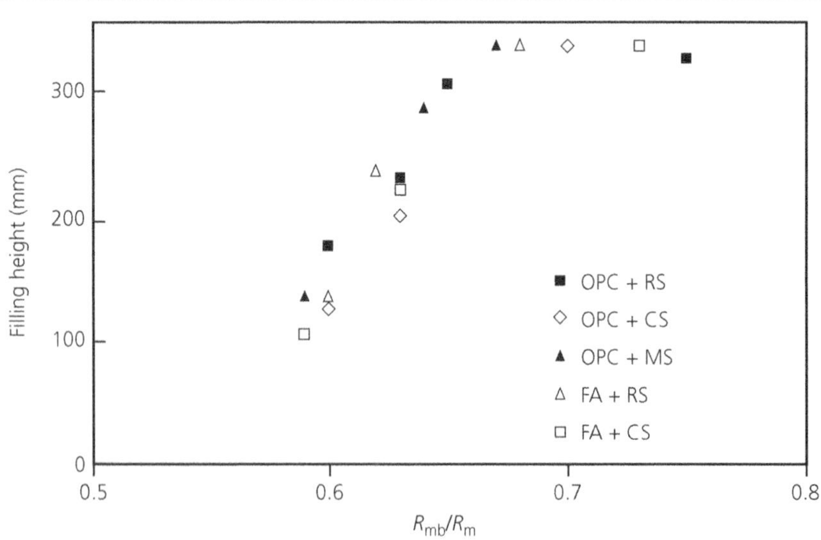

Segregation-inhibiting agent

It has been found possible to manufacture self-compacting concrete with constant quality, especially self-compactability. However, any variation in material characteristics can affect the self-compactability. The most influential variant is the water content of the fine aggregate, which results in variations in the water content of the concrete itself. To solve this problem, some general construction companies employ a segregation-inhibiting agent. This type of agent is effective in making self-compactability less sensitive to variations in the water content. Various agents have been proposed and are available in Japan.[23–28]

Acceptance test on site

Since the degree of compaction in a structure depends mainly on the self-compactability of the concrete, and poor self-compactability cannot be compensated by the construction work, self-compactability must be checked for all concrete just before casting. However, conventional testing methods for self-compactability require sampling, and this can be extremely laborious if the self-compactability acceptance test is to be carried out for the whole batch of concrete. A suitable acceptance test method for self-compactability has been developed by Ouchi and coworkers:[8]

1. The testing apparatus is installed between the agitator truck and the pump at the site. The whole of the concrete batch is poured into the apparatus.
2. If the concrete flows through the apparatus, the concrete is considered to be self-compactable. If the concrete is stopped by the apparatus, the concrete is considered as having insufficient self-compactability, and mix proportion has to be adjusted.

This apparatus was successfully used at the construction site of the Osaka Gas LNG tank, and saved a considerable amount of acceptance test work (Figure 26).[16]

New structural design and construction systems

By employing self-compacting concrete, the cost of compaction can be saved and the compaction of the concrete in the structure can be assured. However, the total construction cost cannot always be reduced, except for large-scale constructions. This is because the conventional construction system is strongly based on the necessity of the vibrating compaction of concrete.

Self-compacting concrete can greatly improve construction systems previously based on conventional concrete requiring vibrating compaction. This sort of compaction, which can easily cause segregation, has been an obstacle to the rationalization of construction work. Once this obstacle has been eliminated, concrete construction can be rationalized, and a new construction system, including formwork, reinforcement, support and structural design, can be developed

One example of a novel design is the so-called sandwich structure, where a steel shell is filled with concrete. This type of structure has already been built in Kobe, and could not have been achieved without the development of self-compacting concrete (Figure 27).[29]

Fig. 26 Acceptance testing apparatus used on site (dimensions in mm)

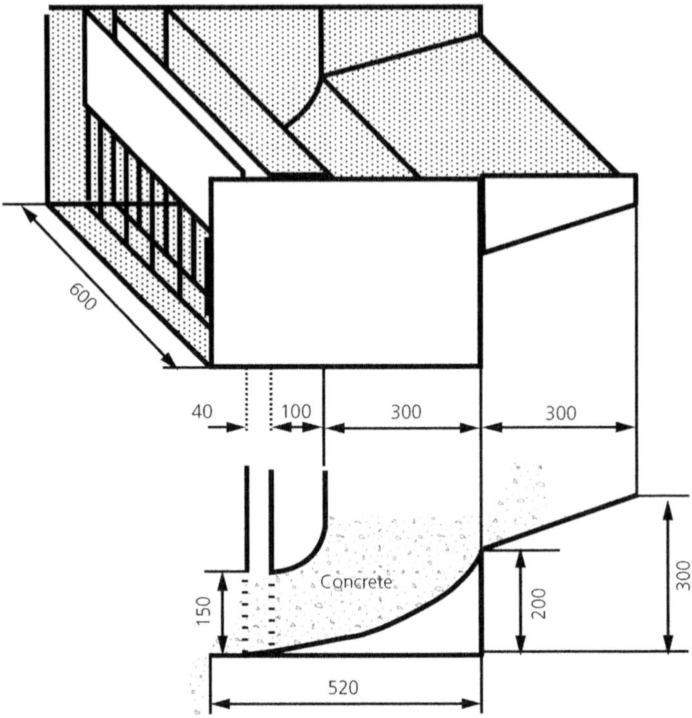

Fig. 27 Sandwich structure for an immersed tunnel

Recommendations and manuals

A set of manuals on self-compacting concrete has been published by different organizations in Japan. At present, the following are available:

- *Recommendations for Mix Design and Construction Practice of Highly Fluidity Concrete* (The Architectural Institute of Japan).
- *Recommendations for Self-Compacting Concrete* (The Japan Society of Civil Engineers).
- *Manual for Manufacturing Self-Compacting Concrete* (The National Ready-Mixed Concrete Industry Association, Japan).

All of these are available in the proceedings of the International Workshop on Self-Compacting Concrete held in Kochi in March 1999.[8]

Conclusions

Since both a rational mix design method and an appropriate acceptance testing method at the job site are close to being established for self-compacting concrete, it is considered that the main obstacles to the wide use of self-compacting concrete have been removed. The next task is to distribute the techniques for manufacturing and construction of self-compacting concrete rapidly. Rational training and qualification systems for engineers should be introduced. In addition, new structural design and construction systems making full use of the benefits of self-compacting concrete should be introduced.

Once self-compacting concrete becomes so widely used that it is seen as a 'standard concrete' rather than as a 'special concrete', we will have succeeded in creating durable and reliable concrete structures requiring very little maintenance work.

REFERENCES

1. Ozawa, K., Maekawa, K., Kunishima, M. and Okamura, H. Development of high performance concrete based on the durability design of concrete structures. *Proceedings of the 2nd East-Asia and Pacific Conference on Structural Engineering and Construction (EASEC-2)*, 1989, Vol. 1, pp. 445–450.
2. Maekawa, K. and Ozawa, K. *Development of SCC's prototype.* Self-Compacting High Performance Concrete, Social System Institute, Tokyo, 1999, pp. 20–32 [in Japanese].
3. Gagne, R., Pigeon, M. and Aitcin, P. C. Decier salt scaling resistance of high performance concrete. *Paul Klieger Symposium on Performance of Concrete*, SP-122. American Concrete Institute, Detroit, 1989.
4. Okamura, H. and Ozawa, K. Mix design for self-compacting concrete. *Concrete Library of the JSCE*, 1995, No. 25, pp. 107–120.
5. Ozawa, K., Tagtermsirikul, S. and Maekawa, K. Role of materials on the filling capacity of fresh concrete. *Proceedings of the 4th CANMET and ACI International Conference on Fly Ash, Silica Fume, Slag and Natural Pozzolans in Concrete.* Istanbul, 1992, pp. 212–237.
6. Okamura, H. and Ozawa, K. *Self-compactable High-performance Concrete In Japan*, SP-169. American Concrete Institute, Detroit, 1994, pp. 31–44.
7. Okamura, H. Ferguson lecture for 1996: Self-compacting high-performance concrete. *Concrete International*, 1997, **19**, No. 7, 50–54.

8. Ozawa, K. and Ouchi, M. (eds). *Proceedings of the International Workshop on Self-Compacting Concrete*, CD-ROM. Kochi, 1999. [Includes state-of-the-art report on self-compactability evaluation, materials and design, construction, manufacturing and concrete products and summary of recommendations and manuals for self-compacting concrete in Japan. Also available from *Concrete Engineering Series*, No. 30, Japan Society of Civil Engineers, 1999.]
9. Yamada, K. *Challenge for Application of SCC to Practical Structures*. Self-compacting High Performance Concrete, Social System Institute, 1999, pp. 34–80 [in Japanese].
10. Tanaka, K. *Development of Low-heat SCC for Anchorage 4A of Akashi-Kaikyo Bridge*. Self-compacting High Performance Concrete, Social System Institute, Tokyo, 1999, pp. 82–110 [in Japanese].
11. Matsuoka, Y. *Progress to Application of SCC to Practical Structures*. Self-compacting High Performance Concrete, Social System Institute, 1999, Tokyo, pp. 112–160 [in Japanese].
12. Sakamoto, J., Matsuoka, Y., Shindoh, T. and Tangtermsirikul, S. An application of super workable concrete to construction of actual structures. *Transactions of the Japan Concrete Institute*, 1991, **13**.
13. Ohno, H., Kawai, T., Kuroda, Y. and Ozawa, K. Development of vibration-free high strength lightweight concrete and its application. *Proceeding of FIP Symposium*. Kyoto, 1993, vol. 1, pp. 297–304.
14. Kodama, Y. Current condition of SCC's application. *Cement Shimbun*, No. 2304, 1997 [in Japanese].
15. Kashima, S., Kanazawa, K., Okada, R. and Yoshikawa, S. Application of self-compacting concrete made with low-heat cement for bridge substructures of Honshu-Shikoku Bridge Authority. *Proceedings of the International Workshop on Self-Compacting Concrete*, 1999, pp. 255–261.
16. Kitamura, H., Nishizaki, T., Ito H., Chikamatsu, R., Kamada, F. and Okudate, M. Construction of prestressed concrete outer tank for LNG storage using high-strength self-compacting concrete. *Proceedings of the International Workshop on Self-Compacting Concrete*, 1999, pp. 262–291.
17. Uno, Y. State-of-the art report on concrete products made of SCC. *Proceedings of the International Workshop on Self-Compacting Concrete*, 1999, pp. 262–291.
18. Chern, J. C. The recent developments and applications of SQC in Taiwan. *Proceedings of The Symposium on Super Quality Concrete*, 1998, pp. 1–16.
19. Okamura, H., Maekawa, K. and Ozawa, K. *High Performance Concrete*. Gihodo, 1993.
20. Hayakawa, M., Matsuoka, Y. and Shindoh, T. Development and application of super workable concrete. *RILEM International Workshop on Special Concretes: Workability and Mixing*. Paisley, 1993.
21. Ouchi, M., Hibino, M., Ozawa, K. and Okamura, H. A rational mix-design method for mortar in self-compacting concrete. *Proceedings of the 6th East Asia-Pacific Conference on Structural Engineering and Construction*. Taipei, ROC, 1998, vol. 2, pp. 1307–1312.
22. Ouchi, M. and Edamatsu, Y. A simple evaluation method for interaction between coarse aggregate and mortar particles in self-compacting concrete. *Proceedings of RILEM SCC Symposium*. Stockholm, 1999.
23. Hibino, M., Okuma, M. and Ozawa, K. Role of viscosity agent in self-compactability of fresh concrete. *Proceedings of the 6th East-Asia Conference on Structural Engineering and Construction*. Taipei, 1998, vol. 2, pp. 1313–1318.

24. Kawai, T. and Okada, T. Effect of superplasticizer and viscosity-increasing admixture on properties of lightweight aggregate concrete. *3rd CANMET/ACI International Conference on Super-plasticizers and Other Chemical Admixture in Concrete*, SP-119. American Concrete Institute, Detroit, 1989, pp. 583–604.
25. Shindoh, T., Matsuoka, Y., Tangtermsirikul, S. and Sakamoto, J. Effect of variation of material quality on properties of super workable concrete. *Transactions of the Japan Concrete Institute*, 1992, **14**.
26. Kawai, T. *Integrated Design and Environmental Issues in Concrete Technology*. Spon, London, 1996.
27. Sakata, N., Maruyama, K. and Minami, M. *Production Methods and Workability of Concrete*. Spon, London, 1996.
28. Matsuoka, Y., Shindoh, T., Yokota, K. and Kusui, S. *Property of Beta-1,3-gurcan (Curdlan) as a Viscosity Agent for Super-workable Concrete*, SP-173. American Concrete Institute, Detroit, 1997.
29. Shishido, T., Shiraiwa, S., Ishii, J., Yamamoto, S. and Kume, H. Pouring works of highly fluidity concrete for immersed tunnel by steel-concrete composite structure. *Proceedings of the International Workshop on Self-Compacting Concrete*, CD-ROM. Kochi, 1999, pp. 328–346.

ice Publishing

Dhir and Lye
ISBN 978-0-7277-6461-4
https://doi.org/10.1680/icetscc.64614.029

Chapter 2
Development of self-compacting concrete

C. I. Goodier
Senior Consultant, Centre for Concrete Construction, Building Research Establishment

Self-compacting concrete (SCC) can be defined as a fresh concrete which possesses superior flowability under maintained stability (i.e. no segregation) thus allowing self-compaction – that is, material consolidation without addition of energy. It was first developed in Japan in 1988 in order to achieve durable concrete structures by improving quality in the construction process. This was also partly in response to the reduction in the numbers of skilled workers available in the industry. This paper outlines a brief history of SCC from its origins in Japan to the development of the material throughout Europe. Research and development into SCC in the UK and Europe are discussed, together with a look at the future for the material in Europe and the rest of the world. Research and development of SCC is being conducted by private companies (mainly product development), by universities (mainly pure research into the material's properties), by national bodies and working groups (mainly the production of national guidelines and specifications) and at European level (Brite-EuRam and RILEM projects on test methods and the casting of SCC, respectively). Although SCC is not expected to ever completely replace conventionally vibrated concrete, the use of the material in both the precast and ready-mix markets in the UK, Europe and the rest of the world is expected to continue to increase as the experience and technology improves, the clients demand a higher-quality finished product and the availability of skilled labour continues to decrease.

1. Introduction

Self-compacting concrete (SCC) can be defined as a fresh concrete which possesses superior flowability under maintained stability (i.e. no segregation), thus allowing self-compaction – that is, material consolidation without addition of energy. The three properties that characterise a concrete as self-compacting are

- flowing ability – the ability to completely fill all areas and corners of the formwork into which it is placed (e.g. Fig. 1)
- passing ability – the ability to pass through congested reinforcement (e.g. Fig. 2) without separation of the constituents or blocking
- resistance to segregation – the ability to retain the coarse components of the mix in suspension in order to maintain a homogeneous material.

These properties must all be satisfied in order to design an adequate SCC, together with other requirements including those for hardened performance.

Fig. 1 Placement of SCC by pump tremie, demonstrating the flowable nature of the material (Mott MacDonald Ltd)

SCC was first developed in Japan in 1988 in order to achieve durable concrete structures by improving quality in the construction process. It was also found to offer economic, social and environmental benefits over traditional vibrated concrete construction. Research and development work into SCC in Europe began in Sweden in the 1990s and now nearly all the countries in Europe conduct some form of research and development into the material.

In addition to the benefits described above, SCC is also able to provide a more consistent and superior finished product for the client, with less defects. Another advantage is that less skilled labour is required in order for it to be placed, finished and made good after casting. As the shortage of skilled site labour in construction continues to increase in the UK and many other countries, this is an additional advantage of the material which will become increasingly important.

This paper outlines the history of the development of SCC from its origins in Japan to its development in Sweden and the rest of Europe. The development of the material in Europe is then discussed, with particular reference to the UK.

Fig. 2 An example of congested reinforcement: an ideal application for SCC (Mott MacDonald Ltd)

2. Properties of SCC

Most materials suitable for conventional concrete can be used in the production of SCC, although there are significant differences in SCC production compared with conventional concrete production. The most important is that the properties of the SCC can be allowed to vary only within much tighter limits. For conventional concrete, variation can be coped with by altering the degree of compaction work, but this is not possible with SCC. The properties of fresh SCC are also much more sensitive to variations in the quality and consistency of the mix constituents. Because of its greater sensitivity to variation, batching accuracy for all component materials is essential for SCC to be successful.

High powder contents are needed in SCC to increase the cohesiveness. Ground granulated blast furnace slag (GGBS), pulverised fuel ash (PFA), or an inert material such as limestone powder are most commonly used. It is possible, however, to produce an SCC without adding any extra fine material.

The placement of SCC is the aspect of the building process that is changed most by the use of SCC. Once the fully compliant SCC is supplied to the point of application then the final operation of casting requires very little skill or manpower compared with traditional concrete to produce uniformly dense concrete. Because of vibration being unnecessary, the noise is reduced and the risk of developing problems due to the use of vibrating equipment is reduced. Fewer operatives are required, but more time is needed to test the concrete before placing.

In general, the hardened properties of SCC are similar or superior to those of equivalent conventional concrete. Provided the self-compacting properties are verified on site, a less variable and denser concrete is achieved. It could also be said that the compaction, and therefore durability, of the concrete is more guaranteed with the use of SCC as it reduces the potential for human error (in the form of poor compaction). The high fines content and the need for well-graded aggregates also improves the concrete, producing a more dense interfacial transition zone between the aggregate and the cement matrix. All these effects are beneficial for both strength and durability, although because of its relatively recent use, knowledge of long-term durability is restricted to a decade of limited use.

At similar water/cement ratios the characteristic strength of SCC is at least equal to that of conventional concrete, and has a similar strength development for the same grade. Due to the low water/cement ratio used in SCC the compressive strength will generally be above C40 and can be up to C100. The tensile strength is also comparable to the same grade of conventional concrete and drying shrinkage is also similar.

3. The history of SCC

The history and development of SCC can be divided into two key stages: its initial development in Japan in the late 1980s and its subsequent introduction into Europe through Sweden in the mid- to late-1990s.

3.1 Japan

SCC was first developed in Japan in 1988 in order to achieve more durable concrete structures by improving the quality achieved in the construction process and the placed material.[1] The removal of the need for compaction of the concrete reduced the potential for durability defects due to inadequate compaction (e.g. honeycombing). The use of SCC was also found to offer economic, social and environmental benefits over traditional vibrated concrete construction. These benefits included faster construction and the elimination of noise due to vibration. One of the main drivers for the development of the technology was the reduction in the number of skilled site operatives that the Japanese construction industry was experiencing in the 1980s. The use of SCC meant that less skilled labour was required for the placing and finishing of the concrete.

SCC was developed from the existing technology used for high workability and underwater concretes, where additional cohesiveness is required. The first research publications that looked into the principles required for SCC were from Japan around 1989 to 1991. These studies concentrated upon high-performance and super-workable concretes and their fresh properties such as filling capacity, flowability and resistance to segregation.[2–4]

The first significant publication in which 'modern' SCC was identified is thought to be a paper from the University of Tokyo by Ozawa *et al.* in 1992.[5] The term 'self-compacting concrete' is not used within the paper, although a high-performance concrete was produced which possessed all the essential properties of a self-compacting concrete mix.

In the following few years many research papers were published on concretes such as super-workable, self-consolidating, highly-workable, self-placeable and highly-fluidised concretes, all of which had similar properties to what we now know as SCC. These were mainly papers on work into the mix design of what would become 'SCC' and its associated fresh properties. In 1993, research papers were beginning to be published of case studies on the use of these early forms of 'SCC' in actual applications.[6,7] One of the first published references utilising the term 'self-compacting' was in Japan in 1995.[8]

After the development of this prototype SCC, intensive research began in many places in Japan, especially within the research institutes of large construction companies, and as a result, SCC has now been used in many practical applications.

The first significant international workshop dedicated to the material was held at Kochi University of Technology, Japan in August 1998.[9] The majority of these papers focused upon the development of SCC in different countries, including research and development into mix design models, mix constituents and rheology.

In April 1997, the Japanese Society of Civil Engineers (JSCE) set up a research subcommittee with the aim of establishing recommendations for the practical application of SCC. This was subsequently published in English in August 1999.[10]

The 2nd International Symposium on SCC was organised by the University of Tokyo in October 2001.[11] A total of 74 papers plus two keynote and four invited speakers from 20 countries were included. Since the 1st International Workshop had concentrated mainly upon mix design and rheology, the 2nd International Symposium concentrated more on the long-term durability and life-cycle cost of SCC.

A conference was recently held in Japan in October 2002 on *Concrete Structures in the 21st Century*, which contained six papers on SCC, including four from Japan.[12] These papers illustrated that the basic technology of the material in Japan is relatively well understood and that the majority of current efforts in research and development are concentrated on taking this knowledge further into new applications such as composite structures and sheet piling.

3.2 Europe

In the second half of the 1990s, interest and use of SCC spread from Japan to other countries, including Europe. Some of the first research work to be published from Europe was at an International RILEM (International Union of Testing and Research Laboratories of Materials and Structures) Conference in London in 1996. Papers were presented on the design of SCC by University College London,[13] and a mix-design model by the Swedish Cement and Concrete Research Institute (CBI).[14]

A Technical Committee (TC 174-SCC) was set up by RILEM in 1997 with the objective of gathering, analysing and presenting a review of the technology of SCC, as well as looking for unified views on testing and evaluation. Seventeen full members and three corresponding members covering ten countries on four continents took part in the work and a state-of-the-art report was published in 2000.[15]

Sweden was the first country in Europe to begin development of SCC, and in 1993 the CBI organised a seminar in Sweden for contractors and producers, leading to a project aimed at studying SCC for housing.[16] As part of this project, large numbers of half-scale house walls were cast using SCCs which were made with different filler materials. The work from this project contributed to the first European project on SCC which began in January 1997 and was completed in 2000. The main goal of this Brite-EuRam project (BRPR-CT96-0366) was to develop a new vibration-free production system to lower the overall cost of *in-situ*-cast concrete construction. The first part concerned the development of SCC with or without steel fibres and the second part dealt with full-scale experiments in civil engineering and housing. This project included partners from several European countries, including the UK. For further details, including project summaries see Reference 17. During this time, CBI also developed a mix design model for the design and production of SCC (Fig. 3).[16]

Parallel to this Brite-EuRam project, CBI continued work together with the Swedish National Roads Authority (SNRA) into SCC for bridge casting. Laboratory investigations, pre-qualifying tests, half-scale trials and finally full-scale trials were all completed. The first of three bridges wholly cast in SCC was completed in January 1998 and is thought to be the first bridge cast with this material outside of Japan (Fig. 4).[16] This work also showed that the hardened properties of SCC are superior to conventional concrete, including compressive strength, frost resistance, permeability and reinforcement bond strength. This is thought to be mainly due to

Fig. 3 A mix design model for SCC developed by the CBI of Sweden[16]

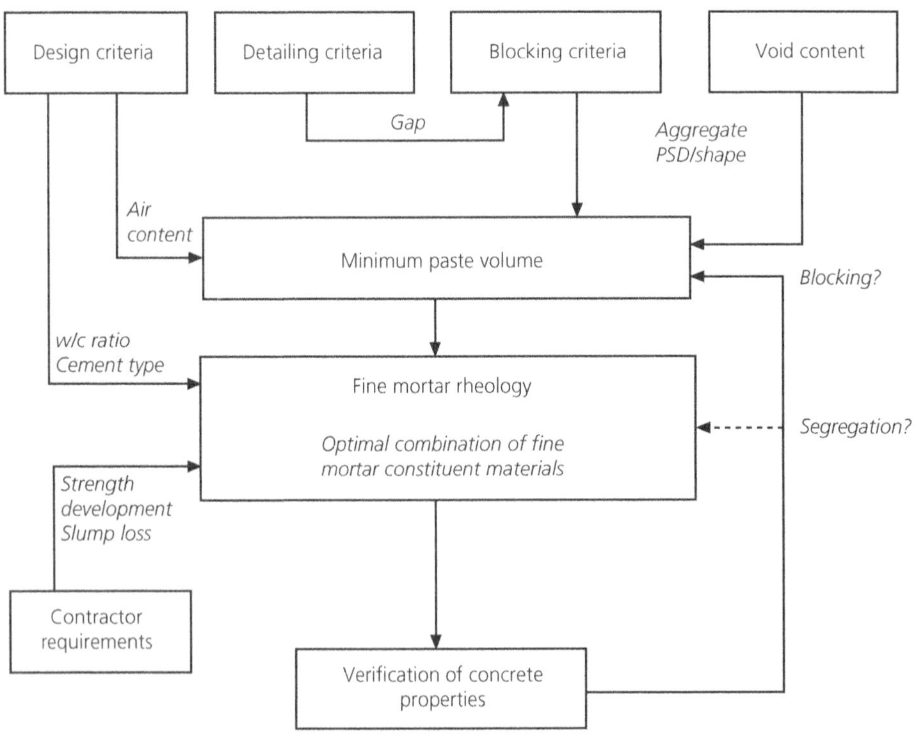

Fig. 4 A Swedish road bridge cast in SCC[16]

both the increased cementitious content generally used and the denser interfacial transition zone between the aggregate and the paste of SCCs.

Studies also showed that *total* bridge building costs can be reduced by as much as 5–15% when using SCC compared with conventional concrete.[16] An earlier report on the same work was also published by the SNRA.[18] This report concluded that as well as reducing overall bridge costs, using SCC has the potential to reduce the energy consumption and emission of greenhouse gases by 20–30% due to the reduced resources required in the construction process and the enhanced durability of the resultant concrete. However, it does not take into account the likely increased cement content used to make the SCC and the additional energy required for its production.

The first major event dedicated to SCC in Europe was the 1st International RILEM Symposium on SCC in Stockholm in 1999.[15] The papers included 23 from Asia, 38 from Europe, five from North America and one from Australia. The symposium attracted 340 participants from 35 countries, which is a ratio of 5 to 1 of participants to papers, indicating the increasing interest in the material in Europe.

4. European development

Development and use of SCC began to quickly spread from Sweden to other Scandinavian countries at the end of the 1990s. The concrete industry in countries such as France and the Netherlands have also recently been developing and using the material, with countries such as Germany and the UK progressing closely behind. All of these countries have now used SCC in both *in situ* and precast applications. Countries such as Greece and the Eastern European countries are still mainly conducting research and development or are at the initial site trial stage.

Two European-wide research projects are currently under way in the field of SCC.

(*a*) Following the completion of the European-wide SCC Brite-EuRam project in 2000, it was clear that the remaining fundamental obstacle to the material's wider use in Europe was the absence of suitable test methods to identify its three key properties (i.e. passing ability, filling ability and resistance to segregation). The EU therefore agreed to support an additional three-year project from 2001 to 2004 through its Growth programme (Growth Project GRD2-2000-30024) to develop test methods for SCC and to prepare the way for European standardisation. The project involves twelve European partners and is led by the University of Paisley, Scotland. For additional details see Reference 19.

(*b*) A RILEM committee was also formed in September 2000 to focus on assembling relevant existing knowledge on various aspects on the casting of SCC.[20] The committee aims to build further on the work of RILEM TC 174-SCC mentioned in the previous section and the objective is to find links between the fundamental basic mechanisms of SCC and the current practical experience regarding SCC.[21] The knowledge will be further developed through technical discussions and a workshop and will result in a report on the existing available knowledge on the casting of SCC.

A specification and guideline document has recently been produced by EFNARC (European Federation of Producers and Contractors of Specialist Products for Structures) which aims to provide a framework for the design and use of high-quality SCC in Europe based on the latest research findings combined with field experience.[22] It is probably the first European guidance document to be universally available, and is freely available on the internet.[23] It is intended that the document be updated as SCC technology evolves and advances.

Most of the major European countries are currently in the process of developing guidelines or specifications for the use of SCC. Table 1 shows a snapshot of the current stage in the development of SCC guidelines or specifications for each of the European countries listed.

Table 1 SCC Guideline and specification development in Europe as of September 2002 (based upon a paper by Dingenouts[24])

Country	Guideline organisation	Acceptance phase	Publication date
Austria	n/a	Draft	2002*
Denmark	n/a	Draft in preparation	n/a
Europe	EFNARC	Guideline	2002
Finland	n/a	Draft	2003*
France	AFGC	Industry recommendation	2000
Germany	Annex to DIN 1045	For comment	2003*
Italy	Annex to EN 206	For comment	2003*
	Italian Ready-mix Assoc.	In preparation	n/a
Netherlands	BRL 1801	Approval	2002
	TC 73/04	Accepted	2001
Norway	Norwegian Concrete Society	Accepted	2002
Sweden	Swedish Concrete Assoc. (SCA)	Accepted	2002

* Expected.

4.1 Scandinavia

As described earlier, Sweden was at the forefront of the development of SCC outside Japan and it is estimated that SCC now accounts for approximately 7–10% of the Swedish ready-mix market,[25] up from approximately 3% in 2000. Currently, the CBI, four universities and the government in Sweden are all conducting research into SCC.

SCC is often used in Sweden today by contractors such as NCC on a commercial basis.[26] NCC and other Swedish contractors are also conducting research into SCC, both internally and with other companies and universities/institutes. The Swedish Concrete Association (SCA) has also recently published recommendations for the use of SCC.[27]

The volume of SCC produced in Norway has increased from approximately 0·5% (or 12 000 m^3) of total concrete volume in 2000 to approximately 1·2% (or 29 500 m^3) in 2001.[28] A Norwegian guideline for the production and use of the material was issued in late 2001 and an English translation has recently been completed.

SCC is used in Finland only to a limited degree, although companies such as Lohja Rudus and Parma Betonila each have experience from approximately 10–20 construction sites.[29] There is also a national project on the practical aspects of SCC led by the Technical University of Helsinki (HUT) and VTT which began in 2001 and will finish in 2003.[29] No SCC standardisation as yet exists.

Universities in Denmark, such as the DTU (Technical University of Denmark), have also recently undertaken research into SCC, as well as holding training courses on SCC.

The first project investigating SCC in Iceland was from 1996 to 1999 and was a collaboration between the Icelandic Building Research Institute (IBRI) and an Icelandic ready-mix concrete company, Steypustö UNKNOWN SYMBOL 240 FONT=Times New Roman in H. F., which showed that it was possible to cast SCC in Iceland with Icelandic materials.[37] IBRI continue to conduct research into SCC and they are hosting the 3rd International Symposium on SCC in August 2003.[38]

4.2 France

France is quite active in the research and development of SCC.[30] A national research project on SCC called BAP (Bétons auto-plaçants) is currently ongoing. French recommendations for the use of the material were established in July 2000[31] and are used as reference on construction sites.

The Lafarge Group have conducted a large amount of research and development at their Laboratoire Central de Recherches (LCR) at L'Isle d'Abeau, near Lyon. Their progress in the development of SCC is approximately two years ahead of the position in the UK. They have spent approximately £2 million on researching and developing the material internationally and currently produce approximately 50 000 m^3/annum of SCC, with this volume increasing almost exponentially at present.[32]

The Lafarge Group wanted to validate the assumption that using SCC generates an overall cost saving. They therefore worked with a contractor to simultaneously construct two identical apartment buildings in Nanterre, France. Conventional concrete methods were used for one building and for the second building the construction process was adapted to utilise SCC materials and processes. The building constructed using SCC materials and processes was completed 2·5 months before the conventionally constructed building and with an overall project cost saving of 21·4%.[33]

4.3 Germany

In Germany, SCC requires technical approval before it can be used on site.[34] The current DIN standards do not allow this type of concrete to be used because the consistency and the fines content do not comply with the standard. Therefore, the DIBt (German Institute of Technical Approvals in Berlin) requires suitability tests from a third-party laboratory, usually universities, who then issue an official approval. Many contractors have obtained approvals and are constructing with SCC.

At least six different universities and research establishments in Germany are also conducting research into SCC. The University of Stuttgart is also involved with the current European-wide SCC test methods project (Growth GRD2-2000-30024).

4.4 Belgium

A Belgian national contact group on SCC exists chaired by Professor De Schutter of the University of Ghent.[35] This group of universities, contractors, suppliers and other interested parties meet several times a year to discuss SCC development in Belgium.

Several national research projects on SCC are also currently under way, funded mainly by the National Fund for Scientific Research, Flanders, and other interested parties. These projects are investigating the transport of potentially aggressive media in SCC, the spalling behaviour of SCC and the integration of SCC into the building industry.

4.5 Spain

SCC production is just beginning in Spain, but the first structures have already been constructed in Malaga, Valencia and Madrid. The current problem is that SCC is not included in the basic Spanish regulations for concrete, and so care has to taken and its use fully justified. Instituto Eduardo Torroja de Ciencias de la Construcción (CSIC) is leading the standardisation group for SCC in the Spanish National Standardisation Agency (AENOR).[36] The target is to issue some Spanish standards on SCC by the end of 2002. They are also collaborating with other Spanish research groups such as UPC (Universitat Politècnica de Catalunya) in Barcelona.

4.6 Holland

The precast concrete industry in the Netherlands first became interested in SCC in 1998. A project was begun to develop SCC precast applications by the Belton Group,[39] an association of 24 precasters, which is a subsidiary of BFBN (Association of the Dutch Concrete and Precasting Industry). Although the Belton project was formally completed in December 1999, intensive collaboration in SCC is still continuing. In 2002 a total volume of 250 000 m^3 of precast SCC is expected to be produced[40] and approximately 30 companies have been certified for the production of SCC pre-cast elements.[24]

Guidelines for ready-mix SCC are finished and are expected to be approved by the end of 2002, at which point the certification of ready-mix plants can begin.[24] Currently, ready-mix SCC is being used, but mainly for demonstration projects.

4.7 Switzerland

SCC currently accounts for approximately 1% of the ready-mix concrete market in Switzerland and the material has largely been developed by trial and error.[41] The use of SCC is currently not limited by Swiss standards because the standards apply to performance and not the composition of concrete. Research into SCC has been conducted at the Swiss Federal Institute of Technology (ETH) in Zurich and at the EMPA (Swiss Federal Institute for Materials Testing and Research) in Dübendorf.[41]

4.8 Italy

In Italy the majority of SCC applications are in the precast market, although SCC has been used for *in situ* applications. The Italian Standards Institute has just completed a document on SCC, which is now being considered as a Standard (as an annex to EN 206-1).[42] The Italian Ready-Mixed Concrete Producers Association is also preparing guidelines on SCC.

4.9 Other European countries

Other countries in Europe such as the Czech Republic and Greece are also beginning to research and construct with SCC. A bridge abutment and an experimental tunnel lining have both been cast with SCC in Prague[43] and a trial was held for the use of SCC in the 2·5 km cable-stayed bridge over the Corinthian Gulf in Greece.[44]

5. UK development

Until about four years ago, interest in SCC in the UK was largely confined to research studies at Paisley University and University College London (UCL). There has, however, been a rapid increase in interest from UK industry in the last three years. Two of the first site applications were in 2000 for the Midsummer Place shopping centre extension in Milton Keynes and the Millennium Point project in Birmingham.

The Concrete Society in the UK formed a working party early in 1999 consisting of 18 experts from universities, suppliers, producers, designers and contractors with the aim to maintain a watching brief on the impact and uptake of SCC in the UK. They have recently produced both an information sheet[45] and Part 1 of a Current Practice Sheet.[46] The Part 1 report deals with materials, properties, production and placing of SCC. At the time of writing, Part 2, which will cover production, placing and optimisation of the construction process, was 90% complete.

A three-year research project into SCC in the UK is also currently being conducted by the Building Research Establishment's (BRE) Centre for Concrete Construction. This work is funded by the UK Department of Trade and Industry's (DTI's) Construction Industry Directorate. The aim of the project is to encourage the wider use of SCC by the UK construction industry, and to demonstrate the economic, social and environmental advantages it offers over traditional vibrated concrete construction. The work involves consulting with UK industry on its perceptions of SCC, researching the production of SCC mixes suited to industry needs and disseminating this and additional information through demonstration events and the publication of case studies and guidance material.

5.1 Academic institutions

The leading academic institution in the UK with regard to SCC is the Advanced Concrete and Masonry Centre at the University of Paisley.[47] This team is led by Professor Bartos and also includes John Gibbs and Dr Mohammed Sonebi. They were key members of the original Brite-EuRam project (BRPR-CT96-0366) mentioned in the previous section and are leading the current European-wide project on test methods. In addition, Professor Bartos is Chairman and John Gibbs the Secretary of the Concrete Society Working Party on SCC. Paisley University also began an 18-month Government-funded research project in December 2001 aimed at reducing the cost of SCC and encouraging its use in general construction.

As mentioned earlier, the other academic institution in the UK with a history of research into SCC is University College London (UCL). The work into SCC is being coordinated by Dr Peter Domone, who is also a member of the Concrete Society Working Party on SCC. UCL is also involved with the European-wide project on the testing of SCC.

Dundee University's Concrete Technology Unit has also recently begun a three-year research project investigating the formwork pressures generated by SCCs and other specialist concretes.[48] This project will determine if the current CIRIA formwork pressure recommendations are applicable to current new materials and types of concrete such as SCC, and if not, safe design pressures will be established.

5.2 Concrete producers

RMC Ready-mix Ltd is the UK's leading supplier of ready-mixed concrete. The company has been undertaking development work on SCC since 1998 with the aim of producing mix design criteria for materials local to any of their 325 batching plants throughout the UK.[49] It was originally thought that the material would be a niche product, but RMC is now concentrating upon developing more economical mixes to enable the material to be more universally used.[50] This development is targeted at customer needs in terms of the application, strength requirement and other technical factors that may be specified.

Lafarge Aggregates Ltd have been conducting research and development into SCC in France for several years, and it is now available at each of their plants in the UK. Lafarge produce their own family of admixtures especially for the production of SCC, called Agilia. These products are fully developed and are all currently available and Lafarge see the next step as getting these existing products accepted and used by the market.[32]

Tarmac Topmix Ltd first began researching and developing SCC in 1998 and first started using it at their precast factory at Tallington in 1999.[51] The first ready-mixed SCC was supplied in 2000 and the eventual aim is to make it available at all plants in the country.

5.3 Admixture suppliers

Sika Ltd launched Sika ViscoCrete in the UK in 1998, which was the first admixture on the UK market specifically for producing SCC. Six different types are available and it can be used with either retarders or accelerators. Other admixture manufactures have now followed this lead. MBT and Grace Construction Products both produce admixtures for the UK SCC market, including superplasticisers, viscosity-modifying admixtures and specialist admixtures for precast applications. The ready-mix companies RMC and Lafarge produce their own admixtures for SCC.

5.4 Consultants

Consultants in the UK do not generally undertake any concrete research or development directly, but do sometimes sponsor or collaborate in research projects at universities and research organisations. Mott MacDonald have an established track record with the development of specifications and guidance notes for SCC and are a member of the Concrete Society Working Party on SCC. They have also worked closely with ready-mix suppliers and contractors to use the material on a variety of projects throughout the UK.[52] However, many UK consultants are cautious when specifying SCC due to the lack of existing guidance, standards and test methods and the lack of an established track record for the material.

5.5 Contractors

Contractors in the UK generally rely upon the ready-mix supplier for any expertise in SCC mix design. However, knowledge and appreciation of the material among contractors is slowly increasing as the use of SCC becomes more widespread.

John Doyle Construction are probably the leading contractor in the UK in the use of SCC and are involved in several research and development projects such as the European SCC test methods project and the project on formwork pressures at Dundee University.[53] They have used the material several times in the UK in actual applications such as column encasement and basements, although its use is still job-specific. Site trials are always conducted before SCC is used in the intended structure.

Other contractors have used SCC in the UK but its use is still job-specific, often to solve a problem during construction such as congested reinforcement.

5.6 UK precasters

Aarsleff Piling's Balderton premises claim to be the UK's first user of SCC in precast pile production.[54] An SCC mix was developed in partnership with RMC with a compressive strength of 26–28 MPa at only 16 h. At nine days the 50 MPa mark had been passed and the piles could be driven. Aarsleff have also won a £4·25 million contract to supply and install approximately 3000 precast concrete piles for the Channel Tunnel Rail Link.[55] SCC was used for these piles and casting began in June 2002 with installation beginning the following month.

Tarmac Precast Concrete Ltd are also now using SCC at their three UK precast plants. SCC currently accounts for approximately 65% of bespoke production at their Tallington plant and their long-term objective is to convert fully to SCC when economics and practicalities allow.[56] Current contracts are for prison units (Fig. 5), double-T beams and columns (Fig. 6).

Trent Concrete Ltd in Nottingham have been experimenting with SCC for the last two years and production using the material first began in August 2002.[57]

Fig. 5 A precast unit being cast in SCC (Tarmac Precast Concrete Ltd)

Fig. 6 A precast column cast in SCC (Tarmac Precast Concrete Ltd)

5.7 Seminars and events

The SCI (Society of Chemical Industry) held a seminar on SCC in January 2001 in London. Approximately 60 delegates attended and the speakers were the main practitioners of SCC in the UK from the areas of research (Professor P. Bartos), design (Dr N. Henderson), ready-mix (R. Gaimster) and contracting (P. Goring). This meeting was one of the first events in the UK to begin to introduce the material to members of the construction industry who had no experience or knowledge of SCC.

A demonstration and international workshop on SCC, lasting a total of two and a half days, was held at the Advanced Concrete and Masonry Centre at Paisley University in May/June, 2001.[58] The event was attended by 70 delegates from eleven countries and helped raise the awareness and knowledge of the material in Scotland and the rest of the UK.

Four demonstration events on SCC around the UK have also recently been held by BRE's Centre for Concrete Construction as part of their DTI-funded project on SCC, which combined practical demonstrations with presentations by industry experts on SCC. The events were aimed at people in the industry with little or no knowledge of the material, with more than 140 people attending the four events.

6. The future

From its origins in Japan in the late 1980s, research, development and use of SCC has spread steadily throughout the world. Approximately 7–10% of the Swedish ready-mix market is now SCC and research is being conducted in virtually every country in Europe. National working groups on the material now exist in some form in most countries in Europe, with the majority of them working towards producing some form of guidance on the use of SCC.

Large amounts of research and development are now also being conducted in Canada and the USA at places such as the University of Sherbrooke and the Université Laval. The first North American conference on SCC was held in November 2002 by Northwestern University. American concrete organisations such as the American Concrete Institute (ACI), American Society for Testing and Materials (ASTM) and PCI (Precast/Prestressed Concrete Institute) have all recently formed committees to produce guidelines, standards and specifications for SCC. The material is slowly gaining acceptance in this huge market and already it is estimated that approximately 4000–5000 m^3 is used per day in precast applications in North America. Nearly 100 000 m^3 of ready-mixed SCC has also so far been used in North America.[59]

Countries such as Argentina, Australia and New Zealand are also all now beginning to conduct research and development work into SCC, with more countries sure to follow. A third International Symposium on SCC is being held in Iceland in August 2003, where much of the research and development work presently under way will be presented. For further details see Reference 60.

The main barrier to the increased use of SCC in the UK and Europe seems to be the lack of experience of the process, and the lack of published guidance, codes and specifications. This situation will improve, however, as experience and knowledge increases and each country begins to produce its own guidance and specifications.

Precasters are currently the overwhelming users of SCC in the UK, in Europe and in the USA. This is partly due to them owning and operating the on-site batching plant, and so they are able to take full advantage of all the potential benefits of SCC. They can also minimise or control the potential disadvantages of the material (such as inconsistency of supply and site acceptance). In the countries where SCC has been adopted relatively quickly, such as Japan and Sweden, the ready-mix concrete producers are owned or operated by the contractors, therefore the increased material costs can be directly offset by savings in the construction process, in a similar way to precast production.

Although SCC is not expected to ever completely replace conventionally vibrated concrete, the use of the material in both the precast and ready-mix markets in the UK, Europe and the rest of the world is expected to continue to increase. The main drivers for this increase in use are expected to include

- an increase in the experience of both producers, contractors, designers and clients
- an increase in available guidance on the production, design and use of SCC
- a decrease in the unit cost of the material as technology and experience improves
- the demand from clients for a higher-quality finished product
- the decrease in skilled labour available in many countries for both the placing and finishing of concrete.

7. Acknowledgements

The author wishes to acknowledge the support provided in collating the information for this paper from the BRE Framework contract on SCC, which is currently being undertaken by BRE for the Construction Industry Directorate, DTI. The author also wishes to thank the industry participants in this programme, including BCA, Concrete Admixtures Association, Concrete Society, John Doyle Construction Ltd, Lafarge UK Ltd, Mott MacDonald Ltd, RMC Readymix Ltd and the University of Paisley, as well as the many individuals who have supplied the information required in the writing of this paper.

REFERENCES

1. Okamura H., Ozawa K. and Ouchi M. Self-compacting concrete. *Structural Concrete*, 2000, **1**, No. 1, Mar., 3–17.
2. Tanigawa Y., Mori H., Yonezawa T., Izumi I. and Mitsui K. Evaluation of the flowability of high-strength concrete by L-flow test. *Proceedings of the Annual Conference of the Architectural Institute of Japan*, 1989/1990.
3. Ozawa K., Maekawa K. and Okamura H. High performance concrete with high filling ability. *Proceedings of the RILEM Symposium, Admixtures for Concrete*, Barcelona, 1990.
4. Tangtermsirikul S., Sakamoto J., Shindoh T. and Matsuoka Y. Evaluation of resistance to segregation of super workable concrete and the role of a new type of viscosity agent. *Reports of the Technical Research Institution*, 1991, Taise: Corporation, Japan, No. 24, pp. 369–376.
5. Ozawa K., Maekawa K. and Okamura H. *Development of High Performance Concrete*. University of Tokyo, 1992, Faculty of Engineering journal.
6. Sakamoto J., Matsuoka Y., Shindoh T. and Tangtermsirikul S. Application of super workable concrete to actual construction. *Proceedings of Concrete 2000 Conference, University of Dundee*, 1993.
7. Kuroiwa S., Matsuoka Y., Hayakawa M. and Shindoh T. *Application of Super Workable Concrete to Construction of a 20-storey Building*. ACI, 1993, SP-140.
8. Ozawa K., Sakata N. and Okamura H. Evaluation of self-compactability of fresh concrete using the funnel test. *Concrete Library of JSCE*, No. 25, 1995.
9. Ouchi M. and Ozawa K. (eds). *Proceedings of the International Workshop on Self-Compacting Concrete*, Kochi, 1998, Japan Society of Civil Engineers, ISBN 4-8106-0310-5.
10. Uomoto T. and Ozawa K. (eds). Recommendation for SCC. *JSCE Concrete Engineering*, Series 31, Tokyo, Japan, 1999.

11. Ozawa K. and Ouchi M. (eds). Proceedings of the 2nd International Symposium on Self-Compacting Concrete, Tokyo, 2001, 743, ISBN 4-90154-04-0 (www.infra.kochitech.ac.jp/sccnet/scc2/).
12. Fib concrete structures in the 21st century. *Proceedings of the 1st Fib Congress*, Osaka, 2002, (www.fib2002.com/e/index.html).
13. Domone P. L. and Chai H. W. Design and testing of SCC: Production methods and workability of concrete. *Proceedings of an International RILEM Conference*. E & F N Spon, London, 1996, ISBN 419 22070 4.
14. Petersson O. and Billberg P. A model for SCC: Production methods and workability of concrete. *International RILEM Conference*. E & F N Spon, London, 1996, ISBN 419 22070 4.
15. Skarendahl A. and Petersson Ö. (eds). *Proceedings of the 1st International RILEM Symposium on SCC*, RILEM Publications, 1999, p. 786, ISBN 2-912143-09-8.
16. Billberg P. *Self-compacting concrete for civil engineering structures—the Swedish experience*. Swedish Cement and Concrete Research Institute, Stockholm, 1999, CBI report 2:99, p. 80.
17. Details available on the internet at: http://scc.ce.luth.se.
18. Nilsson M. *Project on Self-compacting Bridge Concrete*. Swedish National Road Administration (SNRA), 1998, Publication 1998:71 E, ISSN 1401-9612, p. 19.
19. Details available on the internet at: www.acmcentre.com/testing-scc/index.php.
20. RILEM. TC CSC: Casting of SCC, www.rilem.ens-cachan.fr/csc.
21. Billberg P. CBI, Sweden. Personal communication, November 2002.
22. European Federation of Producers and Contractors of Specialist Products for Structures. *Specification and Guidelines for SCC*. EFNARC, 2002, ISBN 0-9539733-4-4, p. 29.
23. European Federation of Producers and Contractors of Specialist Products for Structures (EFNARC). *Specifications and Guidelines for Self-Compacting Concrete*. EFNARC, Surrey, 2002. Also available on the internet at: www.efnarc.org.
24. Dingenouts M. Review of SCC Product Acceptance Standardisation. *Proceedings of Paisley University Workshop*, September 2002.
25. Skarendahl A. Swedish Construction Sector Innovation Centre, BIC. Personal communication, August 2002.
26. Grauers M. NCC Construction Sverige AB, NCC Engineering, Sweden. Personal communication, July 2002.
27. Swedish Concrete Association. *SCC Recommendations for Use*, SCA, 2002, Report series No. 10, p. 84.
28. Johansen K. Sintef, Trondheim, Norway. Personal communication, August 2002.
29. Kukko H. VTT Building and Transport, Finland. Personal communication, September 2002.
30. De Larrard F. Laboratoire Central des Ponts et Chaussées, France. Personal communication, July 2002.
31. Association Française de Genie Civil. *Bétons Auto-plaçants: Recommandations provisoires*, AFGC, Bagneux, France, July 2000.
32. Richardson G. Lafarge Aggregates Ltd. Personal communication, July 2002.
33. Richardson G. *SCC Demonstration*, BRE, Watford, October 2002.
34. Reinhardt H. W. University of Stuttgart. Personal communication, July 2002.
35. De Schutter G. Magnel Laboratory for Concrete Research, Ghent University. Personal communication, August 2002.
36. Revuelta D. Instituto Eduardo Torroja de Ciencias de la Construcción (CSIC), Madrid, July 2002.
37. Níelsson I. Icelandic Building Research Institute (IBRI). Personal communication, July 2002.

38. Icelandic Building Research Institute. For details of 3rd International Symposium see www.ibri.is/scc.
39. Belton Group. For further details of SCC project see www.belton.nl.
40. Bennenk W. and Van Schijndel J. SCC in the daily pre-cast concrete practice. *Proceedings of the 17th International Congress of Pre-cast Concrete Industry*, Istanbul, 2002.
41. Leemann A. Swiss Federal Institute for Materials Testing and Research (EMPA). Personal communication, July 2002.
42. Khurana R. MBT. Personal communication, July 2002.
43. Vitek J. L. Metrostav a.s., Czech Republic. Personal communication, August 2002.
44. Kanakari H. K. Bridging the north–south divide. *Concrete Engineering*, 2001, **5**, No. 2, 6–8, 10–12.
45. Concrete Society. Self-compacting concrete. *Concrete Magazine*, supplement, 2001, **35**, No. 1.
46. Gaimster R. and Gibbs J. Self-compacting concrete current practice sheet No.123. *Concrete*, 2001, July/August, 32–34.
47. Advanced Concrete and Masonry Centre. For more details see www.acmcentre.com.
48. Dundee University Concrete Technology Unit. Details available at: www.dundee.ac.uk/civileng/research/concrete/formwork1.htm.
49. Gaimster R. and Foord C. Self-compacting concrete. *Concrete*, 2000, April, 23–25.
50. Hartley A. RMC Ready-mix Ltd. Personal communication, August 2002.
51. Sealey B. Tarmac. Personal communication, October 2002.
52. Henderson N. Mott MacDonald. Personal communication, July 2002.
53. Goring P. John Doyle Construction. Personal communication, July 2002.
54. Parker D. Star of the cast takes mould breaking role. *New Civil Engineer*, October 2000, pp. 33–34.
55. Concrete Engineering International (CEI). *Pre-cast Piling Contract*. CEI, 2002, 65.
56. Richards C. Tarmac Pre-cast Concrete Ltd. Personal communication, August 2001.
57. Downing M. Trent Concrete. Personal communication, September 2002.
58. Bartos P. Self-compacting concrete demonstration and international workshop. *Concrete*, 2001, July/August, 30–31.
59. Vachon M. and Daczko J. U.S. Regulatory Work on SCC. *Proceedings of the 1st North American Conference on the Design and Use of Self-Consolidating Concrete (SCC), Rosemont, Illinois, USA*, 2002, 423–426.
60. Details available at: www.ibri.is/scc.

Section 2

Designing self-compacting concrete

ice Publishing

Dhir and Lye
ISBN 978-0-7277-6461-4
https://doi.org/10.1680/icetscc.64614.049

Chapter 3
Design of self-compacting concrete with fly ash

Pasla Dinakar
Assistant Professor, Indian Institute of Technology, Bhubaneswar, India

Concrete of a desired strength for any specific application can be obtained by using any of the accepted methods of design. However, the addition of mineral admixtures, such as fly ash, in concrete, requires a systematic design procedure for obtaining the required strength. It is possible to design a fly ash concrete of the required strength through the 'efficiency concept', and the same is also true for the design of self-compacting concrete (SCC). In this paper, a new mix design methodology is proposed for proportioning of self-compacting fly ash concretes based on the efficiency concept. The methodology has already been successfully verified through a proper experimental investigation and the self-compacting fly ash concretes were evaluated for their self compactability and strength characteristics. The results indicate that the proposed method is capable of producing high-quality SCC.

Introduction

The development of self-compacting concrete (SCC), also referred to as 'self-consolidating concrete', has recently been one of the most important developments in the building industry (Brouwers and Radix, 2005). SCC can flow through and fill gaps in reinforcement and corners of moulds without any need for vibration and compaction during the pouring process. Owing to its rheological properties, SCC has many advantages. SCC may contribute to a significant improvement in the quality of concrete structures and open up a new field for the application of concrete. It has the ability to compact itself only by means of its own weight without the requirement of vibration (Okamura, 1997). It fills all recesses, reinforcement spaces and voids, even in highly reinforced concrete members. While flowing in the formwork SCC is able to deaerate almost completely. The use of SCC offers many benefits to construction practice: the elimination of the compacting work results in reduced costs of placement, a shortening of the construction time and, therefore, in an improved productivity. The application of SCC also leads to a reduction of noise during casting, better working conditions and the possibility of expanding the placing times in inner-city areas (Persson, 2001; Sonebi, 2004).

However, SCC has also some disadvantages: SCC needs a high level of quality control of the materials being used (sand, coarse aggregate, filler, superplasticitiser and cement) and of the adopted mix, since the concrete properties in fresh state can be altered, presenting excessive fluidity, segregation or excess of cohesion. The stricter requirements on measuring, adjustments and control is, however, demanding, possibly excluding some of the less qualified and less well-equipped concrete producers (EFNARC, 2005). Often the material costs of SCC will be higher than the equivalent material costs of normal vibrated concrete. However, when SCC is sensibly utilised, the reduction of costs by better productivity, shorter construction time and improved working conditions will compensate the higher material costs (Okamura, 1997). On the other hand, in densely reinforced sections, such as those in columns and beams in moment-resisting frames in seismic areas and in some repair sections, it is quite difficult to cast normal concrete and ensure that all spaces in the formwork are filled with concrete. The use of SCC is necessary in these cases.

The basic components for the mix composition of SCC are the same as those used in conventional concrete. However, to obtain the required properties of fresh SCC, a higher proportion of ultra fine materials and the incorporation of chemical admixtures, effective superplasticisers in particular are necessary. Filler materials commonly used are fly ash, limestone powder, blast furnace slag, silica fume and quartzite powder (Goodier, 2003; Nagamoto and Ozava, 1997; Okamura and Onchi, 1999). Of all these filler materials, fly ash has the highest potential to be used in SCC because of its availability and particle shape. Moreover, it is inexpensive. This paper presents a new mix design methodology for the design of SCC with fly ash for any percentage replacement of fly ash.

Approaches for SCC design

Freshly-mixed SCC of good quality is characterised by an excellent deformability (high deformability of paste and low inter-particle friction) and a good stability (minimal segregation and bleeding). Both of these properties are, however, dependent on a large number of highly complex and interactive parameters that are not easily amenable to mathematical representation and thus often are only vaguely defined. In general, SCC requires a manipulation of the concrete constituents and/or an introduction of chemical agents, such as superplasticisers and viscosity modifying admixtures. To achieve and maintain the ability of SCC to flow under its own weight and eliminate the possibility of segregation and blockage, researchers suggest limiting the absolute volume of coarse aggregate used in the mixture and improving the particle size distribution of solid particles to reduce inter-particle friction (Yurugi *et al.*, 1993). Any reduction in the total volume of the coarse aggregate, must be balanced by increasing the volume of the cement mortar, which often leads to a higher material cost and a significant temperature rise, especially in massive concrete structures.

In general, the available concepts for proportioning SCCs were categorised as: the powder type (proportioned through additional powder material at lower water–powder ratios to provide adequate segregation resistance); the viscosity modifying agent (VMA) type (by using a VMA to provide segregation resistance; and the combination type (a combination of both), with the superplasticisers and air-entrainers used for obtaining the desired deformability (Japan Concrete Society, 1999). Also, to meet the structural performance requirements which take into account the shape, dimensions and reinforcement conditions, three levels or ranks of self-compactability are suggested, based on the reinforcement clearance (around 60, 60–200, over 200 mm) or the corresponding steel content (around 350, 100–200, less than 100 kg/m^3) (Japan Concrete Society, 1999). However, there is no National Standard available for the design process of an SCC, unlike conventional concrete. There are guidelines for mixture proportions published in various works, although there is no specific SCC mix design for a specific strength to date. Owing to the fact that SCC can be achieved through various methods,

Table 1 Regulations for self compacting concrete given by EFNARC (2005)

Parameters	EFNARC guidelines
Volume of paste: litres/m^3	300–380
Powder content: kg/m^3	380–600
Water content: kg/m^3	150–210
Fine aggregate in total aggregate: %	48–55
Size of coarse aggregate: mm	≤ 20
Slump flow (SF) class: mm	
SF1	550–650
SF2	660–750
SF3	760–850
Viscosity class (V-funnel time in s)	
VF1	≤ 8
VF2	9 to 25
Passing ability classes (L-box)	
PA1	$\geq 0{\cdot}8$ with 2 rebars
PA2	$\geq 0{\cdot}8$ with 3 rebars

no single mixture design encapsulates all the mixture variables. In general, compared to conventional concrete, SCC needs a higher paste volume and a lower coarse aggregate content in order to make the aggregates float. In this investigation the increase in the powder content is accomplished through the use of fly ash. With the proposed method, all we need to do is to select the required qualified materials, do the calculations, conduct mixing tests and make some adjustments in such a way that SCC with good flowability and segregation resistance can be obtained with self-compacting ability as specified by Experts for Specialised Construction and Concrete Systems (EFNARC) (Table 1).

Fly ash in self-compacting concrete

The development of self-compacting concrete with fly ash is a positive contribution to the sustainability of concrete technology. SCCs acquire a high slump through the addition of a superplasticiser. To maintain cohesion during handling, special attention is needed in terms of the sand and paste content, apart from a viscosity-modifying admixture to enhance stability (Khayat and Guizani, 1997). The use of mineral admixtures, such as fly ash, could increase the slump of the concrete mixture without increasing its cost. The use of fly ash also reduces the demand for cement, fine fillers and sand that are required in SCC (Khurana and Saccone, 2001). Moreover, the incorporation of fly ash also eliminates the need for viscosity-enhancing chemical admixtures. The lower water content of the concrete results in a higher durability, in addition to better mechanical integrity of the material. Also, the use of fly ash improves rheological properties and reduces cracking of concrete owing to the heat of hydration of the cement. It was reported that replacement of cement by about 30% fly ash helped in achieving improved rheological properties and super flowing concretes (Kim *et al.*, 1996; Miura *et al.*, 1993). Researchers also attempted to produce high-volume fly ash SCCs by replacing up to 85% of Portland cement with class F fly ash (Dinakar *et al.*, 2008a, 2008b). Fly ash in SCCs also improved the durability and corrosion properties studied (Dinakar *et al.*, 2008a, 2008b, 2009).

Earlier mix design methods and limitations

A number of procedures such as empirical methods, rheology-based methods, statistical methods and particle packing models have all been proposed for designing SCCs to date. Okamura proposed a mix design method for SCC based on paste and mortar studies for superplasticiser compatibility followed by trial mixes (Okamura *et al.*, 2000). However, it is emphasised that the final product should be tested for passing ability, filling ability, flow and segregation resistance. Sedran and de Larrard (1996) have proposed a method based on a rheometer, which depicts the rheological characteristics of fresh concrete, and software based on a mathematical model for optimising the aggregate–binder granular skeleton. Su *et al.* (2001) proposed a mix design methodology based on a high fines approach. In this procedure the required amount of coarse aggregates is determined first. Then, the binder paste fills the voids between these aggregates to ensure concrete flowability, self compactability and other SCC properties. The amount of coarse and fine aggregates is then determined based on the optimal packing factor (in the range 1·12–1·16) and the volume of sand to mortar is fixed in the range of 54 to 60%. A water content of about 170–176 kg/m^3 is used for obtaining a medium compressive strength in the range 27–48 MPa. The SCC contains more sand and less coarse aggregate. This approach enhances the passing ability and reduces the amount of binder needed. Gettu *et al.* (2002) proposed a methodology for the design of high-strength SCC with fly ash, similar to Okamura (1997). First, the optimum superplasticiser dosage for cement and fly ash was determined using a Marsh cone. In the second step the mini-slump test was used to choose the optimum fly ash dosage and consequently the paste composition with the prescribed water/cement (w/c) ratio. Next the aggregate proportions were obtained based on the minimum void content of the aggregate skeleton. Finally, concretes with paste volumes and the optimum sand/gravel ratio were tested. Their experiments, aimed at strength of 50 MPa at 7 days, had a fly ash/cement ratio of about 0·60 and an optimal paste content of 45% by volume of the concrete. In spite of an understanding of the concepts and requirements of the mix design methodology, literature available today does not suggest any specific procedure for obtaining SCCs of a definite strength, unlike normal concretes.

Proposed method for proportioning fly ash self-compacting concrete

It should be recognised that the inexpensive, abundantly available and low-efficiency fly ash is a ready-to-use powder (unlike the ground granulated blastfurnace slag (GGBS) or limestone powder that needs grinding) that can be directly added as an additional cementitious constituent. Incidentally, Babu and Rao (1993) proposed a methodology for the design of fly ash concretes of a desired strength (20–120 MPa) at a desired fly ash percentage (85–10%) using the efficiency concept. From this point of view, the present investigation attempts to extend this concept and realise the development of high-volume, high-strength, high-performance SCCs. The advantage of this method is that at higher replacement percentages of fly ash the SCCs are more economical than the use of superplasticisers in SCC. However, this would require specific adjustments to all the other ingredients such as sand, coarse aggregate, superplasticisers and water, to arrive at an optimal mix proportion. The procedure of the proposed mix design method can be summarised in the following steps

Step 1: calculation of total cementitious or powder content for SCC

In the mix proportioning of conventional concretes, the water content is fixed based on the maximum size of the aggregate and/or aggregate grading. In the case of SCC, the quantity of total fines (powder) is of importance. In view of this, fix the total cementitious materials (TCM) content (preferable to have this around 550 kg/m^3). To understand the behaviour of SCCs one can choose this in the range of 500 to 600 kg/m^3 (EFNARC, 2005).

Let the TCM = TP kg/m^3

Step 2: calculation of efficiency of fly ash and fly ash content

Babu and Rao (1993) proposed the efficiency concept methodology for the design of normal fly ash concretes. As per this methodology, the fly ash content can be varied between 15–75%. The overall cementing efficiencies for fly ash at 7, 28 and 90 days ages proposed by them are given in Figure 1. The corresponding relationships for the overall efficiencies (k_7, k_{28} and k_{90}) at 7, 28 and 90 days for replacement levels varying from 15–75% given by Babu and Rao (1996) are

$$k_7 = 2{\cdot}67p^2 - 3{\cdot}75p + 1{\cdot}45 \quad (1)$$

$$k_{28} = 2{\cdot}78p^2 - 3{\cdot}80p + 1{\cdot}64 \quad (2)$$

$$k_{90} = 2{\cdot}50p^2 - 3{\cdot}59p + 1{\cdot}73 \quad (3)$$

where p is the percentage replacement of fly ash

Figure 1 Variation of efficiency factor (k) at different ages

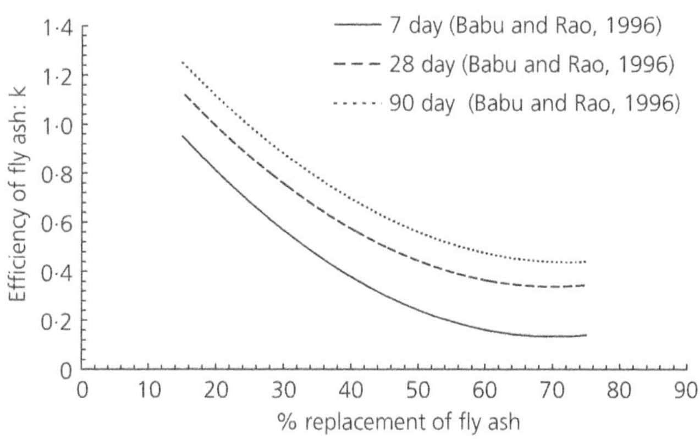

Figure 2 Maximum possible percentage replacement plotted against strength

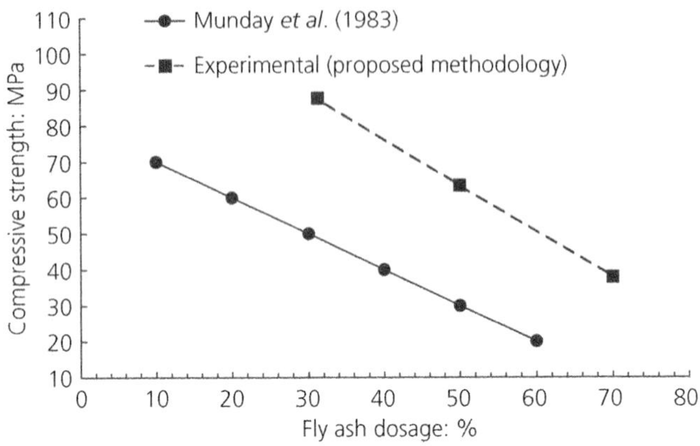

Also, the maximum percentage replacement possible for normal fly ash concretes was decided according to the strength requirements as per Figure 2 (Babu and Dinakar, 2005). This figure presents earlier results proposed by Munday *et al.* (1983). Using the efficiency curve (Figure 1) and the maximum replacement possible for a particular strength (Figure 2) a mix design procedure for the design of concretes with fly ash was proposed by Babu and Rao (1993). The same concept has been extended for the design of SCC with fly ash. In this procedure the 28 day efficiency curve shown in Figure 1 is used for calculating the efficiency of fly ash for any replacements varying between 15 and 75%. The percentage replacement of fly ash is chosen as per the strength requirement using Figure 2. The efficiency of fly ash for this percentage is calculated using Equation 2. However, recent experimental results have shown that it is possible to replace even higher percentages if the aggregate gradings and the filler proportions were to be modulated to minimise the water content needed.

Let the fly ash percentage be p%.
Cement content (c_s) = TP $(1 - p)$ kg/m^3
Fly ash content (f) = TP (p) kg/m^3

The efficiency of fly ash at 28 days (k_{28}) for replacement levels varying between 15–75% is given in Figure 1. For a fly ash replacement of p% the efficiency is calculated using Equation 2.

Step 3: calculation of water content in SCC

Now the water to effective cementitious materials content ratio of SCC with fly ash is calculated using $w_s/(c_s + k_{28}f)$, where w_s is the water content of self-compacting fly ash concrete which needs to be determined.

According to any of the recognised mix design methodologies, the w/c ratio of normal or conventional concretes (w_n/c_n) is chosen based on the compressive strength required. The water content (w) required from the workability consideration is also chosen from the same procedure. In the present investigation the modified ACI relationship (Raja, 1993) was utilised

Figure 3 Strength to water–cement ratio relationship of conventional concrete

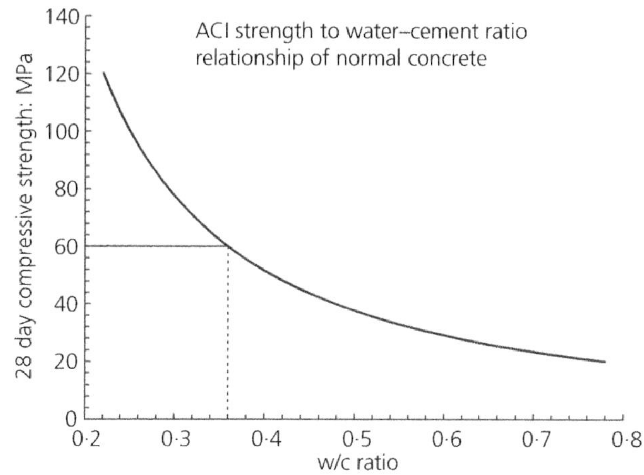

as shown in Figure 3. The water content (w_n) required from the workability consideration is also chosen from the same ACI procedure. From Figure 3 for a desired strength, the w/c ratio (w_n/c_n) is determined. This w/c ratio obtained for normal concrete shall be used to determine the water content of SCC using the following relation

$$w_n/c_n = w_s/(c_s + k_{28}\mathrm{f}) \tag{4}$$

Therefore, $w_s = (w_n/c_n)\,(c_s + k_{28}f)$ kg/m^3

Step 4: calculation of coarse and fine aggregate contents

It is now possible to assess the total aggregate content according to the absolute volume method. The fine aggregate content in the total aggregate is generally recommended to be in the range of 48–55% (EFNARC, 2005). Alternatively one can always follow the continuous grading curves, if required. However, in the present investigation a combined aggregate grading as recommended by the DIN 1045 standards was utilised.

Total volume = 1000 litres;

Assuming air content = 2%, Air = 20 litres

Net concrete volume = 980 litres

Let the cement content be c_s kg/m^3

Fly ash content be f kg/m^3

Water content be w_s kg/m^3

Volume of cement (V_c) = c_s/G_c litres, where G_c is the specific gravity of cement

Volume of fly ash (V_{fly}) = f/G_f litres, where G_f is the specific gravity of fly ash

Volume of water (V_w) = w_s/G_w litres, where G_w is the specific gravity of water

Volume of paste (V_{paste}) = ($c_s/G_c + f/G_f + w_s/G_w$) litres

Volume of total aggregate (V_{agg}) = (980 − V_{paste}) litres

Step 5: calculation of superplasticiser (SP) dosage

The chemical admixtures have the most profound impact on the behaviour of fresh SCC. The dosage of admixtures was adjusted in such a way in order to obtain initial slump–flow values greater than 550 mm, which is necessary for the production of a highly flowable SCC as per EFNARC (2005) guidelines (Table 1). In developing SCCs, polycarboxylate ether (PCE) based admixtures are generally used and, based on experience gained in the current authors' laboratory, it was found that the dosages levels should be between 0·9 to 1·5% of the total cementitious or powder content. Similarly, to attain stability or robustness in the mix, VMAs are also used; the dosage levels of VMAs should be between 0·1 and 0·3% of the total cementitious or powder content. If the dosage of SP used is equal to $n\%$ and that of VMA used is $m\%$ of the total cementitious content (TP), then the dosages can be obtained as follows

$$\text{Dosage of SP used } W_{sp} = n\% \text{ (TP)} \quad (5)$$

$$\text{Dosage of VMA used } W_{vma} = m\% \text{ (TP)} \quad (6)$$

Step 6: adjustment of mix proportion

After obtaining the complete proportions, check once again with the regulations for SCC as presented in Table 1. Once all the calculated proportions fall within the recommendations given in Table 1, go for the development of SCC. If they do not fall within the regulations, redesign the mixture.

Analytical verification – design example

Verification of the mix concept was carried out within the scope of a limited experimental programme. Three different concretes of strengths 30 and 60 and 90 MPa have been designed with the mix design methodology explained above for fly ash replacements varying between 30 and 70%. The mix details are presented in Table 2.

The applied ordinary Portland cement (similar to ASTM type I) and the classified fly ash meet the requirements mentioned in IS:12269 (53 grade) (BIS, 1987) and ASTM C618 (class F) (ASTM, 2002), respectively. The chemical and physical characteristics of cement and fly ash are given in Table 3. Crushed granite with nominal grain size of 20 mm and good-quality well-graded river sand of maximum size 4·75 mm were used as coarse and fine aggregates, respectively. The different size fractions of coarse aggregates (20 mm downgraded, 12 mm downgraded and 6 mm downgraded) were taken in order to obtain a dense concrete. The specific gravities of aggregates were determined experimentally. The coarse aggregates with 20, 12 and 6·0 mm fractions had specific gravities of 2·89, 2·87 and 2·88, whereas the fine aggregate had specific gravity of 2·65, respectively. The high range water reducer (HRWR) used in this study was a commercially available polycarboxylate ether (PCE). Commercially available VMA was also used. As an example, the design procedure is explained for a SCC with design strength of 60 MPa and a cement replacement level of 50%.

Table 2 Mix details of the concretes developed

S. no.	Concrete grade: MPa	Name	TCM: kg/m^3	Fly ash: %	Cement: kg/m^3	(k_{28})	Fly ash (f): kg/m^3	Total aggregate, kg/m^3				Water: kg/m^3	$w/(c + k_{28}f)$	SP: %	VMA: %
								20 mm	12 mm	6 mm	Sand				
1	30	NC30	319	0	319	0	0	722	518	360	368	185	0·58	0	0
2		SCC30	550	70	165	0·34	385	334	448	83	735	175	0·58	1·4	0·3
3	60	NC60	500	0	500	0	0	662	475	330	337	185	0·37	0	0
4		SCC60	550	50	275	0·44	275	362	485	90	796	142	0·37	1·3	0·2
5	90	NC90	552	0	552	0	0	671	481	334	342	160	0·29	1	0
6		SCC90	550	30	385	0·75	165	373	500	93	821	137	0·29	1·5	0·2

TCM – total cementitious materials content (powder content).
k = efficiency of fly ash.
SP – super plasticiser.
VMA – viscosity modifying agent.
NC – normal or conventional concrete.
SCC – self-compacting concrete.

Table 3 Characteristics of cement and fly ash

Chemical composition	Cement: %	Fly ash: %
Silica (SiO_2)	32·9	63·4
Alumina (Al_2O_3)	5·7	25·9
Ferric oxide (Fe_2O_3)	3·9	4·0
Calcium oxide (CaO)	62·5	1·8
Magnesium oxide (MgO)	1·2	0·7
Sodium oxide (Na_2O)	0·1	0·15
Potassium oxide (K_2O)	0·39	1·03
Sulphuric anhydride (SO_3)	2·4	0·20
Loss on ignition (LOI)	1·2	1·1
Blaine (m^2/kg)	370	400
Specific gravity	3·15	2·2

Step 1: calculation of total cementitious or powder content for SCC

Let the TCM = 550 kg/m^3

Step 2: calculation of efficiency of fly ash and fly ash content

For concrete of compressive strength 60 MPa according to Figure 2 the percentage replacement of fly ash should be around 20%, but in the present investigation a higher percentage (50%) was chosen for designing 60 MPa SCC. Similarly for 30 and 90 MPa SCCs percentage replacements offly ash such as 70 and 30 were chosen than the ones given in Figure 2. With the present-day high-grade cements and qualified fly ashes it is possible to realise higher percentages of fly ash. Refer to concrete mixtures given in Table 2.

$$\text{Cement content } (c_s) = 275\ \text{kg/m}^3$$

$$\text{Fly ash content } (f) = 275\ \text{kg/m}^3$$

The efficiency of fly ash at 28 days (k_{28}) for a fly ash replacement of 50% calculated using Equation 2 is 0·44 (k_{28} = 0·44).

Step 3: calculation of water content of SCC

The water to effective cementitious materials content ratio of SCC with fly ash is given by $w_s/(c_s + k_{28}{*}f)$, where w_s is the water content of self-compacting fly ash concrete which is to be determined. From Figure 3 for conventional 60 MPa concrete, the w/c ratio (w_n/c_n) is 0·36. This w/c ratio is used to determine the water content of SCC using Equation 4

$$0{\cdot}36 = w_s/(275 + 0{\cdot}44 \times 275)$$

Therefore, w_s = 142 kg/m^3

Step 4: calculation of coarse and fine aggregate contents

It is now possible to assess the total aggregate content according to the absolute volume method. The fine aggregate content in the total aggregate is generally recommended to be in the range of 48–55% (EFNARC, 2005). However, in the present investigation a combined aggregate grading as recommended by the DIN 1045 standards was utilised. The aggregates were combined in such a way, so that they nearly meet the combined grading specification of DIN 'B' curve. The actual and the standard DIN 'B' combined aggregate curves are presented in Figure 4. For normal vibrated concretes the DIN 'A' curve was utilised and the combined aggregate grading adopted was presented in Figure 5. The percentage fractions of aggregates used are also presented in the same figures.

Total volume = 1000 litres;

Assuming air content = 2%, Air = (2/100) × 1000 = 20 litres

Net concrete volume = 980 litres

From above cement content (c_s) = 275 kg/m^3

Fly ash (f) = 275 kg/m^3

Water (w_s) = 142 kg/m^3

Volume of cement (V_c) = 275/3·15 = 87·3 l, where specific gravity of cement = 3·15

Volume of fly ash (V_{fly}) = 275/2·2 = 125 l, where specific gravity of fly ash = 2·2

Volume of water (V_w) = 142/1 = 142 l, where specific gravity of water = 1·0

Volume of paste (V_{paste}) = 87·3 + 125 + 142 = 354·3 l

Volume of total aggregate (V_{agg}) = 980 – 87·3 – 125 – 142 = 625·7 l

In the combined aggregate grading for SCC it was observed that the percentage of fine aggregate in the total aggregate content is 48% and the coarse aggregate is 52% (20 mm = 20%, 12 mm = 27%

Figure 4 Comparison between the actual and the standard combined aggregate grading used for SCC

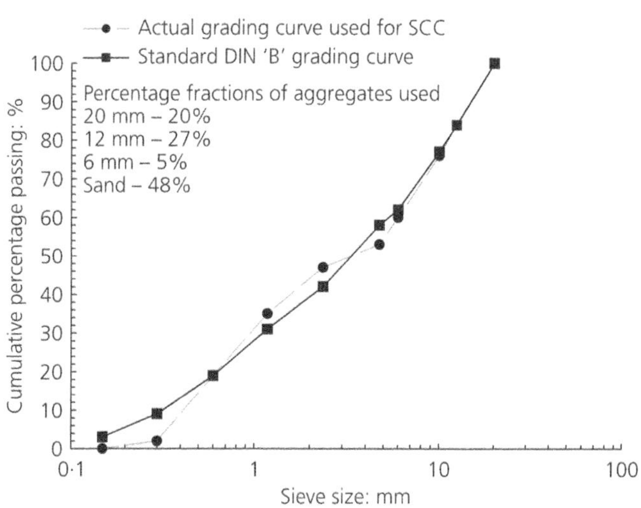

Figure 5 Comparison between the actual and the standard combined aggregate grading used for normal concrete

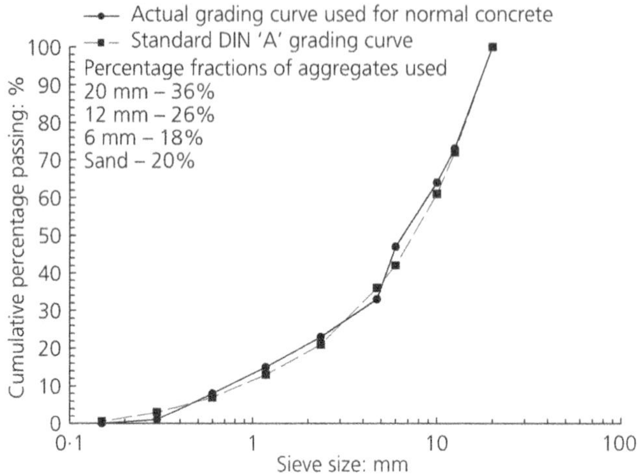

and 6 mm = 5%). This percentage of fine aggregate is in correspondence with the proposed 48–55% range for fine aggregate in SCC.

Volume of fine aggregate (V_{fa}) = 0·48 × 625·7 = 300·33 l

Mass of fine aggregate = 300·33 × 2·65 = 796 kg, where specific gravity of sand = 2·65

Volume of coarse aggregate (V_{ca}) = 0·52 × 625·7 = 325·36 l

Mass of 20 mm aggregate = 0·20 × 625·7 × 2·89 = 362 kg, where specific gravity of 20 mm = 2·89

Mass of 12 mm aggregate = 0·27 × 625·7 × 2·87 = 485 kg, where specific gravity of 12 mm = 2·87

Mass of 6 mm aggregate = 0·05 × 625·7 × 2·88 = 90 kg, where specific gravity of 6 mm = 2·88

Total mass of concrete = coarse aggregate + water + sand + cement + fly ash = 362 + 485 + 90 + 142 + 796 + 275 + 275 = 2425 kg

Summary of volume fractions

$V_{ca} = 0{\cdot}32$

$V_{paste} = 0{\cdot}35$

$V_{fa} = 0{\cdot}31$ $\qquad V_{ca} + V_{fa} = 0{\cdot}63$

Total = 0·98

Step 5: calculation of superplasticiser (SP) dosage

According to previous engineering experience in the current authors' laboratory, it was found that the dosage of SP is 1·3% and that of VMA used is 0·2% of the total cementitious content for meeting the SCC regulations specified in Table 1.

$$W_{sp} = 0{\cdot}013 \times (275 + 275) = 7{\cdot}15\ \text{kg/m}^3$$

$$W_{vma} = 0{\cdot}002 \times (275 + 275) = 1{\cdot}1\ \text{kg/m}^3$$

Step 6: adjustment of mix proportion

Since all the mix proportions met the guidelines of EFNARC specified in Table 1, there is no need of adjusting the mix proportions. Similarly the 30 and 90 MPa concretes were designed as per the methodology outlined above.

Experimental programme

A 120 kg batch has been prepared for all mixtures. The mixing sequence consisted of homogenising the sand, the coarse aggregate, fly ash and cement in a laboratory pan mixer. After incorporation of water, superplasticiser was finally introduced to the wet mixture. The initial mixing time is more critical for polycarboxylate-based admixtures compared to naphthalene-based admixtures, owing to their dispersing mechanism. In order to sustain the equilibrium viscosity, longer mixing times are required. The optimum mixing time and order should be determined by means of pre-tests for each type of plant and concrete composition. The results of pre-tests showed that a total mixing time of 5 min is enough to stabilise the slump flow and V-funnel flow values. Thirty percent of the batch was used for fresh concrete tests. The remaining part was used to prepare 100 mm cube specimens without any vibration in order to determine the strength properties.

The specimens were cured in water at 27°C right up to the testing day. For determining the self-compactibility properties, slump flow, V-flow time and L-box blocking ratio tests were performed. All fresh test measurements were duplicated and the average of measurements was given. In order to reduce the effect of workability loss on variability of test results, the fresh-state properties of mixtures were determined in a period of 30 min after mixing. Before testing, fresh SCC was remixed for 30 s. The order of testing was: (*a*) spread flow test; (*b*) V-flow test; (*c*) L-box test. The tests were performed in accordance with EFNARC (2005) standards. The compressive strength was obtained on 100 mm cube specimens. For SCCs developed with 50 and 30% replacements of fly ash the specimens were demoulded after 24 h of casting, whereas for 70% replacement the specimens were demoulded after 3 days of casting. The specimens were then cured in water at approximately 27°C until testing was carried out at 7, 28, 56 and 90 days. Three specimens of each mixture were tested and the mean values were reported.

Results of fresh concrete are outlined in Table 4. The fresh concrete mixes examined exhibited slump flow values between 670 and 730 mm and showed no signs of segregation. The slump flow values obtained has a direct bearing on the volume of fly ash used. At high volume

Table 4 Fresh properties of the concretes investigated

S. no.	Concrete grade: MPa	Name	Slump: mm	Slump flow: mm	V-funnel flow time: s	L- box ratio for gap of 40 mm
1	30	NC30	80	–	–	–
2		SCC30	–	730	14	0·89
3	60	NC60	80	–	–	–
4		SCC60	–	710	20	0·94
5	90	NC90	75	–	–	–
6		SCC90	–	670	22	0·85

replacements of fly ash SCCs exhibited the highest slump flows. A similar trend was observed in V-funnel flow times also. The obtained funnel times were in the range of 14–22 s. Experimental measurements related with L-box ratio indicate the filling and passing ability of each mixture. L-box test is more sensitive to blocking. There is a risk of blocking of the mixture when the L-box blocking ratio is below 0·8 (Petersson and Billberg, 1999; Tviksta, 2000) The determined L-box ratios of the three SCC mixtures are presented in Table 4. From the results it can be seen that all the three SCC mixtures exhibited L-box ratios of more than 0·80. From the fresh property results it can be concluded that all the SCCs developed have satisfied the norms that were required to qualify them as SCCs according to the EFNARC (2005) regulations given in Table 1.

The compressive strengths were evaluated at 7, 28, 56 and 90 days for SCCs as well as for normal concretes. The normal concretes were designed for strengths of 30, 60 and 90 MPa based on the modified ACI w/c ratio to strength relation (Raja, 1993). The results of these concretes were presented in Table 5. From the results it can be seen that the concretes of above 90 MPa strengths at 28 days can be produced with the available high grade (C53) cements alone at lower water cement ratios. The designed target strengths were easily obtained for all the concretes. In case of 30 MPa concrete strengths even more than the design strengths were obtained.

The self-compacting fly ash concretes were also designed for strengths 30, 60 and 90 MPa for equivalent 28 day strengths with the proposed methodology. The strengths achieved by these concretes were also presented in Table 5. The results of the 30 MPa concrete demonstrate that even at 70% replacement the strength gain rate is almost similar to that of normal concretes. Except for the 70% replacement, the other self-compacting fly ash concretes achieved almost similar results to normal concretes. The results of 60 MPa SCC with 50% replacement of fly ash showed lower strengths compared to the normal concretes, but still achieved the 28 days designed strength. The results of high-strength normal and self-compacting fly ash concretes of 90 MPa indicate that the strength gain rate is comparatively low after 28 days for 90 MPa concrete. It can also be observed that 90 MPa SCC with 30% replacement showed lower strength at early ages, but almost an equal strength of 90 MPa at 28 days.

Table 5 Compressive strengths of the concretes investigated

S. no.	Concrete grade: MPa	Name	Compressive strength: MPa			
			7 day	28 day	56 day	90 day
1	30	NC30	25·50	35·34	39·23	43·4
2		SCC30	25·88	37·92	58·82	67·55
3	60	NC60	53·66	68·50	71·22	76·30
4		SCC60	43·50	63·32	70·00	79·42
5	90	NC90	75·70	91·30	92·80	94·40
6		SCC90	64·30	89·10	92·60	98·73

It is evident from the experimental results that there is a maximum strength that can be achieved at a particular level of fly ash replacement. In general, it was seen that high-volume fly ash replacement is only possible in low-strength SCC mixtures; high-strength concrete mixtures could be made only at the lower percentages of replacement. In order to understand these aspects clearly, the compressive strengths achieved at 28 days were plotted against the percentage replacement of fly ash (Figure 2). It can be distinctly seen that there is only a minor variation of strength at different possible percentages of replacement for any particular strength in these concrete types designed through the efficiency approach. This depicts the limitations on the maximum percentage of replacement possible for a particular strength. The results of this limitation on the maximum percentage replacement that is possible at any strength were compared with the limitation proposed by Munday *et al.* (1983) on the same figure. This comparison shows that the limitations obtained from the present experimental investigations were far above those given by Munday *et al.* (1983). Finally, from this study, the level of replacement of fly ash for making the required strength of self-compacting fly ash concrete can easily be selected. Thus, this approach will lead to an effective utilisation of fly ash in SCC by incorporating the maximum possible replacement of the (low-cost pozzolana) fly ash.

The overall results showed that the proposed mix design method gave good results and strengths of more than 90 MPa can be realised. All the self-compacting fly ash concretes have obtained their design strengths similar to normal concretes. From different ranges of strengths and percentage replacements it can be seen that high volume as well as high-strength self-compacting fly ash concretes can be made by using the proposed mix design methodology. High-volume replacements of up to 70% for 30 MPa concrete were possible. High-strength concretes of more than 90 MPa at 30% fly ash replacement were also possible. Hence, the proposed mix design method can be recommended for the design of high-volume fly ash SCCs for an effective utilisation.

Conclusions

A critical review of the basic approaches and the suggested mix design procedures shows that there is no specific method for obtaining a SCC with a predefined strength. This paper suggested a methodology for the design of SCCs of specific strength with a fly ash content of about 30–70%, which was also verified through a limited experimental investigation.

The experimental investigations on SCC fly ash concretes designed with the proposed mix design method, shows that concretes of very high strengths (more than 90 MPa) can be produced with reasonable ease. The design method also presents a way for obtaining high-volume replacements (up to 70% for 30 MPa).

REFERENCES

ASTM (2002) C618: Standard specification for coal fly ash and raw or calcined natural pozzolan for use in concrete. ASTM International, West Conshohocken, PA, USA.

Babu KG and Dinakar P (2005) Fly ash based high performance cementitious composites. Material Science of Concrete, Special Volume. *Proceedings of the Indo-US Workshop on High-Performance Cement-Based Concrete Composites, Chennai, India*, pp. 171–182.

Babu KG and Rao GSN (1993) Efficiency of fly ash in concrete. *Cement and Concrete Composites* **15(4)**: 223–229.

Babu KG and Rao GSN (1996) Efficiency of fly ash in concrete with age. *Cement and Concrete Research* **26(3)**: 465–474.

BIS (Bureau of Indian Standards) (1987) IS 12269: Specification for 53 grade ordinary Portland cement. Bureau of Indian Standards, New Delhi.

Brouwers HJH and Radix HJ (2005) Self-compacting concrete: theoretical and experimental study. *Cement and Concrete Research* **35(11)**: 2116–2136.

Dinakar P, Babu KG and Santhanam M (2008a) Durability properties of high volume fly ash self compacting concretes. *Cement and Concrete Composites* **30(10)**: 880–886.

Dinakar P, Babu KG and Santhanam M (2008b) Mechanical properties of high volume fly ash self compacting concrete mixtures. *Structural Concrete: Journal of the fib* **9(2)**: 109–116.

Dinakar P, Babu KG and Santhanam M (2009) Corrosion resistance performance of high volume fly ash self compacting concretes. *Magazine of Concrete Research* **61(2)**: 77–85.

EFNARC (Experts for Specialised Construction and Concrete Systems) (2005) Self Compacting Concrete European Project Group. *The European Guidelines for Self-Compacting Concrete*. BIBM, CEMBUREAU, EFCA, EFNARC and ERMCO. See: http://www.efnarc.org (accessed 27/01/2009).

Gettu R, Izquierdo J, Gomes PC and Josa A (2002) Development of high-strength self compacting concrete with fly ash: A four-step experimental methodology. *27th Conference on Our World in Concrete and Structures*, Singapore, pp. 217–224.

Goodier CI (2003) Development of self-compacting concrete. *Proceedings of the Institution of Civil Engineers, Structures and Buildings* **156(SB4)**: 405–414.

Japan Concrete Society (1999) *Recommendations for Self-Compacting Concrete*, Tokyo, Japan.

Khayat KH and Guizani Z (1997) Use of viscosity-modifying admixture to enhance stability of fluid concrete. *ACI Materials Journal* **94(4)**: 332–341.

Khurana R and Saccone R (2001) Fly ash in self-compacting concrete. *Proceedings of Fly Ash, Silica fume, Slag and Natural Pozzolans in Concrete*, Chennai, India. ACI SP-199, pp. 259–274.

Kim JK, Han SH, Park YD, Noh JH, Park CL, Kwon YH and Lee SG (1996) Experimental research on the material properties of super flowing concrete. *Production Methods and Workability of Concrete*, E&FN Spon, London, UK, pp. 271–284.

Miura N, Takeda N, Chikamatsu R and Sogo S (1993) Application of super workable concrete to reinforced concrete structures with difficult construction conditions. *Proceedings of ACI SP-140*, pp. 163–186.

Munday JGL, Ong LT and Dhir RK (1983) Mix proportioning of concrete with PFA: critical review. *Proceedings of Fly Ash, Silica Fume, Slag and Other Mineral by Products in Concrete*, ACI SP-79, pp. 267–288.

Nagamoto N and Ozava K (1997) Mixture properties of self-compacting concrete. *Proceedings of Third CANMET/ACI International Conference on Design and Materials and Recent Advances in Concrete Technology, ACI SP 172, Kuala Lumpur*, American Concrete Institute, Farmington Hills, MI, USA, pp. 623–637.

Okamura H (1997) Self-compacting high-performance concrete. *Concrete International* **19(7)**: 50–54.

Okamura H and Onchi M (1999) Self-compacting concrete: development, present use and future. *Proceedings of the first RILEM International Symposium on Self-Compacting Concrete, Stockholm, Sweden*, pp. 3–14.

Okamura H, Ozawa K and Ouchi M (2000) Self compacting concrete. *Structural Concrete: Journal of the fib* **1(1)**: 3–17.

Persson B (2001) A comparison between mechanical properties of self compacting concrete and the corresponding properties of normal concrete. *Cement and Concrete Research* **31(2)**: 193–198.

Petersson O and Billberg P (1999) Investigation on blocking of self-compacting concrete with different maximum aggregate size and use of viscosity agent instead of filler. *Proceedings of the First RILEM International Symposium on Self-Compacting Concrete, Stockholm, Sweden*, pp. 333–344.

Raja GLV (1993) *Mechanical and Fracture Characteristics of Steel Fibre Reinforced High Strength Concrete*. PhD thesis, Indian Institute of Technology, Madras, India.

Sedran T and de Larrard F (1996) A software to optimize the mix design of high performance concrete. *In Proceedings of Fourth International Symposium on The Utilization of High Strength/High Performance Concrete; BHP 96, RENE–LCPC, Paris, France*, pp. 29–31.

Sonebi M (2004) Medium strength self-compacting concrete containing fly ash: modeling using factorial experimental plans. *Cement and Concrete Research* **34(7)**: 1199–1208.

Su Nan, Kung CH and Chai WH (2001) A simple mix design method for self-compacting concrete. *Cement and Concrete Research* **31(12)**: 1799–1807.

Tviksta LG (2000) *Brite Euram Project: rational production and improved working environment through using self compacting concrete*, Final report: Task 8·4, Quality control, NCC AB.

Yurugi M, Sakata N, Iwai M and Sakai G (1993) Mix proportion for highly workable concrete. *Concrete 2000*, Dundee, UK, pp. 579–590.

Dhir and Lye
ISBN 978-0-7277-6461-4
https://doi.org/10.1680/icetscc.64614.067

Chapter 4
Optimum superplasticiser dosage and aggregate proportions for SCC

A. K. H. Kwan
The University of Hong Kong

I. Y. T. Ng
The University of Hong Kong

An experimental study has been carried out to investigate the effect of superplasticiser dosage on the performance of self-consolidating concrete (SCC). Seven concrete mixes with water/cement (w/c) ratios ranging from 0·35 to 0·45 and fine/total (F/T) aggregate ratios ranging from 0·40 to 0·60 were designed and 42 batches of concrete containing different dosages of superplasticiser were produced. The workability, passing and filling abilities, and segregation stability of the concrete produced were measured using the slump flow, U-box, and sieve segregation tests respectively. For each concrete mix designed, the superplasticiser dosages for maximum slump flow without segregation and maximum filling height were determined. It was found that the maximum performance of the concrete mix and the respective required superplasticiser dosage are dependent on the w/c and F/T aggregate ratios. In general, increasing the F/T aggregate ratio would improve the maximum performance but would also increase the superplasticiser dosage needed. Lastly, the robustness of each concrete mix was evaluated quantitatively as the range of superplasticiser dosage or slump flow satisfying all requirements for SCC. Based on these results, the optimum superplasticiser dosage for maximum robustness of each concrete mix was determined.

Introduction

Self-consolidating concrete (SCC) is a kind of high-performance concrete having excellent ability to flow under its own weight, pass through narrow gaps between reinforcing bars and obstacles, fill itself into far-reaching corners and confined spaces in the mould, and achieve full consolidation without applying compaction. Moreover, it has to have sufficient cohesiveness and stability to avoid segregation after dropping a certain height or flowing a certain distance through narrow gaps along the mould, and sedimentation of coarse aggregate particles after placing at the final position. In other words, SCC must possess, apart from high workability, also high passing ability, filling ability and segregation stability. With these performance attributes, the use of SCC has demonstrated substantial benefits in reducing the number of workers required for concreting, shortening the construction time, improving the productivity and enhancing the quality of the completed concrete structures. Owing to these advantages, SCC has gained popularity in many places[1–4] since its successful development.

SCC is not, however, at all easy to produce. It needs to have both high workability and high cohesiveness, which are difficult to achieve simultaneously. In general, any individual measure (e.g. increase of superplasticiser dosage or decrease of fine/total aggregate ratio) that increases the workability would also decrease the cohesiveness. If the cohesiveness is insufficient, the concrete would tend to segregate during and after placing. Furthermore, when passing through closely spaced reinforcing bars and obstacles, the mortar portion would tend to flow ahead of the coarse aggregate, leaving behind the coarse aggregate particles to block the flow of the remaining concrete and thereby reducing the passing and filling abilities of the concrete. On the other hand, because a concrete with higher cohesiveness tends to be more sticky and reluctant to flow, any individual measure (e.g. decrease of superplasticiser dosage or increase of fine/total aggregate ratio) that increases the cohesiveness would also decrease the workability. If the workability is insufficient, the concrete would not be able to flow far during placing and release the entrapped air after placing. It would also have difficulties in passing through narrow gaps and filling itself into far-reaching corners and confined spaces.

Hence, when designing an SCC, proper balance should be maintained between the workability and cohesiveness that can be achieved at the same time. This is particularly important when determining the superplasticiser (SP) dosage and the fine/total aggregate (F/T) ratio. As the addition of more SP would increase the workability whereas the addition of excessive SP could decrease the cohesiveness to an unacceptable level, for each concrete mix with given materials and mix proportion, there is an optimum SP dosage for best overall performance in terms of both workability and cohesiveness. The determination of the F/T ratio is even more difficult because the F/T ratio affects not only the workability and cohesiveness but also the passing and filling abilities. To ensure good passing and filling abilities, the concrete mix should be designed to have a relatively high F/T ratio, which increases the cohesiveness, improves the shear deformability by reducing particle interaction (also called interparticle collision) between coarse aggregate particles and decreases the amount of coarse aggregate particles that could be entrapped at narrow gaps.[5,6] In fact, the F/T ratios adopted for SCC are usually higher than the common values of 0·35 to 0·45 for conventional concrete.[7] In this regard, Okamura and Ouchi[6] suggested that the coarse aggregate content should not be larger than 50% of the total aggregate content. Likewise, Ng *et al.*[8] recommended that the coarse aggregate content should not be larger than the fine aggregate content (in other words, the F/T ratio should be at least 0·5) whereas the 20 mm aggregate content should not be larger than the 10 mm aggregate content.

Another major difficulty in the design and production of SCC is that the performance of the SCC is often very sensitive to small variations in the quantity and/or quality of the ingredients, especially the water content, the SP dosage and compatibility, and the aggregate grading, shape and moisture content. To ease the pressure on production control and reduce the possibility of having problems on the job site, the SCC should be designed to be tolerant to daily fluctuations of the ingredients or, in other words, to have adequate robustness.[9] To be more specific, the robustness of an SCC is the tolerance of the SCC to expected variations in the ingredients while maintaining the required performance standard. Nkinamubanzi and Aïtcin[10] have found that the chemical compatibility of the SP used is critical to the robustness of a concrete mix. Nunes *et al.*[11] proposed to assess the robustness of an SCC in terms of the frequency of satisfying the acceptance criteria despite daily fluctuations of the ingredients. They found that the mix parameter having the greatest effect on the performance of an SCC is the water-to-powder ratio and that the effect of one mix parameter is dependent on the other mix parameters.

During recent trial mixing to optimise the mix design of SCC, the current authors noticed that apart from the water content, the SP dosage also has great effect on the performance of SCC. Quite often, the addition of more SP to increase the workability would gradually decrease the cohesiveness and, at a certain stage, the passing/filling abilities and segregation stability of the concrete mix could drop drastically. Hence, there is a limited acceptable range of SP dosage within which the concrete would perform satisfactorily. Moreover, the F/T ratio has great effect on such acceptable range of SP dosage, which, in a way, may be taken as a measure of robustness. In order to study the effect of SP dosage on the performance of SCC and the effect of F/T ratio on the robustness of SCC, a comprehensive experimental programme has been launched. In the experimental programme, a total of 42 batches of concrete with water/cement (w/c) ratios ranging from 0·35 to 0·45, F/T ratios ranging from 0·40 to 0·60 and SP dosages ranging from 0·6 to 2·4% by weight of the cement content have been produced for performance evaluation, as reported herein.

Experimental details

Materials

An ordinary Portland cement of strength class 52·5N complying with BS 12: 1996[12] was used as the only cementitious material. Its relative density, fineness and 28-day mortar cube strength (tested as per BS EN 196: Part 1: 2005)[13] were 3·10, 336 m^2/kg and 64·4 MPa respectively. Crushed granite rocks were used for both the fine and coarse aggregates. The fine aggregate had nominal maximum size of 5 mm, relative density of 2·62, fineness modulus of 3·26 and water absorption of 1·5%, while the coarse aggregate had nominal maximum size of 20 mm, relative density of 2·59, fineness modulus of 6·46 and water absorption of 0·8%. Figure 1

Figure 1 Grading curves of the fine and coarse aggregates

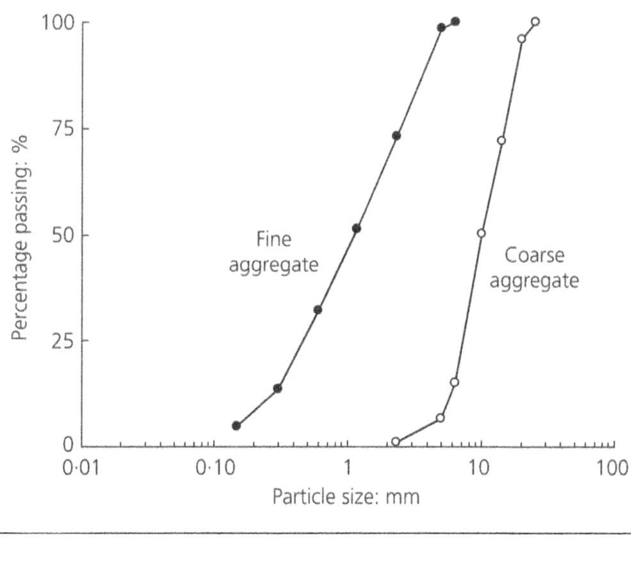

Table 1 Properties of materials

Material	Property	Measured value
Cement	Relative density	3·10
	Fineness	336 m^2/kg
	28-day mortar cube strength	64·4 MPa
Fine aggregate	Relative density	2·62
	Fineness modulus	3·26
	Water absorption	1·5%
Coarse aggregate	Relative density	2·59
	Fineness modulus	6·46
	Water absorption	0·8%
Superplasticiser	Relative density	1·05
	Solid content	20%

depicts the grading curves of the fine and coarse aggregates, which fall within the permissible limits stipulated in BS 882: 1992.[14] The superplasticiser added was a polycarboxylate-based superplasticiser complying with BS 5075: Part 3: 1985.[15] It was a milky white solution with relative density of 1·05 and solid content of 20%. Its molecules have a comb-like structure consisting of a backbone chain and a number of graft chains. The functional group of the backbone chain produces negative charges to disperse the cement grains (this is called electrostatic repulsion) while the graft chains wrap the cement grains to maintain separation and avoid flocculation of the cement grains (this is called steric hindrance). Table 1 summarises the properties of the materials used.

Mix proportions

Seven concrete mixes were designed. They are numbered A to G. The mixes A and B have a w/c ratio of 0·35, mixes C, D and E have a w/c ratio of 0·40, and mixes F and G have a w/c ratio of 0·45. These w/c ratios should have covered the usual range for production of normal- and high-strength concrete. Apart from the w/c ratio, the F/T ratio was also varied to investigate their combined effects. Among the two mixes with w/c = 0·35, A and B have F/T ratios of 0·50 and 0·60 respectively. Among the three mixes with w/c = 0·40, C, D and E have F/T ratios of 0·40, 0·50 and 0·60 respectively. Among the two mixes with w/c = 0·45, F and G have F/T ratios of 0·50 and 0·60, respectively. In all the concrete mixes, the paste volume was set at 35% to strike a balance between workability of fresh concrete and dimensional stability of hardened concrete.[16] The mix parameters and mix proportions of the concrete mixes are summarised in Tables 2 and 3 respectively. For each concrete mix, several batches of concrete having different dosages of SP were produced. Each batch of concrete so produced is assigned an identification code *X–Y*, in which *X* is the concrete mix number and *Y* is the SP dosage expressed in terms of percentage by weight of the cement content.

Table 2 Mix parameters of concrete mixes

Mix no.	w/c ratio	F/T ratio	Range of SP dosage: %
A	0·35	0·50	0·8–1·6
B	0·35	0·60	1·2–2·4
C	0·40	0·40	0·7–1·2
D	0·40	0·50	0·9–1·5
E	0·40	0·60	0·8–1·8
F	0·45	0·50	0·6–1·1
G	0·45	0·60	0·9–1·3

Notes:
(a) The superplasticiser dosage was measured in terms of liquid weight as a percentage by weight of the cement content.
(b) In all concrete mixes, the paste volume was set at 35%.

Table 3 Mix proportions of concrete mixes

Mix no.	Water: kg/m^3	Cement: kg/m^3	Fine aggregate: kg/m^3	Coarse aggregate: kg/m^3	Range of SP content: kg/m^3
A	182	520	859	853	4·2–8·3
B	182	520	1032	683	6·2–12·5
C	194	484	687	1023	3·4–5·8
D	194	484	859	853	4·4–7·3
E	194	484	1032	683	3·9–8·7
F	204	453	859	853	2·7–5·0
G	204	453	1032	683	4·1–5·9

Mixing, testing and casting

An electronic balance was used to weigh the materials and a pan mixer was employed to produce each batch of concrete. During production, the aggregate and cement were first added into the mixer. Then, water was added and mixed with the aggregate and cement for 2 min. After this preliminary mixing, the mixture was left undisturbed for 1 min to allow water absorption of the aggregate particles. Superplasticiser was then added to the mixture and the mixing resumed and continued for a further 2 min. At the same dosage, such delayed addition of superplasticiser has been proven to yield higher fluidity as compared to simultaneous addition of superplasticiser.[17]

Immediately after completion of the mixing process, concrete samples were taken from the mixer for slump flow test, U-box test and sieve segregation test, which were performed concurrently to avoid workability loss with time. After having carried out these tests, the remaining concrete in the mixer was remixed for 30 s and then taken out of the mixer for casting three 150 mm cubes. No tamping or vibration was applied during casting of the cubes; the concrete placed into the moulds was just left to consolidate by itself under its own weight. The top surface of each concrete cube was gently trowelled and then covered with a plastic sheet. After casting, all the cubes were cured in the laboratory with temperature controlled at 24 ± 2°C to allow setting and hardening. At 24 h after casting, the cubes were demoulded and put into a lime-saturated water curing tank controlled at a temperature of 27 ± 2°C until the age of 28 days for cube compression test.

Slump flow test

The slump flow test for SCC was very similar to the slump test for conventional concrete stipulated in BS 1881: Part 102: 1983.[18] In fact, the same apparatus was employed. As for the slump test, the slump flow test was performed on a flat, smooth and level steel plate large enough to contain the concrete patty formed during the test. To measure the slump and slump flow of the concrete, the concrete was first filled into the slump cone, trowelled flat at the top surface and then allowed to deform and flow by lifting the slump cone steadily. Unlike the slump test, no tamping was applied when filling the concrete into the slump cone. After lifting the slump cone, the slump was measured as the drop in height of the concrete and the slump flow was measured as the average diameter of the concrete patty in two perpendicular directions. The slump is a measure of deformability whereas the slump flow is a measure of flowability.

U-box test

A specially designed U-box, comprising a storing compartment, a filling compartment and an opening between the two compartments, was employed for the U-box test. There are two slightly different versions of U-box in common use: one with a round bottom and the other with a flat bottom.[19] The U-box employed in this study was the same as that used by Ng *et al.*,[8] which is of flat bottom design and has three 12 mm diameter plain round bars at the opening between the two compartments serving as an obstacle to test the passing ability of the concrete. At the onset of the U-box test, the sliding gate at the opening was closed and the concrete sample was placed gently into the storing compartment. When the storing compartment was full, the top surface of the concrete was trowelled flat and the sliding gate was opened to allow the concrete to flow under its own weight through the obstacle into the filling compartment. After the flow had ceased, the height of the concrete in the filling compartment was measured and recorded as the filling height of the concrete. This filling height is an integrated measure of the passing and filling abilities of the concrete. The Japan Society of Civil Engineers (JSCE) *Guidelines for Concrete*[20] stipulates that the filling height of SCC should be at least 300 mm.

Sieve segregation test

A 5 mm perforated steel plate sieve was employed for the sieve segregation test. Several slightly different test procedures have been developed.[9,21,22] The test procedure adopted in this study was the same as that used by Ng *et al.*,[8] which followed the recommendation of the European Guidelines for SCC.[9] It started with the placing of a 10 litre concrete sample gently into a 300 mm diameter plastic container. The container was then covered by a lid, put at a level position and left undisturbed to allow sedimentation and bleeding to take place. After 15 min of waiting, the upper 2 litres of the concrete sample, including any bleeding water, was poured gently onto the sieve at a height of 500 mm. The materials dripping through the sieve (the laitance or paste incapable of adhering to the coarse aggregate particles retained by the sieve) were collected by a base receiver placed underneath the sieve. After 2 min of waiting, which was deemed sufficient for the dripping to finish, the sieve was removed without agitation. The weight of the materials collected by the base receiver was then measured and the segregation index of the concrete determined as the ratio of the weight of materials in the base receiver to the weight of concrete poured onto the sieve. This segregation index is an integrated measure of the sedimentation of the coarse aggregate particles and the segregation of the concrete. As both sedimentation and segregation are consequences of the concrete having insufficient cohesiveness, the segregation index may also be taken as a measure of the degree of insufficient cohesiveness. According to the European Guidelines for SCC,[9] the segregation index of SCC should not be higher than 20%.

Experimental results

The experimental results of the 42 batches of concrete mixes produced are summarised in Tables 4, 5 and 6 for mixes A and B with w/c = 0·35, mixes C, D and E with w/c = 0·40, and mixes F and G with w/c = 0·45, respectively. Detailed data analysis of these results is presented in the following subsections.

Effect of superplasticiser on workability

The effect of superplasticiser on the workability of the seven concrete mixes is depicted in Figure 2 by plotting the slump flow against the SP dosage. For clearer illustration, the curves are grouped into three separated graphs according to the w/c ratio of the concrete mix. On the whole, the slump flow increased with the SP dosage at a gradually decreasing rate until the slump flow

Table 4 Experimental results of mixes A and B with w/c = 0·35

Batch code	Slump: mm	Slump flow: mm	Filling height: mm	Segregation index: %	28-day cube strength: MPa
A-0·8	89	217	0	0·0	73·0
A-1·0	232	469	225	0·3	75·2
A-1·2	248	623	205	9·3	78·2
A-1·4	261	685	275	11·3	82·8
A-1·6	270	790	95	27·1	86·0
B-1·2	211	388	273	0·0	69·7
B-1·4	257	577	323	1·0	78·3
B-1·6	266	652	334	2·4	80·7
B-1·8	270	765	338	9·5	85·9
B-2·0	270	792	338	9·5	85·9
B-2·2	271	825	337	15·1	83·1
B-2·4	268	835	336	22·1	83·7

Table 5 Experimental results of mixes C, D and E with w/c = 0·40

Batch code	Slump: mm	Slump flow: mm	Filling height: mm	Segregation index: %	28-day cube strength: MPa
C-0·7	240	574	174	0·5	64·7
C-0·8	256	585	141	5·3	65·5
C-0·9	250	629	162	7·8	67·5
C-1·0	254	635	170	11·3	65·1
C-1·1	250	695	185	15·5	70·9
C-1·2	249	741	91	37·1	72·0
D-0·9	213	416	134	0·0	60·5
D-1·0	248	553	218	0·8	66·0
D-1·1	254	605	246	4·3	66·2
D-1·2	255	648	248	5·2	66·0
D-1·3	264	702	226	9·8	68·6
D-1·4	258	715	188	17·4	68·8
D-1·5	258	748	183	25·2	70·4
E-0·8	77	234	0	0·0	55·8
E-1·0	240	498	318	1·6	57·3
E-1·2	255	538	329	2·5	64·9
E-1·4	263	662	334	6·8	68·2
E-1·6	264	753	336	19·0	72·3
E-1·8	255	810	335	38·8	71·4

increased only marginally when the SP dosage was further increased. However, at a given SP dosage, the slump flow was dependent on the w/c ratio and the F/T ratio. At a fixed w/c ratio, the slump flow at any given SP dosage was lower at a higher F/T ratio, or, in other words, the SP dosage required to achieve any required slump flow was higher at a higher F/T ratio. This was because a higher F/T ratio would increase the total surface area of particles to be adsorbed with superplasticiser for particle dispersion, thereby leading to a higher superplasticiser demand. On the other hand, at a fixed F/T ratio, the slump flow at any given SP dosage was lower at a lower w/c ratio, or, in other words, the SP dosage required to achieve any required slump flow was higher at a lower w/c ratio. This was because a lower w/c ratio would decrease the thickness of the water films coating the particles and increase the particle interaction and internal friction between the particles, thereby leading to a higher superplasticiser demand.

Effect of superplasticiser on passing/filling abilities

The effect of superplasticiser on the passing and filling abilities of the seven concrete mixes is depicted in Figure 3 by plotting the filling height against the SP dosage. As before, the curves are grouped into three separate graphs according to the w/c ratio of the concrete mix.

Table 6 Experimental results of mixes F and G with w/c = 0·45

Batch code	Slump: mm	Slump flow: mm	Filling height: mm	Segregation index: %	28-day cube strength: MPa
F-0·6	225	393	200	1·2	55·2
F-0·7	232	498	163	2·5	56·3
F-0·8	255	588	243	6·4	58·9
F-0·9	255	618	247	8·0	59·9
F-1·0	256	709	185	15·2	64·7
F-1·1	260	736	208	36·5	65·1
G-0·9	230	434	274	0·1	54·9
G-1·0	252	527	310	1·7	56·3
G-1·1	255	588	324	4·8	60·1
G-1·2	266	692	336	12·9	62·0
G-1·3	270	735	306	29·8	64·4

In general, the filling height increased with the SP dosage and then after reaching a certain maximum value decreased drastically when the SP dosage was only slightly further increased. As a result, for each concrete mix, there was a certain optimum SP dosage at which the filling height reached a maximum value. The maximum filling height was dependent on the w/c ratio and the F/T ratio. At a fixed w/c ratio, the maximum filling height was higher at a higher F/T ratio, albeit requiring a higher SP dosage to achieve the maximum filling height. On the other hand, at a fixed F/T ratio, the maximum filling height was higher at a lower w/c ratio, albeit requiring a higher SP dosage to achieve the maximum filling height. Among the various concrete mixes, only those with F/T ratio = 0·60 (coarse aggregate content = 683 kg/m^3) were able to achieve a maximum filling height of at least 300 mm; those with F/T ratio = 0·40 or 0·50 (coarse aggregate content ≥ 853 kg/m^3) were not able to achieve a filling height of at least 300 mm no matter how much SP was added. Hence, apart from optimising the SP dosage, appropriate aggregate proportioning is also important in the design of SCC mixes.

Effect of superplasticiser on segregation stability

The effect of superplasticiser on the segregation stability of the seven concrete mixes is depicted in Figure 4 by plotting the segregation index against the SP dosage. As before, the curves are grouped into three separate graphs according to the w/c ratio of the concrete mix. From these curves, it can be seen that in all the concrete mixes, as the SP dosage was increased, the segregation index steadily increased at a gradually increasing rate until the segregation index abruptly increased to higher than 20% when the SP dosage was only slightly further increased. At a given SP dosage, the segregation index was dependent on the w/c ratio and the F/T ratio. At a fixed w/c ratio, the segregation index at any given SP dosage was lower at a higher F/T ratio while at a fixed F/T ratio, the segregation index at any given SP dosage was lower at a lower w/c ratio. Hence, a higher F/T ratio and/or a lower w/c ratio would render a higher cohesiveness for avoiding sedimentation of the coarse aggregate particles and segregation of the concrete mix.

Figure 2 Slump flow plotted against SP dosage

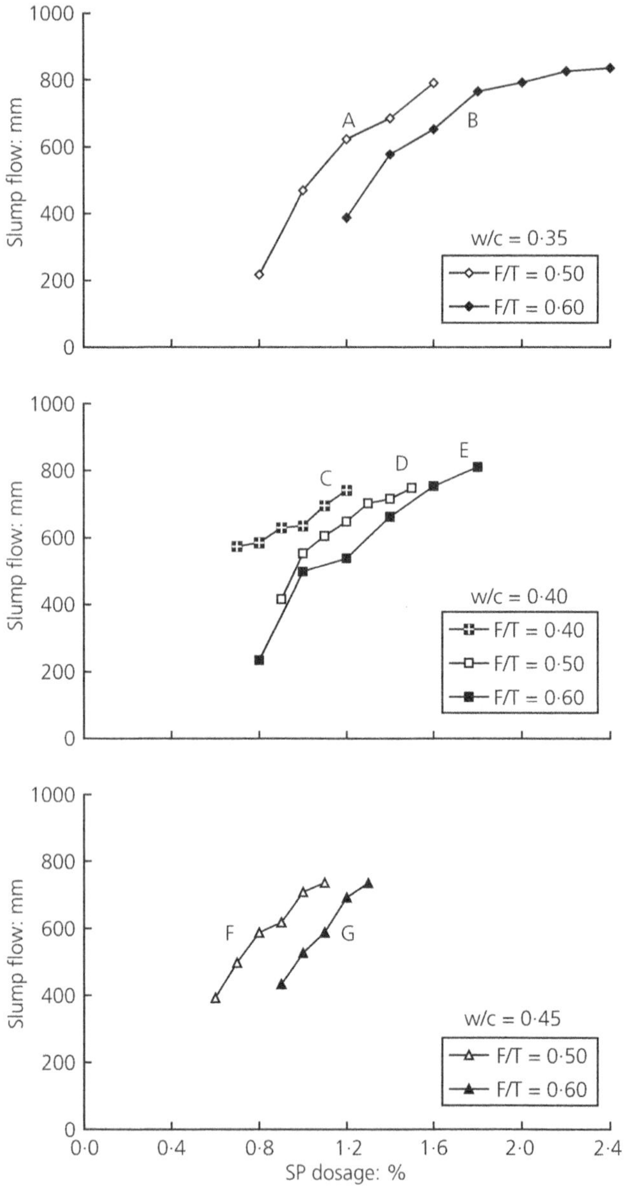

Effect of superplasticiser on strength

The 28-day cube strength results of the 42 batches of concrete produced have been presented in the last column of Tables 4, 5 and 6. Each cube strength result was the mean strength of the three cubes cast from the same batch of concrete and tested at the same time. As the cubes were cast without compaction, these cube strength results represent the uncompacted cube strengths of the concrete.

Figure 3 Filling height plotted against SP dosage

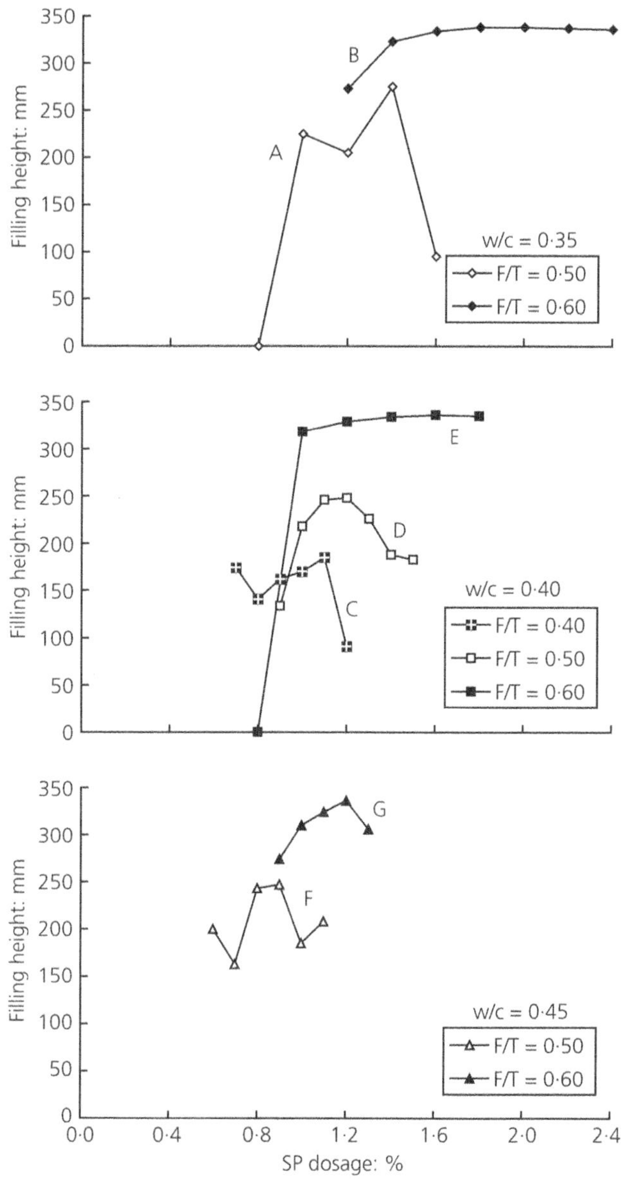

From these results, it can be seen that the maximum 28-day cube strengths achieved by the mixes A and B with w/c = 0·35, mixes C, D and E with w/c = 0·40, and mixes F and G with w/c = 0·45 were 86 MPa, 72 MPa and 65 MPa, respectively. In all the concrete mixes, the cube strength increased with the SP dosage at a decreasing rate until a maximum value was reached at a relatively high SP dosage. This may be explained by the following two effects. First, it was observed during testing

Figure 4 Segregation index plotted against SP dosage

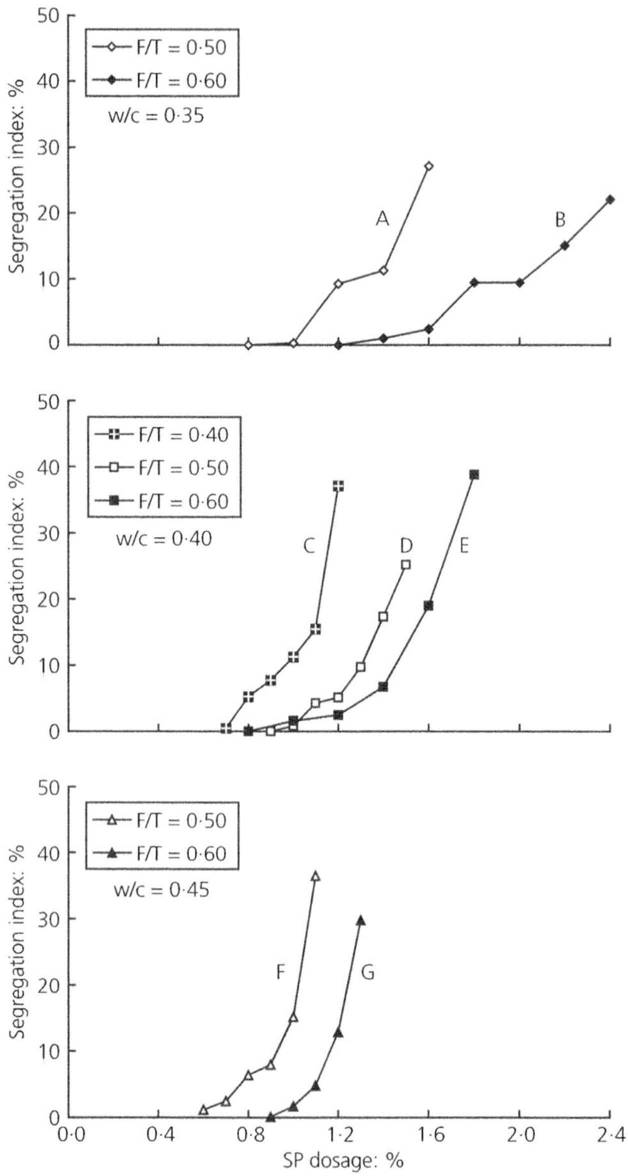

that when the SP dosage was low, the consolidation of the concrete in the cubes was actually incomplete, as indicated by the presence of many air voids on the surfaces of the cubes. However, when the SP dosage was relatively high such that the slump flow of the concrete mix was at least 650 mm, the amount of air voids in each cube became insignificant. Hence, it is advocated that the minimum required slump flow for SCC should be taken as 650 mm. Second, an increased SP dosage would better disperse the cement grains to avoid flocculation, thus resulting in a more uniform distribution of mixing water and a higher degree of hydration.

Maximum performance and required SP dosage

Maximum slump flow without segregation

From the experimental results, it is evident that in all the concrete mixes tested, as the slump flow was increased by adding more superplasticiser, the segregation index was also increased. Hence, generally, when the SP dosage is adjusted to maximise the overall performance, the workability or cohesiveness can only be increased at the expense of the other. It is therefore difficult to achieve both high workability and high cohesiveness at the same time. The variation of the segregation index with the slump flow as the SP dosage changes is illustrated in Figure 5, where the segregation index is plotted against the slump flow for the seven concrete mixes tested. It can be seen from this figure that when the slump flow is relatively low, the segregation index would slowly increase with the slump flow but when the slump flow is increased to beyond a certain limit, the segregation index would drastically increase to higher than 20%, the acceptable upper limit for SCC.[9] From the figure, the maximum slump flow without segregation, defined as the maximum slump flow with segregation index $\leq$ 20%, of each concrete mix may be determined, as listed in the fourth column of Table 7. The required SP dosage for achieving the maximum slump flow without segregation of each concrete mix has also been determined, as listed in the fifth column of Table 7. It is noteworthy that each of the seven concrete mixes has achieved a maximum slump flow without segregation of well above 650 mm, the minimum required for SCC advocated herein.

Maximum filling height

The experimental results also reveal that in all the concrete mixes tested, as the slump flow was increased by adding more superplasticiser, the filling height steadily increased but when the slump flow was increased to such extent that the segregation index became higher than about 15%, the filling height drastically dropped. Hence, generally, for each concrete mix, there is a maximum achievable filling height. The variation of the filling height with the slump flow as the SP dosage changes is illustrated in Figure 6. It can be seen from this figure that the maximum filling height is dependent on both the w/c ratio and the F/T ratio and that the minimum required filling height of 300 mm for SCC[20] is not always achievable. Furthermore, even when achievable, the filling height is higher than or equal to 300 mm only within a limited range of SP dosage or slump flow. The maximum filling height and the corresponding required SP dosage of each concrete mix are presented in the sixth and seventh columns of Table 7. It should be noted that the maximum filling height and the maximum slump flow without segregation do not occur at the same time. Generally, the SP dosage for achieving maximum filling height is lower than the SP dosage for achieving maximum slump flow without segregation. The main reason is that the maximum filling height always occurs when the segregation index is lower than 20% while the maximum slump flow without segregation occurs when the segregation index is equal to 20%.

Effects of w/c and F/T ratios

By nonlinear regression analysis of the data presented in the fourth and fifth columns of Table 7, the maximum slump flow without segregation and the required SP dosage may be expressed as multivariable quadratic functions of the w/c and F/T ratios, based on which the effects of the w/c

Figure 5 Segregation index plotted against slump flow

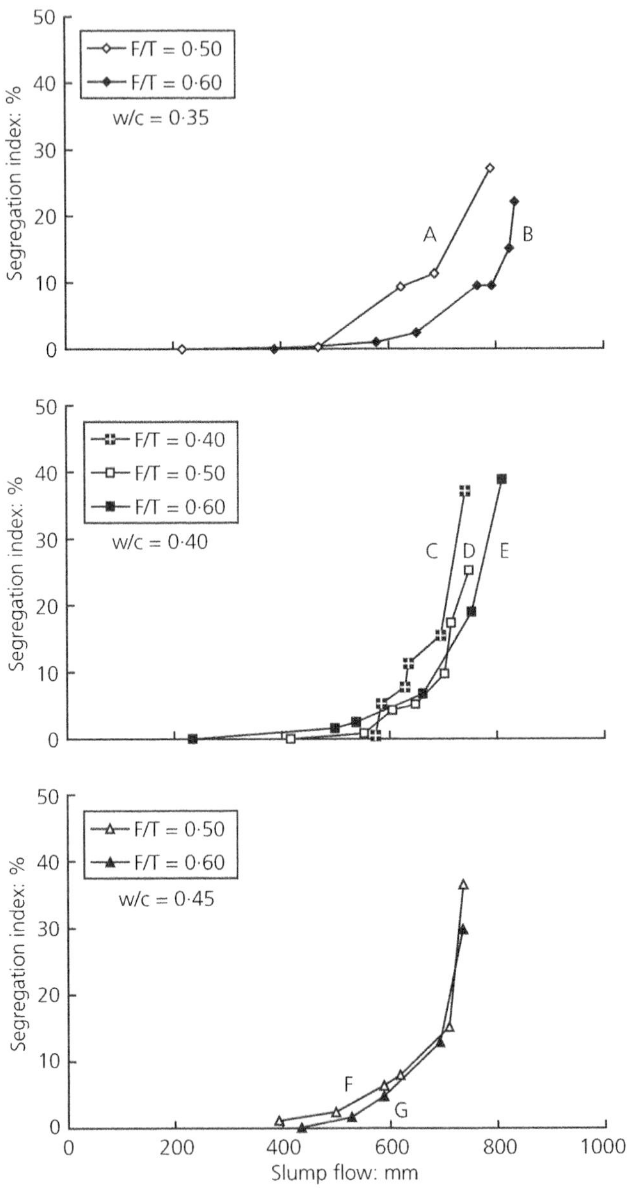

and F/T ratios can be illustrated graphically in the form of contours, as shown in Figure 7. From this figure, it can be seen that the maximum slump flow without segregation is generally higher at a lower w/c ratio or a higher F/T ratio. Hence, although at a given SP dosage, the slump flow is lower at a lower w/c ratio or a higher F/T ratio (see Figure 2), when the SP dosage is allowed to vary in order to achieve maximum performance, the maximum slump flow without segregation

Table 7 Maximum performance and required SP dosage

Mix no.	w/c ratio	F/T ratio	Maximum slump flow with segregation index ≤ 20%		Maximum filling height for best passing/filling abilities	
			Maximum slump flow: mm	Required SP dosage: %	Maximum filling height: mm	Required SP dosage: %
A	0·35	0·50	743	1·51	275	1·40
B		0·60	832	2·34	338	2·00
C	0·40	0·40	705	1·12	185	1·10
D		0·50	726	1·43	248	1·20
E		0·60	756	1·62	336	1·60
F	0·45	0·50	715	1·02	247	0·90
G		0·60	710	1·24	336	1·20

that can be achieved is actually higher at a lower w/c ratio or a higher F/T ratio. There is, however, a price to pay. The required SP dosage for achieving the maximum slump flow without segregation is also generally higher at a lower w/c ratio or a higher F/T ratio.

Likewise, by nonlinear regression analysis of the data presented in the sixth and seventh columns of Table 7, the effects of the w/c and F/T ratios on the maximum filling height and the required SP dosage can be illustrated graphically in the form of contours, as shown in Figure 8. From this figure, it can be seen that the maximum filling height is generally higher at a lower w/c ratio or a higher F/T ratio. Relatively, the F/T ratio has a greater effect than the w/c ratio. In fact, for the ranges of mix parameters covered in this study (more explicitly, for concrete mixes with cement used as the only cementitious material and w/c ratio within 0·35 and 0·45), the minimum required filling height of 300 mm for SCC is achievable only at an F/T ratio higher than 0·55 regardless of the w/c ratio. Hence, if the maximum filling height is lower than the minimum required, the filling height should be more effectively increased by increasing the F/T ratio rather than by decreasing the w/c ratio. As expected, the required SP dosage for achieving the maximum filling height is generally higher at a lower w/c ratio or a higher F/T ratio.

Robustness and optimum SP dosage

Acceptable range of SP dosage and optimum SP dosage

To be qualified as SCC, the concrete mix has to satisfy all the following performance requirements: slump flow ≥ 650 mm; segregation index ≤ 20%; and filling height ≥ 300 mm. From the experimental results, the range of SP dosage satisfying the requirements that slump flow ≥ 650 mm and segregation index ≤ 20% for each concrete mix has been determined and presented in the fourth column of Table 8. Likewise, the range of SP dosage satisfying the requirement that filling height ≥ 300 mm for each concrete mix has been determined and presented in the fifth column of Table 8 (note that since the mixes A, C, D and F do not satisfy this requirement, no range is given for them). Combining these ranges of SP dosage, the acceptable range of SP dosage satisfying all the requirements for SCC can be obtained, as listed in the sixth column of Table 8. The breadth of such acceptable range of SP dosage may be taken as a

Figure 6 Filling height plotted against slump flow

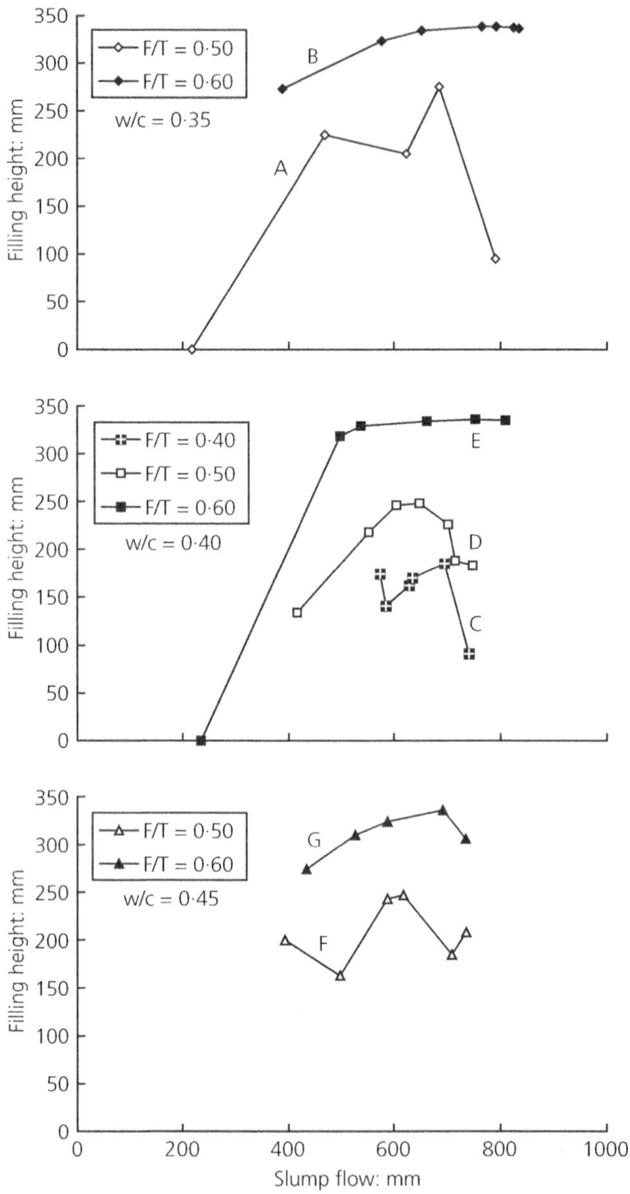

quantitative measure of robustness. The robustness of a concrete mix may be said to be high if the acceptable range of SP dosage satisfying all requirements is relatively wide because a wider acceptable range of SP dosage would render the performance of the concrete mix less sensitive to changes in SP dosage. Comparing the acceptable ranges of SP dosage in Table 8, it can be seen that the robustness is generally higher at a lower w/c ratio or a higher F/T ratio.

Figure 7 Maximum flow without segregation and required SP dosage

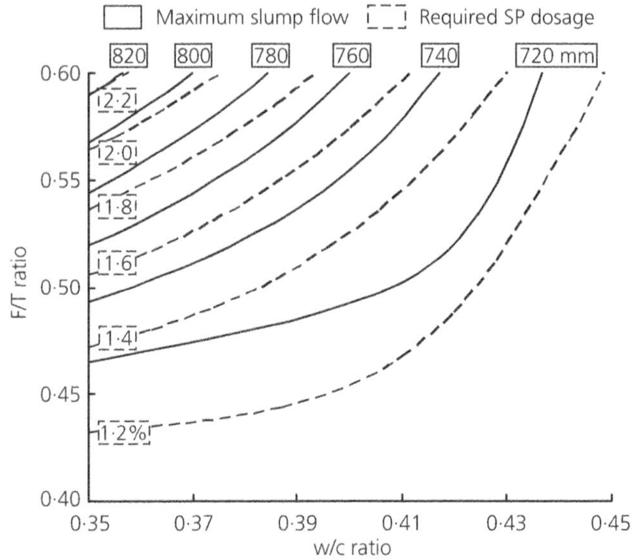

Figure 8 Maximum filling height and required SP dosage

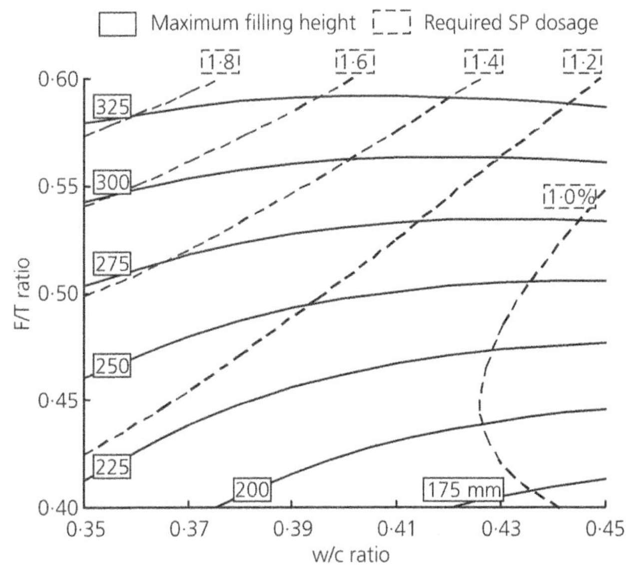

Table 8 Acceptable range of SP dosage and optimum SP dosage

Mix no.	w/c ratio	F/T ratio	Range of SP dosage satisfying following requirements: %		SP dosage satisfying all requirements: %	
			Slump flow ≥ 650 mm and segregation index ≤ 20%	Filling height ≥ 300 mm	Acceptable range of SP dosage	Optimum SP dosage
A	0·35	0·50	1·29–1·51	–	–	–
B		0·60	1·59–2·34	1·31–2·40	1·59–2·34	1·97
C	0·40	0·40	1·03–1·12	–	–	–
D		0·50	1·20–1·43	–	–	–
E		0·60	1·38–1·62	0·99–1·80	1·38–1·62	1·50
F	0·45	0·50	0·94–1·02	–	–	–
G		0·60	1·16–1·24	0·97–1·30	1·16–1·24	1·20

To allow the largest increase or decrease in the SP dosage without causing the concrete mix to fail as an SCC, the design SP dosage should be set as the middle value of the acceptable range of SP dosage. Such a design SP dosage is actually the optimum SP dosage for maximum robustness of the concrete mix. The optimum SP dosage so evaluated for each concrete mix is presented in the last column of Table 8. Comparing the optimum SP dosage of each concrete mix with the respective required SP dosage for maximum filling height (given in last column of Table 7), it is evident that in every concrete mix acceptable as an SCC, the optimum SP dosage is very close to the required SP dosage for maximum filling height. Hence, an alternative is simply to set the optimum SP dosage as the required SP dosage for maximum filling height. However, it should be noted that the design SP dosage should never be set as the required SP dosage for maximum slump flow without segregation because such a design SP dosage would lead to serious segregation (segregation index > 20%) upon the slightest increase in SP dosage and thus a very low robustness.

Acceptable range of slump flow

Following the same procedures, the range of slump flow satisfying the requirements that slump flow ≥ 650 mm and segregation index ≤ 20% for each concrete mix has been determined and presented in the fourth column of Table 9. Likewise, the range of slump flow satisfying the requirement that filling height ≥ 300 mm for each concrete mix has been determined and presented in the fifth column of Table 9. Combining these ranges of slump flow, the acceptable range of slump flow satisfying all the requirements for SCC can be obtained, as listed in the sixth column of Table 9. The width of such acceptable range of slump flow may also be taken as a quantitative measure of robustness, as depicted in the last column of Table 9. In fact, it is the current authors' belief that the width of the acceptable range of slump flow should be a more objective measure of robustness than the width of the acceptable range of SP dosage because the acceptable range of SP dosage is actually dependent on the type and characteristic of the superplasticiser used. As before, the robustness values presented in Table 9, each expressed in terms of the width of the acceptable range of slump flow, reveal that the robustness is generally higher at a lower w/c ratio or a higher F/T ratio.

Table 9 Acceptable range of slump flow and robustness

Mix no.	w/c ratio	F/T ratio	Range of slump flow satisfying following requirements: mm		Acceptable range of slump flow satisfying all requirements: mm	Robustness taken as breadth of acceptable range of slump flow: mm
			Slump flow ≥ 650 mm and segregation index ≤ 20%	Filling height ≥ 300 mm		
A	0·35	0·50	650–743	–	–	0
B		0·60	650–832	490–835	650–832	182
C	0·40	0·40	650–705	–	–	0
D		0·50	650–726	–	–	0
E		0·60	650–756	483–810	650–756	106
F	0·45	0·50	650–715	–	–	0
G		0·60	650–710	501–735	650–710	60

It should be noted that the acceptable range of slump flow is a property of the concrete mix. Depending on the mix parameters, especially the composition of the cementitious materials and the particle size distribution of the aggregate, different concrete mixes may have very different acceptable ranges of slump flow. There is basically no single range of slump flow that is applicable to all concrete mixes. In fact, the relationship between passing/filling abilities and slump flow and the relationship between segregation stability and slump flow are both dependent on the mix parameters. The common practices of predicting the segregation tendency from the slump flow[23] and of recommending a desirable range of slump flow[24] are not universally applicable. In this regard, the authors always object to the idea of setting an upper limit to the slump flow for the purpose of ensuring sufficient cohesiveness. For a given concrete mix, the cohesiveness is generally lower at a higher workability and vice versa. A concrete mix having a higher workability does not, however, necessarily have a lower cohesiveness than another concrete mix having a lower workability. Depending on the mix optimisation technique applied, it may be possible to produce a concrete mix that has a higher slump flow than the specified or recommended upper limit and at the same time satisfies all the other requirements for SCC. The present authors do not see any reason for condemning a concrete mix just because its slump flow is higher than an artificially set upper limit.

Conclusions

A total of 42 batches of concrete, derived from seven basic mix proportions and containing different dosages of superplasticiser, were produced for studying the effects of SP dosage on the workability, passing/filling abilities, segregation stability and strength of concrete mixes with w/c ratios ranging from 0·35 to 0·45 and F/T ratios ranging from 0·40 to 0·60. The slump flow, U-box, sieve segregation and cube compression tests were employed to measure the performance of the concrete mixes. Overall, the results revealed that the increase of SP dosage to increase the workability would cause gradual reduction of the cohesiveness and at a certain stage drastic reduction of the cohesiveness and passing/filling abilities. Hence, the SP dosage has great effects on the performance of SCC and thus should be carefully selected.

The effects of SP dosage were found to be dependent on the w/c and F/T ratios. Regarding the workability and cohesiveness, although at a given SP dosage, the slump flow is lower at a lower w/c ratio or a higher F/T ratio, when the SP dosage is optimised to achieve maximum performance, the maximum slump flow without segregation that can be achieved is actually higher at a lower w/c ratio or a higher F/T ratio. Regarding the passing and filling abilities, the maximum achievable filling height is generally higher at a lower w/c ratio or a higher F/T ratio. Relatively, the F/T ratio has greater effect on the maximum filling height than the w/c ratio and therefore if the maximum filling height is lower than the minimum required, the filling height should be more effectively increased by increasing the F/T ratio rather than by decreasing the w/c ratio. The required SP dosage for achieving the maximum slump flow without segregation or maximum filling height is, however, also generally higher at a lower w/c ratio or a higher F/T ratio.

Based on observation of excessive voids in uncompacted concrete at slump flow < 650 mm and recommendations of JSCE Guidelines[20] and European Guidelines,[9] it is suggested that the requirements for SCC should be: slump flow ≥ 650 mm; segregation index ≤ 20%; and filling height ≥ 300 mm. From the test results, the acceptable range of SP dosage satisfying all such requirements has been determined for each concrete mix. As a wider acceptable range of SP dosage would render the performance of the concrete mix less sensitive to changes in SP dosage, it is proposed to take the width of the acceptable range of SP dosage as a measure of robustness. Furthermore, it is proposed to take the middle value of the acceptable range of SP dosage as the optimum SP dosage for maximum robustness. Likewise, the acceptable range of slump flow satisfying the above requirements has also been determined for each concrete mix. It is proposed that this acceptable range of slump flow may also be taken as a measure of robustness. Taking the acceptable range of either SP dosage or slump flow as the measure of robustness, it can be seen that the robustness is generally higher at a lower w/c ratio or a higher F/T ratio. Hence, the adoption of a higher F/T ratio would increase not only the maximum performance but also the robustness of the concrete mix. Finally, it is advocated that no upper limit should be imposed to the slump flow, as a concrete mix with a high robustness can have a high slump flow while satisfying all other requirements for SCC.

Acknowledgement

The work described in this paper was fully supported by a grant from the Research Grants Council of the Hong Kong Special Administrative Region, China (Project No. HKU 7139/05E).

REFERENCES

1. Okamura H. Self-compacting high-performance concrete. *Concrete International*, 1997, **19**, No. 7, 50–54.
2. Goodier C. I. Development of self-compacting concrete. *Proceedings of Institution of Civil Engineers, Structures and Buildings*, 2003, **156**, No. 4, 405–414.
3. Skarendahl Å. Changing concrete construction through use of self-compacting concrete. In *Design, Performance and Use of Self-Consolidating Concrete (SCC'2005-China)* (Yu Z., Shi C., Khayat K. H. and Xie Y. (eds)). International Union of Testing and Research Laboratories for Materials and Structures (RILEM), Changsha, China, 2005, pp. 17–24.

4. Sonebi M., Grünewald S. and Walraven J. Filling ability and passing ability of self-consolidating concrete. *ACI Materials Journal*, 2007, **104**, No. 2, 162–170.
5. Khayat K. H., Assaad J. and Daczko J. Comparison of field-oriented test methods to assess dynamic stability of self-consolidating concrete. *ACI Materials Journal*, 2004, **101**, No. 2, 168–176.
6. Okamura H. and Ouchi M. Self-compacting concrete. *Journal of Advanced Concrete Technology*, 2003, **1**, No. 1, 5–15.
7. Domone P. L. Self-compacting concrete: an analysis of 11 years of case studies. *Cement and Concrete Composites*, 2006, **28**, No. 2, 197–208.
8. Ng I. Y. T., Wong H. H. C. and Kwan A. K. H. Passing ability and segregation stability of self-consolidating concrete with different aggregate proportions. *Magazine of Concrete Research*, 2006, **58**, No. 7, 447–457.
9. Self Compacting Concrete European Project Group. *The European Guidelines for Self-Compacting Concrete*. BIBM, CEMBUREAU, EFCA, EFNARC and ERMCO, 2005. Available online from: http://www.efnarc.org. Accessed 27/01/2009.
10. Nkinamubanzi P. C. and Aïtcin P. C. Cement and superplasticizer combinations: compatibility and robustness. *Cement, Concrete and Aggregates*, 2004, **26**, No. 2, 102–109.
11. Nunes S., Figueiras H., Oliveira P. M., Coutinho J. S. and Figueiras J. A methodology to assess robustness of SCC mixtures. *Cement and Concrete Research*, 2006, **36**, No. 12, 2115–2122.
12. British Standards Institution. *Specification for Portland Cement*. BSI, London, 1996, BS 12.
13. British Standards Institution. *Methods of Testing Cement. Determination of Strength*. BSI, London, 2005, BS EN 196-1.
14. British Standards Institution. *Specification for Aggregates from Natural Sources for Concrete*. BSI, London, 1992, BS 882.
15. British Standards Institution. *Concrete Admixtures. Specification for Superplasticizing Admixtures*. BSI, London, 1985, BS 5075-3.
16. Mehta P. K. and Aïtcin P. C. Microstructural basis of selection of materials and mix proportions for high-strength concrete. *Proceedings of the 2nd International Symposium on Utilization of High Strength Concrete*. American Concrete Institute, Farmington Hills, MI, USA, 1990, ACI Special Publication SP-121, pp. 265–286.
17. Hanehara S. and Yamada K. Interaction between cement and chemical admixture from point of cement hydration, absorption behaviour of admixture, and paste rheology. *Cement and Concrete Research*, 1999, **29**, No. 8, 1159–1165.
18. British Standards Institution. *Testing Concrete. Method for Determination of Slump*. BSI, London, 1983, BS 1881-102.
19. International Union of Testing and Research Laboratories for Materials and Structures (RILEM). *Self-Compacting Concrete* (Skarendahl Å. and Petersson Ö. (eds)). Technical Committee 174-SCC, RILEM Publications SARL, Bagneux, France, 2002.
20. Japan Society of Civil Engineers. *JSCE Guidelines for Concrete No. 6: Standard Specifications for Concrete Structures – 2002: Materials and Construction*. Japan Society of Civil Engineers, Tokyo, 2005.
21. Fujiwara H. Fundamental study on the self-compacting property of high-fluidity concrete. *Proceedings of Japan Concrete Institute*, 1992, **14**, No. 1, 27–32.
22. El-Chabib H. and Nehdi M. Effect of mixture design parameters on segregation of self-consolidating concrete. *ACI Materials Journal*, 2006, **103**, No. 5, 374–383.
23. Safawi M. I., Iwaki I. and Miura T. A study on the applicability of vibration in fresh high fluidity concrete. *Cement and Concrete Research*, 2005, **35**, No. 9, 1834–1845.
24. Hwang S., Khayat K. H. and Bonneau O. Performance-based specifications of self-consolidating concrete used in structural applications. *ACI Materials Journal*, 2006, **103**, No. 2, 121–129.

ice Publishing

Dhir and Lye
ISBN 978-0-7277-6461-4
https://doi.org/10.1680/icetscc.64614.089

Chapter 5
Effects of paste-to-voids volume ratio on performance of self-consolidating concrete mixtures

Xuhao Wang
Research Assistant, Department of Civil, Construction and Environmental Engineering, Iowa State University, Ames, Iowa, USA

Peter Taylor
Associate Director, National Concrete Pavement Technology Center, Iowa State University, Ames, Iowa, USA

Kejin Wang
Professor, Department of Civil, Construction and Environmental Engineering, Iowa State University, Ames, Iowa, USA

George Morcous
Associate Professor, The Durham School of Architectural Engineering and Construction, University of Nebraska-Lincoln, Omaha, USA

A well-proportioned self-consolidating concrete (SCC) mixture can be achieved by controlling the aggregate system, paste quality and paste quantity. This study aims to apply a concept that is based on paste-to-voids volume ratio (V_{paste}/V_{voids}) to assess the performance of SCC mixtures. The relationship between excess paste theory and V_{paste}/V_{voids} was investigated. Tests were conducted on a large matrix of SCC mixtures that were designed for cast-in-place concrete construction. The mixtures were made with different aggregate types and sizes, and different cementitious materials. The workability, flow properties, compressive strength, shrinkage and surface resistivity of SCC mixtures were determined as indicators of SCC performance. Statistical analyses, response surface models and Tukey honest significant difference (HSD) tests, were conducted to relate the mix design parameters to the concrete performance. Test results revealed that slump flow of the SCC mixtures increased with V_{paste}/V_{voids} at a given viscosity. Strength increases with V_{paste}/V_{voids} up to a point (~150%), after which the strength becomes independent of V_{paste}/V_{voids}. V_{paste}/V_{voids} has little effect on the shrinkage of SCC mixtures, or their surface resistivity. Statistical analysis is an efficient tool to identify the significance of influencing factors on concrete performance.

Notation

D_{av}	the average aggregate diameter
D_{pi}	projected diameter of the particles in size class i

D_{ss}	average spacing between aggregate particle surfaces (particles are assumed to be spherical)
d_i	average size of aggregate fraction i
m_i	percentage of aggregate mass retained between upper and lower sieve sizes in fraction i
n_i	number of particles in size class i
P_e	volume of excess paste
s_i	surface area of particles in size class i
t_{final}	flow time until concrete stops flowing
t_{50}	flow time for SCC mixture to reach a diameter of 500 mm
V_c	total concrete volume
V_p	paste volume
V_{voids}	volume of voids in densely compacted aggregate determined in accordance with ASTM C29
V_{paste}/V_{voids}	paste-to-voids volume ratio
η	plastic viscosity
τ	dynamic yield stress

Introduction

The overall performance of self-consolidating concrete (SCC) 'combines concrete's existing ability to produce a wide range of engineering properties with an increased potential for constructability that exceeds anything possible with conventional concrete (CC)' (Daczko, 2012). Nowadays, how to achieve a quality SCC with minimum impact, while meeting application requirements, has become a key issue in designing mix proportions. A critical aim of research dedicated to mix proportioning is to ensure that SCC mixtures have better performance and sustainability.

A well-proportioned SCC mixture can be achieved by controlling the aggregate system, paste quality and paste quantity. The unique features of SCC mixtures are the fresh properties: flow ability, passing ability, filling capacity and stability. This can be obtained by properly selecting the aggregate system and paste quantity. However, other concrete properties, such as mechanical properties, shrinkage and permeability, need to be assessed as well. These are largely controlled by the paste quality, such as water-to-cementitious material ratio, supplementary cementitious material (SCMs) types and dosages, and use of chemical admixtures.

This study is aimed at applying a concept that is based on a paste-to-voids volume ratio (V_{paste}/V_{voids}) to assess the performance of SCC mixtures. Aggregate system and paste quality are varied in order to comprehensively understand the influences of V_{paste}/V_{voids} in a given paste system.

Background

Excess paste theory and applications on SCC

The 'excess paste theory' was originally developed by Kennedy (1940) and it was built on a two-phase theory, that is, a paste phase is used to fill up the voids between aggregates. Sufficient paste volume is needed to fill the voids and control friction between aggregates to provide desired workability. The 'lubricating' layer of paste around aggregates needs to be sufficient to prevent coarse aggregates from sinking down and segregating and to achieve a good workability (Hu and Wang, 2007; Kennedy, 1940; Koehler and Fowler, 2007; Kosmatka *et al.*, 2008).

Ideally, the excess paste thickness can be approximated by using paste volume divided by the surface area of the aggregates. Heywood (1933) proposed a direct method to measure aggregates in terms of length, width and thickness. However, what he proposed would only work for an individual grain size, and not for a continuous grading of aggregates. Oh *et al.* (1999) modified the equations so that they would allow the total surface area of aggregates to be calculated. Meanwhile, they established the relationship between the relative thickness of excess paste and the relative Bingham parameters for a continuous grading of aggregates in Equation 1.

$$\tau = \frac{P_\mathrm{e}}{\sum_i^n n_i s_i D_{\mathrm{p}i}} \tag{1}$$

where P_e is the volume of excess paste; n_i is the number of particles in size class i; s_i is the surface area of particles in size class i, and $D_{\mathrm{p}i}$ is the projected diameter of the particles in size class i.

This theory has been applied to design SCC mix proportions by Bui *et al.* (2002). The paste volume must be high enough to fill the voids between aggregate particles and create a layer enveloping the particles to achieve deformability and good segregation resistance. The average aggregate spacing is calculated by Equations 2 and 3 and defined as an average distance between surfaces of aggregate particles, or as twice the thickness of paste layer around an aggregate particle, as shown in Figure 1 (Bui *et al.*, 2002).

$$D_\mathrm{ss} = D_\mathrm{av}\left(\sqrt[3]{1 + \frac{V_\mathrm{p} - V_\mathrm{void}}{V_\mathrm{c} - V_\mathrm{p}}} - 1\right) \tag{2}$$

where D_ss is average spacing between aggregate particle surfaces (particles are assumed to be spherical); V_p is paste volume; V_void is volume of voids in densely compacted aggregate determined in accordance with ASTM C29 (ASTM, 1999a); V_c is total concrete volume; and

Figure 1 Spherical aggregate particles, aggregate spacing D_ss, and average aggregate diameter D_av

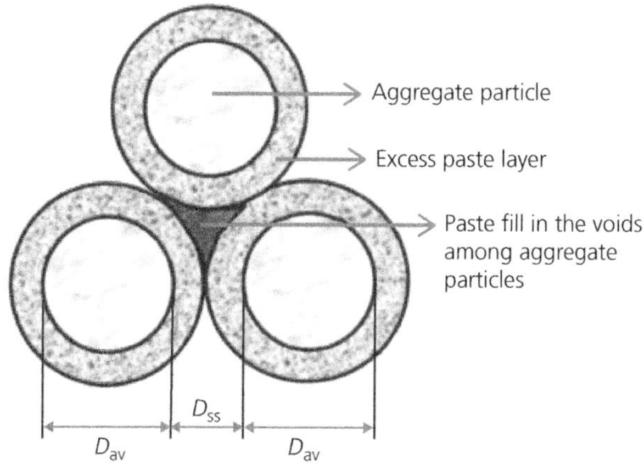

D_{av} is the average aggregate diameter, which is given by

$$D_{av} = \frac{\sum d_i m_i}{\sum m_i} \tag{3}$$

where d_i is the average size of aggregate fraction i; and m_i is the percentage of aggregate mass retained between upper and lower sieve sizes in fraction i.

The relationships among paste rheology, average aggregate spacing and average aggregate size were established and the general trends were found. Some satisfactory zones were defined for different average aggregate spacing, average aggregate diameter, cement contents, water–binder ratios, as well as contents and types of fly ash (Bui *et al.*, 2002).

Paste-to-void volume ratio

An alternative concept, based on the paste-to-voids volume ratio (V_{paste}/V_{voids}), was applied to pavement concrete mixtures by Yurdakul *et al.* (2013). The V_{paste}/V_{voids} can be determined by calculating the paste volume of concrete mixtures and dividing that value by the volume of voids in the consolidated aggregate system determined in accordance with ASTM C29 (ASTM, 1999a). The paste volume comprises the volume of water, the cementitious materials and the measured air in the system. A figure of 100% means that all the space between the aggregates is just filled with paste, with no excess.

The idea of relating performance of a mixture to paste volume for a given aggregate system was initially used to assess SCC mixtures by Koehler and Fowler (2007). The V_{paste}/V_{voids} concept provides a quantitative means to consider the interaction between paste and aggregate system and achieve a quality concrete mixture with minimum impact, while meeting specifications. The approach is believed to be more useful than parameters of 'cementitious content' or 'paste content' because it takes into account differences between aggregate systems (Yurdakul, 2013). In the same way as the excess paste approach, the aim is to

- coat the aggregate particles
- fill the voids between the combined aggregate system
- disperse the aggregate particles to provide the desired workability.

Relationship between parameters from excess paste theory and V_{paste}/V_{voids}

A limitation to the excess paste approach is that it is based on the assumption that aggregate particles are spherical and that they are packed in a cubic lattice, neither of which is true (Yurdakul *et al.*, 2013). The aggregate spacing can be considered as an average paste thickness because the average aggregate diameter is determined based on combined coarse and fine aggregate fractions.

The reason for choosing V_{paste}/V_{voids} to assess SCC mixture performance is that the process includes the effects of aggregate characteristics, such as size, shape and gradation.

For comparison, the aggregate spacing, D_{ss}, and average aggregate diameter, D_{av}, of all the mixtures discussed in this paper were calculated using Equation 2 and Equation 3, respectively. The data are listed in Table 3 (see later). The relationship between average aggregate spacing and V_{paste}/V_{voids} is shown in Figure 2. A linear relationship is found for a given aggregate system, but varies for different coarse/fine combinations.

As shown in Figure 3, the gradations of both limestone and gravel coarse aggregates were similar. Gravel aggregate with particles more spherical in nature tends to have higher V_{paste}/V_{voids} at a given binder content, size and sand-to-aggregate ratio system compared to limestone aggregate with more angular particles.

Figure 2 Relationship between average aggregate spacing and V_{paste}/V_{voids} of the SCC mixtures investigated in the present study

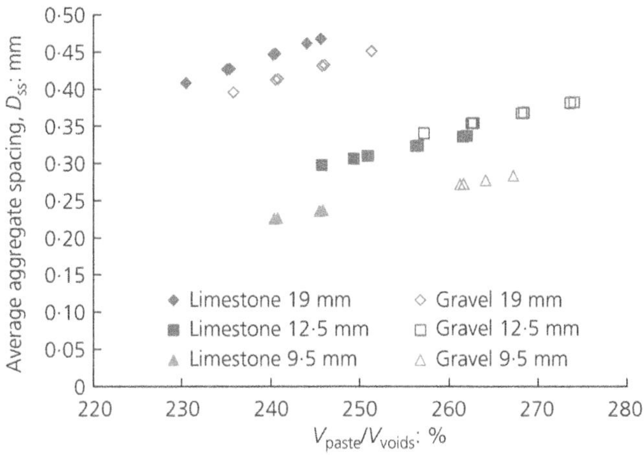

Figure 3 Size distributions of all solid ingredients in this study

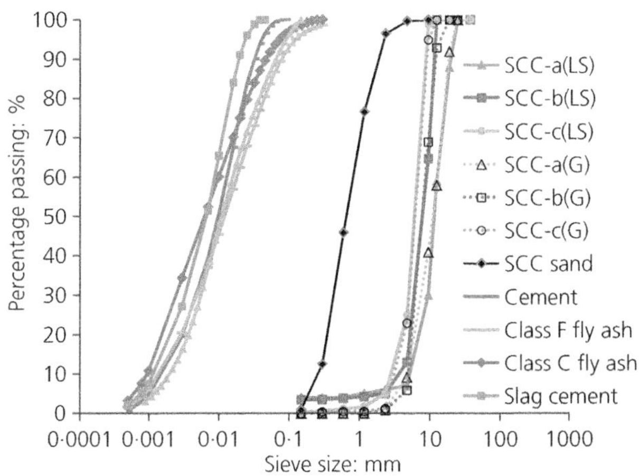

This is because the angular aggregate particles tend to decrease packing density, resulting in higher void content in the combined aggregate system (Quiroga and Fowler, 2003). However, there is an increased effect on packing density with decreased size of aggregate, which is in agreement with the compressible packing model proposed by De Larrard (1999).

Mixture proportion and materials

Forty SCC mixes, designed for cast-in-place construction, were developed with the following targeted parameters

- low slump flow range between 550 and 650 mm or high flow range between 650 and 750 mm
- visual stability index (VSI) ≤ 1
- J-ring ≤ 75 mm

The SCC mixes were made with limestone and river gravel coarse aggregates. Each coarse aggregate was used in three different nominal maximum sizes, 19·0 mm, 12·5 mm and 9·5 mm. The physical properties of the aggregates are shown in Table 1.

For practical purposes, cementitious blends containing 25% Class C fly ash, 25% Class F fly ash, 30% slag cement, or 15% limestone dust with 20% Class F fly ash, were used. Table 2 provides the chemical properties of the cementitious materials. The aggregate gradations and SCMs size distributions are given in Figure 3.

The chemical admixtures used were air-entraining agent (AEA), polycarboxylate-based high-range water reducer (HRWR) and viscosity-modifying admixture (VMA). The mix proportions of all the mixtures are shown in Table 3.

Table 1 The physical properties of aggregates

Aggregates used in the research		Type	Nominal maximum size: mm	Absorption: %	Fineness modulus	Specific gravity
Coarse aggregate	a(LS)	Limestone	19·0	1·3	–	2·66
	b(LS)	Limestone	12·5	1·3	–	2·66
	c(LS)	Limestone	9·5	1·3	–	2·66
	a(G)	Gravel	19·0	1·1	–	2·74
	b(G)	Gravel	12·5	1·4	–	2·68
	c(G)	Gravel	9·5	1·4	–	2·69
Fine aggregate	River sand		–	0·5	2·62	2·68

Note: '–' indicates that the data are not available.

Table 2 Chemical compositions of cementitious materials

Chemical composition: %	Type I/II cement	Class F fly ash	Class C fly ash	Slag cement
Silicon dioxide (SiO_2)	20·10	50·87	42·46	31·63
Aluminium oxide (Al_2O_3)	4·44	20·17	19·46	11·30
Iron (III) oxide (Fe_2O_3)	3·09	5·27	5·51	0·34
Sulfur trioxide (SO_3)	3·18	0·61	1·20	3·30
Calcium oxide (CaO)	62·94	15·78	21·54	41·31
Magnesium oxide (MgO)	2·88	3·19	4·67	10·77
Sodium oxide (Na_2O)	0·10	0·69	1·42	0·19
Potassium oxide (K_2O)	0·61	1·09	0·68	0·36
Phosphorus pentoxide (P_2O_5)	0·06	0·44	0·84	0·02
Titanium dioxide (TiO_2)	0·24	1·29	1·48	0·56
Strontium oxide (SrO)	0·09	0·35	0·32	0·04
Barium oxide (BaO)	–	0·35	0·67	–
Loss on ignition	2·22	0·07	0·19	–
Total	100	100	100	100

Test equipment and procedures

Workability

Slump flow, segregation resistance and blocking assessment were determined in accordance with ASTM C1611 (ASTM, 2005) and ASTM, C1621 (ASTM, 2009). The flow time for SCC mixtures reaching a diameter of 500 mm, t_{50}, and flow time until concrete stopped flowing, t_{final}, were recorded.

The workability retention was determined by the difference between slump flow soon after mixing and 30 min after mixing.

Air content

Air contents of all the mixtures were determined in accordance with ASTM C231 (ASTM, 1997a).

Mortar rheology

Mortar was sieved from the concrete mixtures using a 4·75 mm size sieve. The dynamic yield stress and plastic viscosity were determined using a Brookfield rheometer. The sample was placed in a 50 mm dia. by 100 mm tall cylindrical vessel and sheared with a 15 by 30 mm vane spindle. The loading history employed, based on Lomboy *et al.* (2014), initially ramps the spindle rotation from 0 to 0·2 1/s for 180 s, and then is sustained at 0·2 1/s for 60 s. The spindle rotation was subsequently increased from 0·2 to 100 1/s in the following 60 s and decreased to 0 1/s during the last 60 s, as shown in Figure 4(a). The dynamic yield stress, τ, and plastic viscosity, η, can be captured based on a Bingham model from the downward curve of the plot

Table 3 Mix proportions

ID	C I,II: kg/m^3	SCM: kg/m^3	Filler: kg/m^3	CA: kg/m^3	FA: kg/m^3	Water: kg/m^3	HRWRA: ml/ 100 kg SCMs	VMA: ml/ 100 kg SCMs	AEA: ml/ 100 kg SCMs	Paste volume: %	V_{paste}/V_{voids}: %	D_{av}: mm	D_{ss}: mm
L-a(LS)-C	315	105	0	915	749	166	652·0	0	97·8	36·5	230·5	7·110	0·408
L-a(LS)-F	315	105	0	915	749	166	489·0	0	97·8	37·0	235·4	7·110	0·427
L-a(LS)-S	309	132	0	915	749	166	652·0	0	97·8	37·0	235·4	7·110	0·427
L-a(LS)-FLD	271	83	62	915	749	166	586·8	0	97·8	37·0	235·1	7·110	0·426
H-a(LS)-C	337	112	0	901	737	166	521·6	0	52·2	37·5	240·2	7·110	0·446
H-a(LS)-F	337	112	0	901	737	166	521·6	130	52·2	38·0	245·6	7·110	0·467
H-a(LS)-S	320	137	0	908	743	166	521·6	130	52·2	37·5	240·5	7·110	0·447
H-a(LS)-FLD	290	89	63	901	737	166	782·4	0	97·8	37·8	244·0	7·110	0·461
L-b(LS)-C	317	106	0	867	769	175	521·6	0	65·2	37·5	245·7	4·543	0·297
L-b(LS)-F	317	106	0	867	769	175	391·2	130	97·8	38·0	250·9	4·543	0·310
L-b(LS)-S	311	129	0	874	775	175	521·6	0	97·8	37·8	249·3	4·543	0·306
L-b(LS)-FLD	273	84	63	867	769	175	391·2	0	97·8	38·0	250·8	4·543	0·310
H-b(LS)-C	339	113	0	854	757	175	586·8	196	52·2	38·5	256·2	4·543	0·323
H-b(LS)-F	339	113	0	854	757	175	521·6	196	65·2	39·0	261·9	4·543	0·337
H-b(LS)-S	322	138	0	860	763	175	619·4	196	65·2	38·5	256·5	4·543	0·323
H-b(LS)-FLD	291	90	67	854	757	175	456·4	196	65·2	39·0	261·5	4·543	0·336
H-c(LS)-C	348	116	0	791	791	181	717·2	0	81·5	39·5	240·4	3·534	0·226
H-c(LS)-F	348	116	0	791	791	181	684·6	196	97·8	40·0	245·8	3·534	0·237

Table 3 (*Continued*)

ID	C I,II: kg/m^3	SCM: kg/m^3	Filler: kg/m^3	CA: kg/m^3	FA: kg/m^3	Water: kg/m^3	HRWRA: ml/100 kg SCMs	VMA: ml/100 kg SCMs	AEA: ml/100 kg SCMs	Paste volume: %	V_{paste}/V_{voids}: %	D_{av}: mm	D_{ss}: mm
H-c(LS)-S	331	142	0	798	798	181	782·4	0	97·8	39·5	240·7	3·534	0·226
H-c(LS)-FLD	299	92	69	791	791	181	717·2	0	97·8	40·0	245·5	3·534	0·236
L-a(G)-C	315	105	0	911	746	166	456·4	0	97·8	36·5	235·8	6·593	0·396
L-a(G)-F	315	105	0	911	746	166	391·2	0	97·8	37·0	240·8	6·593	0·413
L-a(G)-S	309	132	0	911	746	166	521·6	0	97·8	37·0	240·8	6·593	0·413
L-a(G)-FLD	271	83	62	911	746	166	391·2	0	97·8	37·0	240·5	6·593	0·412
H-a(G)-C	337	112	0	897	734	166	652·0	196	97·8	37·5	245·8	6·593	0·431
H-a(G)-F	337	112	0	897	734	166	652·0	196	97·8	38·0	251·2	6·593	0·451
H-a(G)-S	320	137	0	904	740	166	652·0	196	97·8	37·5	246·0	6·593	0·432
H-a(G)-FLD	290	89	67	897	734	166	847·6	196	97·8	38·0	251·3	6·593	0·451
L-b(G)-C	317	106	0	864	766	175	456·4	0	97·8	37·5	257·1	4·829	0·340
L-b(G)-F	317	106	0	864	766	175	293·4	0	97·8	38·0	262·5	4·829	0·354
L-b(G)-S	311	133	0	864	766	175	391·2	0	97·8	38·0	262·7	4·829	0·354
L-b(G)-FLD	273	84	63	864	766	175	391·2	0	97·8	38·0	262·4	4·829	0·354
H-b(G)-C	339	113	0	850	754	175	717·2	196	97·8	38·5	268·1	4·829	0·368
H-b(G)-F	339	113	0	850	754	175	652·0	391	97·8	39·0	274·1	4·829	0·383
H-b(G)-S	322	138	0	857	760	175	782·4	326	97·8	38·5	268·3	4·829	0·368
H-b(G)-FLD	291	90	67	850	754	175	652·0	196	97·8	39·0	273·6	4·829	0·381
H-c(G)-C	348	116	0	788	788	181	652·0	0	97·8	39·5	261·2	3·660	0·272

(*continued on next page*)

Table 3 (*Continued*)

ID	C I,II: kg/m^3	SCM: kg/m^3	Filler: kg/m^3	CA: kg/m^3	FA: kg/m^3	Water: kg/m^3	HRWRA: ml/ 100 kg SCMs	VMA: ml/ 100 kg SCMs	AEA: ml/ 100 kg SCMs	Paste volume: %	V_{paste}/V_{voids}: %	D_{av}: mm	D_{ss}: mm
H-c(G)-F	348	116	0	788	788	181	586·8	0	97·8	40·0	267·1	3·660	0·283
H-c(G)-S	331	142	0	795	795	181	782·4	228	97·8	39·5	261·6	3·660	0·272
H-c(G)-FLD	291	92	69	788	788	181	619·4	0	97·8	39·7	264·0	3·660	0·277

Note: C = Class C fly ash; F = Class F fly ash; S = slag cement; FLD = Class F fly ash and limestone dust; a = 19·0 mm NMSA; b = 12·5 mm NMSA; c = 9·5 mm NMSA; H = high slump flow range (i.e. 650–750 mm); L = low slump flow range (i.e. 550–650 mm); C I,II = Type I/II Portland cement; LD = limestone dust; CA = coarse aggregate; FA = fine aggregate; LS = crushed limestone; G = river gravel; SCMs = supplementary cementitious materials.

Figure 4 (a) Loading history with preshear; (b) flow curve of a typical SCC mixture used in this study (Lomboy *et al.*, 2014)

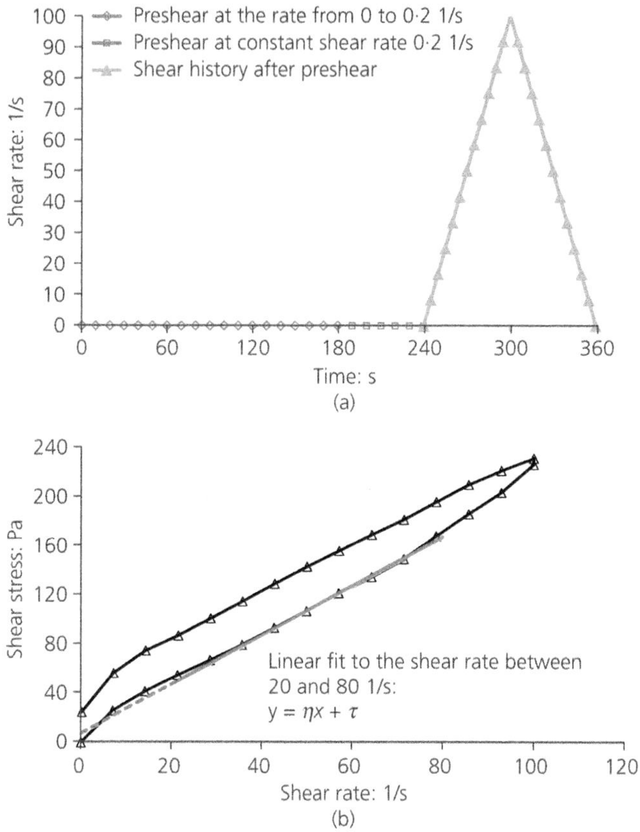

shown in Figure 4(b). The intersection with the *y*-axis and the slope of the linear fit Bingham model represent the yield stress and viscosity, respectively.

Surface resistivity

Research studies have shown that the surface resistivity test is a promising alternative to the rapid chloride penetrability test (RCPT) as a means of indirectly assessing the permeability of concrete mixtures (Aashto TP 95, Aashto, 2011; Chini *et al.*, 2003; Kessler *et al.*, 2008; Rupnow and Icenogle, 2012). Surface resistivity results were determined in accordance with the instructions of the device supplier (Proceq, 2011).

The device comprises four electrodes that are linearly aligned and uniformly spaced (3·8 cm in this study). A potential is applied across the outside probes and the resistivity is measured on the inside probes.

Table 4 LADOTD surface resistivity and permeability classes (LADOTD, 2011)

Permeability class	56-d rapid chloride permeability charge passed: Coulombs	28-d surface resistivity: kΩ cm
High	> 4000	< 12
Moderate	2000–4000	12–21
Low	1000–2000	21–37
Very low	100–1000	37–254
Negligible	< 100	> 254

Three concrete cylindrical specimens for each tested material were prepared so that averaged responses could be measured. The specimens were stored in a fog room at 23°C after casting and tests were conducted at 1, 3, 7 and 28 d.

Four indelible marks were made, oriented 90° apart on the outer edge of one of the end faces, to help place the device evenly about the circumference of the cylinder. Eight readings were taken on each cylinder. A dimensionless correction factor was needed to account for the geometry and size of the specimen.

The qualitative relationship between the charge passed using ASTM C1202 (ASTM, 1997b) and the surface resistivity for 100 × 200 mm cylindrical specimens, as proposed by Louisiana Department of Transportation and Development (LADOTD, 2011), is presented in Table 4:

Compressive strength and free shrinkage

Compressive strength and free shrinkage of the concrete mixtures were measured following ASTM C39 (ASTM, 2001) and C157 (ASTM, 1999b), respectively. The compressive strengths were taken at 56 d and free shrinkage was measured at 3, 7, 28 and 56 d. The concrete prisms for free shrinkage tests were moist cured for 7 d and then transferred to the drying room and maintained at 50% ± 4% relative humidity and a temperature of 23 ± 2°C. Initial readings were taken immediately after demoulding on the first day.

Results and discussion

Workability

The measured fresh properties are summarised in Table 5. The slump flows for all the mixtures fell within the targeted ranges, that is, low flow range between 550 and 650 mm and high flow range between 650 and 750 mm. The t_{50} times of most mixtures were less than 2 s and the t_{final} times ranged from 5·5 to 10·0 s. For the same basic mixture compositions, the lower values of t_{50} and t_{final} correspond to a low viscosity.

In terms of blocking assessment in accordance with ASTM C1621 (ASTM, 2009), all the mixtures fell within ‘no visible blocking’ and ‘minimal to noticeable blocking’ categories.

Table 5 Fresh properties of mixtures

ID	Slump/Slump flow after mixing				J-ring test		Total air: %	Unit weight: kg/m^3	Slump flow after 30 min of mixing		Dynamic yield stress: Pa	Plastic viscosity: Pa s
	t_{50}: s	t_{final}: s	*D*: mm	VSI	Δ*D*: mm	Δ*H*: mm			*D*: mm	Δ Flow: mm		
L-a(LS)-C	1·9	7·9	610	0	32	13	6·2	2297	514	95	78·7	1·07
L-a(LS)-F	1·2	6·8	660	0	51	11	6·0	2262	559	102	62·0	0·93
L-a(LS)-S	1·7	6·5	641	0	38	13	8·0	2223	552	89	53·5	1·17
L-a(LS)-FLD	1·9	7·9	597	0	25	13	5·5	2291	527	70	49·9	0·83
H-a(LS)-C	< 2	6·8	705	1	16	13	5·2	2310	692	13	37·9	1·73
H-a(LS)-F	< 2	7·8	730	1	25	11	3·5	2342	718	13	26·8	0·87
H-a(LS)-S	2	6·2	740	1	3	10	6·5	2262	702	38	10·7	1·95
H-a(LS)-FLD	1·3	7·0	699	1	44	11	6·0	2278	629	70	53·1	1·08
L-b(LS)-C	< 2	6·8	616	0	22	8	4·0	2339	533	83	45·4	0·97
L-b(LS)-F	< 2	7·1	616	0	25	6	5·0	2287	578	38	51·6	1·07
L-b(LS)-S	< 2	6·8	597	0	25	17	6·8	2255	546	51	47·1	1·39
L-b(LS)-FLD	< 2	8·3	629	0	6	6	5·0	2268	572	57	40·7	0·89
H-b(LS)-C	2	9·8	711	1	29	13	3·0	2342	622	89	29·3	0·71
H-b(LS)-F	< 2	6·9	711	1	13	16	6·5	2326	648	64	39·5	1·03
H-b(LS)-S	< 2	6·7	730	1	19	8	7·5	2281	648	83	14·8	1·08
H-b(LS)-FLD	< 2	6·5	667	0	19	11	4·5	2310	610	57	30·9	0·82
H-c(LS)-C	< 2	6·8	737	0	25	6	5·0	2291	679	57	24·6	1·00
H-c(LS)-F	< 2	7·6	699	0	19	8	6·5	2291	629	70	27·6	0·88

(continued on next page)

Table 5 *(Continued)*

ID	Slump/Slump flow after mixing				J-ring test		Total air: %	Unit weight: kg/m^3	Slump flow after 30 min of mixing		Dynamic yield stress: Pa	Plastic viscosity: Pa s
	t_{50}: s	t_{final}: s	D: mm	VSI	ΔD: mm	ΔH: mm			D: mm	Δ Flow: mm		
H-c(LS)-S	2·7	6·5	686	0	13	6	7·0	2249	635	51	20·3	1·73
H-c(LS)-FLD	1·7	8·9	692	0	19	13	6·5	2239	641	51	33·3	1·09
L-a(G)-C	1·6	6·4	616	0	13	13	8·0	2236	527	89	87·6	1·89
L-a(G)-F	1·2	6·8	622	0	32	13	7·4	2287	559	64	50·7	0·79
L-a(G)-S	1·9	6·5	660	0	32	17	7·8	2178	578	83	45·2	1·42
L-a(G)-FLD	1·4	7·7	597	0	19	13	6·6	2265	527	70	43·5	0·64
H-a(G)-C	1·3	7·3	711	1	51	6	5·7	2265	654	57	38·6	1·23
H-a(G)-F	1·2	7·8	762	1	25	3	5·5	2300	705	57	18·6	0·82
H-a(G)-S	2·3	8·7	718	1	25	6	5·7	2166	670	48	13·5	1·98
H-a(G)-FLD	1·6	7·3	718	1	32	6	4·6	2287	641	76	12·8	0·57
L-b(G)-C	1·0	6·0	622	0	13	6	5·6	2252	546	76	55·2	0·77
L-b(G)-F	1·4	5·5	629	0	6	6	6·0	2281	578	51	44·5	0·62
L-b(G)-S	1·8	10·0	603	0	6	13	6·5	2185	546	57	31·1	1·15
L-b(G)-FLD	1·0	5·9	641	0	6	6	5·6	2265	572	70	25·9	0·59
H-b(G)-C	1·5	7·9	730	1	13	8	6·0	2265	622	108	26·0	0·52
H-b(G)-F	1·1	7·4	737	2	6	6	6·0	2284	660	76	8·4	0·40
H-b(G)-S	2·3	9·7	762	2	6	6	7·8	2182	679	83	1·4	0·68
H-b(G)-FLD	0·9	6·2	762	2	13	5	7·6	2233	686	76	15·8	0·46

Table 5 (*Continued*)

ID	Slump/Slump flow after mixing				J-ring test		Total air: %	Unit weight: kg/m^3	Slump flow after 30 min of mixing		Dynamic yield stress: Pa	Plastic viscosity: Pa s
	t_{50}: s	t_{final}: s	D: mm	VSI	ΔD: mm	ΔH: mm			D: mm	Δ Flow: mm		
H-c(G)-C	1·4	7·2	743	0	6	6	6·6	2259	679	64	18·9	0·88
H-c(G)-F	1·3	9·5	737	0	6	6	6·0	2255	648	89	9·8	0·74
H-c(G)-S	1·8	7·6	762	2	6	6	5·4	2220	692	70	0·00	0·53
H-c(G)-FLD	1·1	5·8	737	0	19	6	6·2	2233	660	76	11·4	0·69

Note: t_{50} = time taken for the outer edge of the concrete mass to reach a diameter of 500 mm from the time the mould is first raised; t_{final} = flow time until flow stopping; D = slump flow diameter; ΔD = slump flow diameter – J-ring flow diameter; ΔH = difference between the height of concrete inside the ring and outside the ring at four locations around the ring; Δ Flow = difference of slump flow between after mixing and 30 min after mixing.

For passing ability, the difference between the height of concrete inside the J-ring and outside the J-ring should be less than 13 mm, for acceptable passing ability for most applications (Koehler and Fowler, 2007). Only three mixtures made with 19 mm coarse aggregate and targeted for low flow range exceeded this limit. However, these mixtures should still be acceptable for some applications requiring lower passing ability, such as bridge foundations.

The slump flow of SCC mixtures is considered to be dominated by the quantity of paste and the yield stress and viscosity of the mortar. A commercially available statistical analysis software (JMP, 2005) was used to develop a quadratic response surface model for the mixtures in Figures 5(a) and 5(b). The contour lines developed from the response surfaces are shown on the top surface in two dimensions. The discrete gradients provide visualised ranges for slump flow diameters and the keys indicate the slump flow ranges for each gradient. The prediction model of Equation 4 is valid for the materials and ranges tested in this study.

Figure 5 gives the response surfaces fitted to the data of all SCC mixtures with the prediction equation whose R^2 value was 70%.

$$\begin{aligned} \mathrm{SF} = {} & 1080.46 - 1.31 \times V_{\mathrm{paste}}/V_{\mathrm{voids}} + 16.60 \times \tau - \left(V_{\mathrm{paste}}/V_{\mathrm{voids}} - 251.18\right) \\ & \times (\eta - 0.99) - 105 \times (\eta - 0.99)^2 \end{aligned} \tag{4}$$

where SF stands for slump flow (mm); τ is the dynamic yield stress; η is the plastic viscosity. The t-tests showed that the viscosity does not have statistically significant effects on slump flow, while other parameters in the prediction equations are statistically significant. The general trends that can be observed are

- slump flow increases with increased $V_{\mathrm{paste}}/V_{\mathrm{voids}}$ at a given viscosity of mortar
- at a given $V_{\mathrm{paste}}/V_{\mathrm{voids}}$, it is not surprising that higher viscosity of the mortar gives a lower flow diameter because more viscous paste can result in a higher resistivity of flow
- at a given $V_{\mathrm{paste}}/V_{\mathrm{voids}}$, slump flow tends to decrease with increased yield stress.

Figure 6 shows the relationship between measured and model predicted slump flow. This implies a good slump flow prediction by the models because the data are closely scattered around the line of equality.

Compressive strength

In Figure 7, the correlation between $V_{\mathrm{paste}}/V_{\mathrm{voids}}$ and 28-d compressive strength is graphically depicted, combined with the data for conventional and SCC mixtures reported by Taylor *et al.* (2012a, 2012b), Cook *et al.* (2013), Obla (2012) and the present study. It is noted that an empirical factor, 1·15, was applied to estimate 28-d strength of the SCC mixtures from measured 56-d test results.

A general trend can be observed that the strength increases with $V_{\mathrm{paste}}/V_{\mathrm{voids}}$ up to a point (~150%), above which the strength becomes independent of $V_{\mathrm{paste}}/V_{\mathrm{voids}}$, or even slightly decreases as shown by the envelope in Figure 7. It is also notable that the data from the SCC

Figure 5 Response surface plots show the relationship among V_{paste}/V_{voids}, slump flow, and (a) viscosity; (b) yield stress

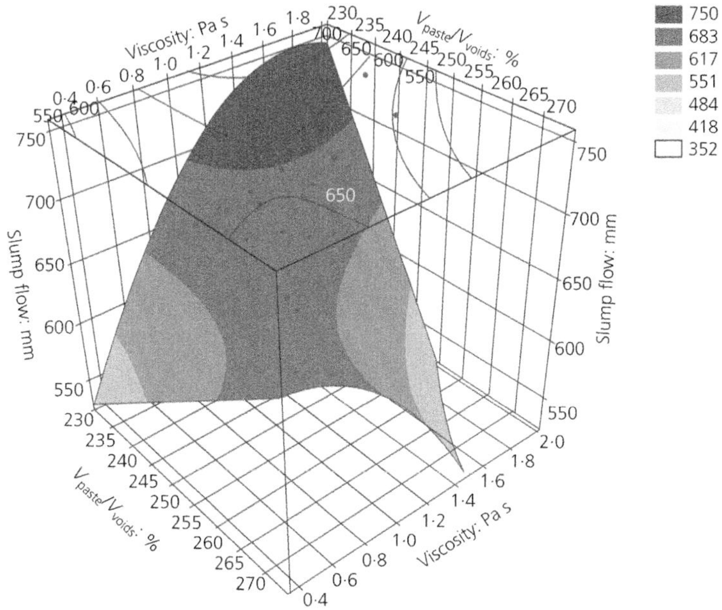

(a)

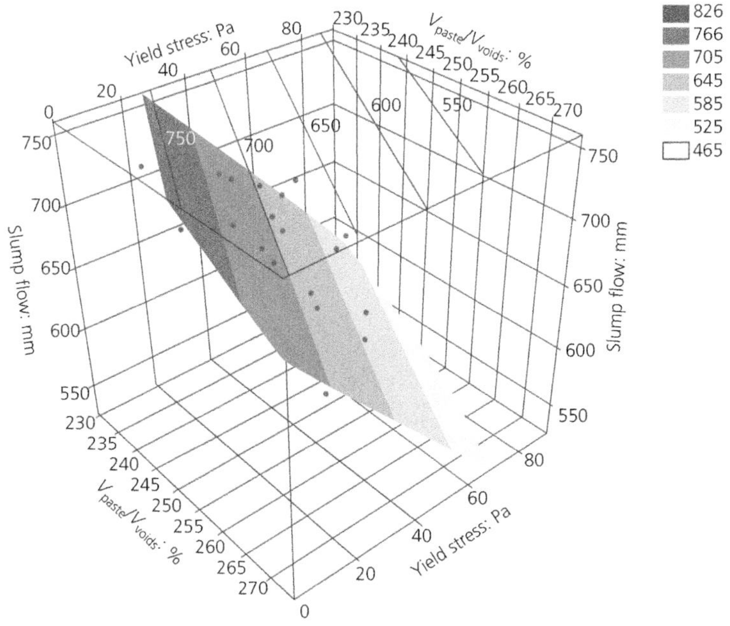

(b)

Figure 6 Measured slump flow plotted against model predicted slump flow

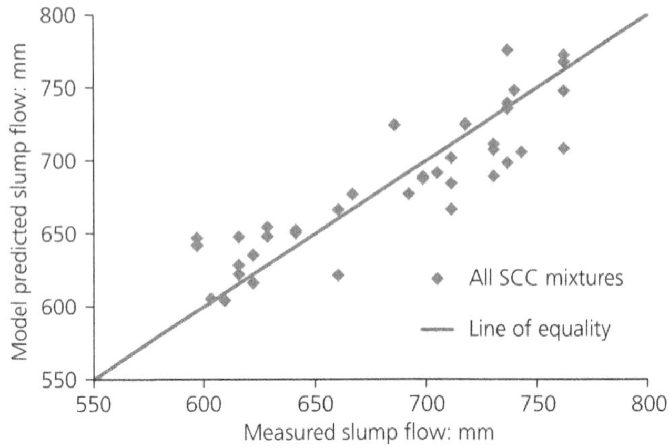

Figure 7 Correlation between V_{paste}/V_{voids} and compressive strength at 28 d based on data retrieved from the literature and the present study (Cook *et al.*, 2013; Obla (2012); Taylor *et al.*, 2012a, 2012b)

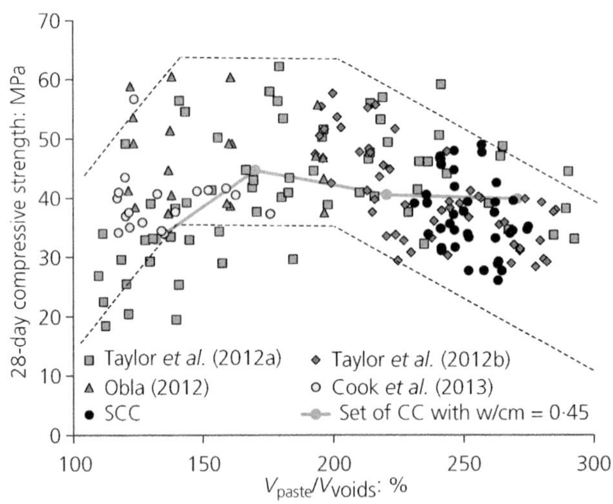

mixtures fall within the same trend as the conventional mixtures, although the paste contents are relatively high.

The trend of reducing strength with increasing paste content may be attributed to crack tortuosity because increased paste content will lead to a shorter path that a crack needs to follow to go from one side of a sample to another. The greater the amount of aggregate, and so greater tortuosity, will mean higher energy required to propagate the crack, and so higher strength (Kolias and Georgiou, 2005).

However, compressive strength will also be affected by other factors, such as SCM type, SCM dosage and water to cementitious materials ratio (w/cm). In particular, a lower w/cm ratio will yield a higher compressive strength in a given mixture and it is considered as one of the most important influencing factors on the strength development (Mehta and Monteiro, 2006).

Free shrinkage

Figure 8 shows the 56-d shrinkage strain plotted against V_{paste}/V_{voids}, using data from SCC mixtures in this study and data reported by Taylor *et al.* (2012a) and NRMCA. It can be observed from Figure 8 that across the range of V_{paste}/V_{voids}, shrinkage is not significantly affected. However, the SCC mixtures are observed to shrink more than the conventional mixtures.

The literature indicates that the size and volume of capillary voids which can be determined by w/cm and degree of hydration drive the mechanism of drying shrinkage behaviour. The paste composition also modifies the microstructure of a paste system (Malhotra and Mehta, 1996; Wang *et al.*, 2012). Shrinkage can be affected by HRWR as well: a higher dosage of HRWR used to achieve a higher workability may result in an increased shrinkage for the SCC mixtures with similar mix proportions (Alsayed, 1998; Kosmatka *et al.*, 2008; Wang, 2011). Therefore, these explain that V_{paste}/V_{voids} is not the only influencing factor related to shrinkage behaviour and SCC mixtures made with similar compositions but higher HRWR dosages compared to CC mixtures tend to have higher shrinkage, as shown in Figure 8.

Figure 8 Overview of the effect of V_{paste}/V_{voids} on free shrinkage at 56 d (using data reported by Taylor *et al.* (2012a) and NRMCA) and SCC mixtures in this study

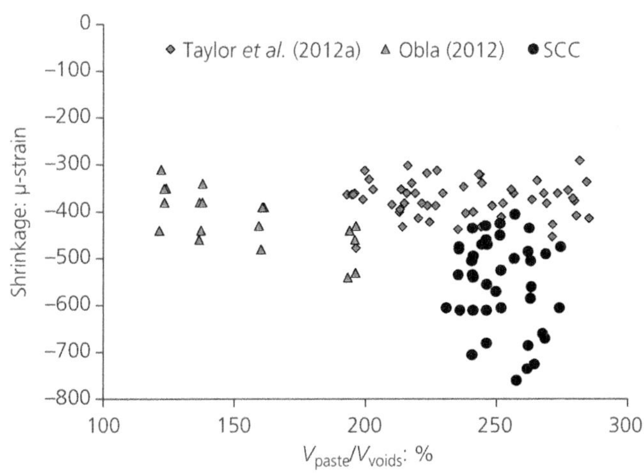

Surface resistivity measurements

The surface resistivity results shown in Figure 9 indicate that the majority of SCC mixtures fall within the 'low' and 'very low' permeability classes in Table 4. The data presented in Figure 9 show the effect of V_{paste}/V_{voids} on surface resistivity categorised by SCM types at 28 d.

The mixtures containing slag cement have significantly higher resistivity than the others, while the mixtures with class F fly ash have similar resistivity to those with both limestone dust and class F fly ash. Slag cement is comparatively quite reactive and may significantly improve the pore structure, and thus increase the resistivity of concrete (Shi, 2004). This is likely to be because the low replacement level of limestone dust has a marginal effect on surface resistivity at 28 d. Class C fly ash mixtures have an unexpectedly low resistivity compared to the other SCMs at 28 d. This is most likely owing to its low reactivity, which can be demonstrated by the X-ray diffractogram shown in Figure 10. The relatively low calcium aluminium oxide (C_3A) and free lime intensity indicate the class C fly ash used in this study is probably slowly reactive.

Full factorial statistical analysis

Compressive strength, shrinkage and surface resistivity are affected by factors other than paste content, including w/cm and system chemistry. Statistically full factorial analyses and Tukey honest significant difference (HSD) tests are therefore appropriate tools to integrate and analyse the 40 mixtures in this study. These tests provide a quantitative method to identify the influence of differences in material on SCC performances from a statistical perspective; these differences may not be easily shown in figures. A commercially available statistical analysis software was employed (JMP, 2005).

As shown in Table 6, input numerical categorical variables, aggregate types, sizes and SCM types, as well as numerical variables, V_{paste}/V_{voids} and w/cm, were applied to perform multiple

Figure 9 The effect of V_{paste}/V_{voids} on surface resistivity with varied SCMs types of SCC mixtures at 28 d

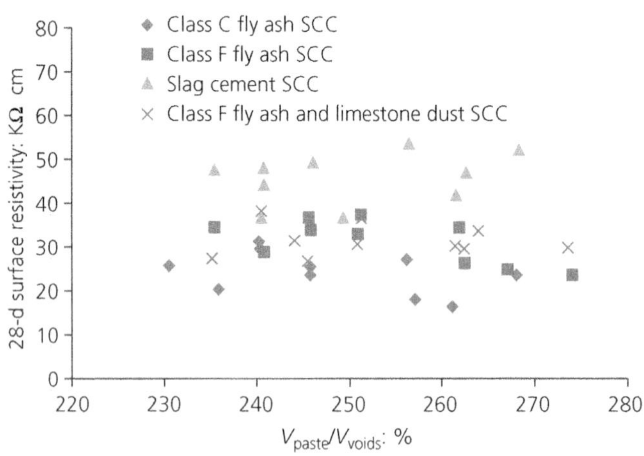

Figure 10 X-ray diffractograms of class C fly ash

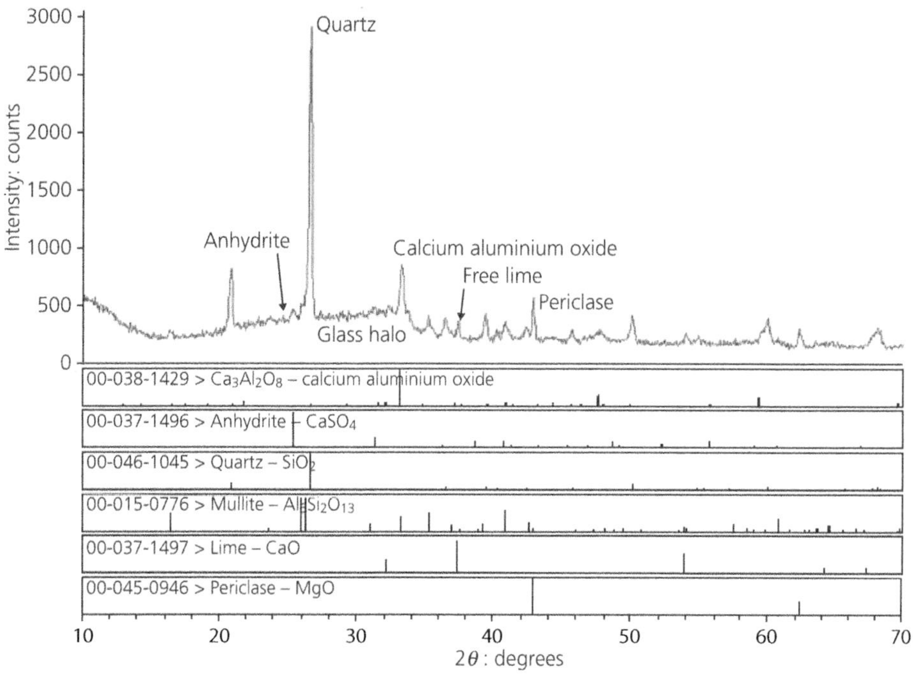

comparisons on compressive strength, shrinkage and surface resistivity results. Only statistically significant factors influencing the properties are listed in the table.

The analysis outputs give an indication of the variables that are statistically significant for 56-d strength, free shrinkage and surface resistivity, respectively. The partition of sums of squares value (i.e. sum of squares/degree of freedom) is a concept that scales for the number of degree of freedom and estimates the variance or spread of the observations about their mean value. In this study, a higher value illustrates that the corresponding variable has a stronger effect on a particular property.

For instance, the partition of sums of squares value of 285·6 for aggregate types is higher than the other variables that are statistically significant in relation to 56-d compressive strength. This indicates that they play the most important role in influencing 56-d compressive strength, followed by SCM types and V_{paste}/V_{voids}. The effect of the aggregate types is likely to be because the angular and rough-textured limestone aggregate tends to improve the quality of the interfacial transition zone and exhibits improved bond to the cement paste compared to gravel aggregate with a round and smooth surface. Moreover, the use of calcareous limestone aggregates may result in increased strength in relation to siliceous gravel aggregate (Mehta and Monteiro, 2006).

Table 6 Input and output parameters for full factorial statistical analysis on strength, shrinkage and surface resistivity

Input numerical and categorical variables				
Aggregate types	Aggregate sizes	SCM types	V_{paste}/V_{voids}	w/cm
Limestone (LS)	19·0 mm	C	Range from 230 to 275%	Range from 0·36 to 0·41
Gravel (G)	12·5 mm	F		
	9·5 mm	S		
		FLD		

Output parameters			
	Variables that are statistically significant	Sum of squares/ Degrees of freedom	Coefficient of correlation, R^2
56-d compressive strength	Aggregate types	285·6	86·1%
	SCM types	165·2	
	V_{paste}/V_{voids}	122·4	
56-d shrinkage	Aggregate sizes	$4{\cdot}6 \times 10^4$	86·0%
	V_{paste}/V_{voids}	$2{\cdot}1 \times 10^4$	
	SCM types	$1{\cdot}8 \times 10^4$	
56-d surface resistivity	SCM types	483·7	97·8%
	Aggregate sizes	66·3	
	Aggregate types × Aggregate sizes	65·2	
	Aggregate types × SCMs type	59·6	
	w/cm	33·4	

For 56-d shrinkage, aggregate sizes have the strongest effect, indicated by the higher value of 46 000, followed by V_{paste}/V_{voids} and SCM types. SCM types are likely to play dominant roles in the surface resistivity of mixtures in this study, as indicated by the higher value of 483·7 compared to the other variables.

A single-step multiple comparison procedure, HSD test, is employed in conjunction with least-square mean to determine the means that are significantly different from each other. The least-square mean and Tukey HSD analysis results for aggregate types (limestone and gravel), aggregate sizes (19·0, 12·5 and 9·5 mm) and SCM types (classes C and F fly ashes, slag cement and class F fly ash plus limestone dust) are shown in Table 7.

Table 7 Statistical least-square means differences, Tukey HSD tests results

Aggregate types	Least square mean: MPa	Aggregate sizes, mm	Least-square mean: MPa	Least-square mean differences, Tukey HSD	SCM types	Least-square mean: MPa	Least-square mean differences, Tukey HSD
56-d compressive strength							
G	33·3	19	49·5	A	C	49·5	A
LS	53·0	12·5	42·4	AB	S	45·5	AB
		9·5	37·6	B	F	40·6	BC
					FLD	37·0	C
56-d shrinkage							
G	−647	19	−456	A	S	−532	A
LS	−489	12·5	−592	B	F	−543	A
		9·5	−655	B	FLD	−559	AB
					C	−637	B
28-d surface resistivity							
G	31·5	19	38·2	A	S	47·4	A
LS	36·8	9·5	34·5	A	F	33·0	B
		12·5	29·7	B	FLD	30·7	B
					C	25·3	C

The levels that are not connected by the same letters are significantly different, namely, A, B and C. For instance, the 56-d compressive strengths of SCC mixtures with 9·5 mm NMSA are significantly different from those of 19·0 mm NMSA mixtures. Generally, concrete mixtures containing larger aggregate particles require less mixing water when the paste composition is similar to those containing smaller size aggregate. This will result in a lower effective w/cm, yielding a higher strength. Contrarily, however, larger aggregates containing more microcracks tend to form a weaker interfacial transition zone, therefore decrease compressive strength (Mehta and Monteiro, 2006). Therefore, the HSD test provides a quantitative manner to identify the net effect of aggregate sizes.

The 56-d shrinkage strains of 19·0 mm mixtures are significantly lower than that of 12·5 and 9·5 mm NMSA mixtures. This can most likely be attributed to the larger size of coarse aggregate yielding a lower paste content in a mixture, which is one of the dominant factors in reducing shrinkage.

Class F fly ash and class F fly ash plus limestone dust mixtures have no significant differences in terms of surface resistivity, whereas slag cement mixtures are significantly higher and class C fly ash mixtures are significantly lower. This is also in agreement with the observations from Figure 9.

Conclusions

The following conclusions are derived from the present study.

(*a*) The V_{paste}/V_{voids} concept can be used in SCC mixtures to assess the following three properties.

- Workability: based on the response surface model, slump flow increases with increased V_{paste}/V_{voids} at a given viscosity of mortar. At a given V_{paste}/V_{voids}, it is not surprising that higher viscosity of the mortar gives a lower flow diameter because more viscous paste can result in a higher resistivity of flow. At a given V_{paste}/V_{voids}, slump flow tends to decrease with increased yield stress.
- Strength: this increases with an increased V_{paste}/V_{voids} up to a point, after which the strength becomes independent of V_{paste}/V_{voids}, and even slightly decreases based on the analysis of CC and SCC mixtures.
- Surface resistivity: V_{paste}/V_{voids} has little effect on surface resistivity of SCC mixtures. The paste quality tends to have the dominant effect.

(*b*) Drying shrinkage: V_{paste}/V_{voids} has little effect on SCC shrinkage, while SCC mixtures tend to have a higher shrinkage than CC for a given V_{paste}/V_{voids}.

(*c*) Statistical analysis, such as response surface models and HSD tests, provides a systematic and quantitative means to predict and assess performance of SCC mixtures. It is also an efficient tool to identify the significance of influencing factors in relation to concrete performance. Aggregate types, sizes and SCM types are statistically evidenced to have most effects on 56-d compressive strength, shrinkage and surface resistivity, respectively. Further research is needed to validate the models.

Acknowledgements

The authors acknowledge the research sponsorship and the collaboration between Iowa State University (ISU), the University of Nebraska – Lincoln (UNL) and Northwestern University (NU). The authors would like to thank Dr Robert Stephenson for his guidance, suggestions and contributions of statistical analysis to this paper. The opinions, findings and conclusions presented in this paper are those of the authors and do not reflect those of the research sponsors.

REFERENCES

Aashto (2011) TP 95: Standard method of test for surface resistivity of concrete's ability to resist chloride ion penetration. American Association of State Highway and Transportation Officials, Washington, DC, USA.

Alsayed SH (1998) Influence of superplasticiser, plasticiser and silica fume on the drying shrinkage of high-strength concrete subjected to hot–dry field conditions. *Cement and Concrete Research* **28(10)**: 1405–1415.

ASTM (1997a) ASTM C231: Standard test method for air content of freshly mixed concrete by the pressure method. ASTM International, West Conshohocken, PA, USA.

ASTM (1997b) ASTM C1202: Standard test method for electrical indication of concrete's ability to resist chloride ion penetration. ASTM International, West Conshohocken, PA, USA.

ASTM (1999a) ASTM C29: Standard test method for bulk density (unit weight) and voids in aggregate. ASTM International, West Conshohocken, PA, USA.

ASTM (1999b) ASTM C157: Standard test method for length change of hardened hydraulic-cement mortar and concrete. ASTM International, West Conshohocken, PA, USA.

ASTM (2001) ASTM C39: Standard test method for compressive strength of cylindrical concrete specimens. ASTM International, West Conshohocken, PA, USA.

ASTM (2005) ASTM C1611: Standard test method for slump flow of self-consolidating concrete. ASTM International, West Conshohocken, PA, USA.

ASTM (2009) ASTM C1621: Standard test method for passing ability of self-consolidating concrete by j-ring. ASTM International, West Conshohocken, PA, USA.

Bui V, Akkaya Y and Shah S (2002) Rheological model for self-consolidating concrete. *ACI Material Journal* **99(6)**: 549–559.

Chini A, Muszynski L and Hicks J (2003) *Determination of Acceptance Permeability Characteristics for Performance-related Specifications for Portland Cement Concrete*. Florida Department of Transportation, Tallahassee, FL, USA, Report No. BC 354-41.

Cook D, Ghaeezadeh A, Ley T and Russell B (2013) *Investigation of Optimized Graded Concrete for Oklahoma –Phase I, Final Report*. Oklahoma Department of Transportation, State Planning and Research Division, USA, ODOT SP&R Item Number 2160.

Daczko J (2012) *Self-consolidating Concrete: Applying What We Know*. Spon, Abingdon, Oxfordshire, UK.

De Larrard F (1999) *Concrete Mixture Proportioning: A Scientific Approach*. CRC Press, London, UK.

Heywood H (1933) Calculation of the specific surface of a powder. *Proceedings of the Institute of Mechanical Engineers* **125(1)**: 383–459.

Hu J and Wang K (2007) Effects of size and uncompacted voids of aggregate on mortar flow ability. *Journal of Advanced Concrete Technology* **5(1)**: 75–85.

JMP (2005) *JMP 8.0.0, Statistical Discovery*. SAS Institute Inc, Cary, NC, USA.

Kennedy C (1940) The design of concrete mixes. *Journal of the American Concrete Institute* **36**: 373–400.

Kessler R, Powers R, Vivas E, Paredes M and Virmani Y (2008) Surface resistivity as an indicator of concrete chloride penetration resistance. *Proceedings of the 2008 Concrete Bridge Conference, St Louis, MO.*

Koehler E and Fowler D (2007) *Aggregates in Self-consolidating Concrete*. International Center for Aggregates Research (ICAR), Austin, TX, USA.

Kolias S and Georgiou C (2005) The effect of paste volume and of water content on the strength and water absorption of concrete. *Cement and Concrete Composites* **27(2)**: 211–216.

Kosmatka S, Kerkhoff B and Panarese W (2008) *Design and Control of Concrete Mixtures*. edn. Portland Cement Association, Skokie, IL, USA.

LADOTD (Louisiana Department of Transportation and Development) (2011) *Test Method for Surface Resistivity Indication of Concrete's Ability to 35 Resist Chloride Ion Penetration*. LADOTD, Baton Rouge, LA, USA, TR 233.

Lomboy G, Wang X and Wang K (2014) Rheological behavior and formwork pressure of SCC, SFSCC, and NC mixtures. *Cement and Concrete Composites* **54(November)**: 110–116.

Malhotra V and Mehta P (1996) *Advances in Concrete Technology, Pozzolanic and Cementitious Materials*. Gordon and Breach Publishers, Amsterdam, the Netherlands, vol. 1.

Mehta P and Monteiro P (2006) *Concrete: Structure, Properties and Materials*. Prentice Hall, Englewood Cliffs, NJ, USA.

Obla KH (2012) Optimizing concrete mixtures for performance and sustainability. *Proceedings of International Concrete Sustainability Conference, Seattle, USA.*

Oh S, Noguchi T and Tomosawa F (1999) Toward mix design for rheology of self-compacting concrete. *Proceedings of the 1st International Rilem Symposium on Self-Compacting Concrete, Stockholm, Sweden*, pp. 361–372.

Proceq (2011) *Resipod Operating Instructions*. Proceq SA, Schewerzenbach, Switzerland.

Quiroga P and Fowler D (2003) *The Effects of Aggregates Characteristics on the Performance of Portland and Cement Concrete*. The University of Texas at Austin, Austin, TX, USA, Project report ICAR 104-1F.

Rupnow T and Icenogle P (2012) Evaluation of surface resistivity measurements as an alternative to the rapid chloride permeability test for quality assurance and acceptance. *Proceedings of the TRB 91st Annual Meeting, Transportation Research Board, Washington, DC.*

Shi C (2004) Effect of mixing proportions of concrete on its electrical conductivity and the rapid chloride permeability test (ASTM C1202 or Aashto T277) results. *Cement and Concrete Research* **34(3)**: 537–545.

Taylor P, Bektas F, Yurdakul E and Ceylan H (2012a) *Optimizing Cementitious Content in Concrete Mixtures for Required Performance*. Federal Highway Administration, Washington DC, USA, Final Report, DTFH61-06-H-00011.

Taylor P, Yurdakul E, Ceylan H and Bektas F (2012b) *Development of Performance Properties of Ternary Mixtures and Concrete Pavement Mixture Design and Analysis (MDA): Effect of Paste Quality on Fresh and Hardened Properties of Ternary Mixtures*. Federal Highway Administration, Washington DC, USA, Technical Report, DTFH61-06-H-00011.

Wang X (2011) *Drying Shrinkage of Ternary Blends in Mortar and Concrete.* Master thesis, Iowa State University, Ames, IA, USA.

Wang X, Wang K, Bektas F and Taylor P (2012) Drying shrinkage of ternary blend concrete in transportation structures. *Journal of Sustainable Cement-Based Materials* **1(1–2)**: 56–66.

Yurdakul E (2013) *Proportioning for Performance-based Concrete Pavement Mixtures.* PhD thesis, *Iowa State University, Ames, IA, USA.*

Yurdakul E, Taylor P, Ceylan H and Bektas F (2013) Effects of paste-to-voids volume ratio on the performance of concrete mixtures. *Journal of Materials in Civil Engineering (ASCE)* **25(12)**: 1840–1851.

ice Publishing

Dhir and Lye
ISBN 978-0-7277-6461-4
https://doi.org/10.1680/icetscc.64614.115

Chapter 6
Mix proportioning of self-compacting normal and high-strength concretes

Rola Deeb
School of Engineering, Cardiff University, Cardiff, UK

Bhushan Lal Karihaloo
School of Engineering, Cardiff University, Cardiff, UK

Over the last two decades, self-compacting concrete mixes of varying strengths and performance have been proportioned on a trial-and-error basis using time-consuming laboratory tests, such as the slump cone, J-ring and L-box tests. An alternative method has recently been proposed for proportioning self-compacting high-performance concrete and self-compacting ultra-high performance concrete mixes with and without steel fibres (with characteristic cube strength between 140 and 160 MPa). This method is based on an extensive investigation of the flow characteristics of such mixes using computational simulations, thus avoiding time-consuming laboratory tests. However, the mixes proportioned using this method contained no coarse aggregate, which has a significant effect on the flow characteristics. This paper extends the method to the proportioning of self-compacting concrete mixes that contain traditional coarse aggregate and whose characteristic cube strengths vary between 35 and 100 MPa.

Notation

$f_i(\phi_i)$	relative plastic viscosity function
i	liquid–solid suspension number
l_d	length to diameter ratio of the fibre
n	total number of solid phases in the mix
P	pressure (Pa)
SP/water	superplasticiser to water ratio
T_{200}	time taken for SCC mixes to reach distance of 200 mm from gate (s)
T_{400}	time taken for SCC mixes to reach distance of 400 mm from gate (s)
T_{500}	time from lifting of slump cone to SCC spreading to a 500 mm diameter (s)
t_i	proportioning factor
u	constant number
v_n	normal velocity (m/s)
η	plastic viscosity of SCC mixes (Pa s)
$[\eta]$	non-dimensional intrinsic viscosity
$\eta_{c0} = \eta_{paste}$	plastic viscosity of cement paste (Pa s)
η_{ci}	plastic viscosity of ith liquid–solid suspension (Pa s)
η_{NF}	plastic viscosity of viscous concrete mix without fibres (Pa s)
η_{NF} (error%)	error in plastic viscosity of base SCC mix without fibres (Pa s)

η_{WF}	plastic viscosity of SCC mix with fibres (Pa s)
τ_y	yield stress of SCC mix (Pa)
ϕ_f	fibre volume fraction
ϕ_i	volume fraction of solid phase
ϕ_m	maximum possible volume fraction of solid phase

Introduction

Over the last two decades, extensive research has been conducted on achieving self-compactability of concrete. Unlike normal vibrated concrete, for which there are different standards around the world for selecting the mix proportions for plain and reinforced concrete (ASTM C938–10, ASTM, 2010; BIS, 2009; DOE, 1988), there are still no clear rules for proportioning self-compacting concrete (SCC) mixes. This lack of rules is intimately connected with the very special flow characteristics of SCC.

The flow behaviour of SCC in its fresh state is strongly influenced by the composition and characteristics of its constituents. Therefore, it is crucial to understand the rheology of SCC, which affects its flow behaviour and has a great influence on its properties in the hardened state.

Development of SCC base mixes with and without fibres

An extensive laboratory investigation was conducted to produce different grades of normal and high-strength SCC mixes (with nominal 28-day cube characteristic compressive strengths of 35, 45, 60, 80 and 100 MPa); on the one hand these were as fluid as possible to ensure that they filled the formwork under their own weight and passed through congested reinforcement cages without blockage, on the other hand they were sufficiently stable to prevent segregation of solids during the flow.

This investigation followed the traditional trial-and-error approach, using the slump cone, J-ring and L-box tests on trial mixes, until the mix was found that met the flowability and passing ability criteria and had no visible signs of segregation. In this manner, the mixes of different grades were proportioned to meet all requirements in the fresh state according to BS EN 206–9 (BSI, 2010). All mixes so proportioned fall into the viscosity class 2 (VS2) with $T_{500} \geq 2$ s.

The mix proportions of these mixes are shown in Table 1. The binder refers to cement plus ground granulated blast furnace slag (GGBS) and/or micro-silica. The trial mixes for normal strength concrete with 28-day cube compressive strengths of 35, 45 and 60 MPa (mixes 1, 2 and 3 in Table 1) were proportioned guided by the corresponding vibrated concrete mix proportions produced in the same laboratory.

For the high-strength SCC mixes (mixes 4 and 5 in Table 1), micro-silica was added to the mix in place of GGBS. The water-to-binder ratio was reduced to 0·23, while superplasticiser to water ratio (SP/water) was increased to ensure good workability. As expected, it was found that the addition of 0·5% by volume of 30 mm long, high-strength steel fibres to the 80 MPa mix 4 also gave a 100 MPa mix (mix 6 in Table 1). However, not surprisingly, the flow characteristics of the 100 MPa mix achieved with the use of steel fibres (mix 6) are somewhat sluggish

Table 1 The mix constituents of normal strength SCC (mix 1, mix 2, mix 3), SCHPC (mix 4, mix 5), SCHPFRC (mix 6), SCUHPC (mix 7) and SCUHPFRC (mix 8) (note in the descriptive mix acronyms, FR represents fibre-reinforced)

Constituents	Mix 1	Mix 2	Mix 3	Mix 4	Mix 5	Mix 6	Mix 7	Mix 8
Cement: kg	210	230	294	500	500	500	543·5	543·5
Micro-silica: kg	–	–	–	75	75	75	214	214
Ground granulated blast furnace slag: kg	151·3	151·9	98	–	–	–	311·5	311·5
Limestone	229·1	210·3	176	200	105	200	–	–
Coarse aggregates: kg (crushed limestone) < 10 mm	839·3	842·6	896·3	833	990	833	–	–
Sand < 2 mm	713·6	716·4	732·6	700	660	700	–	–
Quartz sand 9–300 μm: kg	–	–	–	–	–	–	470	470
Quartz sand 250–600 μm: kg	–	–	–	–	–	–	470	470
Water: kg	203·3	190·9	196	138	133·5	138	188	188
Fibres (30 mm long with crimped ends, vol. fraction)	–	–	–	–	–	0·5%	–	2·5%
Superplasticiser/water	0·009	0·012	0·017	0·14	0·12	0·14	0·28	0·28
Water/binder	0·56	0·55	0·5	0·24	0·23	0·24	0·18	0·18
Flow spread: mm	710	700	680	805	800	760	910	850
T_{500}: s	2·42	2·65	2·35	2·2	2·88	3	3	3
T_{200}: s	0·45	0·51	0·36	0·35	0·5	1·88	0·3	1·9
T_{400}: s	0·95	1·04	0·8	0·76	1·02	5·1	0·75	5·6
Level-off: s	8·36	9·26	6·8	7·23	9·2	38	0·95	54
Compressive strength: MPa	35	45	60	80	100	100[a]	140	160[b]

[a] Mix with 0·5% volume fraction of steel fibres.
[b] Mix with 2·5% volume fraction of steel fibres.

compared with the 100 MPa mix (mix 5) produced without the use of steel fibres. For comparison and completeness, the mix proportions of self-compacting ultra-high-performance concrete mixes with and without steel fibres reported by Karihaloo and Ghanbari (2012) are also reproduced in Table 1 (mixes 7 and 8).

The compressive strength of all mixes that passed the flowability, passing ability and visual cohesiveness tests in the fresh state (as shown in Figures 1–10) was measured in the hardened state at age 28 days. Figure 6 gives an example of a trial mix that had the required flowability, but failed the passing ability test. In such instances, the paste content had to be increased. The results are reported in Table 1.

Figure 1 Horizontal spread of SCC (mix 2)

Figure 2 Horizontal spread of SCC (mix 3)

Figure 3 Horizontal spread of SCC (mix 4)

Figure 4 Horizontal spread of SCHPC (mix 5)

Figure 5 Flow and passing ability of SCC (mix 2)

Figure 6 Fibres and large aggregates are nested around the steel rods in an earlier trial version of mix 4, which met the flowability criterion

Figure 7 Flow and passing ability of SCUHPFRC (mix 8)

Figure 8 Flow and passing ability of SCHPFRC (mix 6)

Figure 9 Flow and passing ability of SCC (mix 2)

Figure 10 Flow and passing ability of SCC (mix 3)

Self-compacting and vibrated concrete mixes

The paste-to-solids ratios by volume in the above SCC, self-compacting high-performance (SCHPC) (Deeb *et al.*, 2012) and self-compacting ultra-high-performance (SCUHPC) (Deeb *et al.*, 2012) mixes (Table 1) are compared in Table 2 with their counterpart normal, high-strength (Tailor, 1997) and ultra-high-strength (Karihaloo, 2012) vibrated concrete (VC) mixes, ranging in 28-day compressive strength from 35 to 160 MPa; these were all developed in the same laboratory at Cardiff university over many years.

Table 2 shows that, for the same compressive strength and almost the same water-to-binder ratio, the paste-to-solids ratio in a SCC mix is much higher than in the counterpart VC mix. However, the difference reduces as the strength of the mix increases.

It should be noted that the compressive strength of concrete can be increased by decreasing the water-to-binder ratio, which can be achieved by either increasing the binder content or

Table 2 Comparison of paste-to-solids ratio by volume between SCC and VC mixes of equal compressive strength

Compressive strength: MPa	35	45	60	80	100	140	162
Paste (SCC)	0·425	0·417	0·403	0·434	0·393	0·646	0·631
Solids (SCC)	0·574	0·583	0·597	0·566	0·607	0·353	0·369
Water/binder (SCC)	0·56	0·55	0·5	0·24	0·23	0·18	0·18
Paste/solids (SCC)	0·74	0·717	0·675	0·767	0·646	1·828	1·708
Paste (VC)	0·353	0·337	0·316	0·309	0·366	0·628	0·613
Solids (VC)	0·647	0·663	0·684	0·691	0·634	0·371	0·387
Water/binder (VC)	0·56	0·5	0·44	0·39	0·26	0·18	0·18
Paste/solids(VC)	0·545	0·509	0·462	0·447	0·576	1·695	1·583

decreasing the water content. These changes have an opposite effect on the volume of paste; an increase in the binder content increases this volume, whereas a decrease in the water content reduces it. This may explain why the paste-to-solids ratio of the 80 MPa SCC mix (Table 2) does not follow the expected trend. This mix contains a large amount of binder (mix 4, Table 1) and therefore a large paste volume compared with the solids.

For SCUHPC and ultra-high-performance vibrated concrete (UHPVC) with and without the fibres, the paste volume is almost double that of the solids (Table 2). This large paste-to-solids ratio distinguishes the ultra-high-performance concrete mixes (both self-compacting and vibrated types) from the traditional normal and high-strength mixes.

Estimation of plastic viscosity

A SCC mix behaves like a viscous non-Newtonian fluid, described by a bilinear Bingham-type rheological model (Dransfield, 2003). This constitutive model contains two material properties, namely the yield stress τ_0 and the plastic viscosity η. However, the yield stress of SCC mixes is very low (around 200 Pa) compared with the normal VC (thousands of Pa), and remains almost constant over a wide range of mixes with different plastic viscosities (Dransfield, 2003).

For a homogeneous viscous fluid such as cement paste, plastic viscosity can be measured accurately; this cannot be said regarding a non-homogeneous viscous fluid such as the SCC. However, the plastic viscosity of a heterogeneous viscous fluid such as SCC, high-performance SCC and ultra-high-performance SCC mixes with and without fibres can be estimated from the known viscosity of the cement paste using a micromechanical model, developed by Ghanbari and Karihaloo (2009).

This model regards concrete as a two-phase suspension in which the solid phase is suspended in a viscous liquid phase. The plastic viscosity of the liquid phase η_{c0} (cement, water and superplasticiser) can be measured accurately using a rheometer. The increase in the plastic viscosity of the paste (cement, water and superplasticiser) induced by the addition of the solid phase (cement-replacement materials, fine and coarse aggregates, and fibres) can be predicted in stages from the two-phase suspension model. In the first stage, the solid phase is the finest solid material, for example micro-silica in the viscous fluid phase (i.e. the paste). In the second stage, the next finest solid (e.g. GGBS) is suspended in the viscous fluid phase formed by the two-phase suspension from the first stage. This procedure is continued until all the solid phase constituents have been added. The viscosity of the two-phase suspension is given by the product of the viscosity of the liquid phase and a function $f_i(\phi_i)$ that depends on the volume fraction of the solid phase ϕ_i and the shape of the particles.

At low concentrations of the solid phase, that is $\phi_i \leq 10\%$, Einstein showed that the addition of the second phase to a liquid suspension leads to an increase in the bulk viscosity proportional to the volume fraction of particles only

$$f_i(\phi_i) = 1 + [\eta]\phi_i \tag{1}$$

where $[\eta]$ is the so-called non-dimensional intrinsic viscosity that measures the effect of individual particles on the viscosity (Struble and Sun, 1995). For rigid spherical particles separated by distances larger than the particle diameter, the value of $[\eta] = 2{\cdot}5$.

At high concentrations of the solid phase, that is $\phi_i > 10\%$, but less than the maximum possible volume fraction ϕ_m which represents the situation in which the particles have the minimum possible separation, the Krieger and Dougherty (1959) formula given below has been found to be accurate for cement-based suspensions. For random hexagonal packing, $\phi_m = 0{\cdot}637$.

$$f_i(\phi_i) = \left(1 - \frac{\phi_i}{\phi_m}\right)^{-[\eta]\phi_m} \tag{2}$$

The intrinsic non-dimensional viscosity $[\eta]$ and the maximum possible volume fraction ϕ_m depend upon the shear rate; whereas the former tends to decrease with increasing shear rate (so-called shear thinning), the latter shows the opposite trend (so-called shear thickening). However, the product $[\eta]\phi_m$ is nearly constant and equal to about 1·9 for rigid spherical particles.

The addition of fibres increases the plastic viscosity further. By treating the fibres as rigid slender bodies that only undergo rigid body translation and rotation in the viscous SCC fluid, Ghanbari and Karihaloo (2009) estimated the effect of a dilute concentration of steel fibres (volume fraction less than 3%) on the plastic viscosity

$$\eta = \eta_{NF}\left[(1 - \phi_f) + \frac{\pi\phi_f l_d^2}{3\ln(2l_d)}\right] \tag{3}$$

Here, η_{NF} is the plastic viscosity of the SCC mix without fibres, estimated as above, l_d is the ratio of length to diameter of fibre and ϕ_f is the volume fraction of fibres.

The plastic viscosity of cement paste, η_{Paste}, which is mainly influenced by water-to-cement ratio and superplasticiser-to-water ratio, has been estimated from literature sources (Grzeszczyk and Lipowski, 1997; Nehdi and Rahman, 2004; Sun *et al.*, 2006).

The effect of cement replacement materials, coarse and fine aggregates has been estimated using the step-wise procedure described above and using the Krieger and Dougherty (1959) formula (Equation 2).

Table 3 shows that the plastic viscosity of the self-compacting normal and high-strength mixes containing large aggregates (mixes 1–5), η_{NF}, varies only marginally from 9·77 Pa s to 12·42 Pa s. However, the plastic viscosity of SCUHPC (mix 7), which contains no coarse aggregates, is much lower (3·1 Pa s).

The addition of steel fibres, however, increases the plastic viscosity η_{WF} significantly (mixes 6 and 8), both in the high-strength and the ultra-high-performance mixes.

Table 3 Plastic viscosity of self-compacting concrete mixes

Compressive strength: MPa	35	45	60	80	100	100[a]	140	162[b]
Paste (SCC)	0·425	0·417	0·403	0·434	0·393	0·432	0·646	0·631
Solids (SCC)	0·574	0·583	0·597	0·566	0·607	0·568	0·353	0·369
Paste/solids (SCC)	0·74	0·717	0·675	0·767	0·646	0·761	1·828	1·708
Water/cement	0·968	0·83	0·667	0·276	0·267	0·276	0·346	0·346
Water/binder	0·56	0·55	0·5	0·24	0·23	0·24	0·18	0·18
SP/water	0·009	0·012	0·017	0·14	0·12	0·14	0·28	0·28
η_{NF}: Pa s	10·55	11·27	9·94	9·77	12·42	9·77	3·1	3·1
η_{WF}: Pa s	–	–	–	–	–	42·1	–	54·3

[a] Mix with 0·5% volume fraction of steel fibres.
[b] Mix with 2·5% volume fraction of steel fibres.
SCC: self-compacting concrete.

Modelling the flow of SCC mixes

We used a Lagrangian particle-based method, the so-called smoothed particle hydrodynamics (SPH) method, to simulate the flow of SCC with and without short steel fibres. The constitutive behaviour of these non-Newtonian viscous fluids is best described by a Bingham-type model. To simplify the use of this model in the SPH simulation, the kink in the shear stress–shear strain relationship has been smoothed out. The basic equations solved in the SPH are the incompressible mass conservation and Navier–Stokes equations.

The solution procedure uses prediction–correction fractional steps with the temporal velocity field integrated forward in time without enforcing incompressibility in the prediction step. The resulting temporal velocity field is used in the mass conservation equation to satisfy incompressibility through a pressure Poisson equation derived from an approximate pressure equation.

The numerical simulations of SCC flow in two different geometrical configurations were carried out. The slump cone test was simulated in an axisymmetric configuration along a vertical plane of symmetry, followed by the flow of SCC with and without fibres in the two-dimensional L-box configuration. The simulated and measured results are compared in Table 4.

It is clear from Table 4 that there is good agreement between the measured and simulated T_{500} times for the slump flow. The large difference in the measured and simulated T_{200}, T_{400} and level-off times in the L-box configuration is a result of several factors. First, because of the error involved in estimating the time from the video recording of the flow, this error decreases as the measured time increases. Second, there is an inevitable delay in opening the gate and in the mix passing through the gaps in the bars, thus increasing the measured times. The delay is greater the larger the content of coarse aggregate and/or fibres in the mix. Finally, the frictional resistance to the flow along the vertical sides of the horizontal part of the L-box cannot be taken into account in the two-dimensional simulation. As a result the simulated times are always shorter than the actual measured times. An exact simulation of flow requires that the full three-dimensional L-box geometry be considered. Such a simulation is currently in progress.

Table 4 Comparison between measured and simulated slump and L-box results

Parameter	Mix 1	Mix 2	Mix 3	Mix 4	Mix 5	Mix 6	Mix 7	Mix 8
Flow spread: mm	710	700	680	805	800	760	910	850
Measured T_{500}: s	2·42	2·65	2·35	2·2	2·88	3	3	3
Simulated T_{500}: s	2·5	2·7	2·2	2·1	2·8	2·8	2·8	2·9
Error: %	3·20	1·85	6·82	4·76	2·86	7·14	7·14	3·45
Measured T_{200}: s	0·45	0·51	0·36	0·35	0·5	1·88	0·3	1·9
Simulated T_{200}: s	0·28	0·31	0·25	0·24	0·29	1·3	0·3	1·7
Error: %	37·78	39·22	30·56	31·43	42·00	30·85	0·00	10·53
Measured T_{400}: s	0·95	1·04	0·8	0·76	1·02	5·1	0·75	5·6
Simulated T_{400}: s	0·7	0·83	0·63	0·6	0·8	3·7	0·7	4·8
Error: %	26·32	20·19	21·25	21·05	21·57	27·45	6·67	14·29
Measured level-off: s	8·36	9·26	6·8	7·23	9·2	38	0·95	54
Simulated level-off: s	7·2	8·1	6·2	5·9	8·1	31	0·9	48·2
Error: %	13·88	12·53	8·82	18·40	11·96	18·42	5·26	10·74
Compressive strength: MPa	35	45	60	80	100	100[a]	140	160[b]

[a] Mix with 0·5% volume fraction of steel fibres.
[b] Mix with 2·5% volume fraction of steel fibres.

Mix proportioning

A method for mix proportioning of SCHPC and SCUHPC mixes with and without short steel fibres, having the right flow and passing ability, was developed by Karihaloo and Ghanbari (2012). These mixes did not contain any conventional coarse aggregates. This method will be extended here to conventional normal and high-strength self-compacting mixes (mixes 1–6 in Table 1). It will provide an alternative tool to the heuristic mix proportioning of SCC, thus avoiding expensive and time-consuming laboratory tests. The mix proportioning procedure involves the following three steps.

(*a*) The slump cone flow is simulated numerically using the technique described above to obtain T_{500} time for a given target plastic viscosity of the mix ranging from 6 Pa s to 14 Pa s because of the large aggregates used in the mixes. It is useful to point out that the range was only from 2 Pa s to 6 Pa s for SCUHPCs, which need to be less viscous to accommodate the large amount of steel fibres. The simulated T_{500} time will be compared with the measured T_{500} from Table 4.

(*b*) The flow of the mixes is simulated numerically in a two-dimensional L-box configuration to obtain the T_{200}, T_{400} and level-off times.

(*c*) Guidance is given for mix proportioning a range of mixes that would attain the target plastic viscosity, and T_{500}, T_{200}, T_{400} and level-off times, and thus satisfy the flow and passing ability criteria.

Slump flow simulation

The slump flow simulation of a series of mixes was carried out using the technique described above to obtain T_{500} that gives an indication of the flowability of the SCC mix. The target plastic viscosity was chosen in the range 6–14 Pa s for mixes 1–6.

Two-dimensional L-box simulation

The next step is the flow simulation of the above mixes in the L-box configuration. The remarks made above on the inaccuracies of the two-dimensional simulation need to be borne in mind.

The data in Table 5 show that T_{500}, T_{200}, T_{400} and level-off times correlate well with the plastic viscosity for all the mixes. This is clearly seen from Figures 11 and 12.

Table 5 Slump flow and L-box flow times (T_{500}, T_{200}, T_{400} and level-off time) for self-compacting concrete (SCC) mixes with different target plastic viscosity

SCC type	Plastic viscosity of mix without fibres: Pa s	Plastic viscosity of mix with fibres: Pa s	T_{500}: s	T_{200}: s	T_{400}: s	Level-off time: s
Mix 1	6	–	1·8	0·2	0·55	5·7
	8	–	2·0	0·23	0·61	6·5
	10	–	2·4	0·27	0·69	7·2
	12	–	2·7	0·31	0·79	8·2
	14	–	3·1	0·38	0·9	9·2
Mix 2	6	–	1·9	0·22	0·57	5·9
	8	–	2·1	0·25	0·61	6·6
	10	–	2·55	0·29	0·7	7·3
	12	–	2·75	0·33	0·81	8·6
	14	–	3·2	0·4	0·91	10·1
Mix 3	6	–	1·6	0·18	0·5	5·3
	8	–	1·8	0·21	0·55	5·4
	10	–	2·1	0·24	0·6	5·9
	12	–	2·4	0·28	0·7	6·7
	14	–	2·8	0·32	0·81	7·1

Table 5 (*Continued*)

SCC type	Plastic viscosity of mix without fibres: Pa s	Plastic viscosity of mix with fibres: Pa s	T_{500}: s	T_{200}: s	T_{400}: s	Level-off time: s
Mix 4	6	–	1·7	0·18	0·53	5·7
	8	–	1·9	0·21	0·6	6·3
	10	–	2·2	0·25	0·65	6·9
	12	–	2·5	0·29	0·74	7·9
	14	–	2·9	0·35	0·87	9·3
Mix 5	6	–	2·1	0·23	0·58	6·1
	8	–	2·3	0·26	0·65	6·8
	10	–	2·6	0·3	0·75	7·7
	12	–	2·8	0·34	0·85	8·9
	14	–	3·3	0·41	0·96	10·8
Mix 6 (0·5% steel fibres)	6	25·89	2·1	0·7	2·8	23·2
	8	34·52	2·4	0·9	3·6	28·1
	10	43·15	3·0	1·2	3·8	31·7
	12	51·78	3·2	1·4	4·9	37
	14	60·32	3·4	1·7	5·2	40·1

Figure 11 Relationship between the plastic viscosity and T_{500} of self-compacting concrete mixes

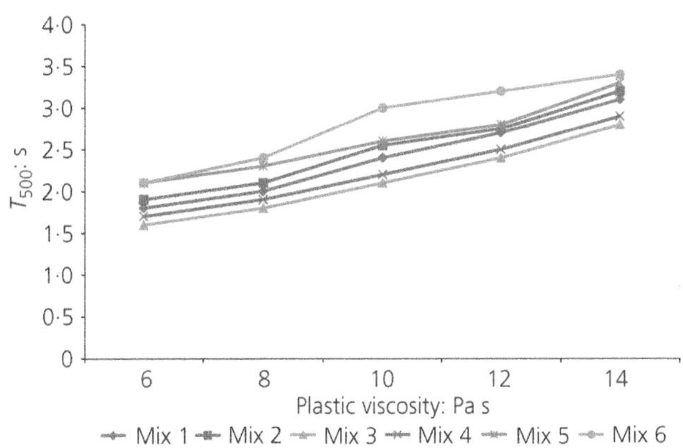

Figure 12 T_{200}, T_{400} plotted against the plastic viscosity of mixes 1–5

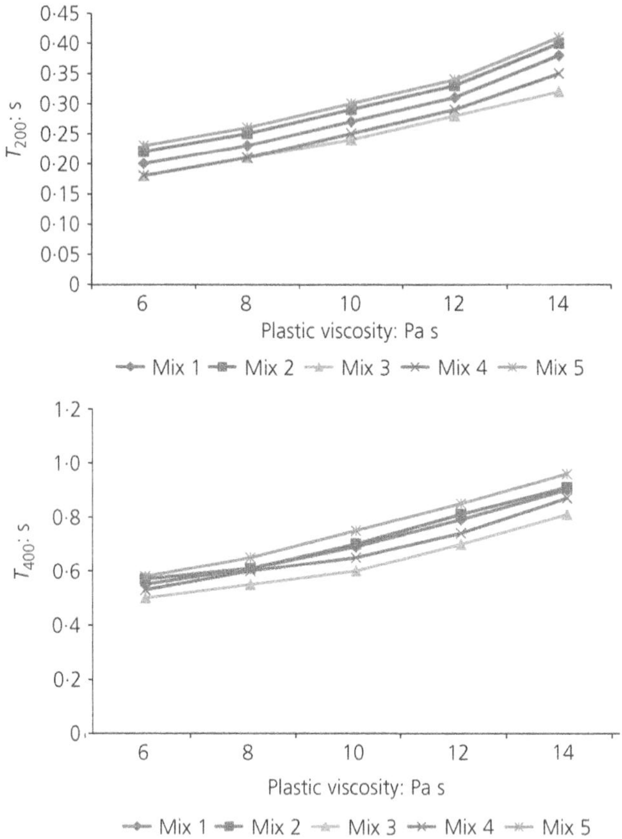

Mix proportioning

The final step is to choose the mix proportions that meet the flowability (slump cone flow) and passing ability (flow in L-box) requirements. This is done as follows.

(*a*) Assume the amount of cement and use the initial water and superplasticiser contents, as indicated in Table 1.
(*b*) Estimate the volume fraction of solid particles (explained below).
(*c*) Calculate the amount of the finest solid particles suspended in the previous liquid suspension (in the first step it is the cement paste) according to its volume fraction.
(*d*) Repeat step (c) until the amounts of all solid constituents in the mix have been calculated.
(*e*) Adjust the water content such that the water-to-binder ratio is the same as in Table 1.
(*f*) Adjust the superplasticiser content to the same value as in Table 1.
(*g*) Estimate the plastic viscosity of the new cement paste (cement, water and superplasticiser) (Sun *et al.*, 2006).

(*h*) Calculate the plastic viscosity of the new mix, based on the two-phase liquid–solid suspension model.
(*i*) Compare the plastic viscosity calculated at step (h) with the one used in the simulations. If the error is more than 5%, repeat steps (c) to (h).
(*j*) Estimate T_{500}, T_{200}, T_{400} and the level-off time from their correlations with the plastic viscosity.

Estimating the volume fractions of solid phases

The plastic viscosity of the SCC mixes without fibres is calculated from

$$f_i(\phi_i) = f_1(\phi_1)f_2(\phi_2)\cdots f_n(\phi_n) \tag{4}$$

where $f_i(\phi_i)$ can be replaced by its equivalent, relative plastic viscosity, that is the plastic viscosity of the SCC mix without fibre (η_{NF}) divided by the plastic viscosity of paste (η_{paste}). Therefore, Equation 4 can be rewritten as

$$\eta_{\mathrm{NF}} = \eta_{\mathrm{paste}}[f_1(\phi_1)f_2(\phi_2)\cdots f_n(\phi_n)] \tag{5}$$

where n is the total number of solid phases in the mix, such as coarse aggregate, sand and cement additives; η_{paste} is the plastic viscosity of the cement paste; η_{NF} is the plastic viscosity of SCC mix without fibres.

The Krieger and Dougherty equation is the most suitable for cement pastes. Therefore, it is used here to obtain $f_i(\phi_i)$. The values of $[\eta] = 2{\cdot}5$ and $\phi_{\mathrm{m}} = 0{\cdot}63$ are used.

For a given η_{NF} and η_{paste}, Equation 5 can be rewritten as

$$(0.63 - \phi_1)(0.63 - \phi_2)(0.63 - \phi_3)\cdots(0.63 - \phi_n) = u \tag{6}$$

where u is a constant number. The value of ϕ_i is given by

$$\phi_i = 0.63 - t_i x \tag{7}$$

where $x = \sqrt[n]{u}$ and t_i is a factor such that

$$t_1 \times t_2 \times \cdots \times t_n = 1 \tag{8}$$

Typical mix proportions of normal and high-strength SCC

Tables 6–8 give proportions of a few selected, typical, normal and high-strength SCC mixes obtained using the procedure described above for prescribed target plastic viscosity. For each strength grade, the plastic viscosity that achieves the T_{500}, T_{200} and T_{400} that are closest to the experimental values (Table 4) is chosen as the target value. For example, for normal strength SCC mix 1, the T_{500}, T_{200} and T_{400} (Table 5) are closest to the experimental values (Table 4) when the plastic viscosity is 10 Pa s. Therefore, many other mixes (theoretically infinitely many because $t_1, t_2, \ldots, t_n$ above can be chosen arbitrarily as long as their product equals unity) can be proportioned with the same plastic viscosity η_{NF} within an acceptable 5% error that will have the same nominal strength (35 MPa) and meet the flowability and passing ability criteria. In Tables 6–8, four typical mix proportions for each strength grade are shown.

Table 6 Typical mix proportions of 35 MPa self-compacting concrete mixes (1 m^3)

Constituents	Mix 1	Mix 2	Mix 3	Mix 4
Cement: kg	220·0	233·0	233·4	207·2
Ground granulated blast furnace slag: kg	134·6	117·9	117·3	152·2
Limestone: kg	508·1	517·5	514·3	484·4
Coarse aggregates: kg	556·1	451·1	547·6	540·2
Sand: kg	708·4	789·7	697·7	709·0
Water: kg	200·0	200·0	197·2	202·4
SP: kg	1·8	1·8	1·8	1·8
Density: kg/m^3	2329·0	2310·9	2309·3	2297·1
SP/water	0·009	0·009	0·009	0·009
Water/binder	0·56	0·56	0·56	0·56
Water/cement	0·91	0·86	0·84	0·98
η_{NF}: Pa s	10·45	10·41	10·51	10·1
η_{NF}: error%	4·39	4·00	4·91	1·05
T_{500} (slump flow test): s	2·5	2·5	2·5	2·45
T_{200} (L-box test): s	0·25	0·25	0·25	0·24
T_{400} (L-box test): s	0·7	0·7	0·7	0·65
Level-off time (L-box test): s	6·7	6·7	6·7	6·1

Table 7 Typical mix proportions of 60 MPa self-compacting concrete mixes (1 m^3)

	Mix 1	Mix 2	Mix 3	Mix 4
Cement: kg	296·1	255·1	222·8	237·6
Ground granulated blast furnace slag: kg	82·8	130·6	174·3	146·8
Limestone: kg	541·5	508·9	421·5	487·8
Coarse aggregates: kg	437·0	693·1	785·6	543·1
Sand: kg	766·7	531·4	517·1	686·5
Water: kg	191·0	193·9	201·0	194·1
SP: kg	3·3	3·3	3·4	3·3
Density: kg/m^3	2318·3	2316·2	2325·6	2299·2
SP/water	0·02	0·02	0·02	0·02
Water/binder	0·5	0·5	0·5	0·5
Water/cement	0·79	0·79	0·79	0·79
η_{NF}: Pa s	9·54	10·19	10·37	10·37
η_{NF}: error%	4·81	1·82	3·58	3·61
T_{500} (slump flow test): s	2·2	2·3	2·3	2·3
T_{200} (L-box test): s	0·25	0·26	0·26	0·26
T_{400} (L-box test): s	0·62	0·65	0·66	0·66
Level-off time (L-box test): s	6·1	6·2	6·3	6·3

Table 8 Typical mix proportions of 100 MPa mixes without steel fibres (1 m^3)

	Mix 1	Mix 2	Mix 3	Mix 4
Cement: kg	428·5	389·0	420·5	430·7
MS: kg	136·8	137·6	137·6	160·4
Limestone: kg	450·2	452·9	452·9	448·4
Coarse aggregates: kg	560·2	563·5	689·9	958·4
Sand: kg	690·6	694·8	564·1	268·4
Water: kg	130·6	137·7	130·0	134·6
SP: kg	15·7	16·5	16·8	17·3
density: kg/m^3	2214·4	2392·3	2407·9	2418·1
SP/water	0·12	0·12	0·12	0·12
Water/binder	0·23	0·23	0·23	0·23
Water/cement	0·3	0·35	0·3	0·31
η_{NF}: Pa s	11·93	11·95	12·46	11·68
η_{NF}: error%	0·58	0·41	3·67	2·78
T_{500} (slump flow test): s	2·7	2·7	2·8	2·7
T_{200} (L-box test): s	0·27	0·27	0·29	0·27
T_{400} (L-box test): s	0·78	0·78	0·80	0·78
Level-off time (L-box test): s	7·9	7·9	8·1	7·9

Concluding remarks

Different grades of normal and high-strength SCC mixes with and without steel fibres that contain traditional coarse aggregate and whose characteristic cube strength varies between 35 and 100 MPa were developed. The maximum size of the coarse aggregate in all the SCC mixes reported above was 10 mm. This was the same as in the corresponding VC mixes prepared in the same laboratory. The procedure described above for proportioning the SCC mixes can, of course, also be applied if larger size coarse aggregate (say, 20 mm) is used, especially in mixes with compressive strength up to 60 MPa. It is, however, necessary to maintain the paste-to-solids ratio given in Table 2. The plastic viscosities of the mixes reported above were estimated from the measured plastic viscosity of the paste using the micromechanical constitutive model proposed by Ghanbari and Karihaloo (2009).

Slump cone flow and L-box numerical simulations were then performed on these mixes to test their flowability and passing ability, as measured by T_{500} for the slump test, and T_{200}, T_{400} and level-off time for the L-box test.

The method for proportioning SCHPC and SCUHPC mixes with and without steel fibres that would attain the target plastic viscosity developed by Karihaloo and Ghanbari (2012) was extended to conventional normal and high-strength self-compacting mixes. The method developed by Karihaloo and Ghanbari (2012) was for mixes containing no conventional coarse aggregate. However, as the latter has a significant effect on the flow characteristics of SCC, the extension proposed here is by no means trivial. It provides an alternative tool to the heuristic mix proportioning of SCC, thus avoiding expensive and time-consuming laboratory tests.

REFERENCES

ASTM (2010) ASTM C938-10: Standard practice for proportioning grout mixtures for preplaced-aggregate concrete. ASTM, West Conshohocken, PA, USA.

BIS (Bureau of Indian Standards) (2009) IS 10262: Concrete mix proportioning – Guidelines. Bureau of Indian Standards, New Delhi, India.

BSI (2010) BS EN 206–9:2010: Concrete. Additional rules for self-compacting concrete (SCC). BSI, London, UK.

Deeb R, Ghanbari A and Karihaloo BL (2012) Development of self-compacting high and ultra high performance concretes with and without steel fibres. *Cement and Concrete Composites* **34(2)**: 185–190.

DOE (Department of the Environment) (1988) *Design for Normal Concrete Mixes*. HMSO, London, UK.

Dransfield J (2003) Admixtures for concrete, mortar and grout. In *Advanced Concrete Technology* (Newman J and Choo BS (eds)). Elsevier, vol. 2, ch. 4.

Ghanbari A and Karihaloo BL (2009) Prediction of the plastic viscosity of self-compacting steel fibre reinforced concrete. *Cement and Concrete Research* **39(12)**: 1209–1216.

Grzeszczyk S and Lipowski G (1997) Effect of content and particle size distribution of high-calcium fly ash on the rheological properties of cement pastes. *Cement and Concrete Research* **27(6)**: 907–916.

Karihaloo BL (2012) Influence of micro-structural parameters and thermal cycling on the properties of CARDIFRC. *Sadhana, Indian Academy of Sciences* **37(1)**: 125–132.

Karihaloo BL and Ghanbari A (2012) Mix proportioning of self-compacting high and ultra-high performance concretes with and without steel fibres. *Magazine Concrete Research* **64(12)**: 1089–1100.

Krieger IM and Dougherty TJ (1959) A mechanism for non-Newtonian flow in suspensions of rigid spheres. *Journal of Rheology* **3(1)**: 137–152.

Nehdi M and Rahman MA (2004) Estimating rheological properties of cement pastes using various rheological models for different test geometry, gap and surface friction. *Cement and Concrete Research* **34(11)**: 1993–2007.

Struble L and Sun GK (1995) Viscosity of Portland cement paste as a function of concentration. *Advanced Cement Based Materials* **2(2)**: 62–69.

Sun Z, Voigt T and Shah SP (2006) Rheometric and ultrasonic investigations of viscoelastic properties of fresh Portland cement pastes. *Cement and Concrete Research* **36(2)**: 278–287.

Tailor M (1997) *Characterisation of Normal and High-Strength Plain and Fibre-Reinforced Concretes by Means of Strength, Fracture and Combined Fracture/Relaxation Tests*. PhD thesis UK Cardiff University, UK.

Section 3

Influence from constituent materials

ice Publishing

Dhir and Lye
ISBN 978-0-7277-6461-4
https://doi.org/10.1680/icetscc.64614.135

Chapter 7
Nanoparticles in self-compacting concrete – a review

Farhad Aslani
Postdoctoral Research Fellow, Centre for Infrastructure Engineering and Safety, School of Civil and Environmental Engineering, University of New South Wales, Sydney, Australia

The development of nanotechnology delivers materials with new properties and over recent years much effort has been made to incorporate nanoparticles (NPs) in concrete technology in order to enhance properties and produce concrete with improved performance. The durability and sustainability of concrete are becoming vitally important for the construction industry and, in this context, self-compacting concrete (SCC) has generated tremendous interest. SCC is a concrete that can be placed and compacted under its own weight with little or no vibration and without segregation or bleeding. It is used to facilitate and ensure proper filling and thus good structural performance of restricted areas and heavily reinforced structural members. It has gained increasing importance in recent years because of the advantages it offers. This paper presents an overview of previous studies on the effect of using NPs in SCC. Properties and methods concerning the use of NPs in SCC mixes, such as mixing procedure, mix proportion, heat of hydration, workability, setting time, mechanical strength and durability, are reviewed.

Introduction

Nanotechnology is the creation of materials and devices by monitoring matter at the levels of atoms, molecules and supramolecular (nanoscale) structures (Roco *et al.*, 1999). In other words, it is the use of very small particles of materials to create new large-scale materials (Mann, 2008).

Concrete has been the most widely used manufactured material; in the last few years, the total annual worldwide production of concrete exceeded 2600 Mt (USGS, 2010). Cement is a major component of concrete, and rapid progress in characterisation techniques has made it possible to characterise cementitious materials at the micro/supramolecular level. Improved understanding of the structure and behaviour of concrete at this level could help to improve concrete properties and thus make it more durable. It is expected that the strengths and durability of concrete would be enhanced if the overall porosity and capillary pore sizes in cement paste were reduced or the diffusion of pore solution and aggressive chemicals were hindered by the introduction of additives with a similar range of capillary pore sizes (Mukhopadhyay, 2011).

Moreover, the cement industry is one of the most energy-intensive industries and every year is responsible for approximately 5% of global manmade carbon dioxide emissions (WBCSD, 2002); 50% of these emissions are caused by chemical manufacturing processes and 40% are due to the burning of fuel. Widespread research efforts have been directed towards reducing the effect of the cement industry on greenhouse gas emissions by improving the efficiency of the cement manufacturing process (Barker *et al.*, 2009; Deja *et al.*, 2010) or by using supplementary cementitious materials to partially replace ordinary cement (Gartner, 2004). It is believed that nanotechnology is one of the most promising research fields that may significantly improve the mix design and the performance and production of cement-based materials.

Among the cementitious materials, self-compacting concrete (SCC) is a concrete that has little resistance to flow so it can be placed and compacted under its own weight with no compaction effort, yet possesses enough viscosity to be handled without segregation or bleeding (Aslani and Bastami, 2014; Aslani *et al.*, 2014a, 2014b). SCC was first developed to improve the durability of concrete structures in Japan in 1988 (Ozawa *et al.*, 1996). Initial studies on the workability of SCC were carried out by Ozawa and Okamura of Tokyo University (Okamura, 1997; Okamura and Ozawa, 1994). SCC is used to facilitate and guarantee proper filling and good structural performance of restricted areas and heavily reinforced structural members and, due to the advantages it offers, has gained increasing significant importance in recent years (Aslani and Nejadi, 2012a, 2012b, 2012c, 2012d; Collepardi *et al.*, 2003). The most important advantage of SCC over conventional concrete is its flowability, but other advantages of using SCC include shorter construction periods, reductions in labour costs and better compaction in a structure, especially in confined zones where compaction is difficult.

One of the differences between SCC and conventional concrete is the incorporation of a mineral admix. In order to avoid the separation of large particles in SCC, viscosity-increasing additives or fillers are used. An additive to increase viscosity is often used when concrete is cast under water and for SCC in tunnels. Mineral admixtures such as limestone powder, natural pozzolans, ground granulated blast-furnace slag (GGBFS), fly ash, glass filler, silica fume or quartzite fillers may be used to increase the viscosity of SCC mixes (Persson, 2001). Researchers have investigated the behaviour of SCC with several types of pozzolanic materials used to replace a proportion of the cement (Aslani, 2013; Aslani and Maia, 2013; Aslani and Natoori, 2013; Khatib, 2008; Siddique *et al.*, 2012; Valcuende *et al.*, 2012). Siddique (2011) reported that the use of mineral admixtures increases the slump of a concrete mix without increasing its cost, while enabling a reduced dosage of superplasticiser. Previous studies have shown that mixes containing fly ash show lower compressive strength at early ages as a result of the slow pozzolanic reaction between the cement and fly ash (Aslani and Nejadi, 2013a, 2013b, 2013c).

El-Dieb (2009) studied the mechanical and durability properties of ultrahigh-strength fibre-reinforced concrete with self-compacting characteristics. According to Fava *et al.* (2003), a strength increase can be achieved in SCCs with added GGBFS. Kulakowski *et al.* (2009) reviewed the influence of silica fume on reinforcement corrosion in concrete and the effect of metakaolin on the transport properties of concrete was investigated by Shekarchi *et al.* (2010).

Craeye *et al.* (2010) studied the effect of mineral filler type on the autogenous shrinkage of SCC and concluded that there were no significant differences in final autogenous shrinkage values using different filler types. According to Barbhuiya (2011), it is possible to manufacture

SCC using fly ash and dolomite powder with acceptable fresh and hardened properties. Another previous study on fly ash was made by Dinakar *et al.* (2013), who observed that fly ash replacements of around 30–50% would be ideal for developing SCCs when Portland pozzolana cement (PPC) was used. They reported that a high percentage of fly ash (more than 50%) could not be used to produce SCC when PPC was used, while a 30% replacement of fly ash resulted in the highest compressive strength, splitting tensile strength and elastic modulus. It can thus be concluded that the use of mineral admixtures reduces the materials cost of SCC and also improves the fresh and hardened properties of the concrete mix (Aslani and Nejadi, 2012e; Bouzoubaa and Lachemi, 2001; Gesoglu *et al.*, 2009).

Research significance

Nanoparticles (NPs) have recently received special attention because of their better performance when compared with traditional mineral additions. The addition of NPs can improve the properties of concrete due to the effect that the increased surface area has on reactivity and through filling the NPs of the cement paste. Some works on incorporating NPs into SCC specimens to achieve improved physical and mechanical properties have been reported. This paper presents a review of the literature on the effect of NPs on properties of SCC mixes such as mixing procedure, mix proportion, heat of hydration, workability, setting time, mechanical strength and durability.

Database of NPs in SCC experimental campaigns

Using experimental results from various published investigations as a database is an effective tool for studying the applicability of NPs in SCC. In order to apply the results to a particular concrete mix accurately, it is necessary to use only the investigations that are adequately consistent with the applied testing methodology. The experimental results included in the database are mainly from presented papers and published articles on NPs in SCC. The database includes information regarding cement type, coarse aggregate type, fine aggregate type, filler type, superplasticiser type, NP properties, percentage cement replacement with NPs, water to binder (w/b) ratio, binder content, curing regime, mechanical properties test age, split tensile strength specimen type, compressive strength specimen type, flexural strength specimen type, durability tests and mixing procedures.

Using experimental results from different sources can, however, be problematic because

- there is often insufficient information regarding the exact composition of the concrete mixes
- the specimen sizes, curing conditions and testing methodologies vary between different investigations and, in some cases, such information is not fully indicated
- in many cases it is difficult to extract the relevant experimental values because the published results are incomplete or are presented in graphical form and data values have to be extrapolated from the graphs given.

Tables 1 to 6 provide a general summary of the experimental results included in the database (Beigi *et al.*, 2013; Jalal *et al.*, 2012, 2013; Khalaj and Nazari, 2012; Madandoust *et al.*, 2011; Nazari and Riahi, 2010, 2011a, 2011b, 2011c, 2011d; Quercia *et al.*, 2014).

Table 1 Properties of cement and coarse and fine aggregates from the experimental database

Reference	Cement	Coarse aggregate	Fine aggregate
Nazari and Riahi (2010)	OPC	Crushed limestone with gravel 4/12	Two types of sand (0/4 and 0/2, with a very high fines content (particle size < 0·063 mm) of 19·2%)
Nazari and Riahi (2011a, 2011b, 2011c, 2011d)	OPC	Crushed limestone with gravel 4/12	Two types of sand (0/4 and 0/2, with a very high fines content (particle size < 0·063 mm) of 19·2%)
Madandoust *et al.* (2011)	OPC	Expanded polystyrene (EPS) crushed gravel	River sand, fineness modulus 2·83
Jalal *et al.* (2012)	ASTM type II Portland cement	Limestone gravel, nominal maximum size 12·5 mm	Silica aggregate sand, maximum size 4·75 mm
Khalaj and Nazari (2012)	OPC	Crushed limestone	Crushed limestone
Beigi *et al.* (2013)	ASTM type II Portland cement	Crushed gravel, size < 12·5 mm	River sand, sieve 4·75 mm
Jalal *et al.* (2013)	ASTM type II Portland cement	Limestone gravel, nominal maximum size 12·5 mm	Silica sand and crushed limestone, maximum size 4·75 mm
Quercia *et al.* (2014)	OPC	Crushed granite, fractions 2–8 mm and 8–16 mm	Dredged river sand (0–4 mm) and microsand (0–1 mm)

Table 2 Filler and superplasticiser properties used in works from the experimental database

Reference	Filler	Superplasticiser
Nazari and Riahi (2010)	Limestone	Polycarboxylate with a polyethylene condensate de-foamed based admix (Glenium C303)
Nazari and Riahi (2011a)	GGBFS	Polycarboxylate with a polyethylene condensate de-foamed based admix (Glenium C303)
Nazari and Riahi (2011b, 2011c)	Limestone	Polycarboxylate with a polyethylene condensate de-foamed based admix (Glenium C303)
Madandoust *et al.* (2011)	Silica fume	Polycarboxylic ether based high range water reducer (HRWR) (Glenium51), density 1·06–1·08 g/cm^3
Nazari and Riahi (2011d)	GGBFS	Polycarboxylate with a polyethylene condensate de-foamed based admix (Glenium C303)
Jalal *et al.* (2012)	Limestone	Polycarboxylic ether type, specific gravity 1·06–1·08 g/cm^3
Khalaj and Nazari (2012)	Limestone	Polycarboxylate with a polyethylene condensate de-foamed based admix (Glenium C303)
Beigi *et al.* (2013)	Limestone	Carboxylic ether (Glenium-110P), specific gravity 1·1 g/cm^3
Jalal *et al.* (2013)	Fly ash (class F)	Polycarboxylic ether type, specific gravity 1·06–1·08 g/cm^3
Quercia *et al.* (2014)	Ground limestone	Based on polycarboxylate ethers

Table 3 Properties of NPs

Reference	Type	Diameter: nm	Density: g/cm^3	Surface volume ratio: m^2/g	Purity: %
Nazari and Riahi (2011a)	Silicon dioxide	15 ± 3	< 0·15	165 ± 17	> 99·9
Nazari and Riahi (2011b)	Copper oxide	15 ± 3	< 0·13	155 ± 12	> 99·9
Nazari and Riahi (2011c)	Zinc peroxide	15 ± 4	< 0·10	166 ± 17	> 99·9
Madandoust *et al.* (2011)	Silicon dioxide	15	1·30	160	> 99·9
Nazari and Riahi (2011d)	Titanium dioxide	15 ± 3	< 0·13	155 ± 12	> 99·9
Jalal *et al.* (2012)	Silicon dioxide	15 ± 3	< 0·15	165 ± 17	> 99·9
Khalaj and Nazari (2012)	Silicon dioxide	80	0·30	> 800	> 99·9
Beigi *et al.* (2013)	Silicon dioxide	15 ± 5	< 0·15	160 ± 20	> 99·9
Jalal *et al.* (2013)	Titanium dioxide	20 ± 5	< 0·15	165 ± 17	> 99·9
Quercia *et al.* (2014)	Colloidal silicon dioxide	–	1·40	50	> 99·9
Quercia *et al.* (2014)	Powder silicon dioxide	–	2·15	56	> 99·9

Table 4 SEM micrographs, XRD analysis and AFM images of silicon dioxide and titanium dioxide NPs in experimental database

Reference

Nazari and Riahi (2010, 2011a)

SEM and XRD analysis of silicon dioxide NPs

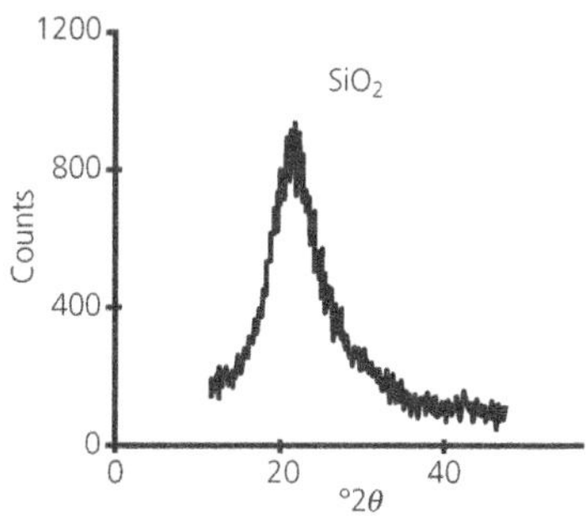

Nazari and Riahi (2011b)

SEM and XRD analysis of copper oxide NPs

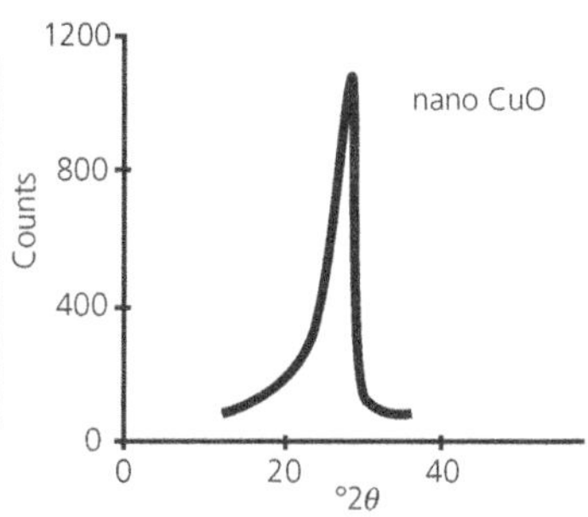

(continued on next page)

Table 4 (*Continued*)

Reference	
Nazari and Riahi (2011c)	SEM and XRD analysis of zinc peroxide NPs 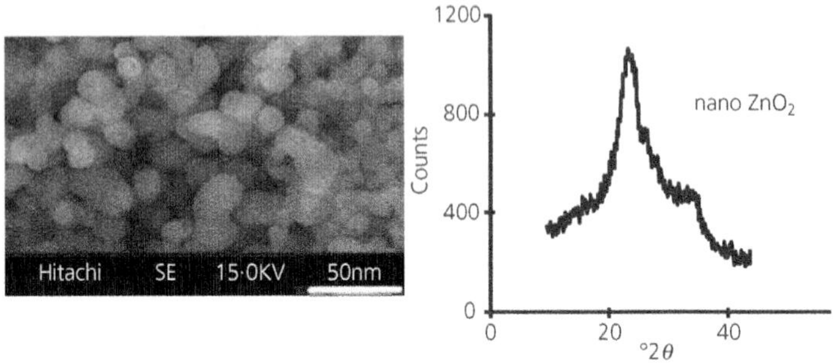
Nazari and Riahi (2011d)	SEM and XRD analysis of titanium dioxide NPs 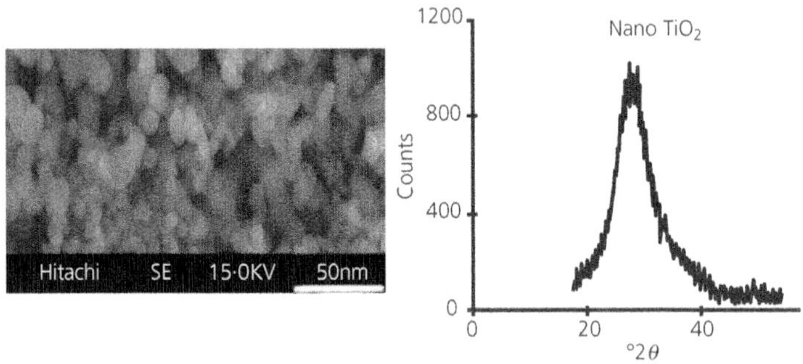
Khalaj and Nazari (2012)	SEM and XRD analysis of silicon dioxide NPs 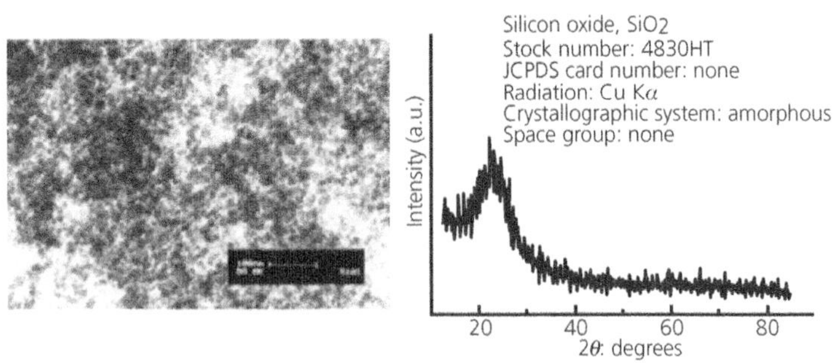

Table 4 (*Continued*)

Reference

Jalal *et al.* (2013) SEM and XRD analysis of titanium dioxide NPs

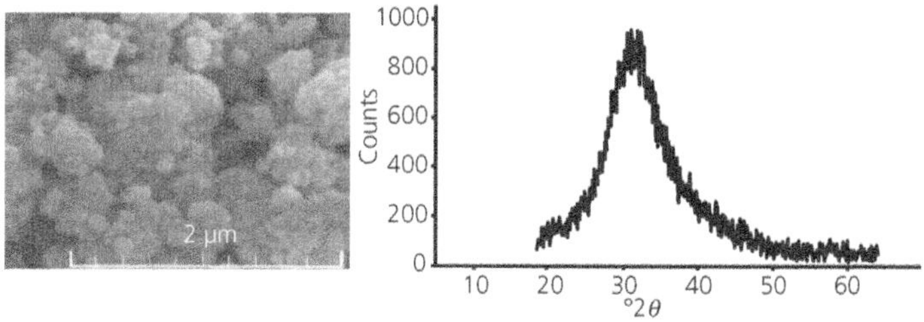

Beigi *et al.* (2013) AFM image of sample with 4 wt% silicon dioxide NPs

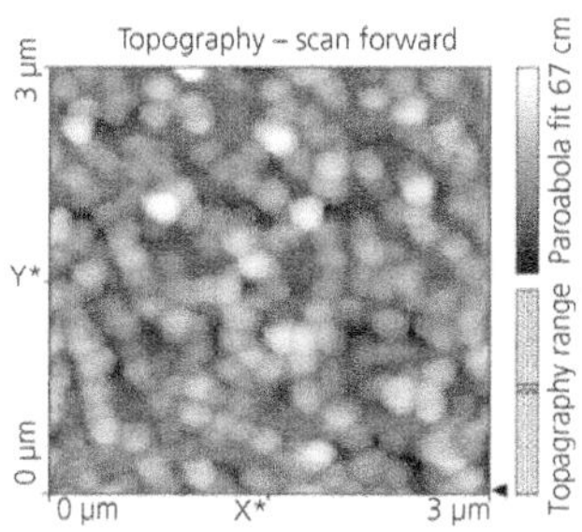

Table 5 Cement replacement by NPs, w/b ratio, binder content and curing type in works in the experimental database

Reference	Cement replacement with NPs: wt%	w/b ratio	Binder content: kg/m^3	Curing type
Nazari and Riahi (2010)	1, 2, 3, 4, 5	0·40	450	Water
Nazari and Riahi (2011a, 2011b, 2011c, 2011d)	1, 2, 3, 4	0·40	450	Water
Madandoust *et al.* (2011)	2·7	0·44	440	Water
Jalal *et al.* (2012)	2	0·38	400, 450, 500	Water
Khalaj and Nazari (2012)	4	0·40	450	Water
Beigi *et al.* (2013)	0, 2, 4, 6	0·39	413·1	Water
Jalal *et al.* (2013)	1, 2, 3, 4, 5	0·38	450	Water
Quercia *et al.* (2014)	1	0·30	519·4	Water

Table 6 Test age and specimen type for mechanical properties of specimens in experimental database

Reference	Test age: d	Specimen type		
		Split tensile strength	Compressive strength	Flexural strength
Nazari and Riahi (2010)	2, 7, 28	–	–	Prism: 200 mm × 50 mm × 50 mm
Nazari and Riahi (2011a)	2, 7, 28	Cylindrical: diameter 150 mm; height 300 mm	–	–
Nazari and Riahi (2011b)	2, 7, 28	–	Cubic: 100 mm edge length	–
Nazari and Riahi (2011c)	2, 7, 28	–	–	Prism: 200 mm × 50 mm × 50 mm
Madandoust *et al.* (2011)	28	–	–	–
Nazari and Riahi (2011d)	2, 7, 28	Cylindrical: diameter 150 mm; height 300 mm	Cubic: 100 mm edge length)	Prism: 200 mm × 50 mm × 50 mm
Jalal *et al.* (2012)	3, 7, 28, 90	Cylindrical: diameter 100 mm; height 200 mm	Cubic: 150 mm edge length	–
Khalaj and Nazari (2012)	28	Cylindrical: diameter 150 mm; height 300 mm	–	–
Beigi *et al.* (2013)	28	Cylindrical: diameter 150 mm; height 300 mm	Cubic: 100 mm edge length	Prism: 500 mm × 100 mm × 100 mm
Jalal *et al.* (2013)	7, 28, 90	Cylindrical: diameter 100 mm; height 200 mm	Cubic: 150 mm edge length	–
Quercia *et al.* (2014)	1, 3, 7, 28, 91	Cubic: 150 mm edge length	Cubic: 150 mm edge length	–

Table 1 shows that the research studies included in the database used ordinary Portland cement (OPC) or ASTM type II Portland cement and the common coarse and fine aggregates were crushed limestone gravel and river sand. Table 2 shows that the common filler types used in the SCC mixes were limestone, GGBFS, silica fume and fly ash and the most common superplasticiser used was polycarboxylic ether. Table 3 lists the NP properties, indicating that silicon dioxide (SiO_2) and titanium dioxide (TiO_2) were the most commonly used. Moreover, Table 4 shows that scanning electron microscopy (SEM), X-ray diffraction (XRD) analyses and atomic force microscopy (AFM) were used to assess the microstructure of the silicon dioxide and titanium dioxide NPs.

Table 5 indicates that cement replacement with NPs was in the range 1–6 wt%, the w/b ratio range used was 0·30–0·44, the binder content range was 400–520 kg/m^3 and water curing was used to assess the mechanical properties of the specimens.

Table 6 shows that the test ages of mechanical properties were 1, 3, 7, 28 and 91 d, but the most common testing age was 28 d. The common split tensile strength specimen type was cylindrical (diameter 150 mm, height 300 mm) and cubic specimens (100 mm or 150 mm edge length) were used to test for compressive strength. The type of specimens used to test flexural strength was not specified in some works. Table 7 indicates the SCC mixing procedures in the experimental database, which generally follow SCC mixing guidelines.

Table 7 Mixing procedures as listed in experimental database

Reference	Mixing procedure
Nazari and Riahi (2010)	'The mixing sequence for specimens was consisted of homogenizing the sand and cementitious materials for one minute in the mixer and then approximately 75% of the mixing water were added. The coarse aggregate was introduced and then the superplasticizer was pre-dissolved in the remaining water and was added at the end of the mixing sequence. The total mixing time including homogenizing was 5 min.'
Nazari and Riahi (2011a, 2011b, 2011c, 2011d)	Same as Nazari and Riahi (2010)
Madandoust *et al.* (2011)	'The mixing sequence for specimens was, natural coarse and fine aggregates followed by EPS aggregates were homogenized for 30 s at normal mixing speed (about 18 rpm). Thereafter, adding about half of the mixing water into the mixer while mixing goes on for 1 min. The mix was rested for 1 min so that the aggregates could absorb the water in the mixer. Then, binder materials including cement and silica fume were added and mixed for one more minute. The remaining water and HRWR were introduced to the wet mix, while mixing was going on for 3 min. Finally, after 2 min resting, mixing sequence resumed for additional 2 min.'
Jalal *et al.* (2012)	'The powder material and the aggregates were mixed in dry form for 2 min. Then half of the water containing the whole amount of super plasticizer was poured and mixed for 3 min. After that, about 1 min rest was allowed and finally rest of the water containing VMA was added into the mix and mixed for 1 min.'
Khalaj and Nazari (2012)	'To improve workability of the fresh concrete, 1 wt% of water was replaced by polycarboxylate admix in all series. The superplasticizer was dissolved in water and stirred at a high speed for 3 min. For N and NSF series, before stirring in water, SiO_2 NPs were added though SiO_2 NPs cannot be dissolved in water and a smaller amount of SiO_2 NPs can be dispersed evenly by the superplasticizer. The total mixing time including homogenizing was 5 min.'
Beigi *et al.* (2013)	–
Jalal *et al.* (2013)	Same as Jalal *et al.* (2012)
Quercia *et al.* (2014)	'These SCCs were mixed for 5 min in total in a compulsory mixer, and subsequently tested for their fresh properties, according to the procedure recommended by the EFNARC committee.'

Influence of NPs on fresh and hardened properties of SCC based on the experimental database

Fresh properties

Madandoust *et al.* (2011), Jalal *et al.* (2012, 2013), Beigi *et al.* (2013) and Quercia *et al.* (2014) observed that the flowability characteristics of SCC mixes incorporating silicon dioxide or titanium dioxide NPs decreased. Since silicon dioxide or titanium dioxide NPs have a large effective surface, they absorb a lot of water and this thus reduces the concrete's workability and would also increase the SCC's shear strength against flowability and viscosity. This property of silicon dioxide or titanium dioxide NPs can be attributed to their high pozzolanic activity.

Quercia *et al.* (2014) observed that only their mix with colloidal silicon dioxide NPs resulted in a spread flow close to the lower limit of the target range.

Madandoust *et al.* (2011) and Jalal *et al.* (2012, 2013) reported that the addition of silicon dioxide or titanium dioxide NPs improved the consistency of the SCC mixes. Less bleeding and segregation were also observed in the SCC mixes containing these NPs. Moreover, it was observed that the addition of silicon dioxide or titanium dioxide NPs led to a higher T_{50} time (the time for concrete to spread to a diameter of 500 mm), which can be expected to improve the segregation resistance. Madandoust *et al.* (2011) also noted that the incorporation of silicon dioxide NPs increased the V-funnel times of SCC mixes.

Quercia *et al.* (2014) found that the SCC mix with silicon dioxide NPs had a higher air content than the reference SCC mix, caused by the higher viscosity of the paste due to the addition of particles with high specific surface area. The assumed air entrainment was also confirmed by comparing the difference between the designed and measured concrete densities.

Hardened properties

Some studies have been conducted on the compressive, tensile and flexural strengths of cementitious composites reinforced with NPs and the following possible reasons have been put forward to explain the increases in strength.

- With a small amount of NPs uniformly dispersed in the cement paste, the NPs act as nuclei to tightly bond with cement hydrates and further promote cement hydration due to their high activity, which is favourable for the strength of cement mortar (Flores-Velez and Dominguez, 2002).
- The NPs among hydrate products will prevent crystals from growing, which again is beneficial for the strength of cement paste (Fernandez *et al.*, 1999; Grzeszczyk and Lipowski, 1997).
- The NPs fill the cement pores, thus increasing the strength. NPs can contribute to the hydration process to generate calcium silicate hydrates through their reaction with calcium hydroxide (Massazza, 1987).

Compressive strength

Silicon dioxide NPs

Beigi *et al.* (2013) reported that, in all the SCC mixes studied, when the silicon dioxide NP content increased to an optimum value of 4·0 wt%, the compressive strength of the sample was significantly improved. This improvement could be attributed to the appropriate positioning of silicon dioxide NPs within the cement matrix. Moreover, Jalal *et al.* (2012) reported that replacement with 2·0 wt% silicon dioxide NPs in SCC mixes with binder contents of 400, 450 and 500 kg/m^3 increased the compressive strength by 5%, 22%, 38% and 43%, 14%, 21%, 55% and 61% and 11%, 22%, 56% and 62% at 3, 7, 28 and 90 d, respectively. Quercia *et al.* (2014) found that, during the development of strength over time, the mix with colloidal silicon dioxide NPs showed the highest increase in compressive strength from 3 d to 91 d. An interesting observation was that, at 7 d, the SCC mix with powder silicon dioxide NPs showed compressive strength similar to that of the reference mix as a result of pozzolanic reaction with calcium hydroxide generated during OPC hydration. At 91 d, the SCC with powder silicon dioxide NPs developed a higher compressive strength than the reference SCC. In the case of SCC with colloidal silicon dioxide NPs, the 91 d compressive strength was almost the same as that at 28 d.

Titanium dioxide NPs

Nazari and Riahi (2011d) reported that compressive strength increased by adding titanium dioxide NPs up to 3·0 wt% replacements and then decreased, although the addition of 4·0 wt% titanium dioxide NPs produced specimens with much higher compressive strength with respect to all the other SCC mixes without NPs. The reduced compressive strength by adding more than 3·0 wt% titanium dioxide NPs may be due to the fact that this quantity of titanium dioxide NPs in the mix was higher than the amount required to combine with lime liberated during the process of hydration, thus leading to excess silica leaching out and causing a deficiency in strength as it replaces part of the cementitious material but does not contribute to strength.

Copper oxide (CuO) NPs

Nazari and Riahi (2011b) noted that compressive strength increased by adding copper oxide NPs up to 4·0 wt% replacement and then it decreased, although adding 5·0 wt% copper oxide NPs produced specimens with much higher compressive strength than all the other SCC mixes without NPs. The reduced compressive strength from adding more than 4·0 wt% copper oxide NPs may be due to the fact that the amount of copper oxide NPs present in the mix was higher than the amount required to combine with liberated lime from the process of hydration.

Split tensile strength

Silicon dioxide or titanium dioxide NPs

Nazari and Riahi (2011a, 2011d) reported that split tensile strength increased by adding silicon dioxide or titanium dioxide NPs up to 3·0 wt% replacements and then decreased.

However, adding 4·0% silicon dioxide or titanium dioxide NPs produced specimens with much higher split tensile strengths than the other SCC mixes without NP addition. In previous work, Nazari and Riahi (2010, 2011b, 2011c) showed that using up to 4·0 wt% of NPs in SCC could improve the split tensile strength of the specimens. However, in the later works (Nazari and Riahi, 2011a, 2011d) an optimum level of 3·0 wt% was achieved by utilising GGBFS. This may be due to the reduction of calcium oxide content when GGBFS is used. Thus, the amount of crystalline calcium hydroxide is reduced and the amount of silicon dioxide NPs assisting the formation of calcium silicate hydrate gel is decreased. Hence, the addition of more than 3·0 wt% of silicon dioxide NPs may decrease the split tensile strength of SCC specimens with NPs. Furthermore, the reduced split tensile strength by adding more than 3·0 wt% silicon dioxide or titanium dioxide NPs may be due to defects generated in the dispersion of NPs, causing weak zones.

Silicon dioxide NPs

From split tensile strength test results, Jalal *et al.* (2012) found that the highest values were obtained from SCC mixes containing both micro and nano silicon dioxide with a binder content of 500 kg/m^3. Compared with control SCC specimens, replacement with 10% micro silicon dioxide and 2% silicon dioxide NPs in ternary mixes increased the splitting tensile strength for binder contents of 400, 450 and 500 kg/m^3 by 19%, 17%, 33% and 25%, 18%, 40%, 8% and 11%, 3·5%, 27%, 2% and 11% at 3, 7, 28 and 90 d, respectively. Quercia *et al.* (2014) reported that the determined average 28 d tensile splitting strength of their reference SCC mix was 4·51 N/mm^2 and the equivalent values for SCC mixes with colloidal and powder silicon dioxide NPs were higher, at 4·92 N/mm^2 and 5·48 N/mm^2, respectively. A possible explanation for the higher tensile splitting strength of the samples containing silicon dioxide NPs is that the bond between the hardened paste and the aggregates was stronger.

Silicon dioxide NPs in SCC fibre mixes

Khalaj and Nazari (2012) investigated the split tensile strength of SCC specimens containing silicon dioxide NPs with an average particle size of 80 nm and randomly oriented steel fibres. The obtained results showed that the addition of the steel fibres up to 12 wt% improved concrete strength. Beigi *et al.* (2013) reported that increases in the content of steel, polypropylene and glass fibres produced 45%, 15% and 33% increases respectively in the tensile strength of SCC samples. However, the addition of silicon dioxide NPs to the samples dramatically improved these increases to 90%, 57% and 77%, respectively. This improvement could be attributed to the larger contact area between fibres and mortar, resulting in more friction and higher performance of the SCC.

Titanium dioxide NPs

Jalal *et al.* (2013) found that the split tensile strength increased by adding titanium dioxide NPs up to 4·0 wt% replacements and then decreased, although adding 5·0 wt% titanium dioxide NPs produced specimens with higher split tensile strengths than SCC specimens with 1·0, 2·0 and 3·0 wt% titanium dioxide NPs. The reduced split tensile strength by adding more than 4·0 wt% titanium dioxide NPs may be due to the fact that the quantity of titanium dioxide NPs in the mix was higher than the amount required to combine with lime liberated during the hydration process.

Flexural strength

Silicon dioxide, titanium dioxide or zinc peroxide (ZnO_2) NPs

Nazari and Riahi (2010, 2011c, 2011d) reported that flexural strength increased by adding silicon dioxide, titanium dioxide or zinc peroxide NPs up to 4·0 wt% replacement and then it decreased, although adding 5·0 wt% of these NPs produced specimens with much higher flexural strengths than the SCC mixes without NPs. The reduced flexural strength by adding more than 4 wt% silicon dioxide, titanium dioxide or zinc peroxide NPs may be due to the fact that the quantity of NPs present in the mix was higher than required to combine with liberated lime from the process of hydration. This phenomenon leads to excess silica leaching out and causes a deficiency in strength as it replaces part of the cementitious material but does not contribute to strength. It may also be due to defects generated in the dispersion of NPs, causing weak zones. The higher flexural strength in SCC mixes with NPs compared with SCC mixes without NPs is due to the rapid consumption of crystalline calcium hydroxide that quickly forms during the hydration of Portland cement, especially at early ages, as a result of the high reactivity of silicon dioxide, titanium dioxide or zinc peroxide NPs. As a consequence, cement hydration is accelerated and larger volumes of reaction products are formed.

Silicon dioxide NPs in SCC fibre mixes

Beigi *et al.* (2013) reported that, in SCC samples without fibres, an increase in silicon dioxide NP content up to 4 wt% continuously increased the flexural strength in a linear manner to a peak at the optimum percentage of NPs (4·0 wt%). Any further increase in silicon dioxide NP content caused no increase in flexural strength. On the other hand, a higher content of silicon dioxide NPs increased the reinforcing fibre-induced flexural strength more significantly. For illustration, 4·0 wt% silicon dioxide NPs in mixes containing the highest content of fibres (0·3% steel fibres, 0·2% polypropylene fibres and 0·2% glass fibres) caused an increase in flexural strength by as much as 76%, 53% and 75%, respectively, compared with the reference SCC sample.

Influence of NPs on durability properties of SCC based on the experimental database

Table 8 summarises the durability tests carried out on SCC mixes incorporating NPs in the experimental database.

Table 8 Durability tests of experimental database

Reference	Durability tests
Nazari and Riahi (2010, 2011b, 2011c)	Mercury intrusion porosimetry, thermogravimetric analysis, conduction calorimetry
Nazari and Riahi (2011b)	Mercury intrusion porosimetry
Madandoust *et al.* (2011)	–
Nazari and Riahi (2011d)	Water permeability, mercury intrusion porosimetry, conduction calorimetry
Jalal *et al.* (2012)	Absorption test, capillary test, chloride ion percentage, resistivity test
Khalaj and Nazari (2012)	–
Beigi *et al.* (2013)	Chloride ion penetration, water absorption
Jalal *et al.* (2013)	Absorption test, capillary test, chloride ion penetration
Quercia *et al.* (2014)	Permeable porosity, penetration of water under pressure, pore size distribution and porosity of the paste, rapid chloride migration, conductivity, chloride diffusion, freeze–thaw resistance

Water absorption

Silicon dioxide NPs

Jalal *et al.* (2012) reported that water absorption decreased in SCC samples containing 2·0 wt% silicon dioxide NPs: 35%, 28% and 32% for binder contents of 400, 450 and 500 kg/m^3, respectively, in the first timestep. Beigi *et al.* (2013) also found that that addition of silicon dioxide NPs significantly reduced the percentage water absorption compared with reference SCC samples. The addition of 4·0 wt% silicon dioxide NPs resulted in reductions in water absorption of 37%, 30% and 40% in SCCs containing steel, polypropylene and glass fibres respectively. This is probably due to the formation of jelly-like silicate hydrate filling the pores.

Titanium dioxide NPs

Nazari and Riahi (2011d) found that the percentage water absorption in SCC mixes without NPs and with GGBFS filler specimens at 7 d of curing was lower than that of SCC mixes with titanium dioxide NPs; at 28 d and 90 d of curing however, the values were lower for SCC mixes with titanium dioxide NPs. This is due to increased formation of hydrated products in the SCC mixes with titanium dioxide NPs at the early ages of curing. The results also showed that the percentage water absorption in SCC mixes with titanium dioxide NPs at 28 d and 90 d of curing decreased with increasing titanium dioxide NP content up to 4·0 wt% and then increased. This is because of the unfavourable dispersion of NPs in the cement paste for NP contents above 3·0 wt%. Also, the increased water requirement due to increasing the NP content up to 3·0 wt% results in a decreased percentage water absorption at 7 d of curing. It can therefore be suggested that, with prolonged curing, an increase in age and the percentage of titanium dioxide NPs can lead to a reduction in permeable voids. This is due to the high action and filler effects of titanium dioxide NPs. Furthermore, Jalal *et al.* (2013) reported that the percentage water absorption decreased by increasing titanium dioxide NP content up to 4·0 wt% and then it increased. Once again, this may be due to poor dispersion of the NPs in the cement paste for NP contents beyond 4·0 wt%.

Capillary water absorption

Silicon dioxide NPs

Jalal *et al.* (2012) reported that the addition of 2·0 wt% silicon dioxide NPs resulted in capillary water absorption of 3·6, 3·3 and 3·0 mm for binder contents of 400, 450 and 500 kg/m^3, respectively, over 72 h. These values represent reductions of 58%, 52% and 54% compared with SCC mixes without any admixes.

Titanium dioxide NPs

Jalal *et al.* (2013) reported that the height of absorbed water in SCC samples decreased by increasing the content of titanium dioxide NPs up to 4·0 wt% and then increased.

Penetration of water under pressure

Silicon dioxide NPs

Quercia *et al.* (2014) found that two SCC mixes with powder or colloidal silicon dioxide NPs showed penetration depths less than 5 mm. The addition of 3·8 wt% silicon dioxide NPs resulted in a SCC that was highly resistant to the penetration of water under a pressure of 0·5 MPa.

Chloride ion penetration resistance

Silicon dioxide NPs

Jalal *et al.* (2012) reported that, at a depth of 0–5 mm, an increase in binder content from 400 to 450 and 500 kg/m^3 in SCC samples without any admix resulted in reductions of chloride ion penetration by 16·6% and 29·0%, respectively. At the same depth, SCC samples containing 2·0 wt% silicon dioxide NPs showed decreases in chloride ion penetration of 43%, 40% and 43% for binder contents of 400, 450 and 500 kg/m^3, respectively. Beigi *et al.* (2013) found that replacement with 4·0 wt% silicon dioxide NPs decreased the chloride ion penetration in SCCs containing steel, polypropylene and glass fibres by 40%, 28% and 20%, respectively. Silicon dioxide NPs can thus reduce the permeability of SCCs dramatically. This property can be attributed to the high activity of silicon dioxide NPs in reactions with calcium hydroxide (which has low resistance to chemical attack). The chemical reactions produce calcium silicate hydrate, which is compacted; this thus improves the microstructure of fibre-reinforced SCC against chloride ion penetration and reduces the associated corrosion in concrete.

Titanium dioxide NPs

Jalal *et al.* (2013) reported that a decrease in chloride ion penetration was observed by increasing the content of titanium dioxide NPs up to 4·0 wt%.

Conductivity

Silicon dioxide NPs

Quercia *et al.* (2014) found that the conductivity of SCC containing silicon dioxide NPs was reduced by more than 50% compared with a SCC reference mix. The SCC with colloidal silicon dioxide NPs presented a slightly lower conductivity than samples with powder silicon dioxide NPs. This behaviour indicates the ability of the water-saturated pore structure of concrete to transport electrical charge. The lower conductivity values shown by the SCC with silicon dioxide NPs are due to pore structure refinement as a result of progressive pozzolanic reactions and a higher hydration degree.

Rapid chloride migration and chloride diffusion

Silicon dioxide NPs

According to the work of Quercia *et al.* (2014), the chloride migration coefficients and chloride diffusion are much lower for mixes containing silicon dioxide NPs. The SCC mix with colloidal silicon dioxide NPs showed the best performance. A finer porosity, greater tortuosity and more precipitated calcium silicate hydrate gel all act to reduce the ingress rate of chlorides into concrete.

Mercury intrusion porosimetry

Silicon dioxide, titanium dioxide, zinc peroxide or copper oxide NPs

Nazari and Riahi (2010, 2011b, 2011c) reported that the total specific pore volumes of concretes decreased with increasing NP contents up to 4·0 wt% in SCC mixes with fly ash or limestone filler; the most probable pore diameters shifted to smaller sizes, in the range of meso capillary pores, which indicates that the addition of Portland cement refines the pore structure of concrete. Nazari and Riahi (2011a, 2011d) found that the total specific pore volumes of concretes decreased with increasing NP contents up to 3·0 wt% in SCC mixes with GGBFS filler.

Silicon dioxide NPs

Quercia *et al.* (2014) found that the addition of 3·8 wt% silicon dioxide NPs slightly increased the volume of pores of less than 20 nm. The addition of silicon dioxide NPs decreased the median pore diameter (by volume and area) and the average pore diameter from 27·8 nm (for the reference specimen) to 24·3 nm and 25·9 nm for specimens containing colloidal silicon dioxide NPs and powder silicon dioxide NPs, respectively.

Permeable porosity

Silicon dioxide NPs

Quercia *et al.* (2014) reported that the reference SCC mix had a slightly lower total porosity (12·1%) than mixes containing colloidal or powder silicon dioxide NPs (12·5%).

Conduction calorimetry

Silicon dioxide, titanium dioxide, zinc peroxide or copper oxide NPs

Nazari and Riahi (2010, 2011b, 2011c) and Jalal *et al.* (2013) reported that increasing the content of NPs up to 4·0 wt% in a SCC mix with fly ash or limestone filler accelerated the peak times and reduced heat release rate values. This is indicative of an acceleration in initial cement hydration. NPs, as pozzolans, can accelerate cement hydration and hence increase the heat release rate. The appearance of the peaks in the conduction calorimetry tests were due to the formation of calcium hydroxide (CH) and calcium fly ash (C_3H) compounds in the cement paste. When NPs are partially added to cement paste, the acceleration in CH and C_3H formation would result in a more rapid appearance of the related peaks. Nazari and Riahi (2011a, 2011d) found that increasing NP content up to 3·0 wt% in a SCC mix with GGBFS filler accelerated the peak times and reduced the heat release rate.

Thermogravimetric analysis

Silicon dioxide, titanium dioxide, zinc peroxide or copper oxide NPs

Nazari and Riahi (2010, 2011b) reported that, after 28 d of curing, loss in specimen weight increased by increasing the NP content up to 4·0 wt% in SCC mixes with fly ash or limestone filler. Again, as with the conduction calorimetry results, the increase in weight loss is due to increased formation of CH and C_3H compounds in the cement paste. Nazari and Riahi (2011a, 2011d) also showed that the weight loss of specimens increased with increasing NPs up to 3·0 wt% in SCC mixes with GGBFS filler.

Freeze–thaw resistance

Silicon dioxide NPs

Quercia *et al.* (2014) noted that SCC mixes with added silicon dioxide NPs resulted in a surface scaling factor lower than the recommended value for non-air-entrained concrete after 28 cycles. This better resistance to freeze–thaw induced damage of SCC with silicon dioxide NPs can be attributed to different factors, such as the formation of very stiff calcium silicate hydrate gel and a refined pore structure, resulting in limited intrusion of water and improved resistance to temperature changes near the concrete surface. SCC with silicon dioxide NPs should result in a mix having a high resistance to freeze–thaw, with an air entrainment mix that guarantees a minimum air content of 4%.

Influence of NPs on the microstructure of SCC

According to the SEM micrographs available in the experimental database, a more refined microstructure and smaller pores may be achieved through the addition of NPs, which can lead to enhanced mechanical, durability and microstructural properties of the SCC mixes. Jalal *et al.* (2013) and Quercia *et al.* (2014) captured good SEM micrographs for clarifying SCC mixes without NPs and with NPs, as shown in Figures 1–4. The XRD patterns available in the experimental database show that the addition of NPs efficiently reduced the formation of calcium hydroxide in the SCC samples.

Figure 1 SEM micrographs of SCC mixtures without NPs (a), with 1·0 wt% NPs (b) and with 4·0 wt% NPs (c) at 7 d (series 1) and 90 d (series 2) (Jalal *et al.*, 2013)

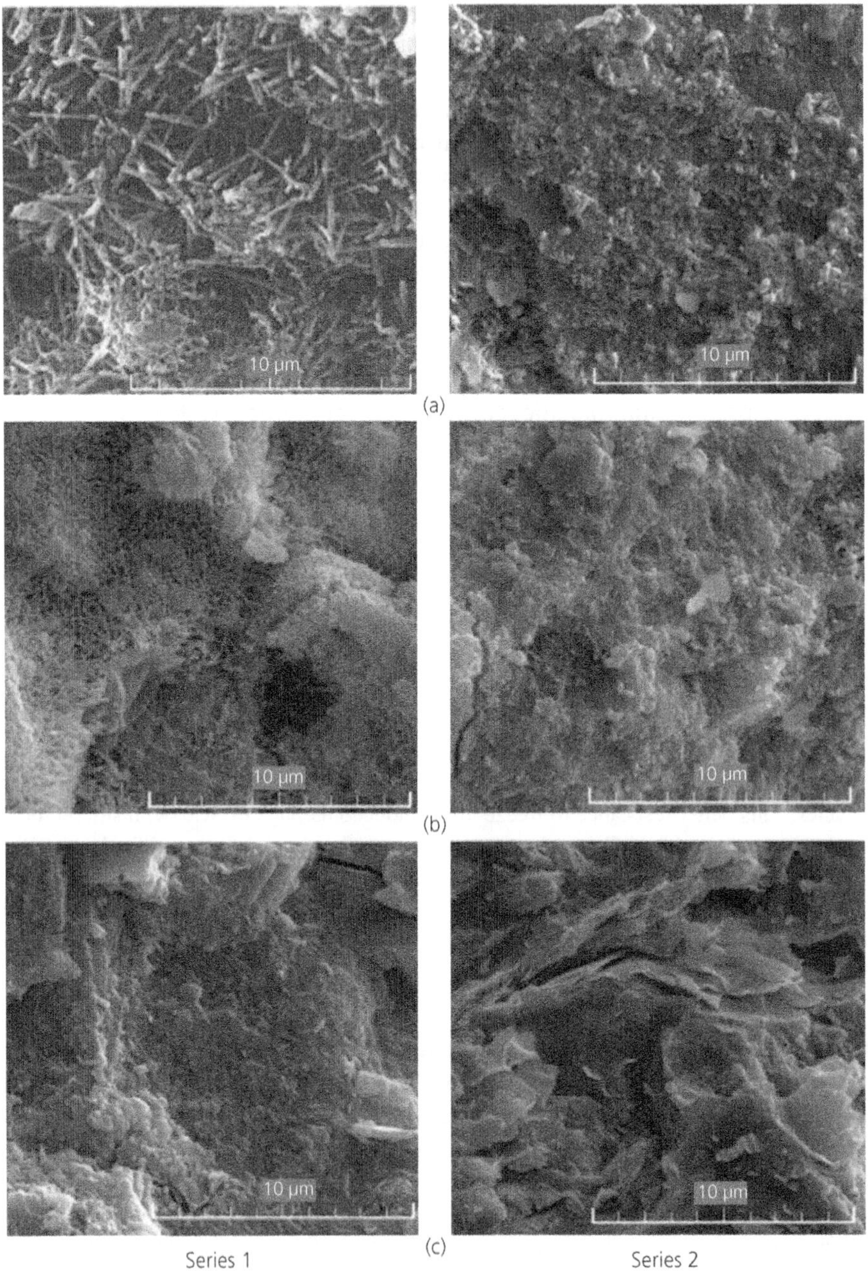

Figure 2 SEM micrographs of SCC reference mix: (a) cement matrix and aggregate interfacial transition zone (ITZ); (b) ettringite and alumina ferric oxide, trisulfate (AFt) needles; (c) precipitated hexagonal portlandite plate crystals (Quercia *et al.*, 2014)

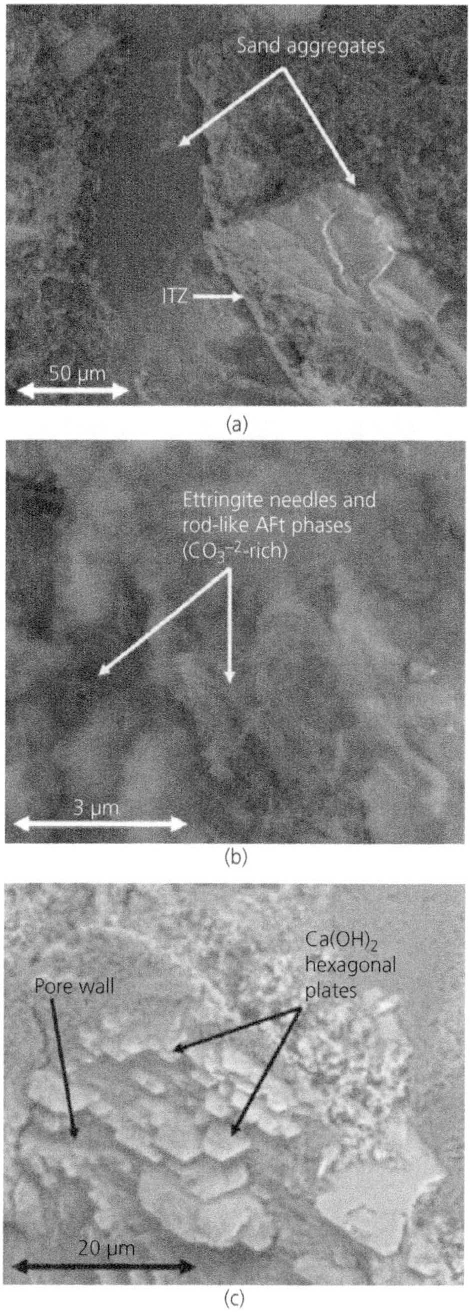

Figure 3 SEM micrographs of SCC with colloidal silicon dioxide NPs: (a) cement matrix; (b) aggregate-dense ITZ; (c) agglomerates in a pore (Quercia *et al.*, 2014)

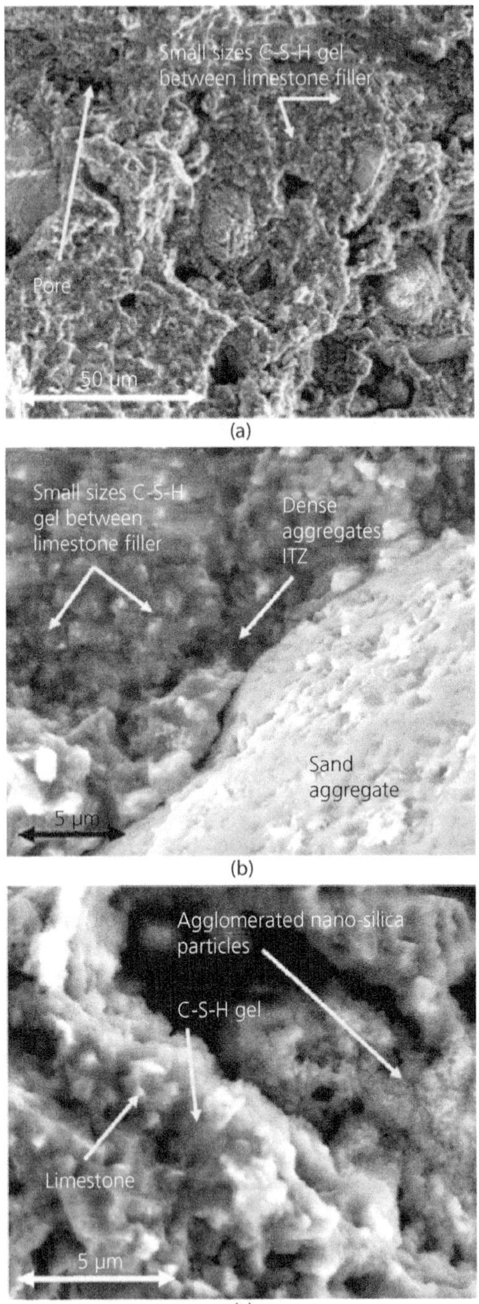

Figure 4 SEM micrographs of SCC with powder silicon dioxide NPs: (a) cement matrix; (b) aggregate ITZ; (c) AFt needle precipitates (Quercia *et al.*, 2014)

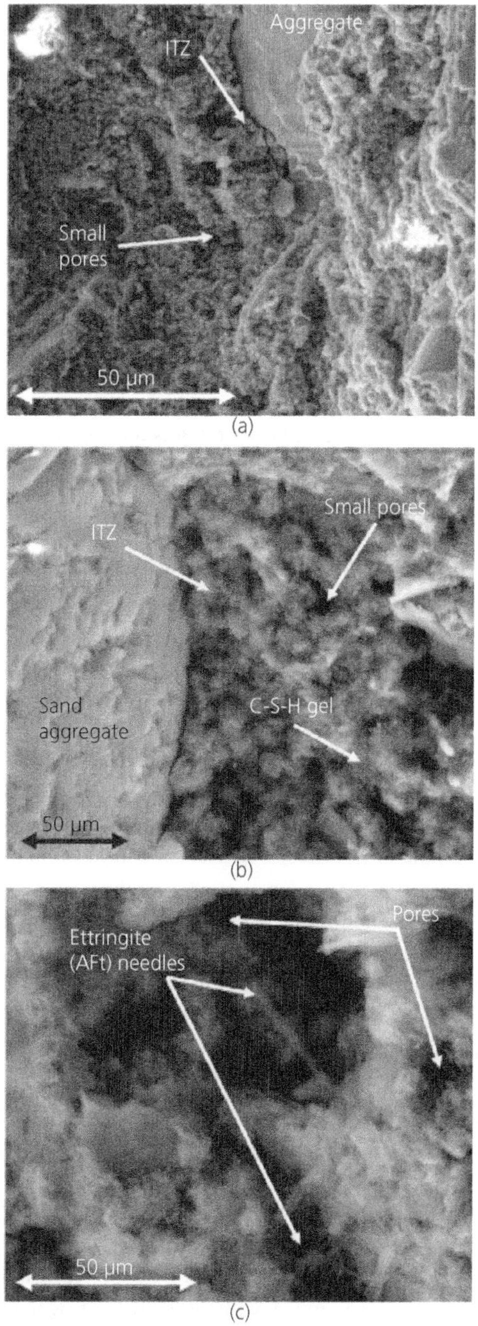

Conclusions

This paper has reviewed previous works investigating the effect of nanoparticles (NPs) on some properties of self-compacting concrete (SCC). The conclusions from this literature review can be summarised as follows.

- The incorporation of NPs in SCC mixes decreases flowability characteristics, which can be attributed to the high pozzolanic activity of the NPs.
- The addition of NPs can improve the consistency of SCC mixes. Less bleeding and segregation were observed in SCC mixes containing NPs.
- Incorporation of NPs increased the V-funnel times of SCC mixes.
- SCC mixes with NPs have higher air content than SCC mixes without NPs.
- In SCC mixes with fly ash or limestone filler, the compressive, split tensile and flexural strengths increased with the addition of silicon dioxide, titanium dioxide, zinc peroxide or copper oxide NPs up to 4·0 wt% replacements, and then they decreased. However, adding 5·0 wt% NPs still produced specimens with much higher strengths than SCC mixes without NPs.
- The compressive, split tensile and flexural strengths increased by adding silicon dioxide, titanium dioxide, zinc peroxide or copper oxide NPs up to 3·0 wt% replacements in SCC mixes with GGBFS filler, and then they decreased; adding 4·0 wt% NPs produced specimens with much higher strengths than SCC mixes without NPs.
- The compressive, split tensile and flexural strengths increased by adding colloidal silicon dioxide NPs in SCC mixes, and these specimens showed better behaviour than SCC mixes with powder silicon dioxide NPs.
- SCC mixes with NPs and fibres showed dramatically improved compressive, split tensile and flexural strengths. This improvement can be attributed to the larger contact area between the fibres and the mortar, resulting in greater friction and higher performance of the SCC.
- The percentage water absorption and capillary water absorption in SCC mixes with NPs at 28 d and 90 d of curing decrease with increasing NP content up to 4·0 wt% and then they increase. This is due to poor dispersion of NPs in the cement paste for NP contents greater than 3·0 wt%.
- The addition of 4·0 wt% NPs to SCC with fly ash or limestone filler results in SCC that is highly resistant to the penetration of water under a pressure of 0·5 MPa.
- NP contents of 4·0 wt% in SCC mixes with fly ash or limestone filler produces decreased chloride ion penetration: the addition of NPs can thus drastically reduce the permeability of SCC.
- The conductivity of SCC with silicon dioxide NPs is reduced by more than 50% compared with SCC without NPs. The lower conductivity of the NP SCC is the result of pore structure refinement due to progressive pozzolanic reactions and a higher hydration degree.
- Chloride migration coefficients and chloride diffusion are much reduced in SCC mixes containing NPs.
- Increasing the content of NPs up to 4·0 wt% in SCC mixes with fly ash or limestone filler and up to 3·0 wt% in SCC mixes with GGBFS filler decreases the total specific

pore volumes of concrete. The most probable concrete pore diameters shift to smaller sizes and fall in the range of meso capillary pores.

- Increasing the percentage of NPs up to 4·0 wt% in SCC mixes with fly ash or limestone filler and up to 3·0 wt% in SCC mixes with GGBFS filler accelerates peak times and reduces the heat release rate. This is indicative of accelerated initial cement hydration due to the higher content of NPs. NPs, as pozzolans, can accelerate cement hydration and hence increase the heat release rate.
- The weight loss of specimens increases by increasing NP contents up to 4·0 wt% in SCC mixes with fly ash or limestone filler or up to 3·0 wt% in SCC mix with GGBFS filler.
- SCC mixes with silicon dioxide NPs produced a surface scaling factor lower than the recommended value for non-air-entrained concrete after 28 cycles. The improved resistance to freeze–thaw induced damage of SCC with silicon dioxide NPs can be attributed to different factors, such as the formation of stiff calcium silicate hydrate gel and a refined pore structure.

REFERENCES

Aslani F (2013) Effects of specimen size and shape on compressive and tensile strengths of self-compacting concrete with or without fibers. *Magazine of Concrete Research* **65(15)**: 914–929, http://dx.doi.org/10.1680/macr.13.00016.

Aslani F and Bastami M (2014) Relationship between deflection and crack mouth opening displacement of self-compacting concrete beams with and without fibres. *Mechanics of Advanced Materials and Structures* in press, http://dx.doi.org/10.1080/15376494.2014.906689.

Aslani F and Maia L (2013) Creep and shrinkage of high strength self-compacting concrete: experimental and numerical analysis. *Magazine of Concrete Research* **65(17)**: 1044–1058, http://dx.doi.org/10.1680/macr.13.00048.

Aslani F and Natoori M (2013) Stress-strain relationships for steel fibre reinforced self-compacting concrete. *Structural Engineering and Mechanics* **46(2)**: 295–322.

Aslani F and Nejadi S (2012a) Mechanical properties of conventional and self-compacting concrete: an analytical study. *Construction and Building Materials* **36**: 330–347.

Aslani F and Nejadi S (2012b) Bond behavior of reinforcement in conventional and self-compacting concrete. *Advances in Structural Engineering* **15(12)**: 2033–2051.

Aslani F and Nejadi S (2012c) Bond characteristics of steel fibre reinforced self-compacting concrete. *Canadian Journal of Civil Engineering* **39(7)**: 834–848.

Aslani F and Nejadi S (2012d) Shrinkage behavior of self-compacting concrete. *Journal of Zhejiang University: Science A* **13(6)**: 407–419.

Aslani F and Nejadi S (2012e) Bond characteristics of reinforcing steel bars embedded in self-compacting concrete. *Australian Journal of Structural Engineering* **13(3)**: 279–295.

Aslani F and Nejadi S (2013a) Self-compacting concrete incorporating steel and polypropylene fibers: compressive and tensile strengths, moduli of elasticity and rupture, compressive stress–strain curve, and energy dissipated under compression. *Composites Part B: Engineering* **53**: 121–133.

Aslani F and Nejadi S (2013b) Creep and shrinkage of self-compacting concrete with and without fibers. *Journal of Advanced Concrete Technology* **11(10)**: 251–265.

Aslani F and Nejadi S (2013c) Mechanical characteristics of self-compacting concrete with and without fibres. *Magazine of Concrete Research* **65(10)**: 608–622, http://dx.doi.org/10.1680/macr.12.00153.

Aslani F, Nejadi S and Samali B (2014a) Short term bond shear stress and cracking control of reinforced self-compacting concrete one way slabs under flexural loading. *Computers and Concrete* **13(6)**: 709–737.

Aslani F, Nejadi S and Samali B (2014b) Long-term flexural cracking control of reinforced self-compacting concrete one way slabs with and without fibres. *Computers and Concrete* **14(4)**: 419–443.

Barbhuiya S (2011) Effects of fly ash and dolomite powder on the properties of self-compacting concrete. *Construction and Building Materials* **25(8)**: 3301–3305.

Barker D, Turner S, Napier-Moore P, Clark M and Davison J (2009) CO_2 capture in the cement industry. *Energy Procedia* **1(1)**: 87–94.

Beigi MH, Berenjian J, Omran OL, Nik AS and Nikbin IM (2013) An experimental survey on combined effects of fibers and nanosilica on the mechanical, rheological, and durability properties of self-compacting concrete. *Materials and Design* **50**: 1019–1029.

Bouzoubaa N and Lachemi M (2001) Self-compacting concrete incorporating high volumes of class F fly ash preliminary results. *Cement and Concrete Research* **31(3)**: 413–420.

Collepardi M, Collepardi S, Ogoumah Olagat JJ and Troli R (2003) Laboratory-test and filled-experience SCCs. *Proceedings of 3rd International Symposium on Self Compacting Concrete, Reykjavik, Iceland*, pp. 912–912.

Craeye B, De Schutter G and Desmet B *et al.* (2010) Effect of mineral filler type on autogenous shrinkage of self-compacting concrete. *Cement and Concrete Research* **40(6)**: 908–913.

Deja J, Uliasz-Bochenczyk A and Mokrzycki E (2010) CO_2 emissions from polish cement industry. *International Journal of Greenhouse Gas Control* **4(4)**: 583–588.

Dinakar P, Reddy MK and Sharma M (2013) Behaviour of self-compacting concrete using Portland pozzolana cement with different levels of fly ash. *Materials and Design* **46**: 609–616.

El-Dieb AS (2009) Mechanical, durability and microstructural characteristics of ultrahigh-strength self-compacting concrete incorporating steel fibers. *Materials and Design* **30(10)**: 4286–4292.

Fava C, Bergol L and Fornasier G (2003) Fracture behaviour of self-compacting concrete. In *Proceedings of 3rd International Rilem Symposium on Self-compacting Concrete* (Wallevik O and Nielsson I (eds)). Rilem Publications Sarl, Reykjavik, Iceland, pp. 636–636.

Fernandez E, Gil FJ, Ginebra MP, Driessens FCM and Planell JA (1999) Best SM production and characterisation of new calcium phosphate bone cements in the $CaHPO_4$-a-$Ca_3(PO_4)_2$ system: pH, workability and setting times. *Journal of Materials Science: Materials in Medicine* **10**: 223–230.

Flores-Velez LM and Dominguez O (2002) Characterization and properties of Portland cement composites incorporating zinc-iron oxide nanoparticles. *Journal of Materials Science* **37(5)**: 983–988.

Gartner E (2004) Industrially interesting approaches to 'low-CO2' contents. *Cement and Concrete Research* **34(9)**: 1489–1498.

Gesoglu M, Güneyisi E and Özbay E (2009) Properties of self-compacting concretes made with binary, ternary, and quaternary cementitious blends of fly ash, blast furnace slag, and silica fume. *Construction and Building Materials* **23(5)**: 1847–1854.

Grzeszczyk S and Lipowski G (1997) Effect of content and particle size distribution of high calcium fly ash on the rheological properties of cement pastes. *Cement and Concrete Research* **27(6)**: 907–916.

Jalal M, Mansouri E, Sharifipour M and Pouladkhan AR (2012) Mechanical, rheological, durability and microstructural properties of high performance self-compacting concrete containing SiO_2 micro and nanoparticles. *Materials and Design* **34**: 389–400.

Jalal M, Ramezanianpour AA and Pool MK (2013) Split tensile strength of binary blended self compacting concrete containing low volume fly ash and TiO_2 nanoparticles. *Composites Part B: Engineering* **55**: 324–337.

Khalaj G and Nazari A (2012) Modeling split tensile strength of high strength self compacting concrete incorporating randomly oriented steel fibers and SiO_2 nanoparticles. *Composites Part B: Engineering* **43(4)**: 1887–1892.

Khatib JM (2008) Performance of self-compacting concrete containing fly ash. *Construction and Building Materials* **22(9)**: 1963–1971.

Kulakowski MP, Pereira FM and Dal Molin DCC (2009) Carbonation induced reinforcement corrosion in silica fume concrete. *Construction and Building Materials* **23(3)**: 1189–1195.

Madandoust R, Ranjbar MM and Mousavi SY (2011) An investigation on the fresh properties of self-compacted lightweight concrete containing expanded polystyrene. *Construction and Building Materials* **25(9)**: 3721–3731.

Mann S (2008) *Nanotechnology and Construction*. Institute of Nanotechnology, Glasgow, UK, European Nanotechnology Gateway, Nanoforum Report.

Massazza F (1987) The role of the additions to cement in the concrete durability. *Cemento* **84**: 359–382.

Mukhopadhyay AK (2011) Next-generation nano-based concrete construction products: a review. *Nanotechnology in Civil Infrastructure* 207–223.

Nazari A and Riahi S (2010) Microstructural, thermal, physical and mechanical behavior of the self compacting concrete containing SiO_2 nanoparticles. *Materials Science and Engineering: A* **527(29–30)**: 7663–7672.

Nazari A and Riahi S (2011a) Splitting tensile strength of concrete using ground granulated blast furnace slag and SiO_2 nano-particles as binder. *Energy and Buildings* **43(4)**: 864–872.

Nazari A and Riahi S (2011b) Effects of CuO nanoparticles on compressive strength of self-compacting concrete. *Indian Academy of Sciences* **36(3)**: 371–391.

Nazari A and Riahi S (2011c) The effects of zinc dioxide nanoparticles on flexural strength of self-compacting concrete. *Composites Part B: Engineering* **42(2)**: 167–175.

Nazari A and Riahi S (2011d) TiO_2 nanoparticles effects on physical, thermal and mechanical properties of self compacting concrete with ground granulated blast furnace slag as binder. *Energy and Buildings* **43(11)**: 864–872.

Okamura H (1997) Self compacting high-performance concrete. *Concrete International* **19(7)**: 50–54.

Okamura H and Ozawa K (1994) Self-compactable high performance concrete in Japan. *Proceedings of International Workshop on High Performance Concrete*. American Concrete Institute, Farmington Hills, MI, USA, SP 169, pp. 31–44.

Ozawa K, Maekawa K and Okamura H (1996) *Self-compacting High-performance Concrete*. Collected papers, Department of Civil Engineering, University of Tokyo, Japan, vol. 34, pp. 135–149.

Persson B (2001) A comparison between mechanical properties of self-compacting concrete and the corresponding properties of normal concrete. *Cement and Concrete Research* **31(2)**: 193–198.

Quercia G, Spiesz P, Hüsken G and Brouwers HJH (2014) SCC modification by use of amorphous nano-silica. *Cement and Concrete Composites* **45**: 69–81.

Roco MC, Williams RS and Alivisatos P (1999) *Nanotechnology Research Directions: IWGN Research Report*. Committee on Technology, Interagency Working Group on Nanoscience, Engineering and Technology (IWGN), National Science and Technology Council, Washington, DC, USA.

Shekarchi M, Bonakdar A, Bakhshi M, Mirdamadi A and Mobasher B (2010) Transport properties in metakaolin blended concrete. *Construction and Building Materials* **24(11)**: 2217–2223.

Siddique R (2011) Properties of self-compacting concrete containing class F fly ash. *Materials and Design* **32(3)**: 1501–1507.

Siddique R, Aggarwal P and Aggarwal Y (2012) Influence of water/powder ratio on strength properties of self-compacting concrete containing coal fly ash and bottom ash. *Construction and Building Materials* **29**: 73–81.

USGS (United States Geological Survey) (2010) *Cement.* See http://minerals.usgs.gov/minerals/pubs/commodity/cement/mcs-2010-cemen.pdf (accessed 07/02/2015).

Valcuende M, Parra C and Marco E *et al.* (2012) Influence of limestone filler and viscosity-modifying admixture on the porous structure of self-compacting concrete. *Construction and Building Materials* **28(1)**: 122–128.

WBCSD (World Business Council for Sustainable Development) (2002) *The Cement Sustainability Initiative: Our Agenda for Action.* WBCSD, Geneva, Switzerland.

ice Publishing

Dhir and Lye
ISBN 978-0-7277-6461-4
https://doi.org/10.1680/icetscc.64614.161

Chapter 8
Influence of aggregate type on workability of self-consolidating lightweight concrete

T. Uygunoğlu
Afyon Kocatepe University, Technical Education Faculty, Construction Department, 03200 Afyonkarahisar, Turkey

i. B. Topçu
Eskişehir Osmangazi University, Civil Engineering Department, Eskişehir, Turkey

In this study, the effects of aggregate type on the workability of self-consolidating concrete produced with lightweight aggregate (self-consolidating lightweight concrete – SCLC) were investigated. In experiments, three coarse lightweight aggregate (LWA) types, namely pumice, volcanic tuff and diatomite, were used. Different combinations of water/powder ratio and superplasticiser dosage levels were prepared for the self-consolidating lightweight concrete mixtures. The total powder content (cement and mineral additives) was constant in the experiments. Workability tests such as slump flow, restricted slump flow, filling-box, V-funnel, GTM screen stability, L-box and U-box were performed to characterise accurately the flow ability, passing ability and segregation resistance of self-consolidating lightweight concrete in fresh condition. The results, in general, showed that self-consolidating lightweight concrete with lightweight aggregate in low unit weight has higher flow ability and passing ability. The segregation of self-consolidating lightweight concrete was prevented by use of air-entraining admixture in high water/powder ratios. As a result, it was observed that the aggregate type and water/powder ratio has significant influence on the workability properties of self-consolidating lightweight concrete.

Introduction

Self-consolidating concrete (SCC) is used for specific applications requiring high deformability, high flow ability and high passing ability (Okamura, 1997; Topçu *et al.*, 2009). Numerous studies have been performed for new testing methods on SCC in the past decade. One of them was the use of lightweight aggregates in the production of SCC (Choi *et al.*, 2006; Hwang and Hung, 2005; Lo *et al.*, 2007; Topçu *et al.*, 2007). Lightweight aggregates such as natural or artificial are available in many parts of the world and can be used in producing concrete in a wide range of unit weights and suitable strength values for different fields of application such as internal and external walls, inner leaves of external cavity walls, fill panels and isolation of roof decks and floors (Topçu, 1997; Ünal *et al.*, 2007).

In practice, lightweight aggregates often have lower particle densities than the density of the mortar matrix and natural aggregates in concrete. The coarse aggregates move upwards with vibration of plain concrete, and thus segregation may take place during the transporting and placing of the fresh concrete (Chia *et al.*, 2005). This is in contrast with the normal-weight aggregate concrete where coarse aggregates may sink to the bottom if the workability is not appropriate. Self-compactability of concrete is a unique property for prevention of upwards movement of lightweight aggregates (LWA) owing to vibration (Topçu and Uygunoğlu, 2008; Uygunoğlu, 2008; Wu *et al.*, 2009). However, when LWAs are used in SCC, aggregate particles are required to have uniform distribution and the minimum segregation risk in self-consolidating lightweight concrete (SCLC) owing to the low density of LWAs (Topçu and Uygunoğlu, 2010; Türkmen and Kantarcı, 2007; Wu *et al.*, 2009).

Aggregates generally constitute about 60% by volume of SCLC. Owing to the large volume fraction they occupy in SCLC, aggregates exert a major influence on the workability of fresh SCLC, and can be expected to have an important influence on other properties as well (Su *et al.*, 2001). There are few studies on workability of SCLC in the literature. This paper presents the extensive test results from the workability of SCLC produced with the different types of LWA. In the study, three types of aggregate – pumice, volcanic tuff and diatomite – were used as LWAs as they are found in large deposits in Afyonkarahisar (Turkey) and the surrounding area.

Experimental programme

Materials used

The cement used was ordinary Portland cement (CEM I 42.5R), which complies with the requirement of European Standard EN 197-1 (Turkish Standard Institute, 2008). Limestone powder and fly ash were used as fine material (or filler) to balance the viscosity. The fly ash (F class) was supplied from Seyitömer/Turkey Thermal Power. Table 1 gives the chemical, physical and mechanical properties of the fine materials that composed the cement, limestone powder and fly ash with a maximum size of 0·125 mm. The admixture was polycarboxylate-based superplasticiser (SP). Its specific gravity and solid content were 1·1 and 20% respectively. Natural river sand was used in all the mixtures. The specific gravity, fineness modulus and water absorption by weight of sand are 2·59, 2·07 and 3·73% respectively. Coarse aggregates were pumice, volcanic tuff and diatomite with a maximum size of 16 mm. The chemical and characteristic properties of these aggregates are presented in Table 1 and Table 2, respectively.

Concrete mixes and tests

The constituents of the SCLC mix were proportioned to achieve maximum packing of the particles and, thus, minimum voids. SCLC mixtures were designed with different water/powder (w/p) ratios, 0·28, 0·31, 0·34 and 0·36, to obtain different fresh-state properties. The powder ratio in the natural sand was also considered in determining the w/p ratio. When the water content increased, the superplasticiser dosage was decreased by ratios of 2·6, 2·3, 2·0 and 1·7% by weight of fine material at the same time. The fine material content, which includes cement, fly ash and limestone powder, and aggregate grading, were kept constant in all mixtures. The aggregate grade was designed as between Turkish Standard curves A-16 and B-16 (Table 3). During the production of SCLCs,

Table 1 Properties of cement fly ash, limestone powder, pumice, tuff and diatomite

	Cement: %	Fly ash: %	Limestone: %	Pumice: %	Tuff: %	Diatomite: %
Chemical properties						
Calcium oxide (CaO)	63·6	0·67	55·0	5·4	0·5	1·4
Silicon dioxide (SiO_2)	19·3	46·5	0·01	56·9	73·0	67·2
Aluminium oxide (Al_2O_3)	5·6	25·5	0·2	16·7	14·1	10·1
Iron (III) oxide (Fe_2O_3)	3·5	4·9	1·0	4·7	1·0	2·7
$SiO_2 + Al_2O_3 + Fe_2O_3$ ($\geq$ 70)	28·3	76·9	0·2	78·2	88·2	80·0
Magnesium oxide (MgO)	0·9	5·9	0·6	1·8	1·0	0·6
Sulfur trioxide (SO_3)	3·0	1·7	–	–	–	–
Potassium oxide (K_2O)	0·8	1·4	–	5·2	3·7	0·7
Sodium oxide (Na_2O)	0·1	0·4	–	4·6	1·2	0·4
Loss of ignition	2·0	2·8	43·7	2·8	6·3	8·0
Insoluble residue	0·7	–	–	–	–	–
Physical properties						
Fineness by Blaine: cm^2/g	3294	6445	856	–	–	–
Specific gravity	3·07	2·13	2·72	1·84	2·04	2·23
Mechanical properties						
Compressive strength: 7 days	37·5	–	–	–	–	–
MPa: 28 days	46·0	–	–	–	–	–

Table 2 Characteristic properties of pumice, tuff and diatomite

Properties	4/16 mm coarse lightweight aggregates			Standard
	Pumice	Tuff	Diatomite	
Loose bulk density: kg/m^3	739	762	435	TS EN 1097-3
Water absorption: %	28·75	23·82	94·24	TS EN 1097-6
Partical shape (flakiness ındex)	FI_{15}	FI_{35}	FI_{15}	TS 9582 EN 933-3

Table 3 Sieve analyses of lightweight aggregates

Sieve: mm	16	8	4	2	1	0·5	0·25
Passing: %	100	59	46	39	27	17	9

Table 4 Constituent of concretes for per cubic meter

Series	Materials: kg/m^3										w/p ratio
	Cement	Fly ash	Limestone powder	Sand	Pumice	Tuff	Diatomite	Water	SP*	AE*	
SCLC 1	450	50	100	722	492	–	–	180	15·6	3·0	0·28
SCLC 2	450	50	100	700	478	–	–	198	13·8	3·0	0·31
SCLC 3	450	50	100	678	462	–	–	216	11·4	3·0	0·34
SCLC 4	450	50	100	664	452	–	–	228	10·2	3·0	0·36
SCLC 5	450	50	100	649	443	–	–	240	9·0	3·0	0·38
SCLC 6	450	50	100	722	–	534	–	180	15·6	3·0	0·28
SCLC 7	450	50	100	700	–	518	–	198	13·8	3·0	0·31
SCLC 8	450	50	100	678	–	502	–	216	11·4	3·0	0·34
SCLC 9	450	50	100	664	–	491	–	228	10·2	3·0	0·36
SCLC 10	450	50	100	649	–	480	–	240	9·0	3·0	0·38
SCLC 11	450	50	100	722	–	–	436	180	15·6	3·0	0·28
SCLC 12	450	50	100	700	–	–	423	198	13·8	3·0	0·31
SCLC 13	450	50	100	678	–	–	410	216	11·4	3·0	0·34
SCLC 14	450	50	100	664	–	–	401	228	10·2	3·0	0·36
SCLC 15	450	50	100	649	–	–	392	240	9·0	3·0	0·38

* SP: superplasticiser; AE: air-entraining admixture.

all the aggregates were used as saturated and dry surface. An air-entrained admixture was used in the production of SCLC to prevent segregation. LWAs were used in the mixes as saturated and dry surface. The composition of the mixes is presented in Table 4.

Experiments

On the fresh SCLC produced with three different aggregates, experiments, presented in Table 5, were carried out to determine the workability of the SCLC according to the EFNARC Committee (EFNARC, 2002). All tests were performed at 20°C. The slump flow test is used to determine the ability of the concrete to flow in a non-restricted condition. To determine the slump flow the traditional hollow Abrams cone is placed on a slump-flow plate with an edge length of 1000 × 1000 mm and is filled with SCC and lifted slowly. In order to prevent any thixotropic effect, the Abrams cone was lifted immediately after having been filled with the SCC. The slump-flow of the final deformed, or slumped, SCC was measured from two perpendicular diameters 2 min after cone lifting (Figure 1). During the slump-flow test, the time required for SCC to reach a slump-flow radius (T_{500}) of 500 mm length was measured.

Table 5 Experiments for evaluating the workability of SCLC

SCLC property	Experiment	Limit*
Filling ability	Slump-flow	650–800 mm
	Time to T_{500} at slump-flow	2–5 s
	V-box flow time	6–12 s
Passing ability	J-ring	0–10 mm
	U-box	0–30 mm
	Filling box	90–100%
	L-box	(h_2/h_1) 0·8–1·0
Segregation resistance	V-box flow at T_{5min}	+3 s
	Sieve segregation (GTM screen stability)	0–15%

* According to EFNARC (2002).

Figure 1 Slump-flow tests

The flow time of SCC was measured with a V-funnel (Figure 2). With the outlet flap closed the funnel is filled to the top with the sample of SCC. The flap is then opened. The time taken by the concrete to flow out of the funnel is measured. Filling ability of SCLC was evaluated with the V-funnel flow time. Then, the passing ability of SCLCs was defined with a V-funnel, again, as $V_{T5minutes}$ value. With the outlet flap closed the funnel is filled with the sample of SCC to the top. SCLC was kept in the V-funnel for 5 min. Then, the flap was opened, and, the flow time was measured as the $V_{T5minutes}$ value.

A J-ring test was used to asses the passing ability of the concrete in restricted condition. To determine the slump flow in restricted condition the traditional hollow Abrams cone was

Figure 2 V-funnel test apparatus

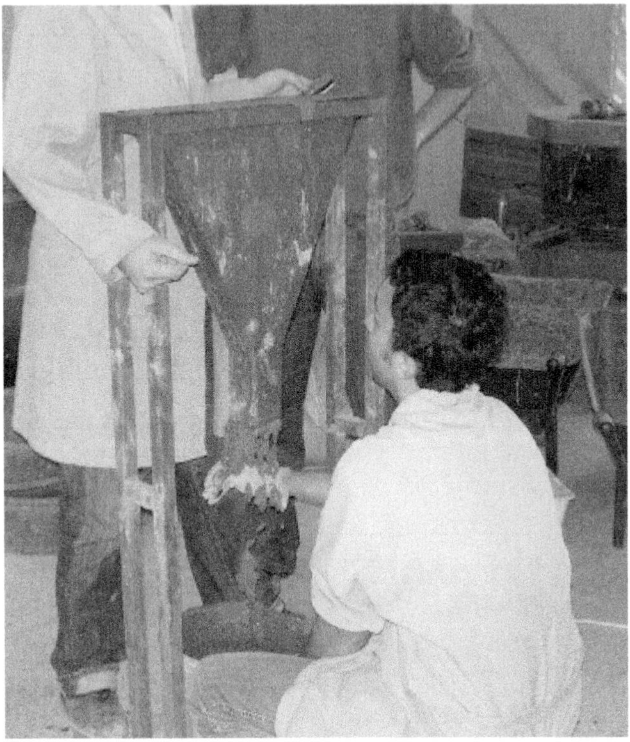

placed centrally inside a J-ring on the slump flow plate with an edge length of 1000 × 1000 mm, filled with SCC and lifted slowly. The diameter of the ring of vertical bars was 300 mm, and the height 100 mm. Similar to the slump-flow test, T_{500} time was measured during the slump-flow of SCLC in restricted condition. The slump-flow of the final deformed SCLC was measured from two perpendicular diameters 2 min after cone lifting. Furthermore, the difference in height between the concrete just inside the bars and that just outside the bars was measured (Figure 3(a)). Four measurements were taken on the outside of the J-ring as k_1, k_2, k_3, and k_4 around the ring at a perpendicular angle (Figure 3(b)). The arithmetic average of k values (k_{av}) was used for the outside height ($h_{outside}$) in the J-ring. Then, step-height (st_j) was determined with the difference of outside ($h_{outside}$) and inside (h_{inside}) concrete height in the J-ring by Equation 1 (Figure 3(c))

$$st_j = (h_{outside} - h_{inside}) = \left[\left(\frac{k_1 + k_2 + k_3 + k_4}{4}\right) - h_{inside}\right] \quad (1)$$

Figure 3 Measuring of (a) h_{inside} value and (b) $h_{outside}$ value; (c) determination of step-height ratio in J-ring

Fill-box test apparatus was used to measure the passing ability of SCLC with a different type of LWA. There are 35 steel obstacles with a diameter of 16 mm and a distance centre to centre of 50 mm. SCLC was poured into the filling pipe at the top side (Figure 4(a)). The apparatus was filled with about 40 litres SCLC through the filling pipe, and the difference in height between two sides of the container was measured to determine the passing ratio according to EFNARC (Figure 4(b)).

To assess the passing ability of SCLC, the L-box test method was used. The height of the fresh SCC after passing through the specified gaps of steel bars and flowing within

Figure 4 A view of fill-box

(a)

(b)

a defined flow distance in the L-box (Figure 5) was measured. Once the concrete flow stopped, the height of concrete in the chimney was recorded as h_1, and the height of the concrete at the end of the horizontal section was recorded as h_2. With this reached height, the blocking behaviour of SCLC was determined by h_2/h_1. During the L-box test, the time required for SCLC to reach 200 (T_{200}), 400 (T_{400}) and 600 (T_{600}) mm length flow time was measured.

The U-box test was used to measure the filling ability of SCLC. The U-box consists of a vessel that is divided by a middle wall into two compartments. An opening with a sliding gate was fitted between the two sections. Reinforcing bars were installed at the gate with centre-to-centre spacing of 50 mm. One section is filled with about 20 litres of concrete then the gate lifted and concrete flows upwards into the other section. The height of the concrete in both sections is measured at the end of the test.

Figure 5 L-box test apparatus

Figure 6 Stability test

In order to check the screen stability self-compactability, segregation resistance tests were performed with a sieve that conformed with the standard procedures given by EFNARC (2002). A sieve with square holes of 5 mm, frame diameter 350 mm and height 75 mm was used through stability experiments (Figure 6). Segregation resistance was determined by Equation 2

$$S = (W_h/W_b) \times 100 \quad (2)$$

where S is segregation resistance; W_h is concrete passing through the sieve; and W_b, is the weight of concrete poured onto the sieve.

Results and discussions

This study focused on those fresh properties of SCLC with different types of lightweight aggregates. However, it should be mentioned that the mechanical properties of hardened SCLC decreased as the w/p ratio increased for each series regardless of aggregate type. When the compressive strength of SCLC with pumice varied between 18 and 25 MPa, the compressive strength varied between 16 and 22 MPa and between 12 and 20 MPa for the SCLC series with tuff and diatomite, respectively, at age 28 days depending on the w/p ratio. The hardened concrete properties of the SCLC series with pumice, tuff and diatomite can be found in the study by Topçu and Uygunoğlu (2010).

Fresh unit weight

The fresh unit weights of SCLCs were measured after the workability tests. It can be seen from Table 6 that the fresh unit weight of the concretes decreased as the w/p ratios increased. While the unit weights of SCLC with pumice were changed between 1750 and 1670 kg/m^3, the unit weight of SCLCs of those made up with tuff ranged between 1810 and 1720 kg/m^3 for 0·28–0·38 w/p ratio, respectively. On the other hand, the unit weights of SCLC with diatomite varied between 1750 and 1640 kg/m^3 for the same w/p ratios. This is attributable to the increase of porous structure in SCLC with high water content. Furthermore, Table 6 shows that the unit weights of SCLCs decreased with increasing density of lightweight aggregates for mixtures with the same w/p ratio. The fresh unit weight reached the maximum value when tuff was used in the production of SCLC. The bulk density of tuff is higher than that of diatomite and pumice; therefore the unit weight of the mix increased when the tuff took the place of pumice and diatomite lightweight aggregates. The density is much lower when compared with the density of normal SCC, namely 2400 kg/m^3.

Filling ability of SCLC

The slump-flow test is a value system for the ability of SCLC to deform under its own weight against the friction of the surface with no external restraint present. The slump flow results in Table 6 indicate that the diameter of SCLC slump flow was 525–840 mm. It is clearly seen that the slump flow diameter increased as the w/p ratio increased for each series regardless of

Table 6 Fresh properties of SCLC

Series	Unit weight: kg/m^3	Slump-flow: mm	V-funnel flow time: s	V_{T5}-flow time: s	V_{T5}-V: s	U-box height difference: mm
SCLC 1	1747	618	6·07	6·31	0·24	22·0
SCLC 2	1701	683	5·85	6·56	0·71	16·0
SCLC 3	1695	723	3·78	4·80	1·02	15·0
SCLC 4	1685	765	2·93	3·66	0·73	9·0
SCLC 5	1657	767	3·19	3·72	0·53	4·0
SCLC 6	1814	529	5·78	6·32	0·54	48·0
SCLC 7	1811	578	4·67	6·06	1·39	48·0
SCLC 8	1754	670	3·48	3·87	0·39	16·0
SCLC 9	1727	728	2·72	3·82	1·10	12·0
SCLC 10	1723	755	2·42	3·68	1·26	4·0
SCLC 11	1737	655	4·19	4·66	0·47	13·0
SCLC 12	1716	770	4·14	4·59	0·45	10·0
SCLC 13	1673	778	2·92	3·35	0·43	5·0
SCLC 14	1655	817	1·87	2·33	0·46	5·0
SCLC 15	1640	834	1·32	1·75	0·43	2·0

aggregate type. This was attributed to the increase in lubrication between the fine particles with an increase of w/p ratio (Chia and Zhang, 2004). The slump-flow value of tuff, pumice and diatomite was 525, 620 and 650 mm, respectively, with a 0·28 w/p ratio. When the w/p ratio was 0·38, the slump-flow value reached 750, 760 and 840 mm for SCLC with tuff, pumice and diatomite respectively. This is because the higher water content in mixes increased the lubrication between the aggregate particles and thus decreased the yield stress and eased the flow ability of concrete (Roussel *et al.*, 2005). The slump-flow reached the maximum value when diatomite was used in the production of SCLC regardless of the w/p ratio. This is due to the lighter weight structure of diatomite than pumice and tuff LWA in SCLC. When the slump-flow values given by EFNARC were considered (650–800 mm), a good filling ability was obtained on SCLCs in 0·28–0·35 w/p ratio, 0·30–0·38 w/p ratio, and 0·34–0·38 w/p ratio for diatomite, pumice and tuff respectively.

The change of the V-funnel flow time of SCLCs with different types of LWA is shown in Table 6 depending on the w/p ratio. The V-funnel flow time is an indicator of viscosity. In other words, viscosity increases with increase of V-funnel flow time. However, it must be borne in mind that V-funnel flow time is often used to estimate the apparent viscosity of a mixture. Many factors influence the results of the V-funnel test: the amount, shape and size distribution of aggregates and also the viscosity and amount of paste and so on. This means that the V-funnel flow time does not necessarily correspond with the viscosity of a mix measured, for example, by a rheometer. The results indicate that the lowest V-funnel flow time was obtained on SCLC with diatomite regardless of the w/p ratio. On the other hand, the highest viscosity was observed on SCLC with pumice in all the w/p ratios. Even though air-entrained admixture was used, the viscosity of SCLCs with diatomite, pumice and tuff in all the w/p ratios was too low according to the limit values given by EFNARC (6–12 s); the exception was SCLC with pumice produced in w/p ratio of 0·28.

Passing ability of SCLC

The variations of slump-flow and T_{500} in free and restricted conditions are compared in Figures 7(a)–(c) for SCLC with pumice, tuff and diatomite respectively. As seen, the T_{500} time value was higher in the J-ring slump flow test than free slump-flow owing to obstacles. It can be noted from Figure 7 that the greatest T_{500} time difference between free and J-ring slump-flow was obtained in the lowest w/p ratio, regardless of aggregate type. The passing ability of SCLC with all the aggregate types was too high in the w/p ratio of 0·38. However, the T_{500} time difference with and without restriction increased as the w/p ratio decreased for each series regardless of aggregate type. This was because the SCLC in the low w/p ratio was blocked between the obstacles because of low viscosity. When aggregate type is considered, it can be clearly seen that the greatest T_{500} time difference was obtained in SCLC with tuff aggregates due to the shape and bulk density of tuff. Even though diatomite has a lower bulk density than other LWAs, the lowest T_{500} time differences were observed in SCLC with pumice in each w/p ratio because pumice has a relatively spherical shape when compared with diatomite and tuff. The shape of crushed aggregates is often more angular with a rougher surface than natural aggregates. Plain and self-consolidating concrete with crushed aggregate often requires higher cement and water content in order to reach the same workability as a concrete with spherical aggregates. When the T_{500} time difference values of SCLC with pumice have changed between 0·7 and 0·1 s, it was ranged between 5 and 0·7 s, and 1·4–0·1 s in SCLC with tuff and diatomite, respectively, depending on the w/p ratio.

Figure 7 Comparison of free and restricted slump flow of SCLC with (a) pumice, (b) tuff and (c) diatomite

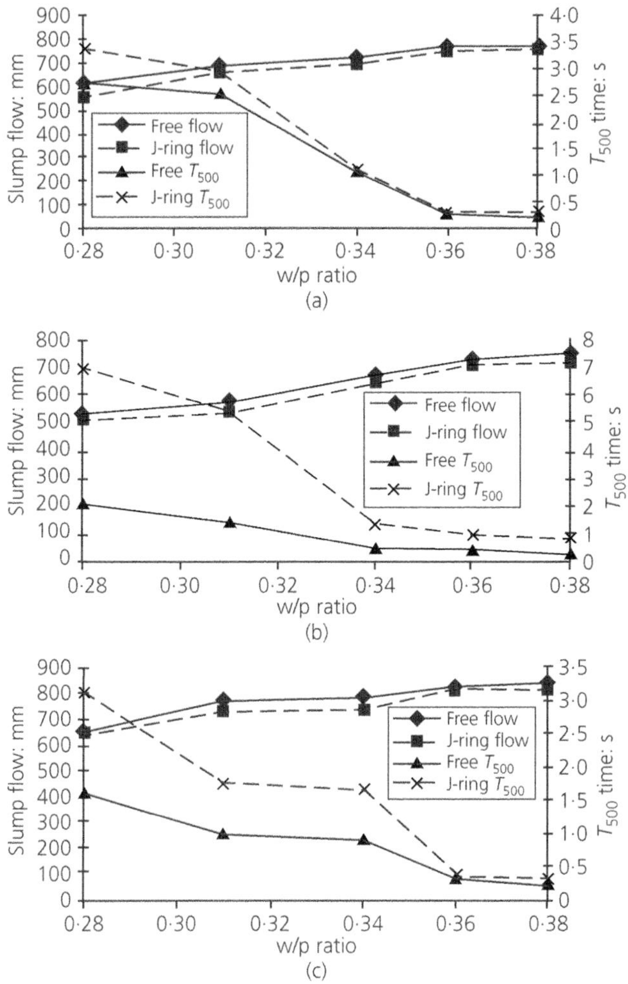

An SCLC mixture with 10 mm slump flow difference between free and restricted condition can be desirable in a good SCLC mixture from the passing ability point of view (Chia and Zhang, 2004). In general, the slump flow difference in the free and J-ring decreased as the w/p ratio increased from 0·28 to 0·38. However, this difference was restricted. It ranged between 63–13 mm, 38–19 mm and 47–9 mm in SCLC with pumice, tuff and diatomite respectively, depending on the w/p ratio.

Since the flow time to T_{500} in free slump-flow and V-funnel flow time is an indicator of viscosity, the relationship between these parameters was examined by statistical methods (Figure 8), and a good correlation coefficient was found ($R = 0{\cdot}927$). The flow time to T_{500} increases by increase of the V-funnel flow time.

Figure 8 Relationship between V-funnel flow time and flow time to T_{500} in slump flow

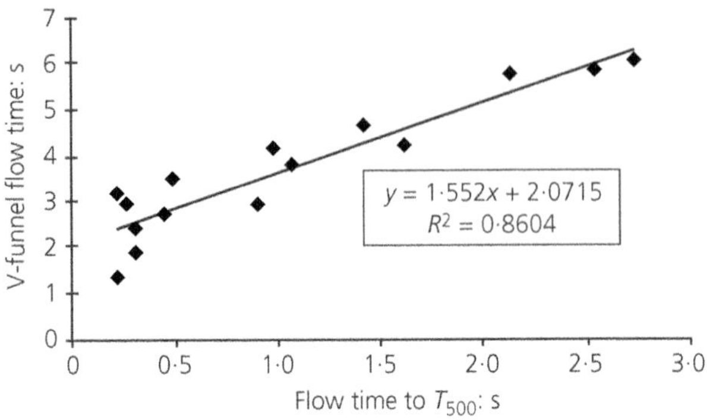

The influence of aggregate type on step-height in the J-ring of SCLCs depending on w/p ratio is shown in Figure 9. There is a substantial decrease in step-height with the increase in w/p ratio, and this decrease is substantial between 0·34 and 0·38 w/p ratios. It is evident that for a given aggregate type, a lower w/p ratio leads to a lower passing ability of SCLC, owing to the low lubrication between the aggregate particles. The high step-height value is an indication of the blocking risk of SCLC. When aggregate type is considered, it can clearly be seen from Figure 9 that the step-height (st_j) of SCLC with diatomite was lower than that of SCLC with pumice and tuff for all w/p ratios. The st_j value of SCLC with diatomite with a w/p ratio of 0·28, 0·31, 0·34, 0·36 and 0·38 was 0·9, 0·6, 0·3, 0·3 and 0 respectively. The highest height value was observed in SCLC with tuff. It was ranged from 3 to 0·3 depending on the w/p ratio. The SCLC mortar moved the diatomite aggregates as its bulk density is lower than that of tuff, and thus SCLC was not blocked. The blocking risk of SCLC with tuff is three times greater than that of SCLC produced with tuff in the lowest w/p ratio. Another blocking factor – the blocking ratio – was measured in the L-box; test results are presented in Figure 10. Similar to st_j, the highest blocking risk occurred in the lowest w/p ratio for SCLC with all types of

Figure 9 Step-height of SCLC in J-ring

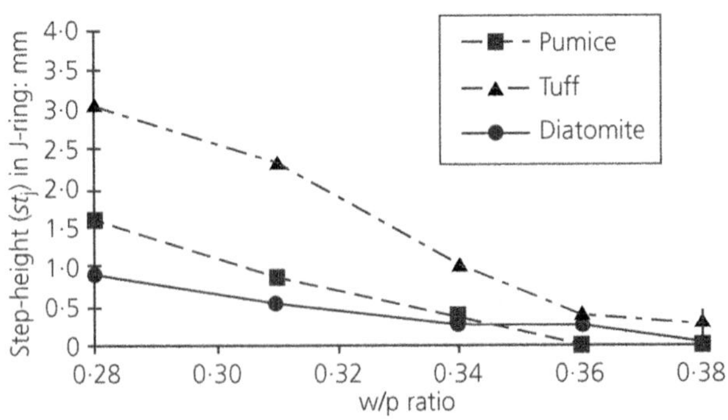

LWA. For a good passing ability of SCLC, the L-box blocking ratio must be between 0·8 and 1 according to EFNARC (2002). As seen from Figure 10, the most appropriate w/p ratio for passing ability of SCLC with pumice, tuff and diatomite is 0·31, 0·36 and 0·28 respectively, in the L-box test.

The effect of a different LWA and w/p ratio on the time required for SCLC to reach T_{200}, T_{400} and T_{600} length flow time during the L-box test is shown in Figures 11(a) to (c) for pumice, tuff and diatomite respectively. Actually, the flow time of SCLC can be associated with viscosity. It is also an indicator of passing ability. There is a substantial increase in the L-box flow time of SCLC produced with all types of LWA with the increase in L-box length from 200 to 600 mm, regardless of the w/p ratio. While the flow time of SCLC with pumice for 0·28 w/p ratio ranged between 1·13 and 7·19 s, the unit weight of SCLC comprising tuff and diatomite for 0·28 w/p ratio was in the range 1·26–10·29 and 0·6–3·89 s respectively. For a w/p ratio of 0·38, the L-box flow time of SCLC with pumice, tuff and diatomite was in the range 0·26–1·08, 0·44–1·64 and 0·24–1·33 respectively. This is probably due to the increase of viscosity in SCLC with increase of flow length, and thus the flow of SCLC stops in a shorter distance, and the passing ability of SCLC mortar decreases. This phenomenon shows that aggregates with low unit weight (or low bulk density) can be moved easily by SCLC mortar. It can be clearly seen that the lowest L-box flow time for lengths of 200, 400 and 600 mm were obtained on SCLC with diatomite which has a lower unit weight than pumice and tuff. On the other hand, the highest flow time in the L-box was noted on SCLC with tuff which has a higher bulk density than pumice and diatomite.

When the effect of the w/p ratio was considered on L-box flow time with length, it could also clearly seen that there was a substantial decrease in T_{200}, T_{400} and T_{600}, respectively, with the increase in w/p ratio from 0·28 to 0·38. It is evident that for a given SCLC type, a lower w/p leads to a higher flow time, owing to the lack of lubrication between the aggregates. The short flow time is an indication of good passing ability achieved by the concrete self-weight.

Figure 10 Blocking ratio of SCLC in L-box

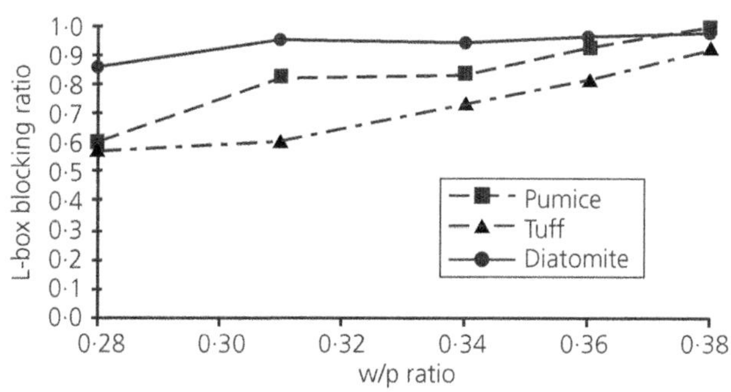

Figure 11 T_{200}, T_{400} and T_{600} flow time of SCLC with (a) pumice, (b) tuff and (c) diatomite

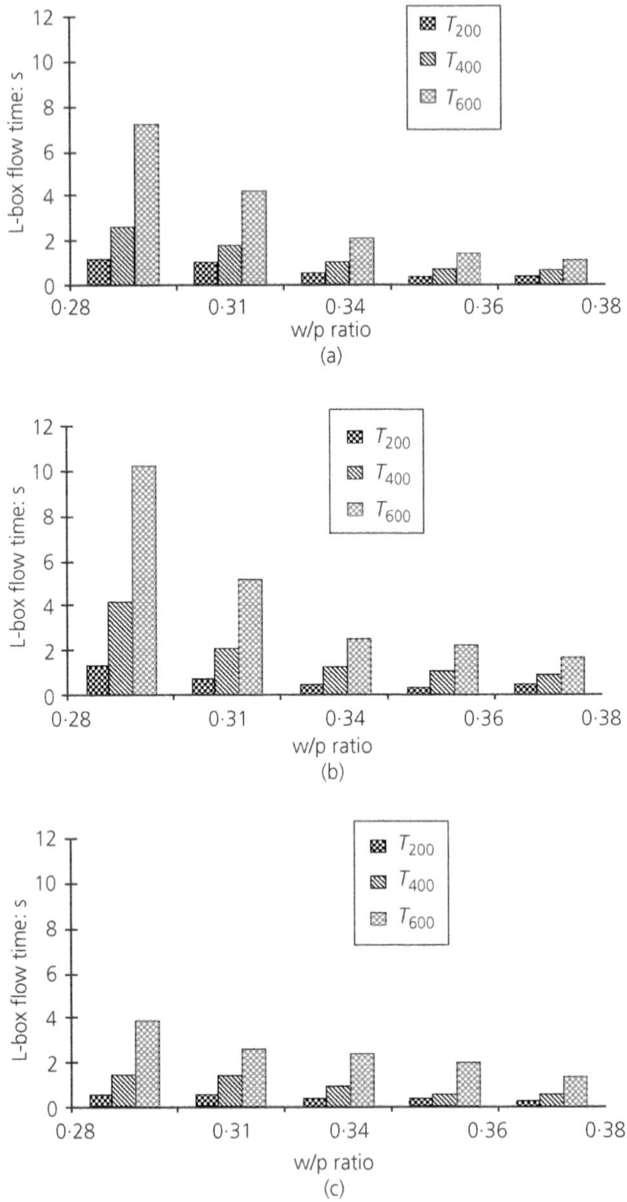

The relationship between L-box flow time (T_{600}) and J-ring flow time to T_{500} is given in Figure 12. It can be noted that there is a good relationship between both parameters with a high correlation coefficient ($R = 0{\cdot}928$). Moreover, the L-box flow time (T_{600}) and J-ring flow time are directly proportional to T_{500}. In other words, the blocking risk of SCLC decreased with the increase of flow time in both the L-box and J-ring.

Figure 12 Relationship between L-box flow time (T_{600}) and J-ring flow time to T_{500}

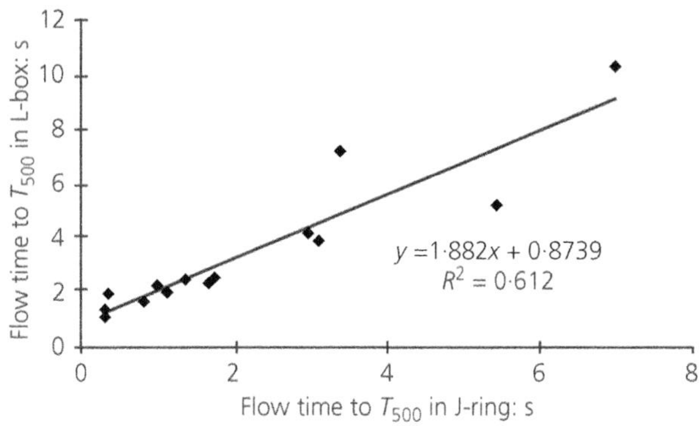

The passing ratio of SCLC with different LWAs was determined with a filling box, and results, depending on the w/p ratio, are given in Figure 13. The passing ratio of SCLC produced with tuff and pumice is 70%, whereas it was 90% on SCLC with diatomite. When the w/p ratio was increased to 0·38, the passing ratio was 100% for all types of SCLC. For good passing ability, the target range is in the range 90–100% in the filling box. SCLC with diatomite has a high passing ability in all of the w/p ratios. However, SCLC produced with pumice and tuff LWA has a good passing ability after w/p ratios of 0·31 and 0·34 respectively. Another passing ability indicator is the height difference in the U-box. The highest height difference was observed in SCLC with tuff in the lowest w/p ratio as 48 mm (Table 6). On the other hand, the height difference was 12 and 22 mm for SCLC with diatomite and pumice, respectively, for the same w/p ratio. An acceptable range for height difference in the U-box test is between 0 and 30 mm. While all the w/p ratios are appropriate for SCLC with pumice and diatomite for good passing ability, it was over 0·33 for SCLC with tuff. The results show that the importance of the aggregate characteristics, such as the shape, surface texture and bulk density, becomes less important with increasing w/p ratio.

Figure 13 Passing ratio of SCLC in filling-box

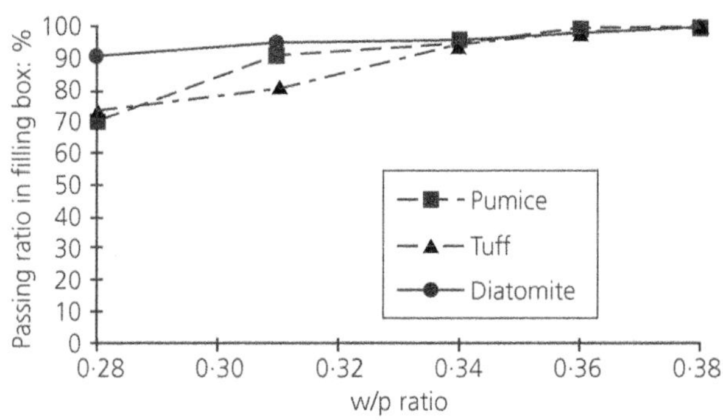

Segregation resistance of SCLC

Another important factor for workability of SCLC is segregation resistance. The segregation resistance of SCLC was evaluated using the wet sieve test for different w/p ratios (from 0·28 to 0·38); the test results are presented in Figure 14. The segregation resistance of the SCLC incorporating the pumice, tuff and diatomite with a w/p ratio of 0·28 was about 0%. The target range was between 0 and 15% for the segregation ratio. The segregation resistance of SCLC containing all the types of LWA was under 15% in the case of a w/p ratio up to 0·34. For a given w/p ratio, it can be noted that the segregation ratio values were close to each other for all the types of SCLC. SCLC with pumice, tuff and diatomite has a segregation ratio of 41, 38 and 33% respectively, in the case of 0·36 w/p ratios. However, for a 0·38 w/p ratio, the segregation ratio of SCLCs was 25, 18 and 20%, respectively, for pumice, tuff and diatomite. It was interesting that SCLC incorporating the diatomite aggregate has a lower segregation value than pumice and tuff in the highest w/p ratio. This can be explained by the increase of the air-entraining effect in high w/p ratios. Air bubbles restricted the exit of LWAs, specifically lower unit weight aggregates, from the SCLC mortars, thus segregation in highly flowable mixtures was prevented.

The V-funnel in the T_5 minutes test is used to determine the segregation resistance of the SCLC. The increase in V-funnel flow time after 5 min is an indicator of the presence of segregation in SCLC. Table 6 shows the difference between V-funnel flow time and V_{T5} flow time. This difference must not exceed 3 s for an acceptable segregation resistance. All the SCLC types had a desirable segregation resistance according to Table 6. The highest difference was obtained on SCLC with tuff as 1·25 s in the highest w/p ratio. The V-funnel time difference was not changed in SCLC with diatomite. On the other hand, in SCLC with pumice, the V-funnel time difference was increased with the increase of w/p ratio up to 0·34. It was then reduced because of air-entraining admixture effect. It must be borne in mind that the air-entraining admixture has a more significant effect on lighter aggregates than weight aggregates for the prevention of segregation. It can be clearly seen that the V-funnel flow time difference in SCLC incorporating the tuff was higher than other SCLC types owing to weightier bulk density of tuff than pumice and diatomite.

In Table 7, statistical analyses of workability test result were presented. For each test, another test was plotted, and correlation and equations for the relationship between the test results were

Figure 14 Segregation ratio of SCLC with different aggregates

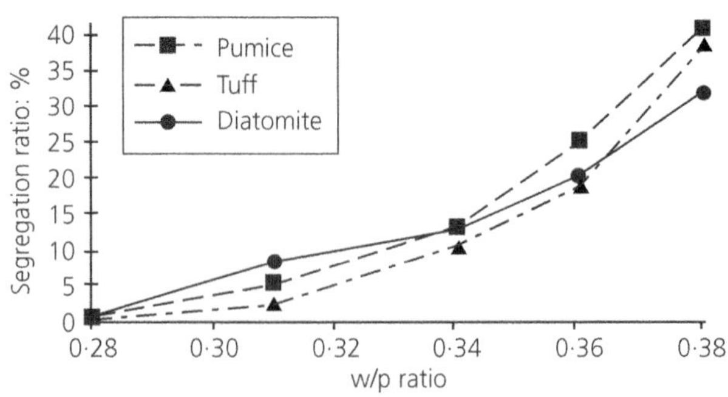

Table 7 Statistical analyses of SCLC workability results

X axes *Y* axes	Slump-flow: mm	V-funnel flow time: s	Segregation ratio: %	L-box (h_2/h_1)	Filling-box: %
Filling-box: %	$y = 0{\cdot}095x + 24{\cdot}42$	$y = -5{\cdot}56x + 112{\cdot}7$	$y = 6{\cdot}12 \ln(x) + 79{\cdot}34$	$y = 60{\cdot}06x + 41{\cdot}8$	–
Correlation, *R*	0·86	0·82	0·90	0·89	–
U-box: mm	$y = -0{\cdot}15x + 121{\cdot}6$	$y = 6{\cdot}9x - 10{\cdot}16$	$y = 24{\cdot}33e^{-0{\cdot}054x}$	$y = -69{\cdot}6 \ln(x) + 1{\cdot}94$	$y = -1{\cdot}2x + 126$
Correlation, *R*	0·91	0·67	0·81	0·90	0·81
L-box (h_2/h_1)	$y = 0{\cdot}0015x - 0{\cdot}25$	$y = -0{\cdot}074x + 1{\cdot}11$	$y = 0{\cdot}67x^{0{\cdot}103}$	–	–
Correlation, *R*	0·93	0·74	0·79	–	–
J-ring step-height: mm	$y = -0{\cdot}0096x + 7{\cdot}62$	$y = 0{\cdot}46x - 0{\cdot}89$	$y = -5{\cdot}3 \ln(x) + 1{\cdot}9$	$y = 12{\cdot}36x^2 - 25{\cdot}12x - 12{\cdot}93$	$y = -0{\cdot}0026x^2 + 0{\cdot}366x - 10{\cdot}5$
Correlation, *R*	0·93	0·73	0·84	0·93	0·91
V-funnel flow time: s	$y = 0{\cdot}014x + 13{\cdot}3$	–	$y = -0{\cdot}83 \ln(x) + 5{\cdot}43$	–	–
Correlation, *R*	0·82	–	0·83	–	–

obtained. Each column shows the '*x* axes' values; however, each line shows the '*y* axes' values for equations that are given in Table 7. The lowest relation was obtained between the V-funnel flow time and U-box height difference with a correlation coefficient (R) of 0·67.

Conclusions

In this paper, results from a study of the influence of aggregate type on the workability of self-consolidating lightweight aggregate concrete have been presented. The experimental results showed that the workability of SCLC strongly depends on characteristic properties of lightweight aggregate such as unit weight (or bulk density), shape and surface texture. However, the importance of the aggregate characteristics becomes less important as the w/p ratio increases. The highest passing ability and filling ability were obtained in SCLC with diatomite aggregates owing to the shape and low bulk density of diatomite in low w/p ratios. On the other hand, segregation of SCLC incorporating diatomite aggregates was prevented by using air entraining admixtures in the high w/p ratios. A high ratio of w/p also contributes to the increase of passing ability and filling ability by increasing the lubrication between aggregate particles and decreasing the viscosity. However, an increase in the w/p ratio resulted in a decrease in segregation resistance. In general, it can be deduced that the optimum w/p ratio is 0·28, 0·31 and 0·34 for good workability of SCLC containing the diatomite, pumice and tuff lightweight aggregates respectively.

REFERENCES

Chia KS and Zhang MH (2004) Effect of chemical admixtures on rheological parameters and stability of fresh lightweight aggregate concrete. *Magazine of Concrete Research* **56(8)**: 465–473.

Chia KS, Kho CC and Zhang MH (2005) Stability of fresh lightweight aggregate concrete under vibration. *ACI Materials Journal* **102(5)**: 347–354.

Choi YW, Kim YJ, Shin HC and Moon HY (2006) An experimental research on the fluidity and mechanical properties of high-strength lightweight self-compacting concrete. *Cement and Concrete Research* **36(9)**: 1595–1602.

EFNARC (2002) *Specification and Guidelines for Self-compacting Concrete*. Association House, Farnham, UK, pp. 1–40.

Hwang CL and Hung MF (2005) Durability design and performance of self-consolidating lightweight concrete. *Construction and Building Materials* **19(8)**: 619–626.

Lo TY, Tang PWC, Cui HZ and Nadeem A (2007) Comparison of workability and mechanical properties of self-compacting lightweight concrete and normal self-compacting concrete. *Materials Research Innovation* **11(1)**: 45–50.

Okamura H (1997) Self-compacting high performance concrete. *Concrete International* **19(7)**: 50–54.

Roussel N, Stefani C and Leroy R (2005) From mini-cone test to Abrams cone test: measurement of cement-based materials yield stress using slump tests. *Cement and Concrete Research* **35(5)**: 817–822.

Su N, Hsu KC and Chai HW (2001) A simple mix design method for self-compacting concrete. *Cement and Concrete Research* **31(12)**: 1799–1807.

Topçu İB (1997) Semi-lightweight concretes produced by volcanic slags. *Cement and Concrete Research* **27(1)**: 15–21.

Topçu İB and Uygunoğlu T (2008) Investigation of use of pumice lightweight aggregate in self-consolidating concrete. *Journal of Concrete Prefabrication* **85(1)**: 5–14, in Turkish.

Topçu İB and Uygunoğlu T (2010) Effect of aggregate type on properties of hardened self-consolidating lightweight concrete (SCLC). *Construction and Building Materials* **24(7)**: 1286–1295.

Topçu İB, Uygunoğlu T and Ünal O (2007) Properties of self-consolidating self-weight concrete. *Proceedings of the 3rd International Congress on Sustainability of Cement and Concrete, Istanbul, Turkey*, pp. 833–844.

Topçu İB, Bilir T and Uygunoğlu T (2009) Effect of waste marble dust content as filler on properties of self-compacting concrete. *Construction and Building Materials* **23(5)**: 1947–1953.

Turkish Standard Institute (2008) TS EN 197-1/A3. Cement – part 1: composition, specifications and conformity criteria for common cements. Turkish Standard Institute, Ankara, in Turkish.

Türkmen İ and Kantarcı A (2007) Effects of expanded perlite aggregate and different curing conditions on the physical and mechanical properties of self-compacting concrete. *Building and Environment* **42(6)**: 2378–2383.

Ünal O, Uygunoğlu T and Yıldız A (2007) Investigation of properties of low strength lightweight concrete for thermal insulating. *Building and Environment* **42(2)**: 584–590.

Uygunoğlu T (2008) *Properties of Self-Consolidating Lightweight Aggregate Concrete*. PhD thesis Süleyman Demirel University in Turkish.

Wu Z, Zhang Y, Zheng J and Ding Y (2009) An experimental study on the workability of self-compacting lightweight concrete. *Construction and Building Materials* **23(5)**: 2087–2092.

ice Publishing

Dhir and Lye
ISBN 978-0-7277-6461-4
https://doi.org/10.1680/icetscc.64614.183

Chapter 9

An experimental study on the performance of self-compacting lightweight concrete exposed to elevated temperature

Xi Wu
PhD student, State Key Laboratory of Coastal and Offshore Engineering, Dalian University of Technology, Dalian, People's Republic of China

Zhi-min Wu
Professor, State Key Laboratory of Coastal and Offshore Engineering, Dalian University of Technology, Dalian, People's Republic of China

Jian-jun Zheng
Professor, School of Civil Engineering and Architecture, Zhejiang University of Technology, Hangzhou, People's Republic of China

Tamon Ueda
Professor, Laboratory of Engineering for Maintenance System, Faculty of Engineering, Hokkaido University, Sapporo, Japan

Sheng-hui Yi
PhD student, Department of Civil and Architectural Engineering, City University of Hong Kong, Hong Kong

This paper presents an experimental study on the spalling characteristics, mass loss ratio, ultrasonic pulse velocity, compressive strength and flexural strength of self-compacting lightweight concrete (SCLC) at elevated temperature. Four types of SCLC specimens with and without polypropylene fibres (PPFs) and one type of normal concrete (NC) specimens were cast and tested. Based on the experimental observations and results, it was found that, compared with NC, SCLC spalls at a lower temperature but maintains a higher residual strength. For all the concretes the peak mass loss ratio increased but the relative ultrasonic pulse velocity decreased with the rise of temperature. At a given elevated temperature, the relative compressive strength and flexural strength of SCLC was larger than those of NC. The addition of PPFs greatly reduced the risk of spalling of SCLC. The thermal damage and the loss in residual mechanical properties of SCLC with PPFs were smaller compared with that without PPFs.

Introduction

Self-compacting lightweight aggregate concrete (SCLC) is a new building material developed by combining the favourable properties of lightweight aggregate concrete (LWAC) and self-compacting concrete (SCC). SCLC needs no external vibration, and can spread into place, fill the form work and encapsulate reinforcement without any bleeding or segregation. In view of the characteristics of being lightweight and self-flowing, SCLC has been used in reinforcement-intensified structural members and earthquake-resistant elements, in particular.

Although there are various tests conducted on the mix design and mechanical properties of SCLC (Choi *et al.*, 2006; Hwang and Hung, 2005; Kim *et al.*, 2010; Madandoust *et al.*, 2011; Shi and Wu, 2005; Topcu and Uygunoğlu, 2010; Turkmen and Kantarci, 2007; Wang, 2009), little research has been devoted to its performance under elevated temperature. It is well known that the spalling characteristics and mechanical properties are two major concerns for concrete under high temperature. Spalling is a common phenomenon occurring in concrete subjected to high temperature, resulting in the severe loss of stability and integrity, and threatening the safety of structures. An efficient way to reduce the risk of spalling is to add polypropylene fibres (PPFs) or steel fibres into concrete. The mechanism of spalling prevention by the two kinds of fibres has been explored experimentally (Aydın *et al.*, 2008; Bentz, 2000; Kalifa *et al.*, 2001; Lau and Anson, 2006; Noumowe, 2005; Poon *et al.*, 2004; Sideris *et al.*, 2009). However, concrete with steel fibres was found to be less spalling resistant than that with PPFs (Aydın *et al.*, 2008; Peng *et al.*, 2006; Sideris *et al.*, 2009), and no agreement has been achieved so far on the effect of PPFs on the residual mechanical properties of concrete.

Some progress on the high-temperature performance of LWAC has been made, with the focus on the spalling and residual mechanical properties. Lindgård and Hammer (1998) carried out a literature survey on the fire resistance of LWAC. They concluded that LWAC has a higher risk of spalling compared with normal concrete (NC) owing to the high moisture content caused by aggregate and the low permeability. Tanyildizi and Coskun (2008a, 2008b) studied the compressive strength and splitting strength of LWAC containing different amounts of fly ash or silica fume after high temperature. It was found that a high residual strength can be achieved by adding 20% silica fume or 30% fly ash. It was also confirmed that PPFs can improve the mechanical properties of LWAC after high temperature (Harun, 2009). Bilodeau *et al.* (2004) investigated the optimum amount of PPFs to prevent spalling of LWAC. It was suggested that 1·5 kg/m^3 of PPFs with a length of 12·5 mm would be sufficient to prevent spalling for concrete with low water/cement ratios.

SCC at high temperatures has also been studied in recent years. It has been demonstrated that the dense matrix owing to fine mineral materials can improve the tendency of spalling. By comparing NC columns with SCC ones, Persson (2004) found that explosive spalling only occurs in SCC columns. Similar conclusions were drawn by Reinhardt and Stegmaier (2006). Although SCC with mineral powders shows the worst behaviour in fire, there is no significant difference in residual compressive strength between NC and SCC. Owing to the severe damage to SCC caused by spalling, PPFs were incorporated to reduce the risk of spalling. To examine the thermal stability after introducing PPFs, Noumowé *et al.* (2006) performed a wide investigation on high-strength SCC and concrete with and without PPFs. No spalling was observed in concrete with PPFs, but the effect of PPFs on the residual mechanical properties of SCC and concrete is negligibly small. Liu *et al.* (2008) and Ye *et al.* (2007) examined SCC paste with and without PPFs to identify the microstructural change of SCC paste at elevated temperature and the mechanism of PPFs in preventing spalling. Their study showed that SCC suffers similar damage to high-performance concrete under fire exposure, but 0·5 and 1 kg/m^3 of PPFs can avoid damage efficiently. Apart from PPFs, steel fibres and hybrid fibres were also introduced by Khaliq and Kodur (2011) to quantify the effect on the thermal stability and residual mechanical properties. It was concluded that steel fibres can increase the tensile strength, which may lead to a reduction in fire-induced spalling, but no obvious effect was observed on the compressive strength of SCC at high temperature. Uysal and Tanyildizi (2012) and Uysal *et al.* (2012) gave an explanation on the effects of PPFs on the high-temperature properties of SCC produced with filler additives, such as granulated ground blast furnace slag, fly ash and limestone powder. They showed that a lower weight loss was observed in SCC with PPFs. However, a higher reduction in ultrasonic pulse velocity for SCC with PPFs suggested that more microcracks be induced after high temperature. The addition of PPFs was also reported to have a negative effect on the hot mechanical properties of concrete (Tao *et al.*, 2010).

In view of the unstable behaviour of LWAC and SCC at high temperature, SCLC has a higher potential risk of spalling, probably owing to the dense matrix and the aggregates with a high moisture content. Uygunoğlu and Topcu (2009) investigated the thermal expansion of SCLC with pumice as coarse aggregate. However, the spalling behaviour and residual mechanical properties of SCLC with and without PPFs after high temperature have not been examined.

The intention of this study was to investigate the behaviour of SCLC subjected to elevated temperature. Based on the experimental results, the spalling characteristics, mass loss ratio, ultrasonic pulse velocity, compressive strength and flexural strength of SCLC at elevated temperature are evaluated. To develop SCLC with good fire spalling resistance, PPFs are introduced and their effects on the spalling and residual mechanical properties of SCLC are examined.

Experimental programme

Materials

An ordinary Portland cement produced in Dalian, China, was used in this study. Class I fly ash was added as a viscous admixture. Sica 3301E polycarboxylate-based high-range water reducer was adopted to adjust the workability of SCLC. A type of high-performance shale was used as coarse aggregate, and its physical properties are shown in Table 1. The maximum aggregate size was 20 mm. PPFs with a length of 45 mm were introduced to improve the resistance to high temperature. The physical properties of PPFs are listed in Table 2.

Table 1 Physical properties and grading of expanded shale lightweight aggregate

Apparent density: kg/m^3	24-hour water absorption: %	Crushing strength: MPa	Volume fraction: %			
			< 5: mm	5–10: mm	10–16: mm	> 16: mm
1363	4	6·0	16·6	43·6	25·7	14·1

Table 2 Physical properties of polypropylene fibre

Tensile strength: MPa	Specific gravity: g/cm^3	Elastic modulus: GPa	Melting temperature: °C	Burning temperature: °C
456	0·91	5·6	165	580

Specimens

Five types of concrete, labelled as SCLC1, SCLC2, SCLCPPF1, SCLCPPF2 and NC, were designed for specimens in this study. SCLC1 and SCLC2 had different water/binder ratios and NC was used for comparison. Two different amounts of fibres, 2 and 3 kg per cubic meter of concrete, were mixed in SCLCPPF1 and SCLCPPF2 respectively. Concrete with PPFs had the same water/binder ratio as SCLC1. The mix proportions and slump flow spread of each type of concrete are listed in Table 3. As PPFs were added to the mixture, more high-range water reducer was required to maintain the flowability of SCLCPPF1 and SCLCPPF2. It can be seen from Table 3 that the workability of fresh SCLC with and without PPFs satisfies the requirements for self-compacting concrete. The distribution of coarse aggregates on the cross-section is shown in Figure 1, which indicates that coarse aggregates are uniformly distributed within SCLC.

For each type of concrete, cubes with a side of 100 mm were cast for measuring the mass loss ratio and compressive strength, and prisms with sizes of 100 × 100 × 400 mm were cast for measuring the flexural strength and ultrasonic pulse velocity. After 24 h of casting, specimens were demoulded and subsequently cured in a standard curing room at 95% relative humidity and 20 ± 3°C. After 28-day curing they were taken out of the standard curing room.

Table 3 Mix proportions of concrete

Type	SCLC1	SCLC2	SCLCPPF1	SCLCPPF2	NC
Cement: kg/m^3	397	441	397	397	462
Aggregate: kg/m^3	417	424	417	417	1200
Sand: kg/m^3	806	836	806	806	496
Water: kg/m^3	170	169	170	170	208·6
HRWR: kg/m^3	5·5	9·65	8	10	–
Fly ash: kg/m^3	170	182	170	170	–
PPF: kg/m^3	–	–	2	3	–
Water to binder ratio	0·3	0·27	0·3	0·3	0·452
Bulk density: kg/m^3	1945	1955	1961	1998	2435
28-day compressive strength: MPa	45	51	60	55	53
Slump flow spread: cm	70	82	71	73	–

SCLC, self-compacting lightweight aggregate concrete; NC, normal concrete; HRWER, high-range water reducer; PPF, polypropylene fibre.

Figure 1 Distribution of coarse aggregates on cross-section

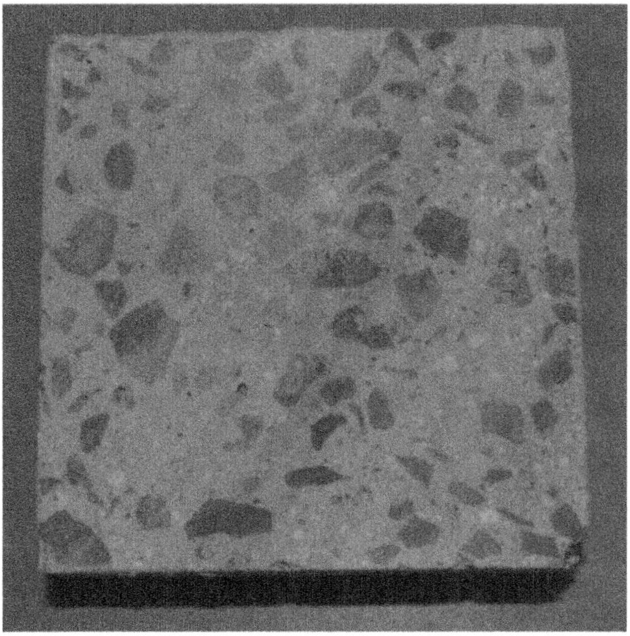

Before testing at high temperature, the specimens were dried at 100°C to constant mass. Non-heated specimens were tested at room temperature for comparison.

Testing procedure

Specimens were subjected to a heating rate of 10°C/min up to 200°C, 400°C, 450°C, 500°C and 600°C respectively. The specimens were maintained at a given temperature for 4 h so that stable thermal properties could be achieved. The specimens were left in the furnace to cool down to room temperature naturally and then tested for the mass loss, ultrasonic pulse velocity, compressive strength and flexural strength.

Results and discussion

Spalling

During the whole heating process, the colour change on the concrete surface can be primarily used to assess heat-damage to concrete. As temperature rose, the colour on the concrete surface changed from straw bluish at room temperature, straw white at 200°C, slight grey at 400°C, slight red at 450°C to dark red at 500°C. On a cross-section of the concrete specimen, the colour in the peripheral region of width 10 mm was darker than that in the central region, indicating that there exists a temperature gradient on the cross-section. The presence of PPFs exhibits no significant influence on the colour on the concrete surface.

With the increase of temperature, cracks were initiated owing to the thermal incompatibility between the cement paste and aggregates, and the building up of vapour pressure (Xu *et al.*, 2003).

Figure 2 Spalling of self-compacting lightweight aggregate concrete at 450°C

Before the temperature reached 400°C, there were no cracks on the concrete surface. Some small cracks occurred on the surface of SCLC1 and SCLC2 specimens at a temperature close to or slightly beyond 400°C, but no cracks were found on the surface of NC or SCLCPPF specimens. These visible cracks subsequently opened rapidly and the edges of SCLC specimens began to spall at a temperature around 450°C. Explosive spalling suddenly happened for SCLC1 and SCLC2 specimens at 470°C and 500°C, respectively, as shown in Figure 2, but there was no sign of spalling for NC and SCLCPPF specimens. The early spalling of SCLC specimens could be attributed to lightweight aggregates and fine additives. The water-absorbing characteristics of lightweight aggregates and the dense hardened matrix caused by mineral additives lead to a greater water vapour pressure and therefore caused earlier spalling than NC. When the temperature reached 600°C, NC specimens began to spall. However, the majority of the SCLCPPF specimens still maintained their integrity. Only a few SCLCPPF specimens showed spalling deterioration at the peripheral surface of the specimens, as confirmed in previous studies (Bilodeau *et al.*, 2004; Noumowé *et al.*, 2006; Persson, 2004). It is suggested that the melting of PPFs around 160°C could create small channels to release the water vapour pressure, thus efficiently preventing the early spalling (Lindgård and Hammer, 1998; Liu *et al.*, 2008; Noumowe, 2005). In this study, it appeared that 2 kg/m^3 PPFs was sufficient to prevent SCLC from spalling.

Mass loss ratio

The mass loss ratio is defined as the ratio of the difference between the original weight and the weight at a given temperature to the original weight. It is usually associated with the process of water evaporation from concrete and the decomposition of chemical hydrates.

In this study, the mass loss ratio at a given temperature was measured from three specimens. The effect of elevated temperature on the mass loss ratio is shown in Figure 3. The mass loss ratio was in the range of 3% to 9% when the temperature increased from 200°C to 600°C.

Figure 3 Relationship between mass loss ratio and temperature. SCLC, self-compacting lightweight aggregate concrete; NC, normal concrete

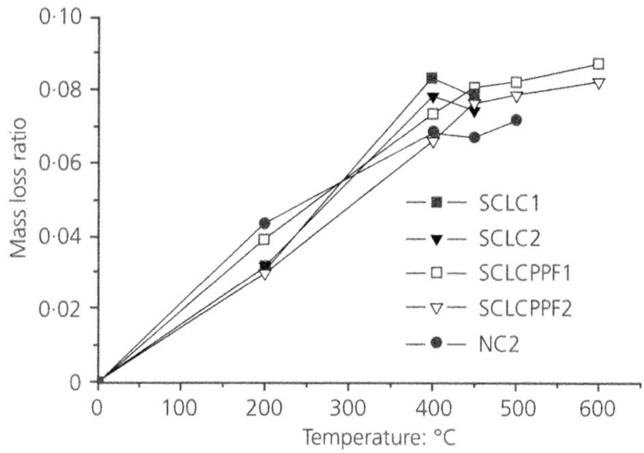

The mass loss ratio curves were slightly different for SCLC and SCLCPPF. When the temperature was in the range of 20°C to 400°C the mass loss ratio increased steadily with the rise in temperature. When the temperature was higher than 400°C, the mass loss ratio remained basically invariable for SCLC, but continued increasing for SCLCPPF. The effects of the water/binder ratio and PPFs on the mass loss ratio were found to be negligibly small.

It is also seen from Figure 3 that the peak mass loss ratios of SCLC1 and SCLC2 were larger than that of NC. This can be explained as follows. As temperature rises, the mass loss ratio increases owing to the release of capillary and gel water. For SCLC1 and SCLC2, the water originally absorbed into lightweight aggregate pores was also released, resulting in a higher peak mass loss ratio. At 400°C, almost all the capillary and gel water was released and therefore the mass loss ratio stayed invariable.

Relative ultrasonic pulse velocity

The ultrasonic pulse velocity test was conducted to assess the effect of elevated temperature on concrete. The relative ultrasonic pulse velocity is defined as the ratio of the ultrasonic pulse velocity at a given temperature to that at room temperature. The experimental results are shown in Figure 4. The smaller the relative ultrasonic pulse velocity, the more severe the damage to concrete specimens is.

From Figure 4, it can be seen that the relative ultrasonic pulse velocity decreased continuously with the rise of temperature owing to heat-induced damage. Above 450°C, a sharp reduction in relative ultrasonic pulse velocity was observed for all the concrete specimens, which implies severe damage beyond that temperature. At a given temperature, the relative ultrasonic pulse velocities for SCLC1 and SCLC2 were larger than that for NC, indicating that SCLC has a higher heat resistance than NC. The reason for this is that lightweight aggregates have better insulation properties than natural ones, owing to the lower heat conductivity, thereby reducing thermal stresses within SCLC specimens. However, the experimental results showed that

Figure 4 Relationship between relative ultrasonic pulse velocity and temperature. SCLC, self-compacting lightweight aggregate concrete; NC, normal concrete

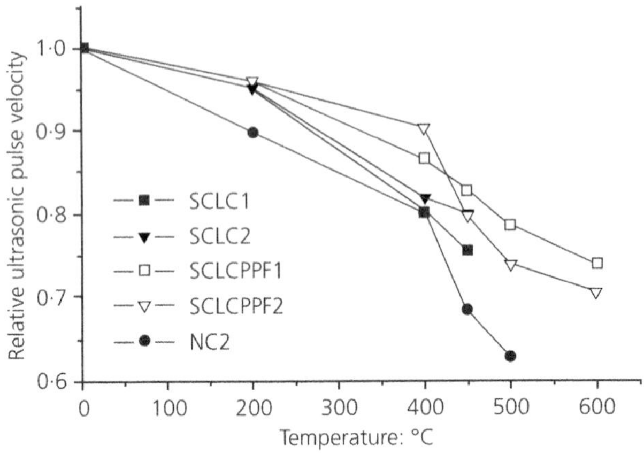

SCLCPPF had the highest ultrasonic pulse velocity, which indicates that PPFs are more efficient in reducing damage, as the additional porosity and small channels created by melting PPFs may lower internal vapour pressure and mitigate damage to concrete (Noumowe, 2005).

Residual compressive strength

The relative compressive strength is defined as the ratio of the compressive strength at a given temperature to that at room temperature. The measured relative compressive strength is plotted against temperature as shown in Figure 5, which indicates that the relative compressive strength generally decreased with the rise of temperature. Initially, the compressive strength dropped at 200°C for all the concrete specimens. When temperature increased to 400°C, the compressive strength kept decreasing for SCLC, but a recovery in strength was observed for SCLCPPF and NC. The recovery was associated with the general stiffening of cement gel or the increase in surface forces between gel particles owing to the removal of absorbed moisture, as explained for high-strength concrete (Castillo and Durrani, 1990). The loss rate in compressive strength increased beyond 400°C, which is consistent with previous studies (Castillo and Durrani, 1990; Xiao and Falkner, 2006). The significant strength reduction at higher temperature is attributable to the phase transformation of cement paste, such as C-S-H gel dehydration, which causes the shrinkage of cement paste and induces cracks in concrete (Omer, 2007).

The strength reduction range varies with the concrete mix. For a given temperature, the relative compressive strengths of SCLC1 and SCLC2 were, in general, slightly larger than that of NC. The reason for this is that the lower heat conductivity of lightweight aggregate reduces thermal stresses within SCLC. Hence, it can be concluded that SCLC maintains a higher residual compressive strength than NC. In addition, Figure 5 also shows that, before spalling, the relative compressive strength of SCLC2 was much larger than that of SCLC1. A smaller water/binder ratio induced a higher residual compressive strength, which is in line with high-strength concrete (Chan *et al.*, 1996; Phan *et al.*, 2001). Furthermore, Figure 5 shows that PPFs helped to maintain the strength after high temperature. The strength reduction of SCLCPPF was

Figure 5 Relationship between relative compressive strength and temperature. SCLC, self-compacting lightweight aggregate concrete; NC, normal concrete

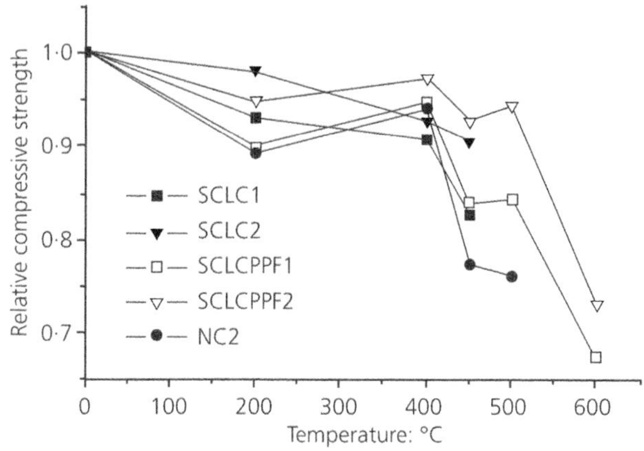

markedly less than that of SCLC and NC, which demonstrates a positive effect of PPFs. The larger the content of PPFs, the higher is the residual compressive strength.

Relative flexure strength

A four-point bending beam test, as shown in Figure 6, was conducted to measure the flexural strength of concrete. The relative flexure strength is defined as the ratio of the flexure strength at a given temperature to that at room temperature. The experimental results are shown in Figure 7, which indicate that all the relative flexural strength curves were similar in shape. In the range of 20°C to 200°C, the relative flexural strength increased slightly with the rise in temperature. A continuous increase in strength was observed even at 400°C for SCLCPPF. The gain in strength can be explained as in the previous section. However, after 450°C, the residual flexure strength of SCLC decreased rather rapidly, with only 40% strength maintained at around 600°C. The rapid drop in residual compressive strength and flexure strength indicates that 400°C to 600°C is a critical range to the strength loss of concrete. However, it is of interest to note that the sharp loss in flexure strength is clearly different from the gradual loss in compressive strength, which implies that the flexure strength is more sensitive to high temperature.

Figure 6 Schematic of flexure test

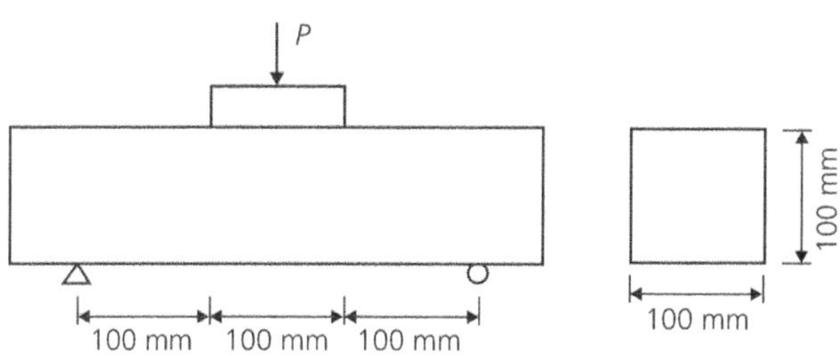

Figure 7 Relationship between relative flexure strength and temperature. SCLC, self-compacting lightweight aggregate concrete; NC, normal concrete

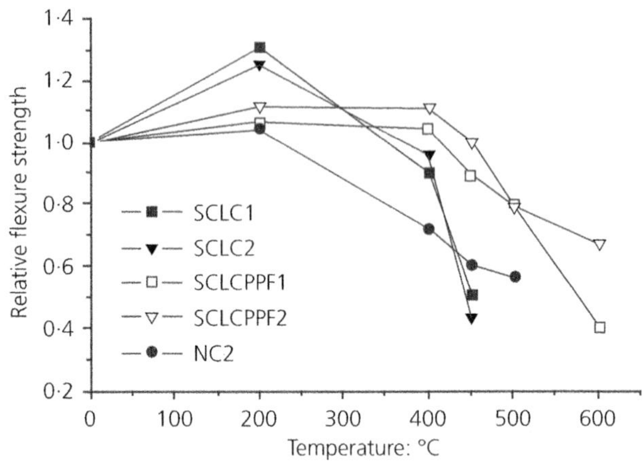

It can be also seen from Figure 7 that, at a given temperature, the relative flexural strengths of SCLC1 and SCLC2 were larger than that of NC, but were significantly lower than that of SCLCPPF. The positive effect of PPFs is again confirmed. The similarity between SCLC1 and SCLC2 in flexure strength reduction indicates that the water/binder ratio has a less significant influence on the residual flexure strength than on the residual compressive strength.

Conclusion

In this study, four types of SCLC specimens with and without PPFs and one type of NC specimens were tested. Based on the experimental results, the fire performance of SCLC has been evaluated in a quantitative manner. The conclusions of this study are summarised as follows.

- With the rise in temperature, the colour change and crack development on the specimen surface were observed. The spalling temperature of SCLC was much lower than that of NC.
- The mass loss ratio basically increased with the rise in temperature. The water/binder ratio and PPFs showed no significant influence on the mass loss ratio.
- The ultrasonic pulse velocity decreased continuously with the rise in temperature. SCLCPPF maintains the highest ultrasonic pulse velocity, indicating less thermal damage than SCLC and NC.
- Except for a noticeable strength recovery around 200°C, the relative compressive and flexure strengths decreased with the rise in temperature. At a given elevated temperature, the relative residual strength of SCLC was larger than that of NC, but smaller than that of SCLCPPF.
- The addition of PPFs is an efficient way to reduce the spalling of SCLC: 2 kg/m^3 of PPFs is sufficient to prevent SCLC from spalling at a lower temperature.
- SCLCPPF had a higher residual compressive strength and flexure strength. PPFs showed a positive effect on the residual mechanical properties of SCLC.

Acknowledgements

The financial support from the National Natural Science Foundation through grant no. 51121005 and the Natural Science Foundation of Zhejiang Province through grant no. LY12E08022, of the People's Republic of China, is greatly acknowledged.

REFERENCES

Aydın S, Yazıcı H and Baradan B (2008) High temperature resistance of normal strength and autoclaved high strength mortars incorporated polypropylene and steel fibers. *Construction and Building Materials* **22(4)**: 504–512.

Bentz DP (2000) Fibers, percolation, and spalling of high-performance concrete. *ACI Materials Journal* **97(3)**: 351–359.

Bilodeau A, Kodur VKR and Hoff GC (2004) Optimization of the type and amount of polypropylene fibres for preventing the spalling of lightweight concrete subjected to hydrocarbon fire. *Cement and Concrete Composites* **26(2)**: 163–174.

Castillo C and Durrani AJ (1990) Effect of transient high temperature on high-strength concrete. *ACI Materials Journal* **87(1)**: 47–53.

Chan S, Peng G and Chan J (1996) Comparison between high strength concrete and normal strength concrete subjected to high temperature. *Materials and Structures* **29(10)**: 616–619.

Choi YW, Kim YJ, Shin HC and Moon HY (2006) An experimental research on the fluidity and mechanical properties of high-strength lightweight self-compacting concrete. *Cement and Concrete Research* **36(9)**: 1595–1602.

Harun T (2009) Statistical analysis for mechanical properties of polypropylene fiber reinforced lightweight concrete containing silica fume exposed to high temperature. *Materials and Design* **30(8)**: 3252–3258.

Hwang C and Hung M (2005) Durability design and performance of self-consolidating lightweight concrete. *Construction and Building Materials* **19(8)**: 619–626.

Kalifa P, Chéné G and Gallé C (2001) High-temperature behaviour of HPC with polypropylene fibres: From spalling to microstructure. *Cement and Concrete Research* **31(10)**: 1487–1499.

Khaliq W and Kodur V (2011) Thermal and mechanical properties of fiber reinforced high performance self-consolidating concrete at elevated temperatures. *Cement and Concrete Research* **41(11)**: 1112–1122.

Kim YJ, Choi YW and Lachemi M (2010) Characteristics of self-consolidating concrete using two types of lightweight coarse aggregates. *Construction and Building Materials* **24(1)**: 11–16.

Lau A and Anson M (2006) Effect of high temperatures on high performance steel fibre reinforced concrete. *Cement and Concrete Research* **36(9)**: 1698–1707.

Lindgård J and Hammer TA (1998) Fire resistance of structural lightweight aggregate concrete a literature review survey with focus on spalling, supplementary papers. *Proceedings of Fourth CANMET/ACI/JCI International Conference on Recent Advances in Concrete Technology, Tokushima, Japan*, pp. 25–39.

Liu X, Ye G, De Schutter G, Yuan Y and Taerwe L (2008) On the mechanism of polypropylene fibres in preventing fire spalling in self-compacting and high-performance cement paste. *Cement and Concrete Research* **38(4)**: 487–499.

Madandoust R, Ranjbar MM and Mousavi SY (2011) An investigation on the fresh properties of self-compacted lightweight concrete containing expanded polystyrene. *Construction and Building Materials* **25(9)**: 3721–3731.

Noumowe A (2005) Mechanical properties and microstructure of high strength concrete containing polypropylene fibres exposed to temperatures up to 200 °C. *Cement and Concrete Research* **35(11)**: 2192–2198.

Noumowé A, Carré H, Daoud A and Toutanji H (2006) High-strength self-compacting concrete exposed to fire test. *Journal of Materials in Civil Engineering* **18(6)**: 754–758.

Omer A (2007) Effects of elevated temperatures on properties of concrete. *Fire Safety Journal* **42(8)**: 516–522.

Peng G, Yang W and Zhao J *et al.* (2006) Explosive spalling and residual mechanical properties of fiber-toughened high-performance concrete subjected to high temperatures. *Cement and Concrete Research* **36(4)**: 723–727.

Persson B (2004) Fire resistance of self-compacting concrete, SCC. *Materials and Structures* **37(9)**: 575–584.

Phan L, Lawson J and Davis F (2001) Effects of elevated temperature exposure on heating characteristics, spalling, and residual properties of high performance concrete. *Materials and Structures* **34(2)**: 83–91.

Poon CS, Shui ZH and Lam L (2004) Compressive behavior of fiber reinforced high-performance concrete subjected to elevated temperatures. *Cement and Concrete Research* **34(12)**: 2215–2222.

Reinhardt HW and Stegmaier M (2006) Self-consolidating concrete in fire. *ACI Materials Journal* **103(2)**: 130–135.

Shi C and Wu Y (2005) Mixture proportioning and properties of self-consolidating lightweight concrete containing glass powder. *ACI Materials Journal* **102(5)**: 355–363.

Sideris KK, Manita P and Chaniotakis E (2009) Performance of thermally damaged fibre reinforced concretes. *Construction and Building Materials* **23(3)**: 1232–1239.

Tanyildizi H and Coskun A (2008a) Performance of lightweight concrete with silica fume after high temperature. *Construction and Building Materials* **22(10)**: 2124–2129.

Tanyildizi H and Coskun A (2008b) The effect of high temperature on compressive strength and splitting tensile strength of structural lightweight concrete containing fly ash. *Construction and Building Materials* **22(11)**: 2269–2275.

Tao J, Yuan Y and Taerwe L (2010) Compressive strength of self-compacting concrete during high-temperature exposure. *Journal of Materials in Civil Engineering* **22(10)**: 1005–1011.

Topcu IB and Uygunoğlu T (2010) Effect of aggregate type on properties of hardened self-consolidating lightweight concrete (SCLC). *Construction and Building Materials* **24(7)**: 1286–1295.

Turkmen I and Kantarci A (2007) Effects of expanded perlite aggregate and different curing conditions on the physical and mechanical properties of self-compacting concrete. *Building and Environment* **42(6)**: 2378–2383.

Uygunoğlu T and Topcu IB (2009) Thermal expansion of self-consolidating normal and lightweight aggregate concrete at elevated temperature. *Construction and Building Materials* **23(9)**: 3063–3069.

Uysal M and Tanyildizi H (2012) Estimation of compressive strength of self compacting concrete containing polypropylene fiber and mineral additives exposed to high temperature using artificial neural network. *Construction and Building Materials* **27(1)**: 404–414.

Uysal M, Yilmaz K and Ipek M (2012) Properties and behavior of self-compacting concrete produced with GBFS and FA additives subjected to high temperatures. *Construction and Building Materials* **28(1)**: 321–326.

Wang H (2009) Durability of self-consolidating lightweight aggregate concrete using dredged silt. *Construction and Building Materials* **23(6)**: 2332–2337.

Xiao J and Falkner H (2006) On residual strength of high-performance concrete with and without polypropylene fibres at elevated temperatures. *Fire Safety Journal* **41(2)**: 115–121.

Xu Y, Wong YL, Poon CS and Anson M (2003) Influence of PFA on cracking of concrete and cement paste after exposure to high temperatures. *Cement and Concrete Research* **33(12)**: 2009–2016.

Ye G, Liu X, De Schutter G, Taerwe L and Vandevelde P (2007) Phase distribution and microstructural changes of self-compacting cement paste at elevated temperature. *Cement and Concrete Research* **37(6)**: 978–987.

ice Publishing

Dhir and Lye
ISBN 978-0-7277-6461-4
https://doi.org/10.1680/icetscc.64614.197

Chapter 10
Mechanical characteristics of self-compacting concrete with and without fibres

Farhad Aslani
Centre for Built Infrastructure Research, School of Civil and Environmental Engineering, University of Technology Sydney, Australia

Shami Nejadi
Centre for Built Infrastructure Research, School of Civil and Environmental Engineering, University of Technology Sydney, Australia

Fibre-reinforced self-compacting concrete (FRSCC) is a high-performance building material that combines positive aspects of fresh properties of self-compacting concrete (SCC) with improved characteristics of hardened concrete as a result of fibre addition. Considering these properties, the application ranges of both FRSCC and SCC can be covered. A test program is carried out to develop information about the mechanical properties of FRSCC. For this purpose, four SCC mixes – plain SCC, steel, polypropylene and hybrid FRSCC – are considered in the test program. The properties include compressive and splitting tensile strengths, modulus of elasticity, modulus of rupture, and compressive stress–strain curve. These properties are tested at 3, 7, 14, 28, 56 and 91 days. Relationships are established to predict the compressive and splitting tensile strengths, modulus of elasticity, modulus of rupture, and compressive stress–strain curve. The models provide predictions matching the measurements.

Introduction

Self-compacting concrete (SCC) can be placed and compacted under its own weight with little or no vibration, and without segregation or bleeding. SCC is used to facilitate and ensure proper filling, and good structural performance of restricted areas and heavily reinforced structural members. It has gained significant importance in recent years because of its advantages. Recently, this concrete has gained wider use in many countries for different applications and structural configurations. SCC can also provide a better working environment by eliminating the vibration noise. Such concrete requires a high slump that can be achieved easily by superplasticiser addition to a concrete mix and special attention to the mix proportions. SCC often contains a large quantity of powdered materials that are required to maintain sufficient yield value and viscosity of the fresh mix, thus reducing bleeding, segregation and settlement. As the use of a large quantity of cement increases costs and results in higher temperatures, the use of additions, such as fly ash, blast furnace slag or limestone filler, could increase the slump of the concrete mix without increasing its cost (Aslani and Nejadi, 2012a, 2012b, 2012c, 2012d).

The incorporation of fibres improves engineering performance of structural and non-structural concrete. The use of fibre-reinforced concrete (FRC) is also of special interest for retrofit and seismic design. The incorporation of metallic fibres can be problematic on some situations, especially when the fibre volume is high and the FRC is cast in sections with a moderate-to-high degree of reinforcement. The fibre content, length, aspect ratio and shape play an important role in controlling workability of FRC. Such concrete presents greater difficulty in handling, and requires more deliberate planning and workmanship than established concrete construction procedures. The additional compaction effort required for such concrete contributes to the increase in construction cost. In order to provide sufficient compaction, improve fibre dispersion and reduce the risk of entrapping voids, the FRC is often proportioned to be fluid enough to reduce the need for vibration consolidation and facilitate placement. An extension of this approach can involve the use of SCC to eliminate, or greatly reduce, the need for vibration and further facilitate placement. A truly fibre-reinforced self-compacting concrete (FRSCC) should spread into place under its own weight and achieve consolidation without internal or external vibration, ensure proper dispersion of fibres, and undergo minimum entrapment of air voids and loss of homogeneity until hardening. Lack of proper self-compaction or intentional vibration and compaction can result in macro- and micro-structural defects that can affect mechanical performance and durability (Khayat and Roussel, 1999).

FRSCC is a relatively recent composite material that combines the benefits of SCC technology with the advantages of the fibre addition to a brittle cementitious matrix. It is a ductile material that, in its fresh state, flows into the interior of the formwork, filling it in a natural manner, passing through the obstacles, and flowing and consolidating under the action of its own weight. FRSCC can mitigate two opposing weaknesses: poor workability in FRC and cracking resistance in plain concrete. A few studies have been carried out on optimisation of the mix proportion for the addition of steel or polypropylene fibres to SCC. Meanwhile, there is insufficient research on the mechanical properties of FRSCC. In mechanical terms, the greatest disadvantage of cementitious material is its vulnerability to cracking, which generally occurs at an early age in concrete structures or members. Cracking may potentially reduce the lifetime of concrete structures, and cause serious durability and serviceability problems. The addition of fibres into SCC mixtures has been studied by a number of researchers (Bui *et al.*, 2003; Busterud *et al.*, 2005; Corinaldesi and Moriconi, 2004; Cunha, 2006; Groth and Nemegeer, 1999; Grünewald, 2006; Grünewald and Walraven, 2001; Khayat and Roussel, 1999; Liao *et al.*, 2006; Massicotte *et al.*, 2000; Sahmaran *et al.*, 2005; Sahmaran and Yaman, 2007; Schumacher, 2008).

The most beneficial properties with fibre addition to the concrete in the hardened state are the impact strength, the toughness and the energy absorption capacity. A detailed description of the benefits provided by fibre addition to concrete can be found elsewhere (ACI 544.2R (ACI, 1999); Balaguru and Shah, 1992). The fibre addition might also improve the fire resistance of cement-based materials, as well as their shear resistance. The possible applications of FRSCC include highways; industrial and airfield pavements; hydraulic structures; tunnel segments; bridges components; and concrete structures of complex geometry that present high difficulties in being reinforced by conventional steel bars, especially those that have a high degree of support redundancy.

The research presented aims at finding the properties of SCC and FRSCC in the fresh and hardened stages. An experimental program is carried out to investigate the mechanical properties of four mixes of SCC. The mechanical properties included in this study are compressive and splitting tensile strengths, modulus of elasticity, modulus of rupture and compressive stress–strain curve. These properties are tested at 3, 7, 14, 28, 56 and 91 days. The developments of mechanical properties with time are investigated.

As only a few correlations among the mechanical properties of FRSCC have been reported and are unclear, regression analyses are conducted on existing experimental data to propose splitting tensile strength, modulus of elasticity and modulus of rupture models on compressive strength and age of concrete. Also, compressive stress–strain relationships for SCC and FRSCC are compared with the test results.

Experimental study

Materials

Cement

In this experimental study, shrinkage limited cement (SLC) corresponding to the AS 3972 (SA, 2010) standard was used. SLC is manufactured from specially prepared Portland cement clinker and gypsum. It may contain up to 5% of AS 3972-approved additions. The chemical, physical and mechanical properties of the cement used in the experiments are shown in Table 1. The chemical, physical and mechanical properties adhere to the limit values specified in AS 2350.2, 3, 4, 5, 8 and 11 (SA, 2006).

Table 1 Properties of cement

Chemical properties	
Calcium oxide (CaO)	64·5%
Silicon dioxide (SiO_2)	19·3%
Aluminium oxide (Al_2O_3)	5·2%
Iron (III) oxide (Fe_2O_3)	2·9%
Magnesium oxide (MgO)	1·1%
Sulfur trioxide (SO_3)	2·9%
Potassium oxide (K_2O)	0·56%
Sodium oxide (Na_2O)	< 0·01%
Chloride (Cl)	0·02%
Loss on ignition	2·8%
Physical properties	
Autoclave expansion	0·05%
Fineness index	405 m^2/kg
Mechanical properties	
Initial setting time	90 min
Final setting time	135 min
Soundness	1·0 mm
Drying shrinkage	590 μstrain
f'_c (3 days)	37·2 MPa
f'_c (7 days)	47·3 MPa
f'_c (28 days)	60·8 MPa

Fly ash

It is important to increase the amount of paste in SCC because it is an agent to carry the aggregates. As a consequence, Eraring fly ash (EFA) has been used to increase the amount of paste. EFA is a natural pozzolan. It is a fine cream/grey powder that is low in lime content. However, in its finely divided form and in the presence of moisture, it will react chemically with calcium hydroxide (e.g., from lime or cement hydration) at ordinary temperatures to form insoluble compounds possessing cementitious properties. The chemical and physical properties of EFA used in the experimental study are given in Table 2. The chemical, physical and mechanical properties of the EFA used adhere to the limit values specified in AS 2350.2 (SA, 2006), and AS 3583.1, 2, 3, 5, 6, 12 and 13 (SA, 1998).

Ground granulated blast furnace slag

Granulated blast furnace slag (GGBFS) is another supplementary cementitious material that is used in combination with SLC. GGBFS used in the experiment originated in Boral, Sydney, Australia, and it conformed to AS 3582.2 (SA, 2001) specifications. The chemical and physical properties of GGBFS are given in Table 3.

Table 2 Properties of fly ash

Chemical properties	
Aluminium oxide (Al_2O_3)	26·40%
Calcium oxide (CaO)	2·40%
Iron (III) oxide (Fe_2O_3)	3·20%
Potassium oxide (K_2O)	1·55%
Magnesium oxide (MgO)	0·60%
Manganese (III) oxide (Mn_2O_3)	<0·1%
Sodium oxide (Na_2O)	0·47%
Phosphorous pentoxide (P_2O_5)	0·20%
Silicon dioxide (SiO_2)	61·40%
Sulfur trioxide (SO_3)	0·20%
Strontium oxide (SrO)	<0·1%
Titanium dioxide (TiO_2)	1·00%
Physical properties	
Moisture	<0·1%
Fineness 45 μm	78% passed
Loss on ignition	2·30%
Sulfuric anhydride	0·20%
Alkali content	0·50%
Chloride ion	<0·001%
Relative density	2·02%
Relative water requirement	97%
Relative strength 28 days	88%

Table 3 Properties of ground granulated blast furnace slag

Chemical properties	
Aluminium oxide (Al_2O_3)	14·30%
Iron (III) oxide (Fe_2O_3)	1·20%
Magnesium oxide (MgO)	5·40%
Manganese (III) oxide (Mn_2O_3)	1·50%
Sulfur trioxide (SO_3)	0·20%
Chloride (Cl)	0·01%
Insoluble residue	0·50%
Loss on ignition	−1·10%
Physical properties	
Fineness index	435 m^2/kg

Aggregate

In this study, crushed volcanic rock (i.e. latite) coarse aggregate was used with a maximum aggregate size of 10 mm. Nepean river gravel with a maximum size of 5 mm and Kurnell natural river sand fine aggregates were also used. Methods for sampling and testing aggregates were determined in accordance with AS 1141 (SA, 2011) and RTA (2006), and the results for coarse and fine aggregates are shown in Tables 4–6 respectively.

Table 4 Properties of crushed latite volcanic rock coarse aggregate

Characteristics	Results
Sieve size	Passing: %
13·2 mm	100
9·5 mm	89
6·7 mm	40
4·75 mm	7
2·36 mm	1
1·18 mm	1
Material finer than 75 μm: %	1
Mis-shapen particles: %	
Ratio 2:1	13
Ratio 3:1	1
Flakiness index: %	20
Uncompacted bulk density: t/m^3	1·36
Compacted bulk density: t/m^3	1·54
Moisture condition of the aggregate: %	1·3
Particle density (Dry): t/m^3	2·65
Particle density (SSD): t/m^3	2·70
Apparent particle density: t/m^3	2·79

(continued on next page)

Table 4 (*Continued*)

Characteristics	Results
Water absorption: %	1·9
Ave. dry strength: kN	391
Ave. wet strength: kN	293
Wet/dry strength variation: %	25
Test fraction: mm	−9·5 + 6·7
The amount of significant breakdown: %	<0·2
The size of testing cylinder = 150 mm diam.	
Los Angeles Value Grd. 'K': %Loss	13

SSD, saturated surface dry.

Table 5 Properties of Nepean river gravel fine aggregate

Characteristics	Results
Sieve size	Passing: %
6·7 mm	100
4·75 mm	99
2·36 mm	83
1·18 mm	64
600 μm	42
425 μm	28
300 μm	19
150 μm	8
Material finer than 75 μm: %	3
Uncompacted bulk density: t/m^3	1·52
Compacted bulk density: t/m^3	1·64
Particle density (dry): t/m^3	2·58
Particle density (SSD): t/m^3	2·60
Apparent particle density: t/m^3	2·63
Water absorption: %	0·7
Silt content: %	7
Degradation factor of fine aggregate	90
The wash water after using permitted 500 ml was: CLEAR	
Moisture content:%	5·5
Method of determining voids content	
% voids	41·7
The mean flow time: s	26·5

SSD, saturated surface dry.

Table 6 Properties of Kurnell natural river sand fine aggregate

Characteristics	Results
Sieve size	Passing: %
1·18 mm	100
600 μm	98
425 μm	87
300 μm	46
150 μm	1
Material finer than 75 μm in aggregate by washing: %	Nil
Uncompacted bulk density: t/m^3	1·39
Compacted bulk density: t/m^3	1·54
Particle density (dry): t/m^3	2·58
Particle density (SSD): t/m^3	2·59
Apparent particle density: t/m^3	2·62
Water absorption: %	0·6
Silt content: %	4

SSD, saturated surface dry.

Admixtures

The superplasticiser, viscosity-modifying admixture and high-range water-reducing agent admixture were used in this study. The new superplasticiser generation Glenium 27 complies with AS 1478.1 (SA, 2000a) type high range water reducer and ASTM C494 (ASTM, 2000) types A and F are used. The Rheomac VMA 362 viscosity modifying admixture that was used in this study is a ready-to-use liquid admixture that is specially developed for producing concrete with enhanced viscosity and controlled rheological properties. Pozzolith 80 was used as a high-range water-reducing agent admixture in the mixes. It reduces the quantity of water required to produce concrete of a given consistency, with greater economy, of a given strength. It meets and exceeds AS 1478 (SA, 2000c) Type WRRe, requirements for admixtures.

Fibres

In this study, two commercially available fibres, Dramix RC-80/60-BN type steel fibres and Synmix 65 type polypropylene (PP) fibres were used. The mechanical, elastic and surface structure properties of the steel and PP fibres are summarised in Table 7.

Mixture proportions

One control SCC mixture (N-SCC) and three FRSCC mixtures were used in this study. FRSCC mixtures contained steel (D-SCC), PP (S-SCC) and hybrid (steel + PP) (DS-SCC) fibres. The content proportions of these mixtures are given in Table 8. These contents were chosen to attempt to keep compressive strength to a level applicable to construction.

Table 7 The physical and mechanical properties of fibres

Fibre type	Fibre name	Density: kg/m^3	Length (*l*)	Diameter (*d*)	Aspect ratio (*l*/*d*)	Tensile strength: MPa	Modulus of elasticity: GPa	Cross-section form	Surface structure
Steel	Dramix RC-80/60-BN	7850	60	0·75	80·0	1050	200	Circular	Hooked end
Polypropylene	Synmix 65	905	65	0·85	76·5	250	3	Square	Rough

Table 8 The proportions of the concrete mixtures (based on saturated surface dry condition)

Constituents	N-SCC	D-SCC	S-SCC	DS-SCC
Cement: kg/m^3	160	160	160	160
Fly ash: kg/m^3	130	130	130	130
GGBFS: kg/m^3	110	110	110	110
Cementitious content: kg/m^3	400	400	400	400
Water: lit/m^3	208	208	208	208
Water cementitious ratio	0·52	0·52	0·52	0·52
Fine aggregate: kg/m^3				
Coarse sand	660	660	660	660
Fine sand	221	221	221	221
Coarse aggregate: kg/m^3	820	820	820	820
Admixtures: lit/m^3				
Superplasticiser	4	4·86	4·73	4·5
VMA	1·3	1·3	1·3	1·3
High range water reducing agent	1·6	1·6	1·6	1·6
Fibre content: kg/m^3				
Steel	–	30	–	15
PP	–	–	5	3

N-SCC, control self-compacting concrete (SCC) mixture; D-SCC, fibre-reinforced SCC (FRSCC) with steel fibres; S-SCC, FRSCC containing polypropylene (PP) fibres; DS-SCC, FRSCC containing hybrid steel and PP fibres; GGBFS, granulated blast furnace slag; VMA, viscosity-modifying admixtures.

Samples' preparation and curing conditions

We used ϕ150 mm × 300 mm molds for the determination of compressive and splitting tensile strengths, and cylindrical molds ϕ150 mm × 300 mm were used for the determination of the modulus of elasticity. Meanwhile, 100 mm × 100 mm × 350 mm molds were used for the determination of modulus of rupture. Specimens for testing the hardened properties were prepared by direct pouring of concrete into molds without compaction. The specimens were kept covered in a controlled chamber at 20 ± 2°C for 24 h until demolding. Thereafter, the specimens were placed in water presaturated with lime at 20°C. These specimens were tested at 3, 7, 14, 28, 56 and 91 days.

Samples' test methods

The compressive strength test, performed on ϕ150 mm × 300 mm cylinders, followed AS 1012.14 (SA, 1991) and ASTM C39 (ASTM, 2000) tests for compressive strength of cylindrical concrete specimens. The cylinders were loaded in a testing machine under load control at the rate of 0·3 MPa/s until failure. The splitting tensile test, run on ϕ150m m × 300 mm cylinders, was in accordance with the AS 1012.10 (SA, 2000a) and ASTM C496 (ASTM, 2000) tests for splitting tensile strength of cylindrical concrete specimens, although ACI committee 544.2R (ACI, 1999) hardly recommends the use of the test on FRC. The running arose because the ratio offibre length to cylinder diameter took a low value of 0·23 in the work and because some investigators have shown that the ASTM C496 test is applicable to FRC specimens.

The modulus of elasticity test that followed the AS 1012.17 (SA, 1997) and ASTM C469 was done to ϕ150 mm × 300 mm cylinders. The flexural strength (modulus of rupture) test, conducted using 100 mm × 100 mm × 350 mm test beams under third-point loading, followed the AS 1012.11 (SA, 2000b) and ASTM C1018 test for flexural toughness and first-crack strength of fibre-reinforced concrete. The mid-span deflection was the average of the ones detected by the transducers through contact with brackets attached to the beam specimen.

Properties of fresh concrete

The experiments required for the SCC are generally carried out worldwide under laboratory conditions. These experiments test the liquidity, segregation, placement and compacting of fresh concrete. Conventional workability experiments are not sufficient for the evaluation of SCC. Some of the experimental methods developed to measure the liquidity, segregation, placement and compaction of SCC are defined in the European guidelines (EFNARC, 2005) and ACI 237R-07 (ACI, 2007) for SCC, including specification, production and use as slump–flow, V-funnel, U-box, L-box and fill-box tests.

This study performed slump flow, T_{50cm} time, J-ring flow, V-funnel flow time and L-box blocking ratio tests. In order to reduce the effect of loss of workability on the variability of test results, the fresh properties of the mixes were determined within 30 min after mixing. The order of testing is as follows: (*a*) slump flow test and measurement of T_{50cm} time; (*b*) J-ring flow test, measurement of difference in height of concrete inside and outside the J-ring, and measurement of T_{50cm} time; (*c*) V-funnel flow tests at 10 s T_{10s} and 5 min T_{5min}; and (*d*) L-box test.

Experimental results

Properties of fresh concrete

The results of various fresh properties tested by the slump flow test (slump flow diameter and T_{50cm}); J-ring test (flow diameter); L-box test (time taken to reach 400 mm distance T_{400mm}, time taken to reach 600 mm distance T_{600mm}, time taken to reach 800 mm distance T_L and ratio of heights at the two edges of L-box [H_2/H_1]); and the V-funnel test (time taken by concrete to flow through V-funnel after 10 s T_{10s}, and time taken by concrete to flow through V-funnel after 5 min T_{5min}); the amount of entrapped air; and the specific gravity of mixes are given in Table 9. The slump flow test judges the capability of concrete to deform under its own weight against the friction of the surface with no restraint present. A slump flow value ranging from 500 to 700 mm for SCC was suggested (EFNARC, 2005). At a slump flow > 700 mm, the concrete might segregate and, at < 500 mm, the concrete might have insufficient flow to pass through highly congested reinforcements. All the mixes in the present study conform to the above range because the slump flow of SCC is in the range of 600–700 mm. The slump flow time for the concrete to reach a diameter of 500 mm for all mixes was less than 4·5 s. The J-ring diameters were in the range of 560–655 mm. In addition to the slump flow test, a V-funnel test was also performed to assess the flowability and stability of SCC. V-funnel flow time is the elapsed time in seconds between the opening of the bottom outlet, depending on when it is opened (T_{10s} and T_{5min}), and the time when light becomes visible at the bottom when observed from the top. A V-funnel time of less than 6 s is recommended for SCC. According to EFNARC (2005), a period ranging from 6 to 12 s is considered adequate for SCC. The V-funnel flow times in the experiment were in the range of 4–10 s. The test results of this investigation indicated that all N-SCC and D-SCC mixes met the requirements of allowable flow time, but the S-SCC and DS-SCC mixes were blocked. The maximum size of coarse aggregates was restricted to 10 mm to avoid a blocking effect in the L-box. The gap between rebars in the L-box test was 35 mm. The L-box ratio H_2/H_1 for the N-SCC mix was above 0·8 which is, according to EFNARC standards and, obviously, for other mixes, blocked.

Table 9 The self-compacting concrete (SCC) mixes workability characteristics

Workability characteristics	N-SCC	D-SCC	S-SCC	DS-SCC
Average spreading diameter: mm	680	670	700	650
Flow time T_{50cm}: s	2·7	3·8	2·5	3·2
Average J-ring diameter: mm	655	580	570	560
Flow time T_{50cm} J-ring: s	3·2	5	6	5
L-box test	0·87	Blocked	Blocked	Blocked
Flow time V-funnel: s	6	7	Blocked	Blocked
V-funnel at $T_{5minutes}$: s	4	5	Blocked	Blocked
Entrapped air: %	1·3	1·2	1·2	1·0
Specific gravity: kg/m^3	2340	2274	2330	2385

N-SCC, control SCC mixture; D-SCC, fibre-reinforced SCC (FRSCC) with steel fibres; S-SCC, FRSCC with polypropylene (PP) fibres; DS-SCC, FRSCC with hybrid steel and PP fibres.

Compressive strength

Figure 1 presents the compressive strength of N-SCC, D-SCC, S-SCC and DS-SCC mixes achieved at different ages. Compressive strength samples with fibre mixes are higher than the N-SCC mix. Samples with the S-SCC mix have lower compressive strength, unlike the D-SCC and DS-SCC mixes. The average compressive strength of the DS-SCC mix is 18·90%, 3·83% and 12·86% higher than the N-SCC, D-SCC and S-SCC mixes respectively. The results show that the D-SCC mix at three days was 32·57%, 26·13% and 22·73% higher than the N-SCC, S-SCC and DS-SCC mixes respectively. Furthermore, the results indicate that the compressive strength of the DS-SCC mix at 91 days is 10·71%, 1·62% and 8·32% higher than the N-SCC, D-SCC and S-SCC mixes respectively.

Tensile strength

Figure 2 shows the splitting tensile strengths of the N-SCC, D-SCC, S-SCC and DS-SCC mixes determined at different ages. The tensile strengths of the D-SCC and DS-SCC samples are higher than those of the N-SCC and S-SCC. The S-SCC mix has a lower tensile strength than N-SCC. The average tensile strength of the D-SCC mix is 23·52%, 27·19% and 15·54% higher than that of the N-SCC, S-SCC and DS-SCC mixes respectively. Moreover, the results indicate that the tensile strength of the D-SCC mix at 91 days is 15·95%, 18·89% and 11·76% higher than that of the N-SCC, S-SCC and DS-SCC mixes respectively.

Modulus of elasticity

Figure 3 presents the modulus of elasticity of the N-SCC, D-SCC, S-SCC and DS-SCC mixes attained at different ages. The average modulus of elasticity of the DS-SCC mix is 2·67%, 4·75% and 3·49% higher than that of the N-SCC, D-SCC and S-SCC mixes respectively. The results show that the N-SCC mix at 14 days is 9·62%, 7·94% and 3·03% higher than the D-SCC, S-SCC and DS-SCC mixes respectively. Additionally, the results indicate that the

Figure 1 Compressive strength of self-compacting concrete (SCC) mixtures at different ages. N-SCC, control SCC mixture; D-SCC, fibre-reinforced SCC (FRSCC) with steel fibres; S-SCC, FRSCC with polypropylene (PP) fibres; DS-SCC, FRSCC with hybrid steel and PP fibres

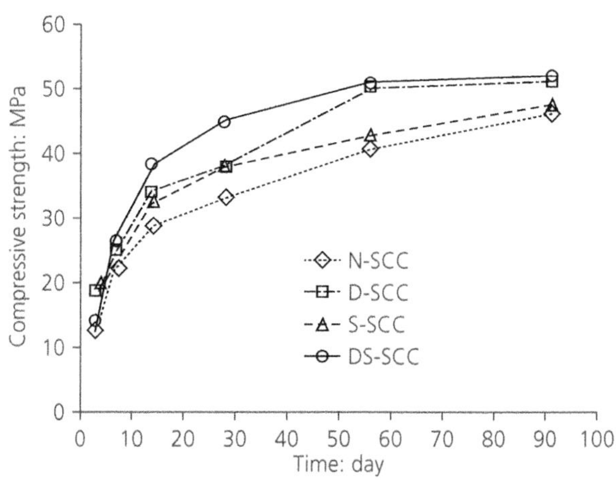

Figure 2 Tensile strength of self-compacting concrete (SCC) mixtures at different ages. N-SCC, control SCC mixture; D-SCC, fibre-reinforced SCC (FRSCC) with steel fibres; S-SCC, FRSCC with polypropylene (PP) fibres; DS-SCC, FRSCC with hybrid steel and PP fibres

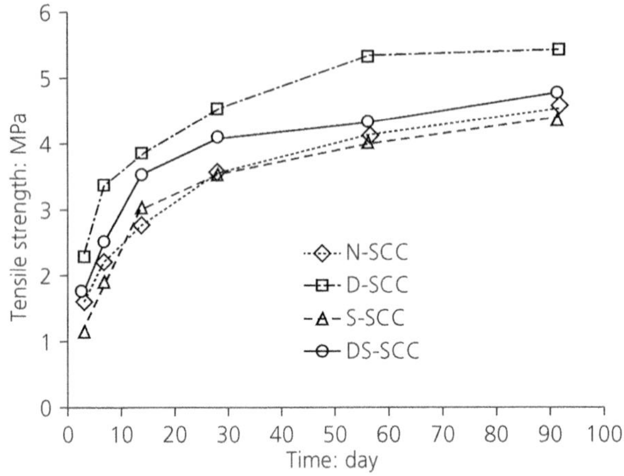

Figure 3 Modulus of elasticity of self-compacting concrete (SCC) mixtures at different ages. N-SCC, control SCC mixture; D-SCC, fibre-reinforced SCC (FRSCC) with steel fibres; S-SCC, FRSCC with polypropylene (PP) fibres; DS-SCC, FRSCC with hybrid steel and PP fibres

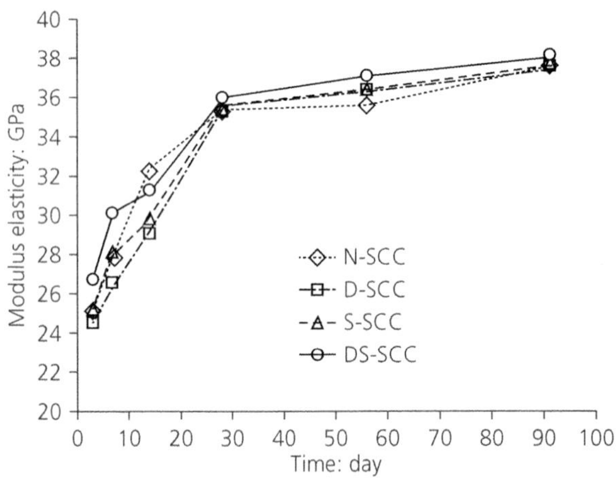

tensile strength of the DS-SCC mix at 91 days is 0·86%, 1·41% and 1·72% higher than that of the N-SCC, D-SCC and S-SCC mixes respectively.

Modulus of rupture (flexural tensile strength)

Figure 4 illustrates the flexural tensile strengths of the N-SCC, D-SCC, S-SCC and DS-SCC mixes determined at different ages. The average flexural tensile strength of the D-SCC mix is 13·96%, 8·80% and 8·89% higher than that of the N-SCC, S-SCC and DS-SCC mixes respectively. The results show that the S-SCC mix at 7 days is 21·30%, 3·97% and 10·52%

Figure 4 Modulus of rupture of self-compacting concrete (SCC) mixtures at different ages. N-SCC, control SCC mixture; D-SCC, fibre-reinforced SCC (FRSCC) with steel fibres; S-SCC, FRSCC with polypropylene (PP) fibres; DS-SCC, FRSCC with hybrid steel and PP fibres

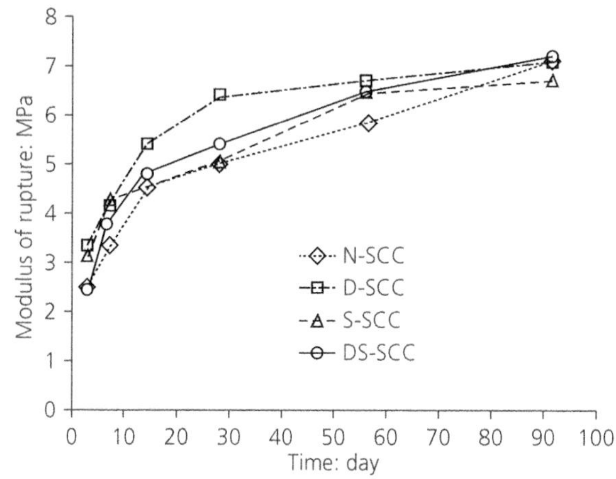

higher than the N-SCC, D-SCC and DS-SCC mixes respectively. Also, the results indicate that flexural tensile strength of the D-SCC mix at 91 days is 1·30%, 6·44% and 0·21% higher than that of the N-SCC, S-SCC and DS-SCC mixes respectively.

Compressive stress–strain curve

Complete stress–strain curves of the concrete specimens were obtained from the compression tests of the cylinders with a controlled displacement rate. For each mix, three cylinders were tested. As the test results reproduced well, each stress–strain curve shown in Figures 5–8 represents the average results of the three tests. It should be noted that the

Figure 5 Compressive stress–strain curve of the control self-compacting concrete mixture mix at different ages

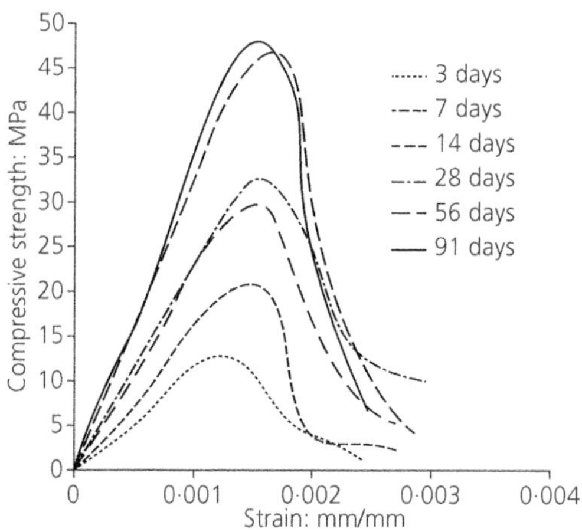

Figure 6 Compressive stress–strain curve of the fibre-reinforced self-compacting concrete containing steel fibres mix at different ages

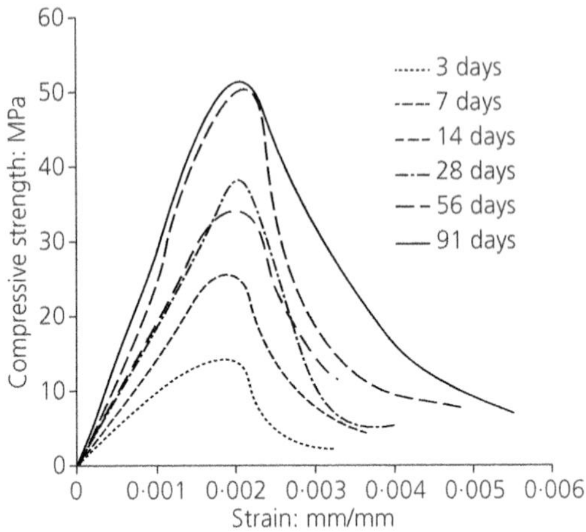

Figure 7 Compressive stress–strain curve of the fibre-reinforced self-compacting concrete containing polypropylene fibres mix at different ages

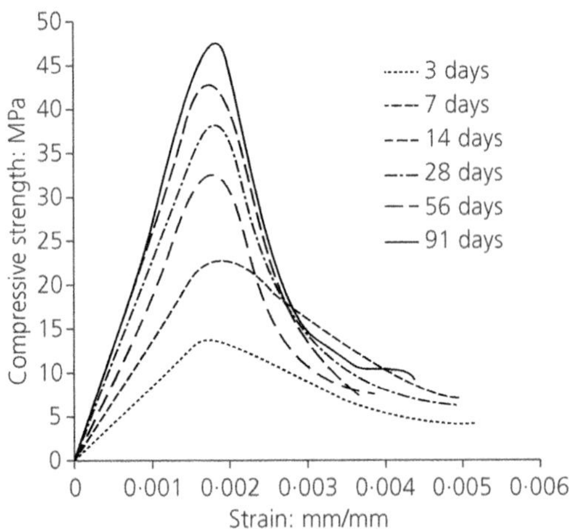

axial strains of the concrete in compression were obtained from the full height shortening of the cylinders using linear variable differential transformers. The compression stress–strain curves at increasing ages of the N-SCC, D-SCC, S-SCC and DS-SCC mixes are shown in Figures 5–8. All the fibrous SCC mixes verified more substantial ductility than the corresponding N-SCC mix. Commonly, the nature of failure in compression for the N-SCC

Figure 8 Compressive stress–strain curve of the fibre-reinforced self-compacting concrete containing hybrid steel and polypropylene fibres mix at different ages

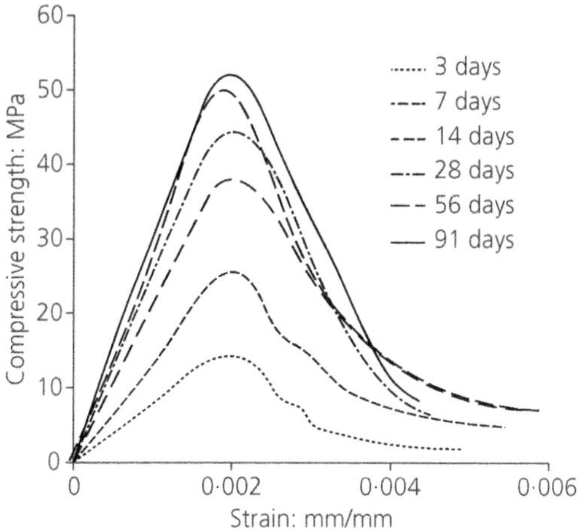

mix tended to be more sudden and brittle as the age of the concrete increased. However, with the increasing age, the majority of the fibrous SCC mixes maintained their ductility and gradual failure mechanism.

Analytical relationships for the mechanical properties

Time-related mechanical properties relationships

To estimate the SCC mixes' compressive strength, tensile strength, modulus of elasticity and modulus of rupture at various ages, Equations 1–4 are proposed based on regression analyses of the experimental data. Figure 9 shows that the proposed time-related relationships of compressive strength, tensile strength, modulus of elasticity and modulus of rupture are in good agreement with the experimental results.

Compressive strength

$$f_{\mathrm{cm}}(t) = \frac{f'_{\mathrm{c}}}{\alpha}\, ln\,(t) + \beta \tag{1}$$

where f'_{cN} is the N-SCC mix compressive strength, f'_{cfD} is the D-SCC mix compressive strength, f'_{cfS} is the S-SCC mix compressive strength, f'_{cfDs} is the DS-SCC mix compressive strength, and α and β are the empirical constants (Table 10).

Tensile strength

$$f_{\mathrm{ctm}}(t) = \frac{f_{\mathrm{ct}}}{\gamma}\, ln\,(t) + \lambda \tag{2}$$

Figure 9 Predicted time-related mechanical properties values versus experimented values (a) compressive strength, (b) tensile strength, (c) modulus of elasticity and (d) modulus of rupture. N-SCC, control SCC mixture; D-SCC, fibre-reinforced SCC (FRSCC) with steel fibres; S-SCC, FRSCC with polypropylene (PP) fibres; DS-SCC, FRSCC with hybrid steel and PP fibres

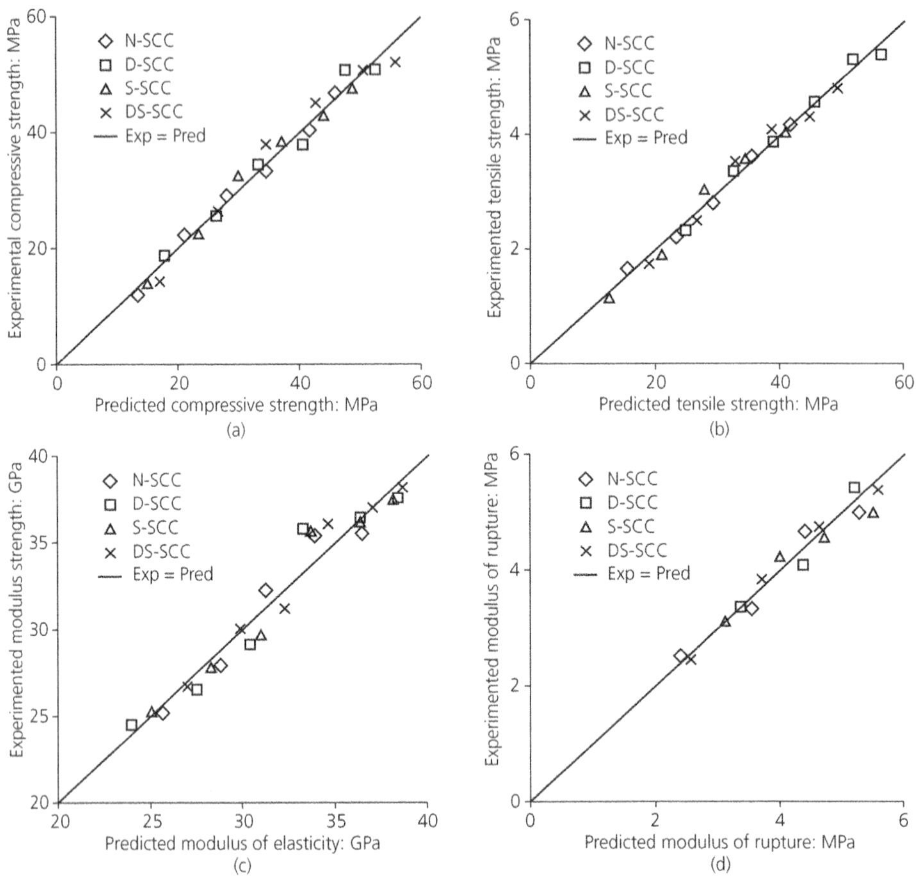

Table 10 Equation 1 constant components for different mixes

Mix	f'_c	α	β
N-SCC	f'_{cN}	3·47	2·54
D-SCC	f'_{cfD}	3·75	6·66
S-SCC	f'_{cfS}	3·84	3·87
DS-SCC	f'_{cfDs}	3·96	4·54

N-SCC, control SCC mixture; D-SCC, fibre-reinforced SCC (FRSCC) with steel fibres; S-SCC, FRSCC with polypropylene (PP) fibres; DS-SCC, FRSCC with hybrid steel and PP fibres.

where f'_{ctN} is the N-SCC mix tensile strength, f'_{ctfD} is the D-SCC mix tensile strength, f'_{ctfS} is the S-SCC mix tensile strength, f'_{ctfDS} is the DS-SCC mix tensile strength, and γ and λ are the empirical constants (Table 11).

Modulus of elasticity

$$E_{\mathrm{cm}}(t) = \frac{E_{\mathrm{c}}}{\eta} \, ln\,(t) + \mu \tag{3}$$

where E_{cN} is the N-SCC mix modulus of elasticity, E_{cfD} is the D-SCC mix modulus of elasticity, E_{cfS} is the S-SCC mix modulus of elasticity, E_{cfDS} is the DS-SCC mix modulus of elasticity, and η and μ are the empirical constants (Table 12).

Modulus of rupture

$$f_{\mathrm{crm}}(t) = \frac{f_{\mathrm{cr}}}{\psi} \, ln\,(t) + \phi \tag{4}$$

Table 11 Equation 2 constant components for different mixes

Mix	f_{ct}	γ	λ
N-SCC	f'_{ctN}	4·09	0·60
D-SCC	f'_{ctfD}	4·87	1·43
S-SCC	f'_{ctfS}	3·69	0·19
DS-SCC	f'_{ctfDS}	4·60	0·91

N-SCC, control SCC mixture; D-SCC, fibre-reinforced SCC (FRSCC) with steel fibres; S-SCC, FRSCC with polypropylene (PP) fibres; DS-SCC, FRSCC with hybrid steel and PP fibres.

Table 12 Equation 3 constant components for different mixes

Mix	E_c	η	μ
N-SCC	E_{cN}	9·47	21·42
D-SCC	E_{cfD}	8·40	19·20
S-SCC	E_{cfS}	9·30	20·83
DS-SCC	E_{cfDS}	10·47	23·15

N-SCC, control SCC mixture; D-SCC, fibre-reinforced SCC (FRSCC) with steel fibres; S-SCC, FRSCC with polypropylene (PP) fibres; DS-SCC, FRSCC with hybrid steel and PP fibres.

Table 13 Equation 4 constant components for different mixes

Mix	f_{cr}	ψ	φ
N-SCC	f_{crN}	3·89	1·00
D-SCC	f_{crfD}	5·39	2·07
S-SCC	f_{crfS}	4·75	1·96
DS-SCC	f_{crfDS}	3·99	1·08

N-SCC, control SCC mixture; D-SCC, fibre-reinforced SCC (FRSCC) with steel fibres; S-SCC, FRSCC with polypropylene (PP) fibres; DS-SCC, FRSCC with hybrid steel and PP fibres.

where f_{crN} is the N-SCC mix modulus of rupture, f_{crfD} is the D-SCC mix modulus of rupture, f_{crfS} is the S-SCC mix modulus of rupture, f_{crfDS} is the DS-SCC mix modulus of rupture, and ψ and ϕ are the empirical constants (Table 13).

Compressive strength-related mechanical properties relationships

Figure 10 illustrates tensile strength, modulus of elasticity and modulus of rupture versus compressive strength. Equations 3–7 are proposed based on regression analyses of the experimental data to predict the SCC mixes' tensile strength, modulus of elasticity and modulus of rupture based on the compressive strength. The bases of the proposed relationships are captured from Aslani and Nejadi's (2012a) study. Figure 11 indicates that the proposed compressive strength-related relationships of tensile strength, modulus of elasticity and modulus of rupture are in good agreement with the experimental results.

Tensile strength

$$f_{ct} = \eta_1 \left(f'_c\right)^{\eta_2} \quad (5)$$

(Table 14)

Modulus of elasticity

$$E_c = \kappa_1 \left(f'_c\right)^{\kappa_2} \quad (6)$$

(Table 15)

Modulus of rupture

$$f_{cr} = \delta_1 \left(f'_c\right)^{\delta_2} \quad (7)$$

(Table 16)

Figure 10 Experimented compressive strength-related mechanical properties (a) tensile strength, (b) modulus of elasticity and (c) modulus of rupture. N-SCC, control SCC mixture; D-SCC, fibre-reinforced SCC (FRSCC) with steel fibres; S-SCC, FRSCC with poly-propylene (PP) fibres; DS-SCC, FRSCC with hybrid steel and PP fibres

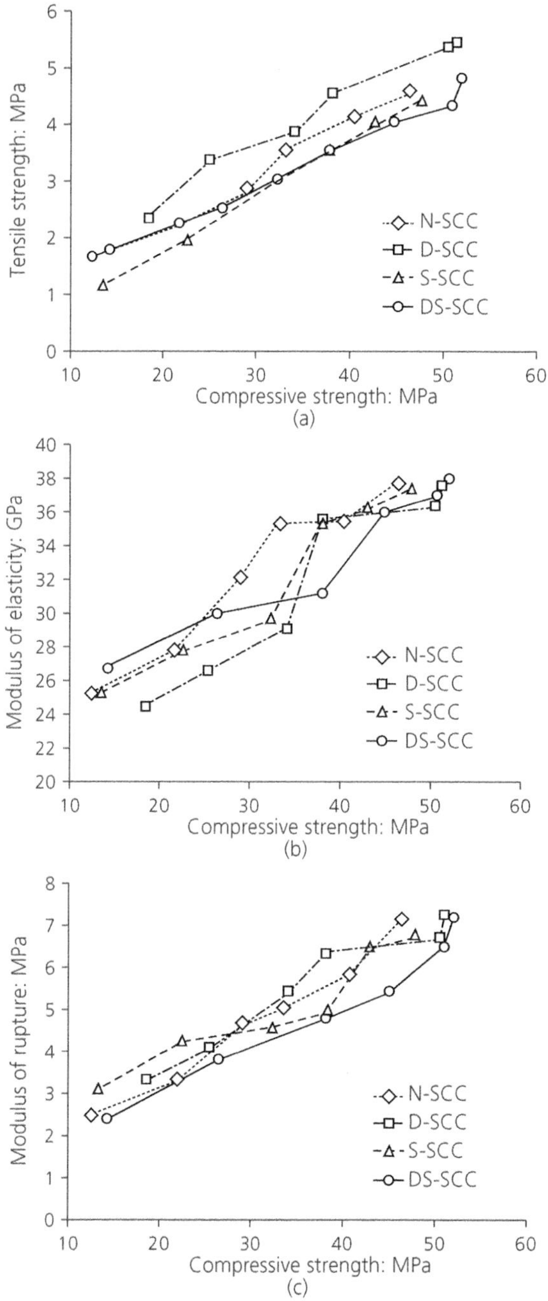

Figure 11 Predicted compressive strength-related mechanical properties values versus experimented values (a) tensile strength, (b) modulus of elasticity and (c) modulus of rupture. N-SCC, control SCC mixture; D-SCC, fibre-reinforced SCC (FRSCC) with steel fibres; S-SCC, FRSCC with polypropylene (PP) fibres; DS-SCC, FRSCC with hybrid steel and PP fibres

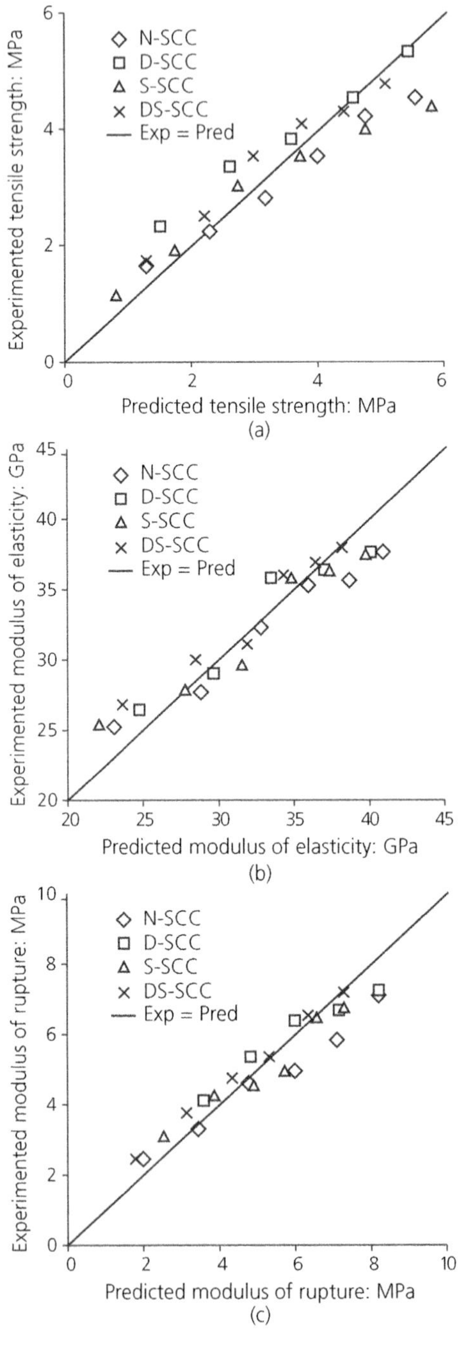

Table 14 Equation 5 constant components for different mixes

Mix	f_{ct}	f'_c	η_1	η_2
N-SCC	f_{ctN}	f'_{cN}	0·204	0·8047
D-SCC	f_{ctfD}	f'_{cfD}	0·237	0·7999
S-SCC	f_{ctfS}	f'_{cfS}	0·067	1·0889
DS-SCC	f_{ctfDS}	f'_{cfDs}	0·226	0·7585

N-SCC, control SCC mixture; D-SCC, fibre-reinforced SCC (FRSCC) with steel fibres; S-SCC, FRSCC with polypropylene (PP) fibres; DS-SCC, FRSCC with hybrid steel and PP fibres.

Table 15 Equation 6 constant components for different mixes

Mix	E_c	f'_c	K_1	K_2
N-SCC	E_{cN}	f'_{cN}	10·913	0·3226
D-SCC	E_{cfD}	f'_{cfD}	6·649	0·4383
S-SCC	E_{cfS}	f'_{cfS}	10·395	0·3271
DS-SCC	E_{cfDS}	f'_{cfDs}	12·895	0·2651

N-SCC, control SCC mixture; D-SCC, fibre-reinforced SCC (FRSCC) with steel fibres; S-SCC, FRSCC with polypropylene (PP) fibres; DS-SCC, FRSCC with hybrid steel and PP fibres.

Table 16 Equation 7 constant components for different mixes

Mix	f_{cr}	f'_c	δ_1	δ_2
N-SCC	f_{crN}	f'_{cN}	0·325	0·7871
D-SCC	f_{crfD}	f'_{cfD}	0·376	0·7511
S-SCC	f_{crfS}	f'_{cfS}	0·670	0·5818
DS-SCC	f_{crfDS}	f'_{cfDS}	0·309	0·7714

N-SCC, control SCC mixture; D-SCC, fibre-reinforced SCC (FRSCC) with steel fibres; S-SCC, FRSCC with polypropylene (PP) fibres; DS-SCC, FRSCC with hybrid steel and PP fibres.

Compressive stress–strain relationship

In this study, a compressive stress–strain relationship (Equations 8–15) for SCC mixes that is based on Aslani and Nejadi's (2012a) model was developed by using the proposed compressive strength (Equation 1) and elastic modulus (Equations 3 and 6) relationships. Figure 12 shows that the proposed stress–strain relationship fits the experimental results well. In Figure 12, typical 91-day age compressive stress–strain curve results are selected to compare with the proposed compressive stress–strain relationship.

$$\frac{\sigma_c}{f'_c} = \frac{n\left(\frac{\varepsilon_c}{\varepsilon'_c}\right)}{n-1+\left(\frac{\varepsilon_c}{\varepsilon'_c}\right)^n} \tag{8}$$

Figure 12 Comparison between experimented compressive stress–strain curve result with proposed relationship (a) control self-compacting concrete (SCC) mixture (N–SCC), (b) fibre-reinforced SCC (FRSCC) containing steel fibres (D–SCC), (c) FRSCC containing polypropylene (PP) fibres (S–SCC) and (d) FRSCC containing hybrid steel and PP fibres (DS-SCC)

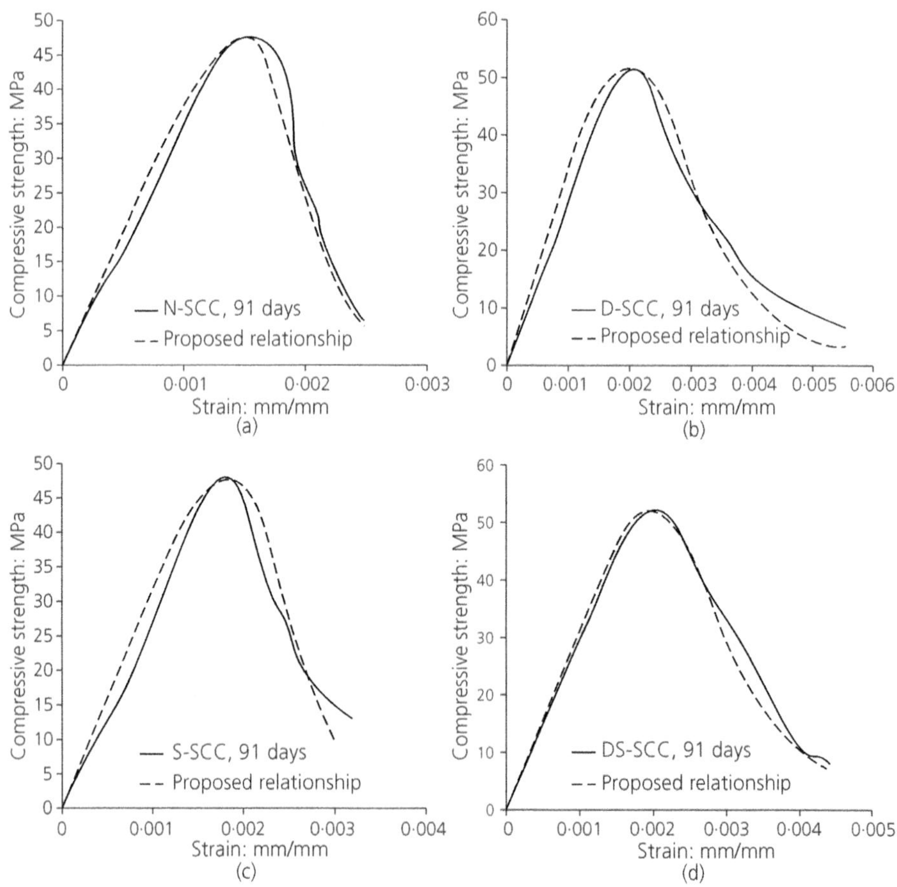

$$n = n_1 = [1{\cdot}02 - 1{\cdot}17\,(E_{sec}/E_c)]^{-0{\cdot}74} \quad \textit{if } \varepsilon_c \le \varepsilon'_c \tag{9}$$

$$n = n_2 = n_1 + (\rho + 28 \times \omega) \quad \text{if } \varepsilon_c \ge \varepsilon'_c \tag{10}$$

$$\rho = (\,135{\cdot}16 - 0{\cdot}1744 f'_c)^{-0{\cdot}46} \tag{11}$$

$$\omega = 0.83 exp\,(-911/f'_c) \tag{12}$$

$$E_{sec} = f'_c/\varepsilon'_c \tag{13}$$

$$\varepsilon\prime_c = \left(\frac{f'_c}{E_c}\right)\left(\frac{v}{v-1}\right) \tag{14}$$

$$v = \frac{f'_c}{17} + 0.8 \tag{15}$$

where σ_c is concrete stress, f'_c is the maximum compressive strength of concrete, n is the material parameter that depends on the shape of the stress–strain curve, ε is the concrete strain, ε'_c is the strain corresponding with the maximum stress f'_c, n_1 is the modified material parameter at the ascending branch, n_2 is the modified material parameter at the descending branch, E_c is the modulus of elasticity, E_{sec} is the secant modulus of elasticity, n_1 is the modified material parameter at the ascending branch, n_2 is the modified material parameter at the descending branch and ρ, ω are the coefficients of linear equation.

Conclusions

The following conclusions can be drawn from this study.

- A novel experimental program was performed. Four different SCC mixes were used in the experiment. These mixes include N-SCC, D-SCC, S-SCC and DS-SCC.
- The compressive strength, tensile strength, modulus of elasticity, modulus of rupture and compressive stress–strain curve are tested at ages of 3, 7, 14, 28, 56 and 91 days. These properties' differences between SCC mixes show that these differences decrease with time.
- The average compressive strength of the DS-SCC mix is higher than that of the N-SCC, D-SCC and S-SCC mixes respectively. The results indicate that compressive strength of the DS-SCC mix at 91 days is 10·71%, 1·62% and 8·32% higher than that of the N-SCC, D-SCC and S-SCC mixes respectively.
- The average tensile strength of the D-SCC mix is higher than that of the DS-SCC, N-SCC and S-SCC mixes respectively. The results show that the tensile strength of D-SCC mix at 91 days is 15·95%, 18·89% and 11·76% higher than that of the N-SCC, S-SCC and DS-SCC mixes respectively.

- The average modulus of elasticity of the DS-SCC mix is higher than that of the N-SCC, D-SCC and S-SCC mixes respectively. Additionally, the results indicate that tensile strength of the DS-SCC mix at 91 days is 0·86%, 1·41% and 1·72% higher than that of the N-SCC, D-SCC and S-SCC mixes respectively.
- The average modulus of rupture of the D-SCC mix is higher than that of the N-SCC, S-SCC and DS-SCC mixes respectively. Also, the results indicate that the modulus of rupture of the D-SCC mix at 91 days is 1·30%, 6·44% and 0·21% higher than that of the N-SCC, S-SCC and DS-SCC mixes respectively.
- Analytical expressions to predict the most significant mechanical properties (i.e., compressive strength, tensile strength, modulus of elasticity and modulus of rupture) of the developed SCC mixes at an age *t* were presented.
- Analytical relationships to calculate the tensile strength, modulus of elasticity and modulus of rupture of the SCC mixes related to compressive strength were proposed.
- A compressive stress–strain relationship was proposed to model the behavior of the SCC mixes from an early age. This relationship was capable of modelling the complete compressive behavior of normal SCC and FRSCC with high accuracy.

Acknowledgements

This work was supported by Centre for Built Infrastructure Research, School of Civil and Environmental Engineering, University of Technology Sydney, Australia. The authors would like to express their sincere gratitude and appreciation to Boral, BOSFA and Concrite companies.

REFERENCES

ACI (1999) ACI544.2R: State-of-the-art report on fiber reinforced concrete. Technical report. ACI, Farmington Hills, MI, USA.

ACI (2007) ACI Committee 237: 237R-07. Self-consolidating concrete. American Concrete Institute, Farmington Hills, MI, USA.

Aslani F and Nejadi S (2012a) Mechanical properties of conventional and self-compacting concrete: An analytical study. *Construction Building Materials* **36**: 330–347.

Aslani F and Nejadi S (2012b) Bond characteristics of steel fibre reinforced self-compacting concrete. *Canadian Journal of Civil Engineering* **39(7)**: 834–848.

Aslani F and Nejadi S (2012c) Bond behaviour of reinforcement in conventional and self-compacting concrete. *Advances in Structural Engineering* **15(12)**: 2033–2051.

Aslani F and Nejadi S (2012d) Shrinkage behaviour of self-compacting concrete. *Journal of Zhejiang University SCIENCE A* **13(6)**: 407–419.

ASTM (2000) *Annual Book of ASTM Standards 2000. Volume 04.02, Concrete and Aggregates.* ASTM International, West Conshohocken, PA, USA.

Balaguru PN and Shah SP (1992) *Fiber Reinforced Cement Composites*. McGraw-Hill, New York, USA.

Bui VK, Shah SP and Geiker MR (2003) *Rheology of Fiber-Reinforced Cementitious Materials. RILEM PRO30 High Performance Fibre Reinforced Composites HPFRCC4*. Reunion Internationale des Laboratoires et Experts des Materiaux, Systemes de Construction et Ouvrages, Bagneux, France, pp. 221–232.

Busterud L, Johansen K and Døssland ÅL (2005) *Production of Fiber Reinforced SCC*. SCC 2005, Session C-3: Fiber-Reinforced SCC.

Corinaldesi V and Moriconi G (2004) Durable fiber reinforced self-compacting concrete. *Cement and Concrete Research* **34(2)**: 249–254.

Cunha V (2006) *Compression behaviour of steel fibre reinforced self-compacting concrete – age influence and modelling*. Report 06-DEC/E-04 University of Minho, Portugal.

EFNARC (2005) *The European Guidelines for Self-Compacting Concrete, Specification, Production and Use*. Experts for Specialised Construction and Concrete Systems, Farnham, UK.

Groth P and Nemegeer D (1999) The use of steel fibres in self-compacting concrete. In *Proceedings of the First International RILEM Symposium on Compacting Concrete, Stockholm, Sweden* (Skarendahl Å and Petersson Ö (eds)), pp. 497–507.

Grünewald S (2006) *Performance-based Design of Self-Compacting Fibre Reinforced Concrete*. PhD thesis, Delft University of Technology, Delft, the Netherlands.

Grünewald S and Walraven JC (2001) Parameter-study on the influence of steel fibers and coarse aggregate content on the fresh properties of self-consolidating concrete. *Cement and Concrete Research* **31(12)**: 1793–1798.

Khayat KH and Roussel Y (1999) Testing and performance of fiber-reinforced, self-consolidating concrete. In *Proceedings of Symposium on Self-compacting Concrete. Stockholm, Sweden* (Skarendahl A (eds)), pp. 509–521.

Liao WC, Chao SH, Park SY and Naaman AE (2006) *Self-Consolidating High Performance Fiber Reinforced Concrete (SCHPFRC) – Preliminary Investigation*. Report UMCEE 06-02 Department of Civil and Environmental Engineering University of Michigan Ann Arbor, USA, p. 68.

Massicotte B, Degrange G and Dzeletovic N (2000) *Mix Design for SFRC Bridge Deck Construction. RILEM PRO15 Fibre-Reinforced Concrete BEFIB*. RILEM, Bagneux, France, pp. 119–128.

RTA (Regional Transportation Authority) (2006) *Materials Test Methods*. vol. 1. RTA, Chicago, IL, USA.

SA (Standards Australia) (1991) AS1012.14: Method for securing and testing from hardened concrete for compressive strength. Standards Australia, Sydney, Australia.

SA (1997) AS1012.17: Determination of the static chord modulus of elasticity and Poisson's ratio of concrete specimens. Standards Australia, Sydney, Australia.

SA (1998) AS3583: Methods of test for supplementary cementitious materials for use with Portland cement. Standards Australia, Sydney, Australia.

SA (2000a) AS1012.10: Determination of indirect tensile strength of concrete cylinders. Standards Australia, Sydney, Australia.

SA (2000b) AS1012.11: Determination of modulus of rupture. Standards Australia, Sydney, Australia.

SA (2000c) AS1478.1: Chemical admixtures for concrete, mortar and grout – Admixtures for concrete. Standards Australia, Sydney, Australia.

SA (2001) AS3582.2: Supplementary cementitious materials for use with Portland and blended cement – slag – ground granulated iron blast-furnace. Standards Australia, Sydney, Australia.

SA (2006) AS2350: Methods of testing Portland and blended cements. Standards Australia, Sydney, Australia.

SA (2010) AS3972: General purpose and blended cements. Standards Australia, Sydney, Australia.

SA (2011) AS1141: Methods for sampling and testing aggregates – Particle size distribution – Sieving method. Standards Australia, Sydney, Australia.

Sahmaran M and Yaman IO (2007) Hybrid fiber reinforced self-compacting concrete with a high-volume coarse fly ash. *Construction and Building Materials* **21(1)**: 150–156.

Sahmaran M, Yurtseven A and Yaman IO (2005) Workability of hybrid fiber reinforced self-compacting. *Building and Environment* **40(12)**: 1672–1677.

Schumacher P (2008) *Rotation Capacity of Self-Compacting Steel Fiber Reinforced Concrete.* PhD thesis, Delft University of Technology, Delft, the Netherlands.

Section 4

Fresh and hardened properties of self-compacting concrete

ice Publishing

Dhir and Lye
ISBN 978-0-7277-6461-4
https://doi.org/10.1680/icetscc.64614.225

Chapter 11
Automatic image analysis process to appraise segregation resistance of self-consolidating concrete

Mahmoud Nili
Associate Professor, Civil Engineering Department, Bu-Ali Sina University, Hamedan, Iran

Mehrdad Razmara
Research Assistant, Civil Engineering Department, Bu-Ali Sina University, Hamedan, Iran
(corresponding author: mehrdaad.razmara@gmail.com)

Maryam Sadeghi
CEO, MetaOptima Technology Inc, Vancouver, BC, Canada

Majid Razmara
CTO, MetaOptima Technology Inc, Vancouver, BC, Canada

The segregation of coarse aggregate in self-consolidating concrete (SCC) can result in some defects in the long-term properties of concrete. In this study, an image processing algorithm has been proposed and implemented, using standard coding languages, to assess segregation of the SCC mixtures. This algorithm has been designed to evaluate the distribution of coarse aggregate and the average thickness of the mortar band layer of SCC. Seven mixtures containing fly ash and silica fume as partial replacements for cement were analysed. Slump, J-ring flow and compressive strength of the samples were also measured. The results revealed that an increase in fly ash replacement resulted in a higher segregation index and a higher ratio of mortar band thickness. However, substitution of silica fume caused a lower segregation index and a lower ratio of mortar band thickness. The results demonstrate that there is a high degree of correlation between the segregation index and the mortar band thickness in all the mixtures.

Introduction

Self-consolidating concrete (SCC) is a new category of high-performance concrete. It behaves as a highly flowable and non-segregating concrete that can spread into place under its own weight, and fill congested structural sections and restricted elements in the absence of mechanical consolidation without exhibiting segregation, bleeding and blockage of the flow (ACI, 2007; Fang and Labia, 2007; Kanellopoulos *et al.*, 2010; Khayat, 1999; Neville, 2006). This type of concrete was first developed in Japan in 1988 (Okamura and Ouchi, 2003). SCC is increasingly becoming an appealing option worldwide because it improves the ease of placement, increases the rate of construction, and has been found to offer a high-quality product with significant reductions in cost through reduced construction time, labour and noise, as well as savings in equipment use. It also improves the environmental quality and workers' health and safety in the construction area (ACBM, 2007; ACI, 2007; Fang and Labia, 2007; Johnson *et al.*, 2010; Khayat, 1999; Nagataki *et al.*, 2010; Neville, 2006; Paultre *et al.*, 2005; Schwarz *et al.*, 2007). Workability requirements for successful casting of SCC include high deformability (low-yield value), passing ability, and proper segregation resistance (moderate viscosity) (EFNARC, 2002; Hwang *et al.*, 2006; Johnson *et al.*, 2010). Owing to the high flowability and relatively low yield stress, and the difference in density between the cement paste and aggregates, SCC makes it susceptible to stability problems during the mixing process, transporting, pumping, spreading into place, and the dormant period (Assaad *et al.*, 2004; Barbosa *et al.*, 2011; Chermant, 2001; Fang and Labia, 2007; Kwan *et al.*, 1999). Segregation is generally defined as separation and settlement of granular particles, resulting in non-uniform distribution of aggregate particles in the hardened concrete. Segregation is still seen as one of the major problems of concrete (Breul *et al.*, 2008; Bui *et al.*, 2002; Johnson *et al.*, 2010; Khayat *et al.*, 2007; Mesbah *et al.*, 2011). Segregation can result in the hardened material developing heterogeneous properties with a direct bearing on the mechanical properties and durability. Beyond the inhomogeneity of the aggregate distribution, instability can also weaken the quality of the interface between the aggregate and cement paste and reduce the bond strength between the reinforcement and cement paste. Segregated SCC has a higher risk of cracking because of aggregate settlement, which leaves a top layer richer in paste, and it may experience more shrinkage, in comparison with unsegregated concrete. Cracking resulting from segregation can reduce the resistance of concrete to the ingress of water and ions. If exposed to freeze–thaw cycles, this can further increase permeation and reduce the mechanical properties. It also impairs the surface finish (Bui *et al.*, 2002; Khayat *et al.*, 1999; 2007; Libre *et al.*, 2010; Mehta and Monteiro, 2006; Mesbah *et al.*, 2011; Nili *et al.*, 2017; Panesar and Shindman, 2012; Pavate *et al.*, 2000). Thus, with the growing use of SCC, it is important to provide a proper assessment of the risk of segregation to ensure the quality and characteristics of the concrete (Khayat *et al.*, 2003).

Image analysis is the process of capturing a scene, digitising it into a pixel image, processing the pixel array, and then extracting information from it. Image analysis techniques are widely used in many disciplines, and in recent years they have been practised in the field of civil engineering, especially in concrete technology (Janaka *et al.*, 2012; Mora *et al.*, 1998). Automatic image analysis was first developed for a granular media study (Breul *et al.*, 2008). Digital image analysis is a quick, convenient, precise, and versatile technique to assess segregation of SCC by measuring the average depth of sedimentation of the upper layer of coarse aggregate particles (mortar band thickness), and the distribution of coarse aggregate as a function of the sample height (Barbosa *et al.*, 2010; Fernlund *et al.*, 2007; Kränkel *et al.*, 2010; Panesar and Shindman, 2011).

In this paper, the algorithms were implemented using standard programming languages to examine the segregation of the SCC specimens by a novel image processing method. The distribution of coarse aggregates and the average thickness of the mortar band layer of SCC in a 200 mm diameter, 600 mm high concrete cylinder were assessed. The effects of the fly ash and silica fume on the results were also examined.

Experimental programme

Materials and mix proportions

Type I Portland cement, in accordance with ASTM C-150-04, was used in this experiment (ASTM, 2004). Table 1 shows the chemical and physical properties of the cement, fly ash and silica fume.

Experimental tests on mix proportion, especially the percentage by mass of superplasticiser, were carried out by trial and error to determine a suitable proportion and the spread diameter of mixtures, in order to facilitate their comparison.

The sieve analysis of gravel, sand and limestone powder is given in Table 2. Table 3 summarises the compositions of the SCC mixtures. Sand with a fineness modulus of 3·37, specific gravity of 2·58 g/cm^3, and water absorption of 1·9% was used. The coarse aggregate for the SCC was crushed to a maximum size of 19 mm, specific gravity of 2·68 g/cm^3, and water absorption of 0·5%. The ratios of gravel to total aggregate and sand to total aggregate were 0·3 and 0·6, respectively. Limestone powder was used as a fine aggregate, at a proportion of 10% of the aggregate.

In the CFA5, CFA15, and CFA25 mixtures, 5%, 15%, and 25% of the cement was substituted with fly ash class F, respectively, and 15% and 25% of the cement was substituted with fly ash in the CSF8FA15 and CSF8FA25 mixtures, respectively. Similarly, 8% of cement was substituted with silica fume in the CSF8, CSF8FA15 and CSF8FA25 mixtures. The superplasticiser was a polycarboxylate-based gelenium 110p, with a specific gravity of 1·08 g/cm^3. The ratio of water to cementitious material (w/cm) and total cementitious materials were kept constant at 0·45 and 400 kg/m^3, respectively. The slump flow was fixed

Table 1 Chemical analysis and physical properties of cement, fly ash and silica fume

Parameters	Cement	Fly ash	Silica fume
SiO_2: %	21·16	61·2	89·83
Al_2O_3: %	4·82		1·31
Fe_2O_3: %	3·9		1·12
$CaO + MgO$: %	65·04	12·2	1·77
$K_2O + Na_2O$: %	0·99	1·92	0·85
SO_3: %	2·51	2·5	0
Loss on ignition: %	1·32	2·6	1·59
Fineness: m^2/kg	295		20 700

Table 2 Sieve analysis of the aggregates

Sieve size: mm	19	12·5	9·5	6·35	4·75	2·38	1·18	0·6	0·3	0·15
Gravel	100	52	6·1	0·8	0	0	0	0	0	0
Sand	100	100	100	97	92	77	52	28	8	3
Limestone powder	100	100	100	100	100	100	100	100	100	100
Mixture of aggregates	100	86	72	68	65	56	41	27	15	12

Table 3 Mix proportioning of concrete mixtures

Mixture	Cement	Water	Fine aggregate	Coarse aggregate: kg/m^3	LP	SF	FA	SP
Reference	400	180	1037	538	174	0	0	3·6
CFA5	380	180	1028	534	173	0	20	3·6
CFA15	340	180	1021	531	172	0	60	3·6
CFA25	300	180	1015	527	170	0	100	3·6
CSF8	368	180	1031	535	173	32	0	4·6
CSF8FA15	308	180	1014	527	170	32	60	4·6
CSF8FA25	268	180	1008	523	169	32	100	4·6

LP, limestone powder; SF, silica fume; FA, fly ash; SP, superplasticiser.

by adding an extra 0·25% high-range water-reducing admixture (HRWRA) in the mixtures CSF8, CSF8FA15, CSF8FA25, which added 8% silica fume by mass. In this experiment, to calculate mix proportions, the absolute volume method was used.

Sample preparation

Cylindrical concrete specimens (200 mm × 600 mm) were used for the image analysis process. Detailing of the column mould is depicted in Figure 1. As shown, a polyvinyl chloride plastic

Figure 1 Image analysis mould used for preparing specimens

pipe was attached to a non-absorbent, 300 mm × 300 mm rigid square base plate, 10 mm thick. To place the column mould firmly, the base plate was lathed by 5 mm, and to form a mortar-tight joint to the base plate, the column was attached to it.

Mixing preparation

The mixing procedure sequence consisted of homogenising the sand, limestone powder and binder for 1 min before adding the water. Next, half of the mixing water and the HRWRA were added and mixed for 2 min. Coarse aggregate and the remainder of the water were then added and mixed for 5 min. After the mixing was completed, the SCC was sampled to determine the slump flow and spread through the J-ring. The concrete was then discharged in one layer into the image analysis mould, without any vibration or tamping. The compressive strength was also determined using 100 × 100 × 100 mm cubes at the ages of 1, 2, 7, 28, 56 and 91 d.

Image processing method

This section explains the authors' image processing method to evaluate the distribution of coarse aggregate and the average thickness of the mortar band layer of the SCC.

Image acquisition

A cylindrical sample of 200 mm diameter × 600 mm height is cured 28 d after demoulding. The hardened samples are then cut vertically into two sections of 200 mm × 600 mm. Phenolphthalein at 1% by weight is sprayed on the saw-cut surface to distinguish the aggregate phase from the cement paste. Phenolphthalein is a weak acid that changes the colour of the paste to purple due to the paste's high alkalinity. Once the colour of the paste changes to purple, the concrete is rinsed with tap water to remove any excess phenolphthalein, and air dried at room temperature. Afterwards, a digital image of the concrete at 600 pixels per inch is captured (Panesar and Shindman, 2012).

An algorithm is a set of steps or rules performed, in this case by a computer, to find a solution for a problem. The surface of the concrete specimen is scanned and the captured image is sent to an algorithm developed by authors in MATLAB software for image analysis. This algorithm defines the set of steps to be followed in calculating the coarse aggregate distribution and mortar band thickness.

Image enhancement

To enhance the quality of the image and to increase the contrast of the pixels for better segmentation, a two-element vector of pixel values that specifies the lower and upper limits can be used for contrast stretching and enhancement. By default, the low and high values specify the bottom 1% and the top 1% of all the pixel values. The grey values returned can be used to increase the contrast of the image by mapping the intensity values in the greyscale image I to new values in image J, such that 1% of the data is saturated at the low and high intensities of image I. This increases the contrast of the output image J.

Particle segmentation and detection

After enhancing the image, a dynamic thresholding method called Otsu (Liao *et al.*, 2001; Otsu, 1979; Sezgin and Sankur, 2004) is used to compute a global threshold (level) to convert the grey-level intensity image to a binary one. This threshold value is a normalised intensity value that lies in the range [0, 1], chosen to minimise the intra-class variance of the black and white pixels. Based on the images used in the dataset, this threshold can be adjusted by multiplying the threshold to a constant number (M), which can be determined using machine learning algorithms. For the database used in this study, the threshold was multiplied by $M=1{\cdot}2$, based on the images in the database.

The binary thresholding approach can result in numerous imperfections, which can be addressed by applying morphological image processing methods to clean up the imperfections (Maragos, 2005). After converting Image I to a binary image B, a morphology method is applied to fill the holes. By definition, a hole is a set of background pixels that cannot be reached by filling in the background from the edge of the image. To remove noise from the binary image, all connected components (objects) that have less than nT pixels are removed (this is known as area opening). To clean the image and remove spur pixels, a morphological closing is performed on the binary image B by using a disc of radius 1 as the structural element. This operation is in fact a dilation followed by an erosion to remove the spur pixels.

Next, to measure the mortar band thickness in this image, non-zero pixels are assigned to objects of interest (the coarse aggregate particles), with zero pixels representing the background region. The following figures (Figures 2–6), which belong to the reference mixture, illustrate these components.

Figure 2 Digital image captured from concrete section

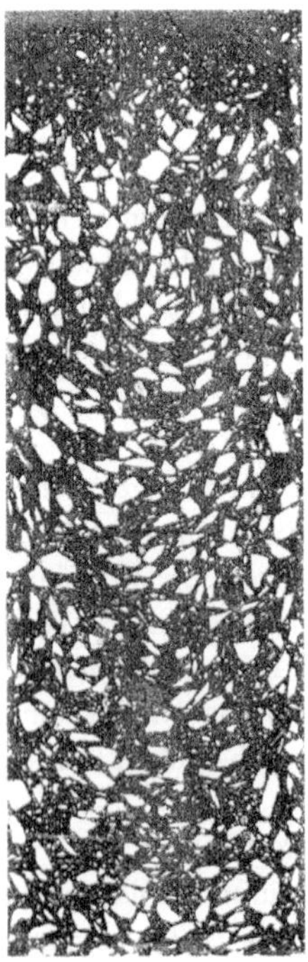

Figure 3 Conversion of image to greyscale

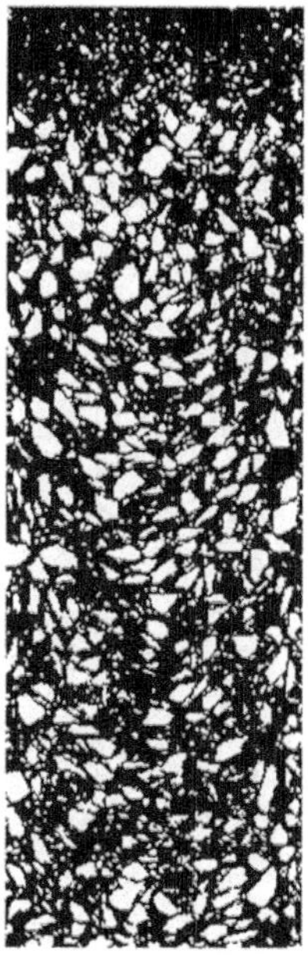

Figure 4 Selection of particles smaller than 5 mm

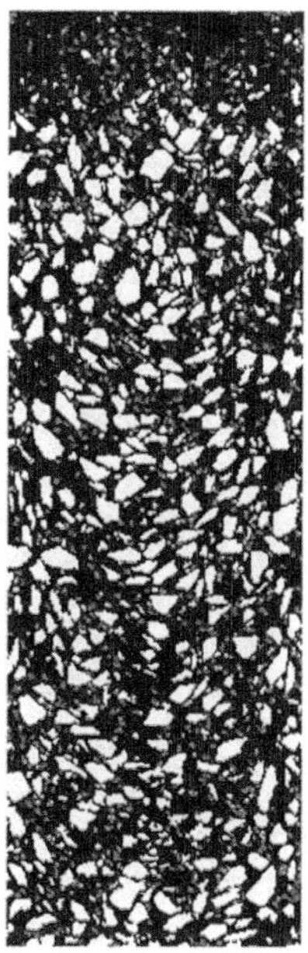

Figure 5 Removal of selected particles smaller than 5 mm

Figure 6 Selection of particles greater than 5 mm

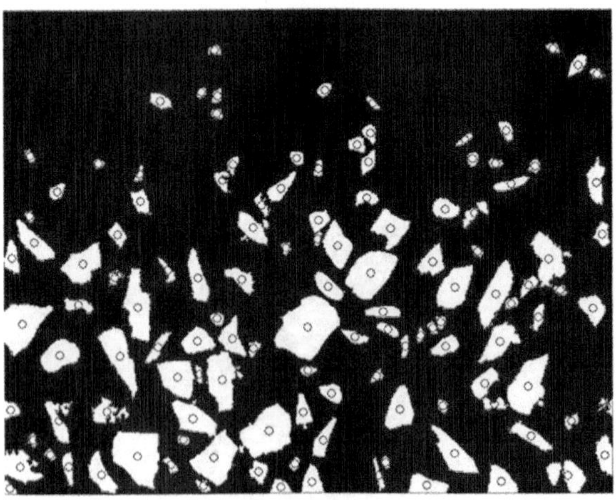

Figure 7 Attribution of the green section to mortar band thickness

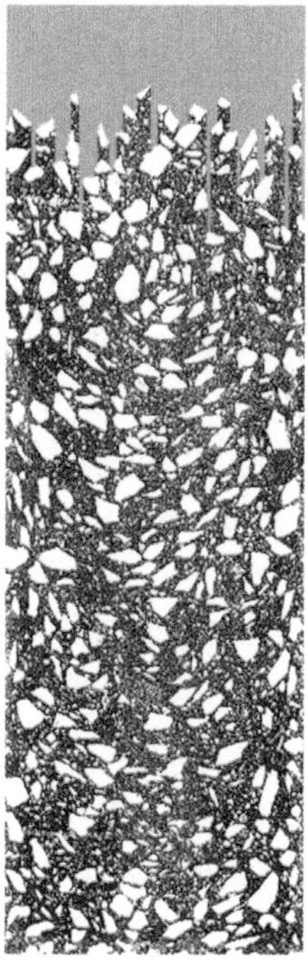

The average mortar band thickness is defined as the average distance between the first row of the pixels and the objects of interest. It is measured by tracing the boundaries of the objects and finding their minimum distance from the top. The image width is scanned to find the pixels of the coarse aggregate particles closest to the top. A green colour is used to separate the area above and below the object. In order to separate mortar band thickness, the pixels are selected vertically from top of the image instead of connecting neighbouring objects (coarse aggregates) as seen in Figure 7.

The flow chart of the digital image analysis procedure is shown in Figure 8.

Coarse aggregate distribution

To evaluate the coarse aggregate distribution of the concrete section, the specimen height is divided into ten sections; that is, the size of each section equals to 200 mm × 60 mm. By computing the cumulative area of the coarse aggregates in each of the ten sections, the distribution of aggregate areas throughout the specimen height can be determined. This is done based on the variance in the aggregate coverage from the top to the bottom of the image. The standard deviation (SD) of each specimen is then calculated and put in the following equation (Equation 1) to determine the segregation index (Is).

$$Is = (1 - \text{Relative coarse aggregate fraction}) \times \text{SD of coarse aggregate distribution} \quad (1)$$

The relative coarse aggregate is defined as the percentage of the coarse aggregate with a minimum dimension greater than 5 mm, which is obtained from mix proportioning. The value

Figure 8 Flow chart of digital image analysis procedure

- Image processing method

1. Concrete is discharged in one layer in the image analysis mould
2. A cylindrical sample is cured after demoulding
3. The hardened sample is cut vertically into two sections of 200 mm × 600 mm
4. Phenolphthalein is sprayed on the saw-cut surface to distinguish the aggregate phase from the cement paste
5. A digital image of the concrete at 600 pixels per inch is captured
6. The quality of the image and the contrast of the pixels are enhanced for a better segmentation
7. The particles which are smaller than 5 mm are selected
8. Selected particles are removed

Figure 9 Flow chart of coarse aggregate distribution procedure

- Coarse aggregate distribution

1. Particles larger than 5 mm are selected
2. The specimen height is divided into ten sections
3. The cumulative area of the coarse aggregates in each of the ten sections is computed
4. The standard deviation of each specimen is calculated and put in the *Is* equation
5. The ratio of cumulative area of coarse aggregate in each section to the area of corresponding section (coarse aggregate distribution) is measured

is computed as the mortar fraction of the concrete, multiplied by the SD of the coarse aggregate along the saw-cut cylinder deduced from image analysis of the ten sections (Khayat *et al.*, 2007). The flow chart for determining the coarse aggregate distribution is depicted in Figure 9.

Mortar band thickness

In this study, mortar is defined as the fraction of coarse aggregate that is smaller than 5 mm. Coarse aggregate is prone to settle at the bottom of fresh concrete, while mortar is prone to remain at the top of the specimen. The mortar band thickness is defined as the average distance from the top edge of the specimen to the first encounter with coarse aggregate. The ratio of the average mortar thickness to the specimen height is an indication of the quantity of concrete segregation; that is, the thicker the mortar band at the top of the specimen, the greater the risk of segregation due to coarse aggregate settlement. Thus, the mortar band thickness is an important

Figure 10 Flow chart of mortar band thickness procedure

- Mortar band thickness

1
- The average area from the top edge of the specimen to the first encounter with coarse aggregate is calculated

2
- The ratio of average depth of the mortar thickness to the specimen height is determined

criterion for assessing the hardened concrete index. The flow chart for the procedure to determine the mortar band thickness is shown in Figure 10. Digital image analysis facilitates and accelerates the determination of mortar band thickness, in order to establish the extent of segregation.

Results and discussions

Slump and J-ring flow

Table 4 summarises the spread diameter of the seven evaluated mixtures. The results showed that for a given w/cm and HRWRA, the increase in fly ash replacement from 0 to 5%, 15% and then to 25% resulted in a consecutively higher slump flow (reference mixture, and mixtures CFA5, CFA15, and CFA25), and resulted in an increase in fly ash replacement from 0 to 15% and 25% in mixtures CSF8, CSF8FA15, and CSF8FA25, respectively. This is due to the lower reactivity of the fly ash compared with cement, which affected the viscosity and adsorbed water. Therefore, it resulted in more free water and a greater level of fluidity (Khayat *et al.*, 2003).

Replacing with 8% silica fume decreased the slump flow and spread in the J-ring test. The intention was to keep the slump flow constant, so excess superplasticiser was added in mixtures CSF8, CSF8FA15 and CSF8FA25.

The maximum values for slump and J-ring flow were found in mixture CFA25 and the minimum values were found in CSF8.

Compressive strength

The compressive strength of the seven mixtures was measured. As expected, the compressive strength, with replacement of 0, 5%, 15% and 25% of fly ash, showed consecutively greater reductions, and increased when 8% silica fume was used as a replacement. Figure 11 depicts the compressive strength of the concrete mixtures. As demonstrated in Figure 11, the compressive strength in mixtures that substituted fly ash and silica fume is lower than the reference mixture at early ages. In other words, when the pozzolanic reaction started, the compressive strength of the mixtures accelerated. Eventually, almost all the compressive strengths of the specimens with various replacement percentages of fly ash were similar by 91 d.

Table 4 Results of the slump and J-ring flow (mm)

Mixture	Ref.	CFA5	CFA15	CFA25	CSF8	CSF8FA15	CSF8FA25
Slump flow	720	730	750	820	700	720	740
J-ring flow	680	680	700	770	680	700	705

Figure 11 Compressive strength of concrete mixtures at ages of 1, 2, 7, 28, 56 and 91 d

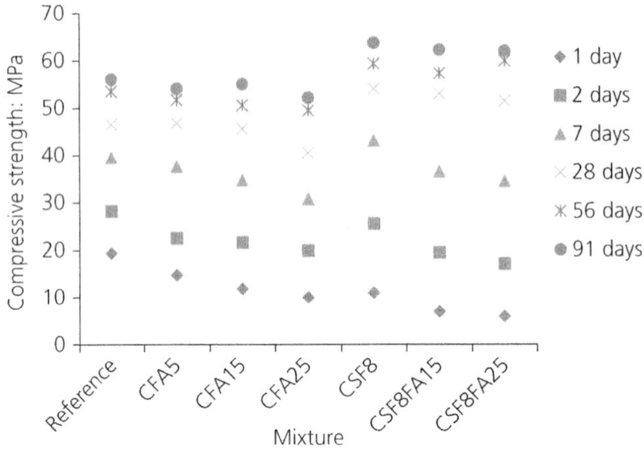

Coarse aggregate distribution

By applying image analysis, it is possible to study the actual segregation that took place in the hardened concrete. The typical curve of the coarse aggregate distribution that belongs to the reference mixture is illustrated in Figure 12. Changes in the distribution of coarse aggregate throughout the height of the test column can be easily observed. If the viscosity of the paste is insufficient, downward migration of coarse aggregate through the column height will occur. Thus, the amount of coarse aggregate in the bottom sections is greater than in the top sections. This causes the coarse aggregate distribution in the top sections to be lower than in the bottom sections. Segregation has a direct effect on the value of *Is*; that is, the greater the segregation, the higher the value of *Is*. Table 5 shows the relative coarse aggregate distribution

Figure 12 Coarse aggregate distribution of reference mixture along the height of column

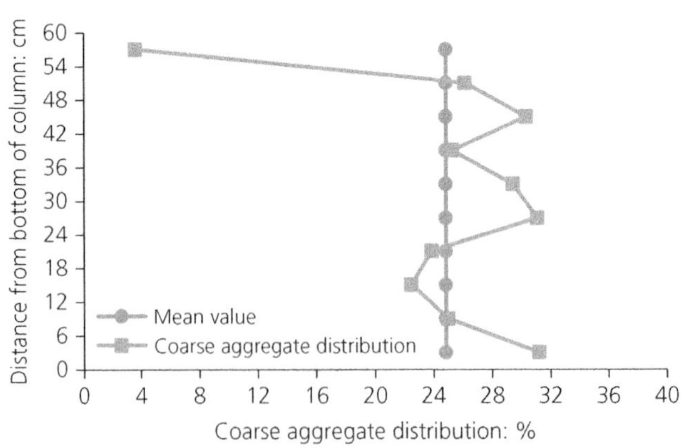

Table 5 Relative coarse aggregate distribution acquired from image analysis

Mixture	Relative coarse aggregate distribution: %												
	Section from top										Mean value: %	SD: %	*Is*
	1	2	3	4	5	6	7	8	9	10			
Reference	3·63	26·12	30·33	25·27	29·40	31·10	23·86	22·48	24·99	31·22	24·84	8·08	5·66
CFA5	1·60	16·62	28·93	25·03	29·34	28·82	28·59	28·74	31·63	28·19	24·75	9·10	6·37
CFA15	3·10	29·40	40·22	32·67	29·03	32·29	36·06	29·17	41·93	36·85	31·07	10·83	7·58
CFA25	0	1·30	26·50	18·46	23·83	29·46	29·66	37·77	37·31	38·12	24·24	13·97	9·78
CSF8	5·16	20·80	18·79	22·75	21·39	29·04	25·49	21·64	25·92	25·53	21·65	3·55	2·48
CSF8FA15	3·21	25·68	34·43	34·72	33·08	37·66	33·18	41·56	38·2	36·55	31·83	10·88	7·62
CSF8FA25	0·41	15·89	16·19	16·64	33·98	35·10	33·04	35·36	34·44	26·57	24·76	11·93	8·35

Figure 13 *Is* of mixtures

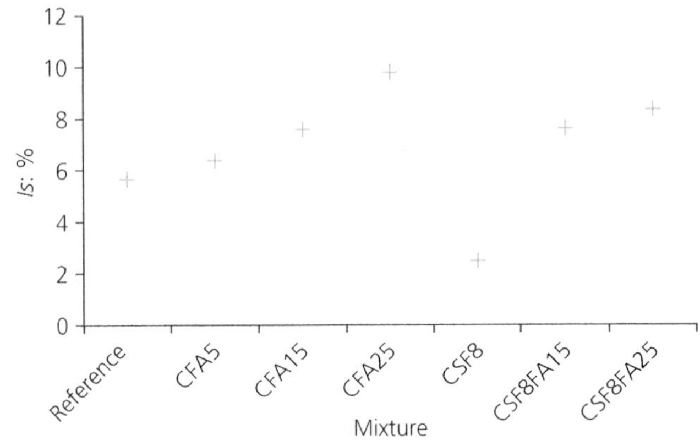

of mixtures acquired from image analysis. Figure 13 demonstrates the *Is* of the mixtures. As shown in Figure 13, the increase in fly ash replacement from 0 to 5%, 15% and then to 25% resulted in higher *Is* (reference mixture, and mixtures CFA5, CFA15, and CFA25 respectively). Similarly, the increase in fly ash replacement from 0 to 15% and 25% resulted in higher *Is* (mixtures CSF8, CSF8FA15, and CSF8FA25 respectively), while the effect of replacing with 8% silica fume reduced the *Is* in mixtures CSF8, CSF8FA15, and CSF8FA25 as defined in Equation 1.

Mortar band thickness

Figure 14 presents the mortar band thickness ratio of the mixtures obtained from the standard coding languages. The ratio of mortar band thickness is defined as the ratio of the average

Figure 14 Ratio of mortar band thickness of mixtures

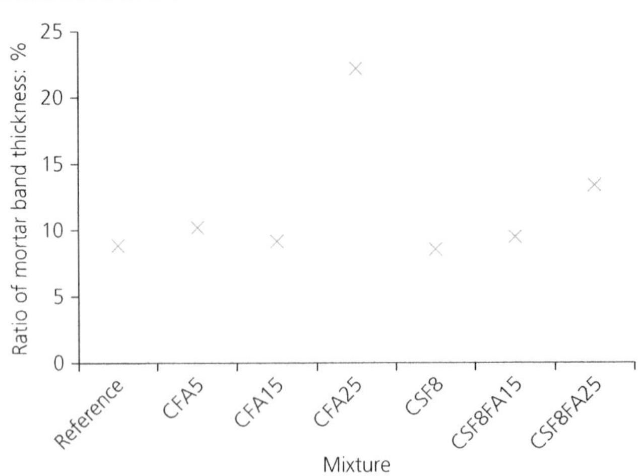

mortar thickness to the specimen height. As illustrated, except in mixture CFA15, the increase in fly ash replacement from 0 to 5%, 15% and then to 25% resulted in a higher ratio of mortar band thickness (reference mixture, and mixtures CFA5, CFA15, and CFA25 respectively). A similar effect occurred with an increase in fly ash replacement from 0 to 15% and 25%, while the effect of replacing with 8% silica fume reduced the ratio of mortar band thickness in mixtures CSF8, CSF8FA15, and CSF8FA25. The general trend of the mortar band thickness ratio is similar to *Is*, and the mortar band thickness ratio is in a good agreement with *Is*.

Conclusions

In this paper, the algorithms were implemented using standard programming languages and trials were undertaken to measure the segregation of the SCC specimens by a new method, applying digital image analysis; that is, the distribution of the coarse aggregates and the average thickness of the mortar band layer of SCC were assessed by this method.

From the results presented in this paper, the following conclusions are offered.

(*a*) For a given w/cm and HRWRA, the increase in fly ash class F replacement from 0 to 5%, 15% and then to 25% (reference mixture, and mixtures CFA5, CFA15, and CFA25 respectively), and similarly the increase in fly ash replacement from 0 to 15% and 25% in mixtures CSF8, CSF8FA15, and CSF8FA25, resulted in consecutively higher slump flows.
(*b*) Replacing with 8% silica fume reduced the slump flow and spread in the J-ring test.
(*c*) The compressive strength of mixtures with a replacement of 0, 5%, 15% and 25% of fly ash decreased, consecutively, and increased with replacement of 8% silica fume.
(*d*) The increase in fly ash replacement from 0 to 5%, 15% and then to 25% resulted in consecutively higher *Is* and SD for the reference mixture and mixtures CFA5, CFA15, and CFA25. Similarly, there was a consecutive increase in fly ash replacement from 0 to 15% and 25%, while replacing with 8% silica fume decreased the *Is* and SD in mixtures CSF8, CSF8FA15, and CSF8FA25.
(*e*) The increase in fly ash substitution makes mixtures segregate more; hence, the value of mortar band thickness increases. Additionally, 8% substitution of silica fume decreases the value of mortar band thickness.
(*f*) The results disclose that regardless of which segregation test method is used (*Is* or ratio of mortar band thickness), the general trend of the mixtures is in good agreement with each other.

Acknowledgements

The authors are grateful for the provision of valuable feedback and suggestions given by Dr. Mohsen Tadayon. The authors would also like to thank Eng. Nader Biglari Joo, Eng. Saeid Khodabandehlou and Eng. Shervin Ilgoli for their assistance in reviewing the paper, preparing mixtures and capturing images for this project.

REFERENCES

ACBM (Advanced Cement Based Materials) (2007) *Self-Consolidating Concrete*. The Center of Advanced Cement Based Materials, Evanston, IN, USA.

ACI (American Concrete Institute) (2007) ACI 237R-07: Self-consolidating concrete. American Concrete Institute, Farmington Hills, MI, USA.

Assaad J, Khayat KH and Daczko J (2004) Evaluation of static stability of self-consolidating concrete. *ACI Materials Journal* **101(3)**: 207–215.

ASTM (2004) C150-04: Standard specification for Portland cement. ASTM International, West Conshohocken, PA, USA.

Barbosa FDS, Faragea MCR, Beaucour AL *et al.* (2010) Evaluation of aggregates particle size distribution in lightweight concrete via image processing. *Mecánica Computacional* **29**: 9711–9718.

Barbosa FDS, Beaucour AL, Farage MCR and Ortola S (2011) Image processing applied to the analysis of segregation in lightweight aggregate concretes. *Construction and Building Materials* **25(8)**: 3375–3381.

Breul P, Geoffray JM and Haddani Y (2008) On-site concrete segregation estimation using image analysis. *Advanced Concrete Technology* **6(1)**: 171–180.

Bui VK, Montgomery D, Hinczak I and Turner K (2002) Rapid testing method for segregation resistance of self-compacting concrete. *Cement and Concrete Research* **32(9)**: 1489–1496.

Chermant JL (2001) Why automatic image analysis? An introduction to the issue. *Cement and Concrete Research* **23(2–3)**: 127–131.

EFNARC (European Federation of National Trade Associations Representing Producers and Applicators of Specialist Building Products for Concrete) (2002) *Specification and Guidelines for Self-Compacting Concrete*. EFNARC, Farnham, UK.

Fang C and Labia S (2007) Image-processing technology to evaluate static segregation resistance of hardened self-consolidating concrete. *Journal of the Transportation Research Board* 1–9.

Fernlund JMR, Zimmerman RW and Kragic D (2007) Influence of volume/mass on grain-size curves and conversion of image-analysis size to sieve size. *Engineering Geology* **90(3–4)**: 124–137.

Hwang SD, Khayat KH and Bonneau O (2006) Performance-based specifications of self-consolidating concrete used in structural applications. *ACI Materials Journal* **103(2)**: 121–129.

Janaka GHA, Kumara J, Hayano K and Ogiwara K (2012) Image analysis techniques on evaluation of particle size distribution of gravel. *International Journal of GEOMATE* **3(1)**: 290–297.

Johnson D, Johnson G and Robertson IN (2010) *Quantifying Segregation in Self-Consolidating Concrete Through Image Analysis*. University of Hawaii, Honolulu, HI, USA, Research Report.

Kanellopoulos A, Petrou MF, Ioannou I and Pantazopoulou S (2010) A study of the quality of interface between SCC and steel reinforcement. *6th International RILEM Symposium on Self-Compacting Concrete and 4th North American Conference on the Design and Use of SCC 2, Montreal, QC, Canada*, pp. 1141–1150.

Khayat KH (1999) Workability, testing, and performance of self-consolidating concrete. *ACI Materials Journal* **96(3)**: 346–354.

Khayat KH, Hu C and Monty H (1999) Stability of self-consolidating concrete, advantages, and potential applications. *First International RILEM Symposium on Self-Compacting Concrete, Stockholm, Sweden*, pp. 143–152.

Khayat KH, Pavate TV, Assaad J and Jolicoeur C (2003) Analysis of variations in electrical conductivity to assess stability of cement-based materials. *ACI Materials Journal* **100(4)**: 302–310.

Khayat KH, Vanhove Y, Pavate TV and Jolicoeur C (2007) Multi-electrode conductivity method to evaluate static stability of flowable and self-consolidating concrete. *ACI Materials Journal* **104 (4)**: 424–433.

Kränkel T, Lowke D and Gehlen C (2010) Direct and indirect determination of the segregation resistance of SCC. *6th International RILEM Symposium on Self-Compacting Concrete and 4th North American Conference on the Design and Use of SCC 2, Montreal, QC, Canada*, pp. 603–610.

Kwan AKH, Mora CF and Chan HC (1999) Particle shape analysis of coarse aggregate using digital image processing. *Cement and Concrete Research* **29(9)**: 1403–1410.

Libre NA, Khoshnazar R and Shekarchi M (2010) Relationship between fluidity and stability of self-consolidating mortar incorporating chemical and mineral admixtures. *Construction and Building Materials* **24(7)**: 1262–1271.

Liao PS, Chen TS and Chung PC (2001) A fast algorithm for multilevel thresholding. *Journal of Information Science Engineering* **17(5)**: 713–727.

Maragos P (2005) Morphological filtering for image enhancement and feasure detection. In *Handbook of Image and Video Processing*(Bovik AL (ed.)), 2nd edn. Elsevier Academic Press, Burlington, MA, USA, pp. 135–156.

Mehta PK and Monteiro PJM (2006) *Concrete Microstructure, Properties, and Materials*. McGraw-Hill, New York, NY, USA.

Mesbah HA, Yahia A and Khayat KH (2011) Electrical conductivity method to assess static stability of self-consolidating concrete. *Cement and Concrete Research* **41(5)**: 451–458.

Mora CF, Kwan AKH and Chan HC (1998) Particle size distribution analysis of coarse aggregate using digital image processing. *Cement and Concrete Research* **28(6)**: 921–932.

Nagataki S, Kawai T and Fujiwara H (2010) State of the art report on SCC in Japan. *6th International RILEM Symposium on Self-Compacting Concrete and 4th North American Conference on the Design and Use of SCC 2, Montreal, QC, Canada*, pp. 3–23.

Neville A (2006) *Concrete, Neville's Insights and Issues*. Thomas Telford Ltd, London, UK.

Nili M, Razmara M, Nili M and Razmara P (2017) Proposing new methods to appraise segregation resistance of self-consolidating concrete based on electrical resistivity. *Construction and Building Materials* **146**: 192–198, https://doi.org/10.1016/j.conbuildmat.2017.04.092.

Okamura H and Ouchi M (2003) Self-compacting concrete. *Advanced Concrete Technology* **1(1)**: 5–15.

Otsu N (1979) A threshold selection method from gray-level histograms. *IEEE Transactions on Systems, Man, and Cybernetics* **9(1)**: 62–66.

Panesar DK and Shindman B (2011) *Comparative Evaluation of Self-Consolidating Concrete Test Methods*. Ministry of Transportation of Ontario, Toronto, Canada, Provincial Highways Management Division Report.

Panesar DK and Shindman B (2012) The effect of segregation on transport and durability properties of self-consolidating concrete. *Cement and Concrete Research* **42(2)**: 252–264.

Paultre P, Khayat KH, Cusson D and Tremblay S (2005) Structural performance of self-consolidating concrete used in confined concrete columns. *ACI Structural Journal* **102(4)**: 560–568.

Pavate TV, Khayat KH and Jolicoeur C (2000) In-situ conductivity method for monitoring segregation, bleeding, and strength development in cement-based materials. *ACI Special Publication* **195**: 535–560.

Schwarz N, DuBois M and Neithalath N (2007) Electrical conductivity based characterization of plain and coarse glass powder modified cement pastes. *Cement and Concrete Composites* **29(9)**: 656–666.

Sezgin M and Sankur B (2004) Survey over image thresholding techniques and quantitative performance evaluation. *Journal of Electronic Imaging* **13(1)**: 146–165.

ice Publishing

Dhir and Lye
ISBN 978-0-7277-6461-4
https://doi.org/10.1680/icetscc.64614.243

Chapter 12
Creep and shrinkage of high-strength self-compacting concrete: experimental and analytical analysis

Farhad Aslani
Centre for Built Infrastructure Research, School of Civil and Environmental Engineering, University of Technology Sydney, Sydney, Australia

Lino Maia
LABEST, Laboratory for the Concrete Technology and Structural Behaviour, Faculdade de Engenharia, Universidade do Porto, Porto, Portugal; Centro de Ciências Exatas e da Engenharia, Universidade da Madeira, Campus Universitário da Penteada, Funchal, Portugal

In the present paper, a numerical and experimental study about creep and shrinkage behavior of a high strength self-compacting concrete is performed. Two new creep and shrinkage prediction models based on the comprehensive analysis on the available models of both conventional concrete and self-compacting concrete are proposed for high strength self-compacting concrete structures. In order to evaluate the predictability of the proposed models, an experimental program was carried out. A concrete which develops 60 MPa within 24 h was used to obtain experimental results. Several specimens were loaded: (i) at different ages and (ii) with different stress-to-strength ratios. Deformation in non-loaded specimens was also measured to assess shrinkage. All specimens were kept under constant stress during at least 600 days in a climatic chamber with temperature and relative humidity of 20°C and 50%, respectively. Results showed that the new models were able to predict deformations with good accuracy, although provided deformations overestimated slightly.

Notation

a and b	constants
c	cement
f'_c	compressive strength
$f_{cm}(t)$	mean value of compressive strength at time t
f'_c, 28d	compressive strength at the age of 28 days
s' and n	parameters that have to be specifically calibrated for each SCC concrete mix by using experimental results
$T(\Delta t_i)$	the temperature (°C) during the time period Δt_i
T_0	1°C
t	temperature-adjusted concrete age
t_0	starting drying concrete age

t_0, t' and t	effective age (days) of concrete at the beginning of drying, at the beginning of loading and during loading respectively
$V_g/V_{g,lim}$	volume of gravel-to-volume limit of gravel ratio
V_s/V_m	volume of sand-to-volume of mortar ratio
V_{sf}/V_s	volume of fine sand-to-volume of sand ratio
V_w/V_p	volume of water-to-volume of powder ratio
w	water
v/s	volume to surface ratio
α	coefficient representing the influence of the cement type
β	represents time dependency of drying shrinkage
γ	coefficient representing the influence of the cement and admixtures type (γ may be 1 when only ordinary Portland cement is used)
Δt_i	the number of days where the temperature T prevails
$\varepsilon'_{as}(t, t_0)$	autogenous shrinkage strain of concrete from the start of setting to age t
$\varepsilon'_{as\infty}$	final value of autogenous shrinkage strain
ε'_{bc}	final value of basic creep strain per unit stress
$\varepsilon'_{cc}(t, t', t_0)$	creep strain
ε'_{cr}	final value of creep strain per unit stress
$\varepsilon'_{cs}(t, t_0)$	shrinkage strain of concrete from age to t
ε'_{ds}	final value of drying creep strain per unit stress
$\varepsilon'_{ds}(t, t_0)$	drying shrinkage strain of concrete from age to t
ε'_{dsp}	final value of drying shrinkage strain
$\varepsilon'_{ds\infty}$	final value of drying shrinkage
$\varepsilon'_{sc}(t, t_0)$	shrinkage strain of concrete from age of t_0 to t
ε'_{sh}	final value of shrinkage strain
η	constant related to compressive strength and water content
κ	conventional scalar damage index
μ, λ and α	parameters to be obtained from a least square minimisation procedure
σ'_{cp}	creep stress unit

Introduction

Self-compacting concrete (SCC) basically consists of the same constituent materials as conventional concrete (CC) (cement, water, aggregates, admixtures and mineral additions), but the final composition of the mixture and its fresh properties are different. In comparison with CC, SCC contains larger quantities of mineral fillers, such as finely crushed limestone or fly ash, and in higher quantities of high-range water-reducing admixtures the maximum size of the coarse aggregate is smaller. These modifications in the composition of the mixture affect the behaviour of the concrete in its hardened state. Using SCC can lead to massive labour and cost savings. It is significant to estimate accurately the crucial mechanical properties of this structural material, including creep and shrinkage deformations, to arrive at a safe and economic analysis and design (Aslani and Nejadi, 2012a).

One critical property is creep of concrete. Creep depends on the characteristics of aggregate stiffness and texture, water/cement (w/c) ratio, volume of paste, volume of coarse aggregate, cement type, admixture type, curing method, ratio of volume to surface area, environmental conditions, magnitude of loads and age of loading. According to Neville (1996) mostly the hydrated cement paste experiences creep, whereas the aggregate is the only portion that resists against creep. Therefore, creep is highly dependent on the stiffness of the chosen aggregate and its proportion within the mixture (Neville, 1996). As a result, as creep mainly occurs in the cement paste, the main concern is that SCC may exhibit higher creep because of its high paste content.

Another essential mechanical parameter is the shrinkage of concrete. The overall shrinkage of concrete corresponds to a combination of several shrinkages: plastic shrinkage, autogenous shrinkage, drying shrinkage, thermal shrinkage and carbonation (chemical) shrinkage. In designing CC, shrinkage is frequently taken as drying shrinkage, which is the strain associated with the loss of moisture from the concrete under drying conditions because with a relatively high w/c ratio (higher than 0·40) CC exhibits a relatively low autogenous shrinkage $<100 \times 10^{-6}$ (Aslani and Nejadi, 2012b). In contrast, the SCC used in the precast industry, namely for prestressed applications, typically has a low w/c ratio (0·32–0·40). These relatively low w/c ratios, coupled with a high content of binder lead to greater autogenous shrinkage. Such shrinkage increases and is notable in SCC because of the use of finely ground supplementary cementitious materials and fillers. Therefore, both drying and autogenous shrinkage deformations have to be considered in the structural detailing of reinforced concrete and prestressed concrete members (Khayat and Long, 2010).

Aware that SCC usually has a higher paste volume and/or higher sand-to-aggregate ratio to achieve high workability, several researchers have claimed relatively large creep and shrinkage of SCC for precast/prestressed concrete, resulting in larger prestress losses (Issa *et al.*, 2005; Naito *et al.*, 2006; Schindler *et al.*, 2007; Suksawang *et al.*, 2006). In fact, although mechanical properties of SCC are superior to those of CC, creep and shrinkage of SCC are significantly high (Issa *et al.*, 2005). Among others, Naito *et al.* (2006) also found that SCC exhibits higher shrinkage and creep than CC, which is probably attributable to the high fine aggregate and paste volume in the SCC.

However, Schindler *et al.* (2007) revealed that the shrinkage of SCC is similar to or less than that of CC. When the shrinkage of SCC is compared with that of CC at 112 days, the sand-to-aggregate ratio effect is not significant for the shrinkage of SCC. The creep coefficients of SCC mixtures were also smaller than those of CC at all loading ages. This was attributed to the low w/c ratio.

The different methodology followed to obtain SCC in different countries (Ouchi *et al.*, 2003) and the limited number of studies concerning its long-term behaviour (Mazzotti and Ceccoli, 2009; Mazzotti *et al.*, 2006; Persson, 2001, 2005; Poppe and De Schutter, 2001; Seng and Shima, 2005) make it still unclear if current international standards can also be applied successfully to SCC (Klug and Holschemaker, 2003; Landsberger and Fernandez-Gomez, 2007; Vidal *et al.*, 2005). Moreover, it has not even been assessed if long-term properties can be predicted with reference to conventional mechanical and physical parameters only (such as strength or w/c ratio), or if the adoption of parameters concerning the mix design is needed.

Research significance

It is vital to investigate whether all the assumed hypotheses used to design structures of CC about creep and shrinkage are also valid for high-strength SCC (HSSCC) structures. Thus, the objectives of this study are

(*a*) to review the accuracies of the CC creep and shrinkage prediction models proposed by international codes of practice, including AASHTO (2004, 2007), ACI (1992), AS (2009), CEB-FIB (2012), BSI (2004) and JSCE (2002)
(*b*) to review the accuracies of the SCC creep and shrinkage prediction models proposed by Cordoba (2007), Khayat and Long (2010), Larson (2006), and Poppe and De Schutter (2005)
(*c*) to propose a new prediction creep and shrinkage model based on the comprehensive analysis of the available models and the experimental results database of both the CC and the SCC
(*d*) to verify the predictability of the proposed models on experimental results conducted in a mix composition previously used in the prefabrication of prestressed bridge girders, i.e. a HSSCC loaded at early ages.

Numerical analysis

Based on recent studies by Aslani and Nejadi (2011a, 2011b), the following procedures are used for comparing available CC creep and shrinkage models

- establish an experimental database for creep and shrinkage results
- establish available creep and shrinkage prediction models database
- compare creep and shrinkage models with SCC experimental results database
- propose SCC creep and shrinkage models based on the previous comparisons
- verify proposed SCC creep and shrinkage models with experimental results tests that have been done in this study.

Creep and shrinkage experimental results database

Tables 1 and 2 present a general summary of the creep and shrinkage concrete mixtures included in the database. The database comprises test results from 11 different investigations, with a total of 52 SCC and 11 CC mixtures for creep tests. Also, the database comprises test results from 14 different investigations, with a total of 165 different SCC mixtures and 21 CC mixtures for shrinkage tests. Tables 1 and 2 also include additional information regarding the applied stress to the creep specimens, age of concrete when shrinkage begins (days), final age of the concrete, relative humidity (RH), type of the specimen, type of cement and filler.

Creep and shrinkage models

This paper also assesses the accuracy of seven commonly used international code-type models that are used to predict creep and shrinkage strains. These empirically-based models, which vary widely in their techniques, require certain intrinsic and/or extrinsic variables, such as mix proportions, material properties and age of loading, as input. The models considered are listed in Table 3, which also shows the factors encountered by each model. In this study the accuracy

Table 1 Creep experimental results database. SCC, self-compacting concrete; CC, conventional concrete; RH, relative humidity; GGBFS, ground granulated blast furnace slag

Reference	No. of SCC mixtures	No. of CC mixtures	Applied stress to the creep specimens	Final age of concrete: days	RH: %	Type of specimen: mm	Type of cement	Type of filler
Chopin *et al.* (2003)	5	1	40% or 60% of the compressive strength at 28 days	365	50	Cylinder (90 × 280)	CEM I	Limestone
Poppe and De Schutter (2005)	6	0	1/3 of the compressive strength at 28 days	1400	60	Prism (150 × 150 × 500)	CEM I 42·5 R, CEM I 52·5	Limestone
Horta (2005)	6	0	40% of the compressive strength at 28 days	70, 200	50	Cylinder (150 × 300)	CEM I, CEM III	Fly ash and GGBFS
Larson (2006)	1	0	40% of the compressive strength at 28 days	520	50	Prism (101·6 × 101·6 × 609·6) and cylinder (114·3 × 609·6)	CEM III	Limestone
Turcry *et al.* (2006)	3	3	20% of the compressive strength at 7 days	65, 100	50	Cylinder (110 × 200)	CEM I 52·5, CEM II 42·5	Limestone
Cordoba (2007)	4	1	30% of the compressive strength at 28 days	365	50	Cylinder (101·6 × 203·2), (101·6 × 1057·8)	CEM I/II	Fly ash and GGBFS
Heirman *et al.* (2008)	7	1	±1/3 of the compressive strength at 28 days	70	60	Cylinder (120 × 300)	CEM I 42·5 R, CEM III/A 42·5 N LA	Limestone

(*continued on next page*)

Table 1 (*Continued*)

Reference	No. of SCC mixtures	No. of CC mixtures	Applied stress to the creep specimens	Final age of concrete: days	RH: %	Type of specimen: mm	Type of cement	Type of filler
Oliva and Cramer (2008)	11	4	40% of the compressive strength at 28 days	495	50	Cylinder (152·4 × 213·6)	CEM I	GGBFS
Kim (2008)	4	4	Changeable for each mixture	150	50	Cylinder (100 × 200)	CEM III	Fly ash and limestone
Zheng *et al.* (2009)	7	1	30% of the compressive strength at loading days	150	60	Prism (100 × 100 × 400)	CEM I	Fly ash
Loser and Leemann (2009)	1	1	Changeable for each mixture	91	70	Prism (120 × 120 × 360)	CEM I 42·5 N, CEM II/A-LL 45·2 N	Fly ash and limestone
Total of 71 mixtures	55	16						

Table 2 Shrinkage experimental results database. SCC, self-compacting concrete; CC, conventional concrete; RH, relative humidity; GGBFS, ground granulated blast furnace slag

Reference	No. of SCC mixtures	No. of CC mixtures	Age of concrete when shrinkage begins: days	Final age of concrete: days	RH: %	Type of specimen (mm)	Type of cement	Type of filler
Chopin *et al.* (2003)	5	1	1	365	50	Cylinder (90 × 280)	CEM I	Limestone
Poppe and De Schutter (2005)	4	0	1	1400	60	Prism (150 × 150 × 500)	CEM I 42, 5 R, CEM I 52,5	Limestone
Horta (2005)	6	0	1	200	50	Cylinder (150 × 300)	CEM I, CEM III	Fly ash and GGBFS
Larson (2006)	1	0	1	520	50	Prism (101·6 × 101·6 × 609·6) and cylinder (114·3 × 609·6)	CEM III	Limestone
Turcry *et al.* (2006)	3	3	1	120, 150, 210	50	Prism (70 × 70 × 280)	CEM I 52·5, CEM II 42·5	Limestone
Cordoba (2007)	4	1	1	365	50	Cylinder (101·6 × 203·2), (101·6 × 1057·8)	CEM I/II	Fly ash and GGBFS
Heirman *et al.* (2008)	7	1	1	98	60	Cylinder (120 × 300)	CEM I 42·5 R, CEM III/A 42·5 N LA	Limestone
Bhattacharya (2008)	6	2	1	90	50	Prism (76·2 × 76·2 × 311·2)	CEM I	Limestone, silica fume and slag
Oliva and Cramer (2008)	11	4	1	350, 495	50	Prism (101·6 × 101·6 × 285·75)	CEM I	GGBFS

(*continued on next page*)

Table 2 *(Continued)*

Reference	No. of SCC mixtures	No. of CC mixtures	Age of concrete when shrinkage begins: days	Final age of concrete: days	RH: %	Type of specimen (mm)	Type of cement	Type of filler
Hwang and Khayat (2010)	10	2	1	56	50	Prism (75 × 75 × 285)	CSA type Gub-F/SF, Gub-S/SF and quaternary blended cement	Fly ash and limestone
Ma *et al.* (2009)	16	0	1	120, 150	60	Prism (100 × 100 × 515)	CEM I	Fly ash
Loser and Leemann (2009)	13	3	1	91	70	Prism (120 × 120 × 360)	CEM I 42·5 N, CEM II/A-LL 45·2 N	Fly ash and silica fume
Güneyisi *et al.* (2010)	63	2	1	50	50	Prism (70 × 70 × 280)	CEM I	Fly ash, GGBFS, silica fume and metakaolin
Khayat and Long (2010)	16	2	1	300	50	Cylinder (150 × 300)	MS and HE (similar to ASTM C150 Type I/II and Type III)	Fly ash
Total of 186 mixtures	165	21						

Table 3 Summary of factors accounted for by different prediction models

Models		CEB-FIB (2012)	ACI (1992)	BSI (2004)	JSCE (2002)	AASHTO (2004)	AASHTO (2007)	AS (2009)
Intrinsic factors	Aggregate type							
	aggregate/ concrete ratio							
	Air content		■					■
	Cement content	■		■	■			
	Cement type							
	Concrete density		■					■
	Fine/total aggregate ratio (mass)		■					■
	Slump		■					■
	w/c ratio				■			
	Water content				■			
Extrinsic factors	Age at first loading	■	■	■	■	■	■	■
	Age of sample				■			
	Applied stress	■	■	■	■			■
	Characteristic strength at loading							
	Cross-section shape				■			
	Curing conditions							
	Compressive strength at 28 days	■	■	■	■	■	■	■
	Duration of load	■	■	■	■			■
	Effective thickness	■	■	■	■	■	■	■
	Elastic modulus at age of loading							
	Elastic modulus at 28 days	■	■	■	■			■
	Relative humidity	■	■	■	■	■	■	■
	Temperature				■			
	Time drying commences							

of the creep and shrinkage prediction models proposed by international codes of practice, including AASHTO (2004, 2007), ACI (1992), AS (2009), CEB-FIB (2012), BSI (2004) and JSCE (2002), are compared with the actual measured creep and shrinkage strains.

As shown in Table 4, the AASHTO (2004, 2007), BSI (2004) and JSCE (2002) models provided better prediction of creep data for CC mixture in the experimental database with a coefficient of correlation factor (R^2) of 0·90, 0·89, 0·89 and 0·86 compared with other models. Also, as shown in Table 4, for the SCC mixture in the experimental database, the AASHTO (2004), ACI (1992) and JSCE (2002) models provided better prediction of creep data with a coefficient of correlation factor (R^2) of 0·87, 0·87 and 0·84 compared with other models.

AASHTO (2004), ACI (1992) and JSCE (2002) CC creep models that have conservative predictions for SCC mixtures in the database are different in the certain intrinsic and/or extrinsic variables. As mentioned in Table 3, the AASHTO (2004) creep model does not have any intrinsic factors, but the JSCE (2002) and ACI (1992) creep models have a good consideration of both intrinsic variables (i.e. aggregate type, aggregates/cement ratio, air content, cement content, cement type, concrete density, fine/total aggregate ratio, slump, w/c ratio and water content) and extrinsic variables (i.e. age at the first loading, age of sample, applied stress, characteristic strength at loading, cross-section shape, curing conditions, compressive strength at 28 days, duration of load, effective thickness, elastic modulus at age of loading, elastic modulus at 28 days, RH, temperature and time drying commences). The modified composition of SCC in comparison with CC influences the creep behaviour of the concrete. Therefore, it is important to include some important variables that have an impact on this behaviour. By considering these variables, the JSCE (2002) creep model has good intrinsic and extrinsic variables.

As show in Table 5, for CC mixture in the experimental database, the AASHTO (2007) and JSCE (2002) models provided a better prediction of drying shrinkage data with a coefficient of correlation factor (R^2) of 0·88 and 0·84 compared with other models. Also, as shown in Table 5, the AASHTO (2007), AS (2009) and JSCE (2002) models provided a better

Table 4 Coefficient of correlation factor (R^2) of creep prediction models for conventional concrete (CC) and self-compacting concrete (SCC)

Creep prediction models	CC	SCC
	R^2	R^2
CEB-FIB (2012)	0·41	0·58
ACI (1992)	0·79	0·84
BSI (2004)	0·89	0·80
JSCE (2002)	0·89	0·87
AASHTO (2004)	0·86	0·87
AASHTO (2007)	0·90	0·80
AS (2009)	0·70	0·75

Table 5 Coefficient of correlation factor (R^2) of shrinkage prediction models for conventional concrete (CC) and self-compacting concrete (SCC)

Shrinkage prediction models	CC	SCC
	R^2	R^2
CEB-FIB (2012)	0·70	0·57
ACI (1992)	0·62	0·66
BSI (2004)	0·72	0·55
JSCE (2002)	0·84	0·83
AASHTO (2004)	0·42	0·47
AASHTO (2007)	0·88	0·86
AS (2009)	0·65	0·80

prediction of SCC mixture in the experimental database drying shrinkage data with a coefficient of correlation factor (R^2) of 0·86, 0·83 and 0·80 compared with other models.

The CC shrinkage models of AASHTO (2007) and JSCE (2002) that have conservative predictions for SCC mixtures in the database are different in the certain intrinsic and/or extrinsic variables. As mentioned in Table 3, the AASHTO (2007) shrinkage model does not have any intrinsic factors, but the JSCE (2002) shrinkage model has a good consideration of both intrinsic and extrinsic variables. When compared with the CC, the modified composition of SCC influences the shrinkage behaviour of concrete. Therefore, it is important to involve some important variables that have an impact on this behaviour. By considering these variables, the JSCE (2002) shrinkage model has good intrinsic and extrinsic variables.

Proposed creep model

The comparison of the different models and the experimental database shows that the ACI (1992), JSCE (2002) and AASHTO (2004) models have conservative creep coefficient predictions. In this study, based on required certain intrinsic and/or extrinsic variables for SCC, the JSCE (2002) creep model gives a good approximation of the creep coefficient. Therefore, with the JSCE (2002) creep model as a basis, an attempt is made to formulate some suggestions to include the cement-to-powder (c/p) ratio into the formulas in order to obtain a better prediction of the time-dependent deformations of normal strength and high strength of SCC. These results are shown in Equations 1–10.

1. For the normal-strength SCC with range of applicability (see the denomination of the parameters in the Notation)

$$45\% \leq \text{RH} \leq 80\%$$

$$120\ \text{kg/m}^3 \leq w \leq 230\ \text{kg/m}^3$$

$$100\ \text{mm} \leq v/s \leq 300\ \text{mm}$$

$$0{\cdot}30 \le w/c \le 0{\cdot}65$$

$$f'_{c,28d} \le 55\,\mathrm{MPa}$$

$$260\,\mathrm{kg/m^3} \le c \le 500\,\mathrm{kg/m^3}$$

$$\begin{aligned} \varepsilon'_{cc}(t,t',t_0) = \sigma'_{cp} & \\ & \times \varepsilon'_{cr}\left[1-\exp\left\{-0{\cdot}09(t-t')^{0{\cdot}54}\right\}\right] \\ & \times (0{\cdot}015 + 1{\cdot}35\,(c/p))^{-1} \\ & \text{for } c/p < 0.65 \end{aligned} \tag{1}$$

$$\begin{aligned} \varepsilon'_{cc}(t,t',t_0) = \sigma'_{cp} & \times \varepsilon'_{cr}\left[1-\exp\left\{-0.09(t-t')^{0{\cdot}54}\right\}\right] \\ & \times (0{\cdot}015 + 1{\cdot}05\,(c/p))^{-1} \\ & \text{for } c/p \ge 0.65 \end{aligned} \tag{2}$$

$$\sigma'_{cp} = \frac{\mu + \lambda\,.\,\sigma(t,t_0)^{\alpha}}{1-\kappa} \tag{3}$$

non-linear creep amplification function

where μ, λ and α are additional parameters to be obtained from a least square minimisation procedure starting from experimental data (Mazzotti and Ceccoli, 2009) $\mu = 0{\cdot}90, \lambda = 1{\cdot}80, \alpha = 2{\cdot}10$; moreover, the stress function $\sigma(t, t_0)$ is the actual stress/strength ratio, being

$$\sigma(t,t_0) = \frac{\sigma(t_0)}{f_{cm}(t)} \tag{4}$$

in the case of constant applied load. In Equation 3, the numerator and denominator indicate the effect of sustained load and the effect of a damage level owing to instantaneous loading. The law $f_{cm}(t)$ representing the evolution with time of compression strength has been defined by modifying the MC90 proposal according to the expression

$$f_{cm}(t) = f'_{c,28}\,\cdot\,\exp\left[s'\left(1-\left(\frac{28}{t}\right)^{n}\right)\right] \tag{5}$$

where parameters s' and n have been specifically calibrated for each SCC concrete mix by using experimental results previously described (see Table 3). According to the available data, parameters s' and n range from 0·2 to 0·6 and 0·28 to 0·35 respectively (Mazzotti and Ceccoli, 2009). The adoption of function $\sigma(t, t_0)$ allows for the variable rate of increase of mechanical properties to be taken into account, which is particularly important for concretes loaded at early ages. Finally, the non-linear behaviour during the load application has been introduced in Equation 3 according to the conventional scalar damage index $\kappa = 1 - E/E_0$, in which E is the secant stiffness at the end of loading and E_0 is the initial tangent stiffness. Usually, damage index κ is about 0·10–0·15 or 0·22–0·35 for low ($0{\cdot}35f_{\mathrm{cm}}(t)$) or medium ($0{\cdot}55f_{\mathrm{cm}}(t)$) applied stress levels respectively

$$\varepsilon'_{\mathrm{cr}} = \varepsilon'_{\mathrm{bc}} + \varepsilon'_{\mathrm{dc}} \tag{6}$$

$$\varepsilon'_{\mathrm{bc}} = \left[17{\cdot}5\,(c+w)^{2{\cdot}0}(w/c)^{2{\cdot}4}\{\ln(t')\}^{-0{\cdot}67}\right] \times 10^{-10} \tag{7}$$

$$\varepsilon'_{\mathrm{dc}} = \left[4500\,(w/c)^{4{\cdot}2}\,(c+w)^{1{\cdot}4}\left[\ln\left(\frac{v/s}{10}\right)\right]^{-2{\cdot}2} \times \left\{1-\frac{\mathrm{RH}}{100}\right\}^{0{\cdot}36} t_0^{-0{\cdot}30}\right] \times 10^{-10} \tag{8}$$

2. For the HS SCC with range of applicability by using Equations 3–5

$45\% \leq \mathrm{RH} \leq 90\,\%$

$120\ \mathrm{kg/m^3} \leq w \leq 230\ \mathrm{kg/m^3}$

$100\,\mathrm{mm} \leq v/s \leq 300\ \mathrm{mm}$

$0{\cdot}30 \leq w/c \leq 0{\cdot}65$

$55 \leq f'_{\mathrm{c,28d}} \leq 100\,\mathrm{MPa}$

$$\varepsilon'_{\mathrm{cc}}(t, t', t_0) = \sigma'_{\mathrm{cp}} \times \left[\frac{4w(1-\mathrm{RH}/100)+350}{12+f'_{\mathrm{c}}(t')}\ln(t-t'+1)\right] \times \left(10 \times (c/p)^{0{\cdot}678}\right)\ \text{for}\ c/p < 0.65 \tag{9}$$

$$\varepsilon'_{\mathrm{cc}}(t, t', t_0) = \sigma'_{\mathrm{cp}} \times \left[\frac{4w(1-\mathrm{RH}/100)+350}{12+f'_{\mathrm{c}}(t')}\ln(t-t'+1)\right] \times \left(13 \times (c/p)^{0{\cdot}701}\right)\ \text{for}\ c/p \geq 0.65 \tag{10}$$

where for t_0, t' and t are replaced by

$$t = \sum_{i=1}^{n} \Delta t_i \exp\left[13{\cdot}65 - \frac{4000}{273+T\,(\Delta t_i)/T_0}\right]$$

Proposed shrinkage model

The comparison of the different models and the experimental database shows that the AASHTO (2004), ACI (1992) and JSCE (2002) models have conservative drying shrinkage predictions. In this study, based on required certain intrinsic and/or extrinsic variables for SCC, the JSCE (2002) drying shrinkage model gives the best approximation of the drying shrinkage strain. Therefore, with the JSCE (2002) model as a basis, an attempt is made to formulate some suggestions to include the c/p ratio into the formulas in order to obtain a better prediction of the time-dependent deformations of normal strength and high strength of SCC. These results are shown in Equations 11–17.

For normal-strength SCC (with range of applicability same as creep proposed model)

$$\varepsilon'_{\mathrm{cs}}(t, t_0) = \varepsilon'_{\mathrm{sh}}\left[1 - \exp\left\{-0{\cdot}1\,(t - t_0)^{(-2{\cdot}4(c/p)+2{\cdot}3)}\right\}\right] \tag{11}$$

$$\begin{aligned}\varepsilon'_{\mathrm{sh}} = &\left[-50 + 78\left\{1 - \exp\left(\frac{\mathrm{RH}}{100}\right)\right\} + 38{\cdot}3\ \ln w\right.\\ &\left.-0.92\ \ln\left(\tfrac{w}{c}\right) - 5\left[\ln\left(\frac{v/s}{10}\right)\right]^2\right] \times \left(10^{-5}\right)\\ &\text{for } c/p < 0.65\end{aligned} \tag{12}$$

For the HS SCC with range of applicability (with range of applicability same as creep proposed model)

$$\varepsilon'_{\mathrm{cs}}(t, t_0) = \varepsilon'_{\mathrm{ds}}(t, t_0) + \varepsilon'_{\mathrm{as}}(t, t_0) \tag{13}$$

$$\varepsilon'_{\mathrm{ds}}(t, t_0) = \frac{\varepsilon'_{\mathrm{ds}\infty}(t - t_0)}{\beta + (t - t_0)} \tag{14}$$

$$\varepsilon'_{\mathrm{ds}\infty} = \frac{\varepsilon'_{\mathrm{ds}p}}{\eta t_0}\left(\times 10^{-6}\right) \tag{15}$$

$$\varepsilon'_{\mathrm{ds}p} = \left[\frac{\alpha(1 - \mathrm{RH}/100)w}{1 + 110\exp\left\{-\dfrac{400}{f'_{\mathrm{c,28d}}}\right\}}\right] \times (0{\cdot}015 + 1{\cdot}35(c/p))^{-1} \quad \text{for } c/p < 0.65 \tag{16}$$

$$\varepsilon'_{\mathrm{ds}p} = \left[\frac{\alpha(1 - \mathrm{RH}/100)w}{1 + 110\exp\left\{-\dfrac{410}{f'_{\mathrm{c,28d}}}\right\}}\right] \times (0{\cdot}015 + 1{\cdot}05(c/p))^{-1} \quad \text{for } c/p \geq 0.65 \tag{17}$$

with

$$\beta = \frac{4w\sqrt{v/s}}{100 + 0{\cdot}7t_0},$$

Table 6 Variations of *a* and *b* constants with water/cement (w/c) ratio

w/c	*a*	*b*
0·20	1·2	0·4
0·23	1·5	0·4
0·30	0·6	0·5
0·40	0·1	0·7
≥0·50	0·03	0·8

$$\eta = [15\exp(0{\cdot}007 f'_{c}(28)) + 0{\cdot}25w] \times 10^{-4},$$

$$\varepsilon'_{as}(t, t_0) = \varepsilon'_{as}(t) - \varepsilon'_{as}(t_0),$$

$$\varepsilon'_{as}(t) = \gamma\varepsilon'_{as\infty}\left[1 - \exp\{-a(t - t_s)^b\}\right] \times 10^{-6},$$

$$\varepsilon'_{as\infty} = 3070\exp\{-7{\cdot}2(w/c)\},$$

$\alpha = 11$ for normal and low heat cement or $\alpha = 15$ for high early strength cement.

where for t_0, t' and t are replaced by the temperature-adjusted concrete age and γ is a coefficient representing the influence of the cement and admixtures type (may be 1 when only ordinary Portland cement is used). The variations of *a* and *b* constants with w/c ratio are given in Table 6.

Experimental programme

In order to validate the proposed models, an experimental programme with shrinkage and creep tests was carried out in a laboratory. A mix composition previously used in the prefabrication of prestressed bridge girders was used. Bearing aware that in precast/prestress industries it is necessary to apply prestress as soon as possible in order to reduce the time of each production cycle, the concrete used is a SCC that reaches a compressive strength higher than 60 MPa at the age of 24 h. As a consequence, loading could be applied at early ages. In order to verify the predictability of the proposed models in a HSSCC loaded at early ages, in this experimental work specimens were loaded at the ages of 12 h, 16 h, 20 h, 24 h, 48 h and 72 h.

Materials and mix proportions

Cement CEM I 42.5R (BSI, 2000a) with specific gravity of 3·10 and limestone filler with a specific gravity of 2·70 were used. A polycarboxylate type superplasticiser was used, having a specific gravity of 1·08 and 40·0% solid content. The aggregates were dried, and the specific gravities of the coarse aggregate, fine sand and coarse sand were 2·68, 2·63 and 2·62, and absorption values were 1·40%, 0·20% and 0·40%, respectively, according to EN 1097–6:2000

Table 7 Mix compositions per cubic meter

Material denomination: kg	
CEM I 42.5R (Portland)	400
Limestone filler	192
Sika viscocrete 20HE	4·7
Natural fine siliceous sand	399
Natural coarse siliceous sand	397
Crushed limestone (washed gravel)	852
Tap water (total)	139·5

(BSI, 2000b). The bulk density of the compacted gravel was 1·38. The water content was recalculated owing to water included in the superplasticiser and the water required to saturate the aggregates.

The constituent materials used and the corresponding mix proportions are reported in Table 7. Note that the w/c ratio was 0·32, superplasticiser-to-powder ratio was 0·80%, volume of water-to-volume of powder ratio (V_w/V_p) was 0·64, volume of sand-to-volume of mortar ratio (V_s/V_m) was 0·48, volume of fine sand-to-volume of sand ratio (V_{sf}/V_s) was 0·50 and volume of gravel-to-volume limit of gravel ratio ($V_g/V_{g,lim}$) was 0·65.

Mixing, fresh testing and casting

Several batches of 57 l were prepared using a mixer with a vertical axis according to the following sequence

- mixing aggregates and 15% of total water during 2·5 min
- stop mixing for 2·5 min
- adding cement, filler, remaining water and superplasticiser and mixing for 5 min
- stop mixing for 1 min (cleaning mixer paddles)
- mixing for 3 min
- evaluation of the self-compacting properties and casting.

Immediately after mixing was finished, self-compacting properties were evaluated: the slump flow, the V-funnel, the L-box and the segregation tests were carried out according to the European guidelines for self-compacting concrete (EFNARC, 2005). All specimens were cast within 15 min after mixing was finished. After casting was finished, specimens were carefully moved in order to be stored up to the demoulding time in a temperature and RH controlled room (20 ± 0·3°C and 50 ± 3% respectively). All tests were performed using cylinder specimens: specimens with 150 mm diameter and 300 mm height for compressive strength and elastic modulus tests; specimens with 141 mm diameter and 500 mm height for deformation measurements. In order to eliminate the effect of the specimens' surface roughness on the uniaxiality, the top and bottom faces of all specimens were smoothed just after demoulding. A thermocouple was located nearly at the centre of two specimens and to record the temperature over a period of 70 h to allow for maturity corrections.

Deformation measurements

The creep tests were carried out by applying a uniaxial load in two superposed specimens with a flat hydraulic jack, which was able to keep the stress constant throughout the test. Measurements were manually acquired by using a digital extensometer. Each specimen had eight pins divided per four generatrices rotated 90° to each other. Three readings per generatrix were taken (the base of measuring was 200 mm), and the specimen deformation calculated using the average of the four generatrices.

The loads corresponding to 20%, 30% and 40% of the stress-to-strength ratio (at the age of loading) were kept constant for at least 600 days. The first measurement was taken immediately before loading and the second one immediately after loading. In order to evaluate shrinkage deformations, two non-loaded specimens per age of loading were also measured.

Concrete fresh and mechanical properties

All the batches produced presented similar self-compactability and pertained to classes SF3, VS2/VF2, PA2 and SR2 (EFNARC, 2005). The following results were observed: diameter for slump flow was 803 ± 9·7 mm; time for V-funnel test was 18·43 ± 1·06 s; ratio for L-box test was 0·94 ± 0·04; and the segregation was 5·9 ± 2·5%.

The evolutions of the elastic modulus and the compressive strength of measured are summarised in Table 8. Bearing in mind that the mix composition was previously used in the prefabrication of prestressed bridge girders, as expected the concrete strength increased very quickly at early ages reaching almost 50 MPa at the age of 12 h and more than 60 MPa at 24 h. Then, it continued to increase, but at lower rates, reaching nearly 70 MPa at 72 h and 90 MPa at 28 days.

Deformation results

Results of the total strain (creep strain + shrinkage strain) for all sets of loaded specimens are presented in Figure 1. Analysing Figure 1, it is noted that, roughly speaking, the total strain is similar for all specimens loaded at 30% of the stress-to-strength ratio (the maximum difference was $\sim 150 \times 10^{-6}$). Therefore, one may conclude that the resistance against the deformation increases with strength (note that to keep the stress-to-strength ratio constant, higher loads are applied for higher strengths).

Table 8 Evolution of mechanic properties measured

Age	Elastic modulus: GPa	Compressive strength: MPa
12 h	35·6	48·4
16 h	–	55·2
20 h	–	59·9
24 h	39·9	62·3
48 h	–	66·3
72 h	–	69·3
28 d	45·5	91·9

Figure 1 (a) Creep strain for all specimens loaded at 12 h and 16 h, (b) creep strain for all specimens loaded at 20 h and 24 h, (c) creep strain for all specimens loaded at 48 h and 72 h, (d) total strain for all specimens

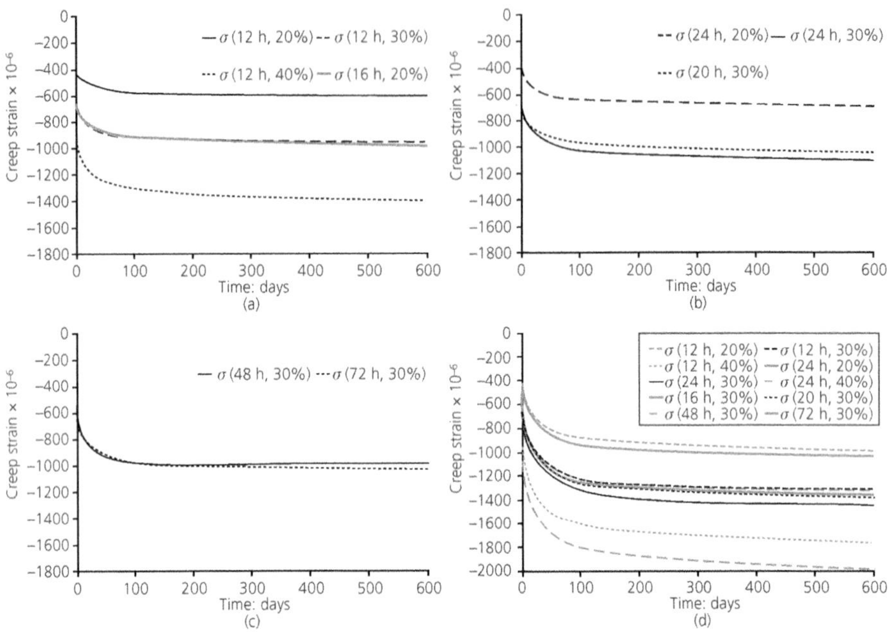

Comparing the total strain of specimens loaded at the age of 12 h with those loaded at the age of 24 h, it was observed that those loaded at the age of 24 h showed higher strain – such a difference being higher for greater loads. Thus, one may conclude that, from the age of 12 h to the age of 24 h, the increase of resistance against the concrete deformation is lower than the increase of strength. Furthermore, looking at specimens loaded at the same age, but with different stress-to-strength ratios, it is observed that there is a near linear increase of strain; that is, the total strain of the specimens loaded at the age of 24 h varies (at the age of 600 days) from approximately 1030×10^{-6} to 1450×10^{-6} and then to 2000×10^{-6} when the stress-to-strength ratio varies from 20% to 30% and to 40%. Analogous trends can be observed for specimens loaded at the age of 12 h.

Regarding the shrinkage strain, Figure 2 presents evolution curves for the deformation of non-loaded specimens. Analysing the measured curves, one observes a rapid evolution for approximately 200 days. However, after that age, the shrinkage increases by only about 10%. According to the results shown in Figure 2, generally speaking, a consistent tendency relative to the age of demoulding (as previously reported in Bissonnette and Pigeon (1995) and as suggested by several models) is not observed. In fact, the specimens demoulded at the age of 12 h and at the age of 72 h are the ones with the lowest shrinkage, while the ones demoulded at the age of 16 h are the ones with the highest.

Figure 2 Shrinkage evolution

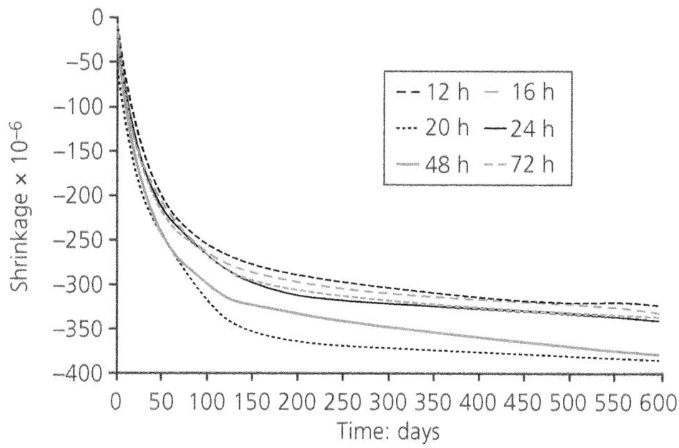

Predictability and discussion of the proposed models

Proposed creep and shrinkage models are useable for lower and higher c/p, and are adjusted to the normal and HSSCC. A non-linear creep amplification function (Equation 3) is added to the creep model, which shows an influential stress function on the creep behaviour. The proposed creep model is adjusted to normal and HSSCC. Furthermore, SCC loading age parameter is included in the creep model as given by Equation 5.

However, before analysing the predictability of the proposed models, it is important to remember that although shrinkage and creep are not totally independent phenomena (Bažant *et al.*, 1994; Neville *et al.*, 1983; Reinhardt and Rinder, 2006), in this experimental programme the total strain was roughly understood as composed of the addition of these independent phenomena. Consequently, experimental results of creep strain used to verify the predictability of the proposed creep model were determined as the difference between the total strain and shrinkage strain (CEB-FIB, 2012; Leemann *et al.*, 2011; Reinhardt and Rinder, 2006). Besides, according to the experimental data reported above, the following parameters were considered in the predictability of the proposed models: RH = 50%, w = 128 kg/m^3, c = 400kg/m^3, w/c = 0·32, v/s = 141 mm, c/p = 67·6% and $f'_{cd,28d}$ = 91.9 MPa.

Figure 3 shows a comparison of the proposed creep model with the available experimental results of creep. In analysing Figure 3 one observes that the proposed model provides an accurate prediction. In fact, most of the creep results predicted by the proposed model were slightly overestimated, but always with a difference lower than 10% to the experimental results.

Figure 3 shows a comparison of the SCC creep from experimental results plotted against calculated values from the proposed model for (*a*) loading at the age of 12 h with different levels of the stress-to-strength ratios loading percentages, (*b*) loading at the age of 24 h with different levels of the stress-to-strength ratios loading percentages, and (*c*) 30% loading at the age of loading rate for 16, 20, 48 and 72 h with a level of 30% of the stress-to-strength ratio.

Figure 3 Comparison of the self-compacting concrete creep from experimental results versus calculated values from proposed model for (a) 12 h with different loading percentages, (b) 24 h with different loading percentages and (c) 30% loading rate for 16, 20, 48, and 72 h

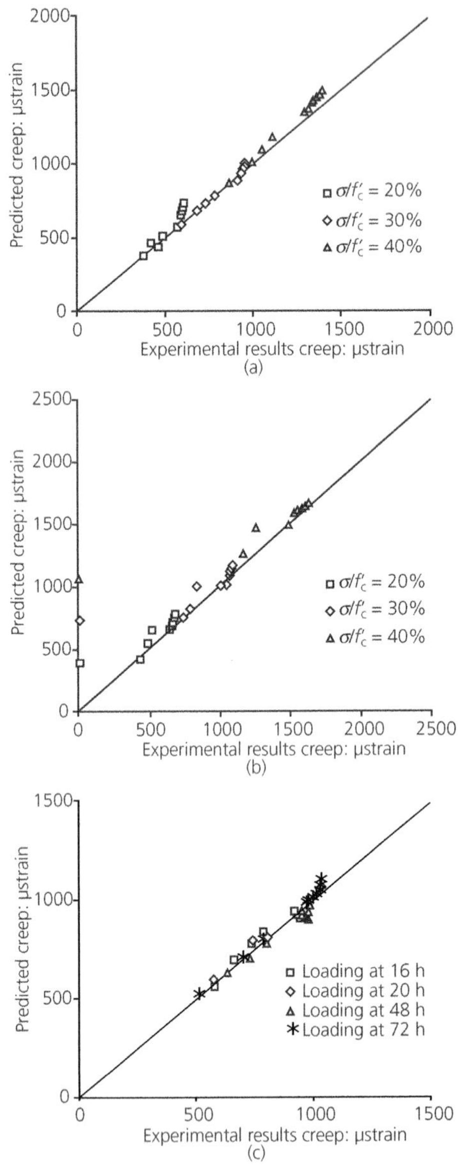

Based on the results and comparisons reported in Figure 3, the proposed creep model was accurate for different levels of stress and different ages at loading time. In fact, most of the creep results predicted by the proposed model were slightly overestimated, but always with the difference being lower than 10% of the experimental results.

In terms of the proposed shrinkage model, Figure 4 shows a comparison of the SCC shrinkage from experimental results against calculated values from the proposed model for (a) 12 h, (b) 16 h, (c) 24 h (d) 48 h and (e) 72 h. Looking at Figure 4, roughly speaking, it can be noted that the proposed model provided a good prediction (especially up to the age of 100 days). The predicted results were mostly conservative (especially after the age of 300 days), the maximum

Figure 4 Comparison of the self-compacting concrete drying shrinkage from experimental results versus calculated values from proposed model for (a) 12 h, (b) 16 h, (c) 24 h, (d) 48 h and (e) 72 h

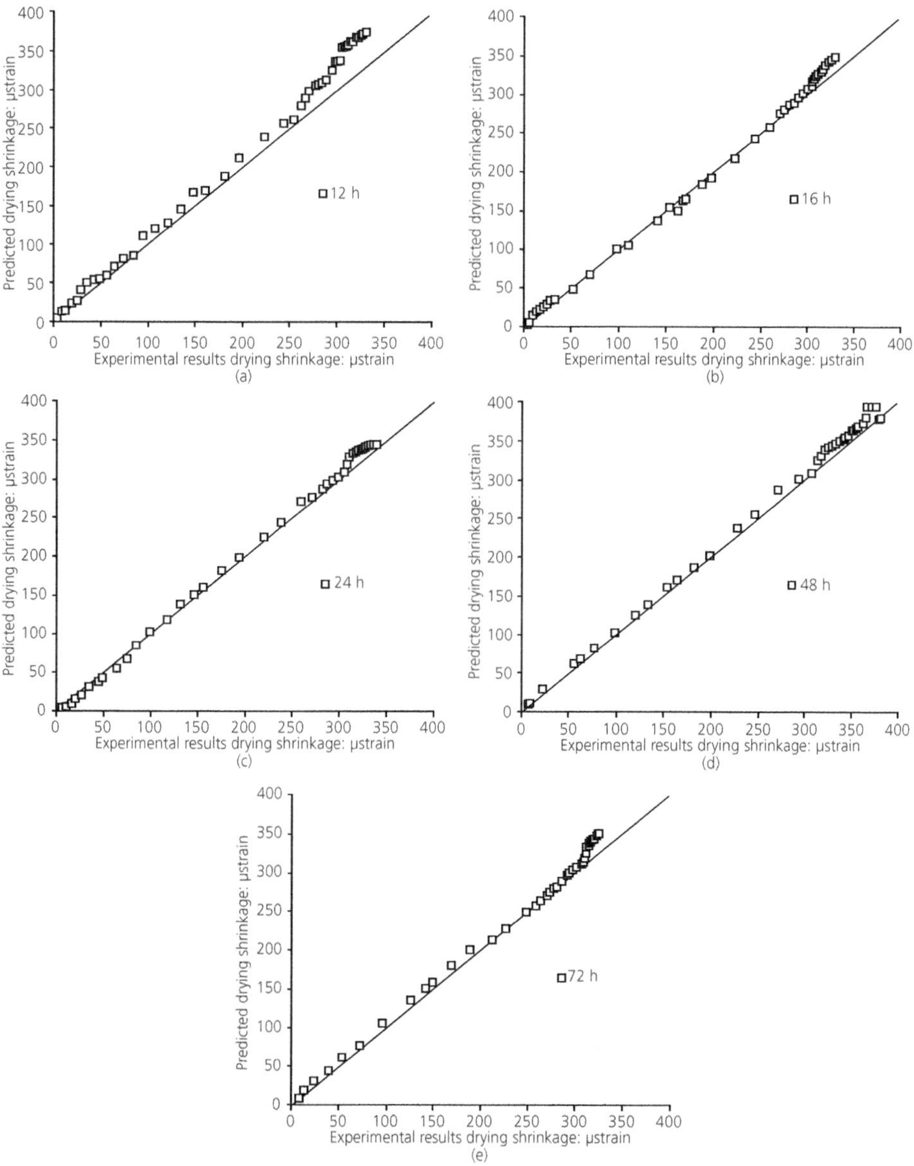

difference (~20%) between predicted and experimental results occurring when specimens were demoulded at the age of 12 h (Figure 4(a)).

Conclusions

The predictability of deformation models have been investigated for CC and SCC in this paper. Two new models are proposed for an accurate prediction of creep and shrinkage for concrete structures made with HSSCC. Based on comparisons between different models and on comparisons to the experimental results, some conclusions can be drawn from this study.

- For the CC mixtures the AASHTO (2007) and JSCE (2002) models provided better predictions of the creep and shrinkage data compared with the other models. Although the BSI (2004) and AASHTO (2004) models had provided suitable predictions for creep, they were not so successful for shrinkage.
- For the SCC mixtures the JSCE (2002) model provided better predictions of the creep and shrinkage data than the other models because of the certain intrinsic and/or extrinsic variables. The AASHTO (2004) and ACI (1992) CC models also provided suitable predictions for creep, and the AASHTO (2007) and AS (2009) CC models also provided suitable predictions for shrinkage.
- The proposed creep and shrinkage models have good predictions for high strength of the SCC mixtures. The comparison between the predicted values and the experimental results conducted in this study showed that the proposed models were able to predict creep and shrinkage with an accuracy of 10% and 20% respectively.

Acknowledgements

Funding provided by the Portuguese Foundation for Science and Project Technology (FCT) and the European Social Fund (ESF), namely for the Research Unit 'LABEST, Laboratory for the Concrete Technology and Structural Behavior, is gratefully acknowledged.

REFERENCES

AASHTO (American Association of Highway and Transportation Officials) (2004) *Bridge design Specifications and Commentary.* AASHTO, Washington, DC, USA.

AASHTO (2007) *Interim Bridge Design Specifications and Commentary.* AASHTO, Washington, DC.

ACI (American Concrete Institute) (1992) ACI 209R-92: Prediction of creep, shrinkage, and temperature effects in concrete structures. ACI, Farmington Hills, MI, USA.

AS (Standards Australia) (2009) AS 3600: Concrete structures. AS, Sydney, Australia.

Aslani F and Nejadi S (2011a) Comparison of creep prediction models for self-compacting and conventional concrete. In *Proceedings of 9th International Symposium on High Performance Concrete, Rotorua, New Zealand* (Khrapko M and Wallevik O (eds)). New Zealand Concrete Society, Auckland, New Zealand, pp. 1–10.

Aslani F and Nejadi S (2011b) Comparison of shrinkage prediction models for self-compacting and conventional concrete. In *Proceedings of 9th International Symposium on High Performance Concrete, Rotorua, New Zealand, August 2011* (Khrapko M and Wallevik O (eds)). New Zealand Concrete Society, Auckland, New Zealand, pp. 1–10.

Aslani F and Nejadi S (2012a) Mechanical properties of conventional and self-compacting concrete: An analytical study. *Construction Building Materials* **36**: 330–347.

Aslani F and Nejadi S (2012b) Shrinkage behavior of self-compacting concrete. *Journal of Zhejiang University Science A* **13(6)**: 407–419.

ASTM (2002) ASTM C 469: Standard test method for static modulus of elasticity and Poisson's ratio of concrete in compression. ASTM International, Philadelphia, PA, USA.

Bažant Z, Huet C and Müller H (1994) Comment on recent analysis of concrete creep linearity and applicability of principle of superposition. *Materials and Structures* **27(6)**: 359–61.

Bhattacharya A (2008) *Effects of Aggregate Grading and Admixtures/Fillers on Fresh and Hardened Properties of Self-Consolidating Concrete.* MSc thesis, West Virginia University, USA.

Bissonnette B and Pigeon M (1995) Tensile creep at early ages of ordinary, silica fume and fiber reinforced concretes. *Cement and Concrete Research* **25(5)**: 1075–1085.

BSI (2000a) BS EN 197-1:2000: Cement – Part 1: Composition, specifications and conformity criteria for common cements. BSI, London, UK.

BSI (2000b) BS EN 1097-6:2000: Tests for mechanical and physical properties of aggregates. Determination of particle density and water absorption. BSI, London, UK.

BSI (2004) BS EN 1992-1-2:2004: Eurocode 2: Design of concrete structures. Part 1.2: General rules–structural fire design. BSI, London, UK.

CEB-FIB (2012) *Model Code 2010 – Final Draft.* FIB, Laussanne, Switzerland, FIB bulletin no, 65, vol. 1.

Chopin D, Francy O, Lebourgeois S and Rougeau P (2003) Creep and shrinkage of heat-cured self-compacting concrete (SCC). *Proceedings of 3rd International Symposium on Self-Compacting Concrete, Reykjavik, Iceland*, pp. 672–683.

Cordoba B (2007) *Creep and Shrinkage of Self-Consolidating Concrete (SCC).* MSc thesis, University of Wyoming, WY, USA.

EFNARC (Experts for Specialised Construction and Concrete Systems) (2005) *The European Guidelines for Self-compacting Concrete.* See http://www.efnarc.org/pdf/SCCGuidelinesMay2005.pdf (accessed 28/06/2013).

Güneyisi E, Gesoğlu M and Özbay E (2010) Strength and drying shrinkage properties of self-compacting concretes incorporating multi-system blended mineral admixtures. *Construction and Building Materials* **24(10)**: 1878–1887.

Heirman G, Vandewalle L and Van Gemerta D *et al.* (2008) Time-dependent deformations of limestone powder type self-compacting concrete. *Engineering Structures* **3(10)**: 2945–2956.

Horta A (2005) *Evaluation of Self-Consolidating Concrete for Bridge Structures Applications.* MSc thesis, Georgia Institute of Technology, Atlanta, GA, USA.

Hwang SD and Khayat K (2010) Effect of mix design on restrained shrinkage of self-consolidating concrete. *Materials and Structures* **43(3)**: 367–380.

Issa M, Alhassan M, Shabila H and Krozel J (2005) Laboratory performance evaluation of self-consolidating concrete. *Proceeding of the Second North American Conference on the Design and Use of Self Consolidating Concrete and the Fourth International RILEM Symposium on Self-Consolidating Concrete, Center for Advanced Cement-Based Materials (ACBM), Chicago, USA*, pp. 857–862.

JSCE (Japanese Society of Civil Engineers) (2002) Standard specifications for concrete structure – 2002. JSCE, Tokyo, Japan.

Khayat KH and Long WJ (2010) Shrinkage of precast, prestressed self-consolidating concrete. *ACI Materials Journal* **107(3)**: 231–238.

Kim YH (2008) *Characterization of Self-consolidating Concrete for the Design of Precast, Pretensioned Bridge Superstructure Elements.* PhD thesis, Texas A&M University TX, USA.

Klug Y and Holschemaker K (2003) Comparison of the hardened properties of self-compacting and normal vibrated concrete. *Proceedings of 3rd RILEM Symposium on Self Compacting Concrete, Reykjavik, Iceland*, pp. 596–605.

Landsberger GA and Fernandez-Gomez J (2007) Evaluation of creep prediction models for self-consolidating concrete. *Proceedings of 5th RILEM Symposium on SCC, Pro 54, Ghent, Austria*, pp. 605–610.

Larson K (2006) *Evaluation the Time-Dependent Deformation and Bond Characteristics of a Self-Consolidating Concrete Mix and the Implication for Pretensioned Bridge Applications*. PhD thesis, Kansas State University, Manhattan, KS, USA.

Leemann A, Lura P and Loser R (2011) Shrinkage and creep of SCC – the influence of paste volume and binder composition. *Construction and Building Materials* **25(5)**: 2283–2289.

Loser R and Leemann A (2009) Shrinkage and restrained shrinkage cracking of self-compacting concrete compared to conventionally vibrated concrete. *Materials and Structures* **42(1)**: 71–82.

Ma K, Xie Y, Long G and Luo Y (2009) Drying shrinkage of medium strength SCC. *Proceedings of Second International Symposium on Design, Performance and Use of Self-Consolidating Concrete SCC'2099, China*, pp. 657–663.

Mazzotti C, Savoia M and Ceccoli C (2006) Creep and Shrinkage of Self Compacting Concrete. Proceedings of 2nd fib Conference, Naples, pp. 1–10 (on CD).

Mazzotti C and Ceccoli C (2009) Creep and shrinkage of self-compacting concrete: Experimental behavior and numerical model. In *Proceedings of the Eighth International Conference on Creep, Shrinkage and Durability Mechanics of Concrete and Concrete Structures, Ise-Shima, Japan* (Tanabe T-A, Sakata K and Mihashi H *et al.* (eds)). CRC Press, Boca Raton, FL, USA, pp. 671–673.

Naito CJ, Parent G and Brunn G (2006) Performance of bulb-tee girders made with self-consolidating concrete. *PCI Journal* **51(6)**: 72–85.

Neville AM (1996) *Properties of Concrete*, 4th edn. Wiley, New York, NY, USA.

Neville AM, Dilger WH and Brooks JJ (1983) *Creep of Plain and Structural Concrete*. Construction Press, London New York.

Oliva MG and Cramer S (2008) *Self-Consolidating Concrete: Creep and Shrinkage Characteristics. Report*. University of Wisconsin, Madison, WI, USA.

Ouchi M, Nakamura S, Osterson T, Hallberg S and Lwin M (2003) *Applications of Self-Compacting Concrete in Japan*. ISHPC, Europe and the United States, pp. 1–20.

Persson B (2001) A comparison between mechanical properties of SCC and the corresponding properties of normal concrete. *Cement and Concrete Research* **31(2)**: 193–198.

Persson B (2005) Creep of self-compacting concrete. *Proceedings of International Conference CONCREEP 7, Nantes, France*, pp. 535–540.

Poppe AM and De Schutter G (2001) Creep and shrinkage of self-compacting concrete. *Proceedings of the Sixth International Conference CONCREEP 6*, pp. 563–568.

Poppe AM and De Schutter G (2005) Creep and shrinkage of self-compacting concrete. *First International Symposium on Design, Performance and Use of Self-Consolidating Concrete, China*, pp. 329–336.

Reinhardt H-W and Rinder T (2006) Tensile creep of high-strength concrete. *Journal of Advanced Concrete Technology* **4(2)**: 277–283.

Schindler AK, Barnes RW, Roberts JB and Rodriguez S (2007) Properties of self-consolidating concrete for prestressed members. *ACI Material Journal* **104(1)**: 53–61.

Seng V and Shima H (2005) Creep and shrinkage of self-compacting concrete with different limestone powder contents. In *Proceedings of 4th RILEM Symposium on Self-Compacting Concrete, Chicago, IL, USA*, pp. 981–987.

Suksawang N, Nassif HH and Najim HS (2006) Evaluation of mechanical properties for self-consolidating, normal, and high-performance concrete. *Transportation Research Record* **1979**: 36–45.

Turcry P, Loukili A, Haidar K, Pijaudier-Cabot G and Belarbi A (2006) cracking tendency of self-compacting concrete subjected to restrained shrinkage: experimental study and modelling. *ASCE, Journal of Materials in Civil Engineering* **18(1)**: 46–54.

Vidal T, Assié S and Pons G (2005) Creep and shrinkage of self-compacting concrete and comparative study with model code. In *Proceedings of International Conference CONCREEP 7, Nantes, France*, pp. 541–546.

Zheng J, Chao P and Luo S (2009) Experimental study on factors influencing creep of self-compacting concrete. *Proceedings of Second International Symposium on Design, Performance and Use of Self-Consolidating Concrete, SCC'2009, China*, pp. 703–709.

ice Publishing

Dhir and Lye
ISBN 978-0-7277-6461-4
https://doi.org/10.1680/icetscc.64614.269

Chapter 13
Microstructure, thermal analysis and chloride penetration of self-compacting concrete under different conditions

Kianoosh Samimi
Director, Department of Research and Development, ImenRah Consulting Engineering Co., Tehran, Iran (corresponding author: samimi.kianoosh@imenrah.com)

Gholam Reza Dehghan Kamaragi
Director, Department of Architecture, ImenRah Consulting Engineering Co., Tehran, Iran

Robert Le Roy
Professor, University of Paris-Est, Navier Laboratory (ENPC/IFSTTAR/CNRS), Ecole des Ponts ParisTech, Marne-la-Vallée, France

This paper concerns the characterisation of the raw materials used in four formulations of self-compacting concrete (SCC) and the study of their fresh state, compressive strength, thermal analysis, transport properties and chloride penetration under immersion and tidal conditions. The control concrete corresponds to a concrete with class 40 MPa based on Portland cement and limestone filler. A second formulation involves the complete substitution of limestone filler by sand filler. In the two other formulations, 10% and 15% of metakaolin are substituted for cement. The results show that SCC with metakaolin exhibits higher resistance to chloride penetration at different states. A higher strength development is obtained when using sand filler. This is true up to 90 d of maturation; beyond this, the compressive strength increases for the control SCC. However, mixtures with metakaolin present compressive strength values close to or higher than the ones of the control mix at both early- and long-term ages. This research shows that the substitution of Portland cement by 15% metakaolin in SCC is affordable in all aspects.

Notation

C_{Ag}	concentration of silver nitrate
D_{nSSm}	migration coefficient (×10–12 m^2/s)
D_{ns} (dif)	apparent diffusion coefficient of chloride ions in the saturated condition (m^2/s)
L	thickness of the sample (mm)
M_{pe}	sampled weight of powder product
S	sorptivity
T	anolyte solution temperature (°C)

t	duration of the test (h)
U	applied voltage (V)
V_e	volume of silver nitrate
V_{fiol}	volume of the filtrate solution
V_p	volume of the solution
W	water content
W_e	evaporable water content
W_{ne}	non-evaporable water content
X_d	depth of penetration of chloride ions (mm)

Introduction

Globally, concrete is one of the most commonly used materials worldwide after water. In structures made of concrete, in order to obtain the necessary strength and reduce the porosity as well as the air inside the concrete and then ensure reliability, the concrete is vibrated in different ways. Most of the time, the vibration effect is not accurate and the mechanical properties of the concrete are not satisfactory. With the development of concrete structures and the prefabricated industry in general, coinciding with the lack of skilled workers, the notion of producing self-compacting concrete (SCC) was presented for the first time by Okamura and Ouchi (2003) in Japan. SCC can move under its own weight. In addition, its consolidation does not require vibration during concreting. SCC, with its specific characteristics, can increase durability and stability in a reinforced concrete structure. A durable concrete must retain its original form, quality and facility of realisation under environmental working conditions. Since the movement of aggressive media fluids from the environment in concrete is the main cause for most deterioration in concrete, the lifetime of reinforced concrete elements and permeability are the two most influential factors in sustainability. To better understand the lifetime of the mechanical and environmental stresses of materials, it is necessary to have sufficient knowledge of the properties of materials, in particular their transport properties. In this respect, the various transport processes of harmful substances through concrete have been distinguished as diffusion, absorption and penetration as a function of the driving force of the process and the nature of the material being transported. The contribution of mineral additions to the activity of the cement binder essentially results from these effects: physicochemical, microstructural and chemical (Bessa-Badreddine, 2004). According to Bessa-Badreddine (2004), the chemical effect allows an improvement in the quality of the bonds in the cementitious matrix with decreasing hydrate volume. In general, the compressive strength increases by the use of pozzolanic additions (Ahmadi *et al.*, 2014; Najimi *et al.*, 2012; Sadrmomtazi *et al.*, 2017), because pozzolan reacts with portlandite from the cement hydration reaction and creates a new hydrated product that could be very useful in increasing the compressive strength. Thereby, the presence of a large number of submicron particles in the cement matrix around the cement grains multiplies the possibilities of germination of the hydrated products and develops the microstructural complexity and the efficiency of the bonds (Bessa-Badreddine, 2004). The binding contribution of the mineral additions thus has a structural role in the connections of the cementitious matrix from the mechanical point of view, which suggests a quantitative role in the reduction of the porosity. In terms of durability, this structural contribution translates to better resistance in the face of severe chemical aggression. The use of fillers in a concrete formulation generates an acceleration of its mechanical resistance at young ages (De Larrard, 1999; Pera *et al.*, 1999). The fine particles of the filler, when well deflocculated by a superplasticiser, promote the hydration of the cement, mainly by a physical effect, and lead to a denser cementitious matrix. These effects have a significant influence on mechanical strength up to 28 d but subsequently become less significant (Assié, 2004). In addition, the fillers may be inert, active or even pozzolanic.

Pozzolans enhance concrete performance by reacting with carbon (C) and hydrogen (H) to form the secondary calcium–silicate–hydrate (C–S–H) gel. Alongside other cementitious materials, such as silica fume, ground-granulated blast-furnace slag and pulverised fly ash, which are frequently used for concrete production, metakaolin is a material that is being used more frequently in the concrete industry. Pure metakaolin has been used successfully as a cementitious additive in concrete since 1990 (Hassan *et al.*, 2012). When the kaolin is heated to 700–900°C, it is calcined and loses up to 14% of hydroxyl water, changing to metakaolin (Caldarone *et al.*, 1994). Metakaolin reacts with the calcium hydroxide formed during the hydration of Portland cement and creates additional cementitious products, modifying the concrete structure via improved mechanical performance as well as overall durability (Hubertova and Hela, 2007).

Indeed, metakaolin improves the behaviour of concrete via two effects:

- A filler effect: the fine particles of metakaolin will be inserted into the empty interstitial spaces between the grains of cement. This modifies the properties in the fresh state and makes it possible to obtain a more compact composition.
- Acceleration of hydration: for the same duration of hydration, a greater quantity of hydrates is formed. Then the pores are gradually filled.

It is well known that metakaolin improves concrete properties, such as compressive strength and resistance to chloride penetration. The application of metakaolin is known to considerably refine the pore structures of concrete and decrease the calcium hydroxide (Dubey and Banthia, 1998). Kim *et al.* (2007) show that the resistance to chloride ion penetration reduces significantly as the proportion of silica fume and metakaolin binders is increased. The results of Kannan and Ganesan (2014) show replacement with 30% metakaolin leads to a 26% improvement in compressive strength, a 98·10% decrement in total charge passed in an RCPT test and a 38·76% decrement in water permeability.

Currently, very little information is available on the properties of SCC containing metakaolin and sand filler, and there is no extensive sustainability data regarding the use of metakaolin and sand filler in SCC. Therefore, the use of SCC containing metakaolin and sand filler requires further study. The studied transport properties are mainly obtained using chloride migration, water penetration, capillary absorption and electrical resistivity tests. However, one can observe a lack in the literature regarding the influence of sand filler on transport properties in SCC and also on the characterisation of the microstructure and the thermal analysis of metakaolin, as well as on their performance in non-saturated conditions such as in the tidal zone in marine environments. This paper investigates and compares the effect of a 10% and 15% metakaolin substitution of Portland cement, and of the total replacement of limestone filler by sand filler, on the rheological behaviour, compressive strength and transport properties of SCC at early ages and up to 180 d. The impact on the resistance to chloride penetration both in immersion and tidal conditions is also investigated. For this purpose, first the chemical and mineralogical composition as well as the microstructure of the cementitious material were studied. Then different tests were performed to study the fresh phase of SCC, and mechanical testing was subsequently done on hardened states to evaluate the compressive strength of the different prepared mixtures. The tests concerning the transport phenomena were also performed at different ages in order to evaluate the durability properties over time. Finally, the resistance to chloride penetration of the different mixtures was evaluated in immersion and tidal conditions.

Experimental programme

Materials and mix proportion

A constant Portland cement type II content (350 kg/m^3), a constant water/cement (w/c) ratio of 0·4 and constant gravel/sand ratio of 1 were used for the preparation of four concrete mixtures. One of the concrete mixtures contained sand filler (SCCS); one contained limestone filler (SCCL); the remaining two mixtures were also formulated with limestone filler, but metakaolin was used as an additive by substituting two different percentages of Portland cement with metakaolin at 10% (SCCM10) and 15% (SCCM15). Portland cement with a specific gravity of 3·15 g/cm^3 and a Blaine fineness of 2900 cm^2/g, in compliance with ASTM C 150 2009 (ASTM, 2009), was used. The metakaolin used in this study has a specific surface area of 3920 cm^2/g and a specific gravity of 2·49 g/cm^3. The chemical analysis and particle size distribution of cementitious materials are shown in Table 1 and Figure 1, respectively.

Table 1 Chemical analysis of Portland cement type II, limestone filler, sand filler and metakaolin

Chemical analysis: % by mass	Cement (type II)	Limestone filler	Sand filler	Metakaolin
Loss on ignition	1·60	44·50	7·11	0·78
SiO_2	18·9	1·93	55·85	65·93
Al_2O_3	4·60	0·67	14·42	13·71
Fe_2O_3	3·19	0·32	7·01	0·23
CaO	62·71	49·91	9·50	5·46
MgO	2·80	1·72	2·83	3·84
SO_3	2·37	–	0·08	–
CO_2	–	–	5·70	–
$CaSO_4$	–	–	0·14	–
Cl	–	–	–	–
Insoluble residue	0·62	–	–	–
Alkalis (Na_2O%+0·658K_2O %)	1	–	–	–
Na_2O+K_2O	–	4·33	–	8·93
Free CaO	1·42	–	–	–
Humidity	–	0·10	–	0·10
C_3S	56·20	–	–	–
C_2S	16·23	–	–	–
C_3A	6·30	–	–	–
C_4AF	12·40	–	–	–

Figure 1 Particle size distribution of fine materials

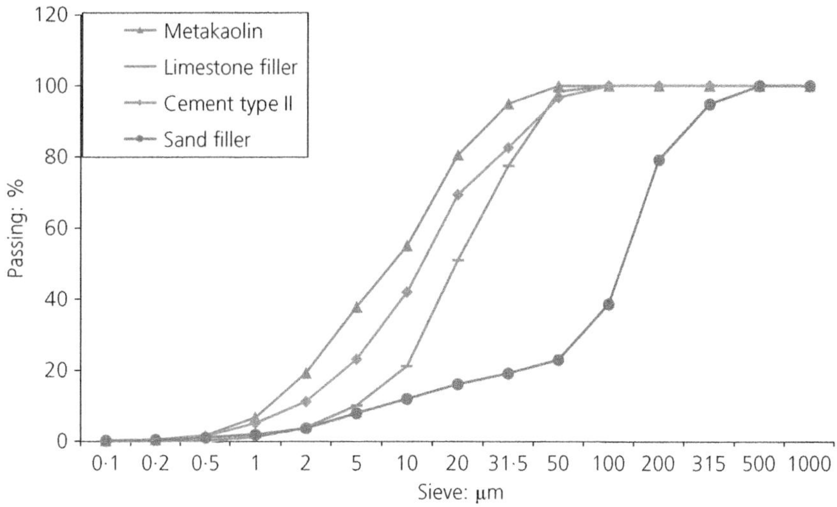

Table 2 Mix design of SCC mixtures (kg/m^3)

Mix name	Water	Cement	w/c	Gravel	Sand	Limestone powder	Superplasticiser (% cement)	Metakaolin
SCCL	140	350	0·4	800	800	160	5·10 (1·46)	0
SCCS	140	350	0·4	800	800	0	5·62 (161)	0
SCCM10	140	315	0·4	800	800	160	4·90 (1·40)	35·0
SCCM15	140	297·5	0·4	800	800	160	6·87 (1·96)	52·5

For all mix designs, crushed angular material of 6–12 mm nominal size was used as a coarse aggregate (gravel), and natural sand with a maximum size of 4 mm was used as a fine aggregate. A high-range water-reducing admixture, with a specific gravity of 1·11 g/cm^3 based on chains of modified polycarboxylate ether (PCE 200), was used in all mixtures to produce SCC. Potable water was also used to prepare concrete mixes. The dosage of superplasticiser was experimentally determined from tests on fresh concrete to obtain a slump flow diameter of 710±20 mm for all SCCs. Table 2 shows the mix proportions of the mixtures. To enhance the stability of SCC mixes, 160 kg/m^3 filler was used in the four mixtures.

Mix design

All concrete mixtures were prepared in a 150-l mixer. The batching sequence consisted of a decant total of the fine and coarse aggregate being placed into the mixer and mixed for 3 min. During this period, two-thirds of the water required was added. Next, cementitious materials were added and mixing was continued for one more minute. After this, the superplasticiser and the remaining water were introduced and the blend was mixed for 2 min. The mixer was covered with a plastic cover to minimise the evaporation of the mixing water.

Casting and curing

Cubic and cylindrical samples were cast in accordance with ASTM C 31 (ASTM, 2012) and ASTM C 511 (ASTM, 2013). After casting, samples were covered with two layers of plastic sheets and placed in a temperature-controlled room at 22±2°C for 24 h. All samples were demoulded after 24 h and cured up to the age of testing in saturated lime solution to prevent possible leaching of calcium hydroxide ($Ca(OH)_2$) from these specimens. The hardened concretes were tested for compressive strength and durability as well as water absorption and permeable voids, capillary absorption, water penetration, chloride diffusion in non-permanent and permanent state, electrical resistivity and determination of the chloride penetration in marine environments.

Test method

Compressive strength

Cubic specimens ($15 \times 15 \times 15$ cm^3) were used to determine the mechanical resistance during compression of the various concretes studied. After demoulding, the concrete samples were stored in the lime-saturated water until the date of the test. The compression test was carried out for several ages as young age and long-term conservation (1, 3, 7, 14, 28, 90 and 180 d). This test was performed in accordance with BS 8110-1:1997 (BSI, 1997).

Analysis by scanning electron microscopy (SEM) and X-ray diffraction (XRD)

SEM observations were performed on pozzolans using a JSM 7100F SEM. Before analysis, the samples were metallised with gold–palladium in order to make their surface conductive. XRD analysis was performed on metakaolin using a Bruker D8 X-ray diffractometer with a monochromatic copper (Cu) source and a fast linear detector, the Lynx Eye.

Evaporable and non-evaporable water content measurement

The thermal measurement of the evaporable and non-evaporable water content was carried out by measurements of mass loss as a result of different thermal loadings. The evaporable water content corresponds to the mass loss of water between 105°C and 20°C (Equation 1). Also, the non-evaporable water content corresponds to the mass loss of water between 1000°C and 105°C (Equation 2). The measurements were carried out on binder pastes at 1, 3, 7, 14, 28, 90 and 180 d of ageing. The binder pastes have the same composition as the binder used in concretes but without limestone, and use a lesser amount of superplasticiser. The measurements were carried out on hardened and crushed binder pastes (about 3 g). The powders obtained were put in crucibles and placed in a furnace (see Figure 2(a)). The samples were carried in the ambient temperature to 1000°C with a rate of 10°C/min and bearing 20 h at 105°C and 2 h at 1000°C (see Figure 2(b)). The evaporable water content (W_e) and non-evaporable water content at 105°C (W_{ne}) are calculated by the following equations

$$W_e = \frac{W_{20^\circ C} - W_{105^\circ C}}{W_{20^\circ C}} \tag{1}$$

$$W_{ne} \text{ at } 105^\circ C = \frac{W_{105^\circ C} - W_{1000^\circ C}}{W_{105^\circ C}} \tag{2}$$

Figure 2 (a) Placement of the crucibles in the furnace; (b) Loading and unloading of the temperature for the measurement of mass charges

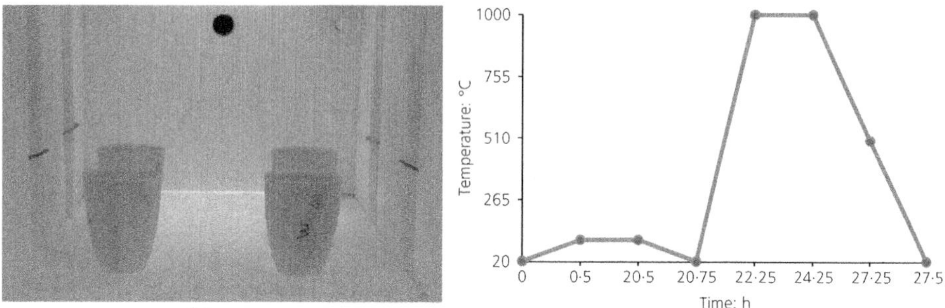

(a)

(b)

Water absorption and permeable voids

Absorption and permeable pore volume measurement was performed according to ASTM C 642-97 (ASTM, 1997). This is determined by weighing: (*a*) mass of a specimen concrete, (*b*) mass in the air (while still soaked in liquid), (*c*) mass in water and (*d*) dry mass. The test was carried out after 7, 14, 28, 90 and 180 d of curing in lime water on slices from cast cylinders with dimensions of 100 × 50 mm. Three samples by concrete formulation were tested.

Capillary absorption (sorptivity)

This test was carried out according to AFPC-AFREM recommendations (Durabilité des Bétons, 1997) on slices from cast cylinders with dimensions of 100 × 50 mm (three samples per formulation). The coefficient of sorptivity (S) after 24 h is measured for each formulation at different curing times (7, 14, 28, 90 and 180 d).

Water penetration

This test was measured on three 150 × 150 × 150 mm^3 cube specimens after 7, 14, 28, 90 and 180 d of cure in lime water by applying water under a pressure of 0·5 MPa for 72 h in accordance with BS EN 12390-8 (BSI, 2009). Afterwards, the samples were split and the depth of penetration of the water in concrete measured.

Resistance to chloride penetration

Figure 3 provides a summary of measurement methods and diffusion coefficients obtained. In this study, diffusion in non-steady-state conditions according to the ASTM C 1543-02 standard (ASTM, 2002) and migration in non-steady-state conditions under electric fields by the NT BUILD 492 method (Nordtest, 1999) is investigated.

Non-steady-state diffusion test

This test makes it possible to characterise the material with respect to its resistance to the simple diffusion of chloride ions in the non-steady state. To carry out this test, ASTM C 1543-02 (ASTM, 2002) was used on three concrete specimens (Ø100 × 50 mm) per mix design with ageing of 28 d. They are sealed by an epoxy resin on all the surfaces except on one of the two bases, through which the diffusion of the chlorides will be effected. This sealing is necessary

Figure 3 Chart method of chloride diffusion (Deby, 2010)

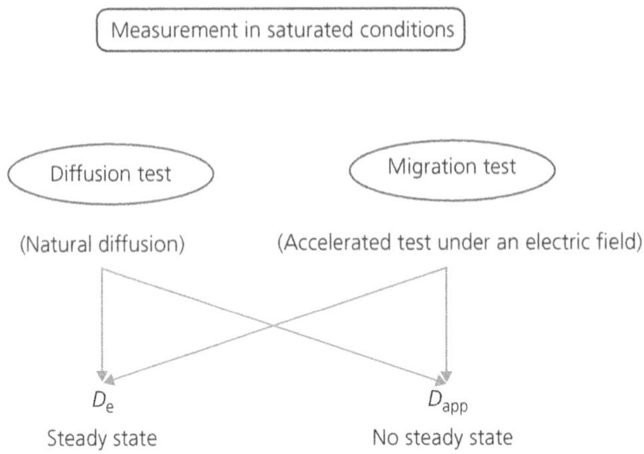

to ensure a one-dimensional penetration of the chlorides (Deby, 2010). The accessible surface is exposed to a saline solution for a specified period of time. The concrete specimens are maintained for 90 and 180 d in the saline solution containing 165 g/l of sodium chloride (NaCl).

At the end of the exposure period, the depth of penetration of chloride ions was measured using silver nitrate ($AgNO_3$) solution. The diffusion coefficient is then deduced using the formula proposed by Baroghel-Bouny *et al.* (2007)

$$D_{ns}(\text{dif}) = \frac{x^2}{4t} \tag{3}$$

where D_{ns} (dif) is the apparent diffusion coefficient of chloride ions in the saturated condition (m^2/s), x is the depth of penetration of chloride ions (m) and t is the immersion duration of the specimens in the saline solution (s).

In order to measure the free-chloride concentration profile in the tested specimen, 3-mm slices of concrete (with a precision of 0·2 mm) were cut from the unsealed surface using a fully automatic device manufactured for this study (Figure 4). It should be noted that this apparatus was used in previous work carried out by the first author.

The dissolution and measurement of the free chlorides was obtained by filtration of a solution made by mixing the powder product (almost 5 g) and distilled water. Figure 5 illustrates the apparatus for filtering the solution in order to measure the concentration of free chlorides by silver nitrate solution. The chloride ion concentration is determined by adding the solution of silver nitrate (with concentration of 0·01 M) until the solution turned red. This test was performed according to the standards laid out by APFC-AFREM (Durabilité des Bétons, 1997).

Figure 4 The device used to obtain concrete powder samples at different depths from the exposure surface (Samimi *et al.*, 2017)

Figure 5 The filtration apparatus

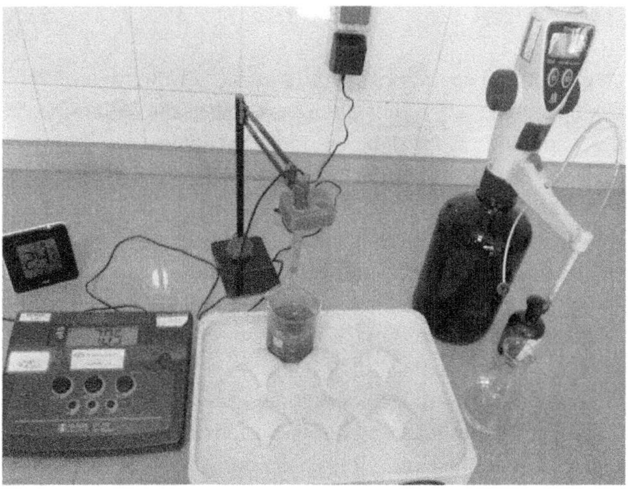

The content of free chlorides expressed in grams per 100 g of sample is given by the following relation

$$C_1 = \frac{(35{\cdot}5 \times C_{\mathrm{Ag}} \times V_{\mathrm{e}} \times V_{\mathrm{p}})}{(10 \times M_{\mathrm{pe}} \times V_{\mathrm{fiol}})} \tag{4}$$

where C_{Ag} is the concentration of silver nitrate (≈0·01 M), V_{e} is the volume of silver nitrate, M_{pe} is the sampled 5 g of powder product, V_{fiol} is the volume of the filtrate solution up to 250 ml in a volumetric flask, and V_{p} is the volume of the solution taken (≈50 ml) to mix with 2 ml concentrated nitric acid in order to acidify the medium and stabilise the chlorides in solution.

Migration test in non-steady-state conditions under electric fields

For the qualification of the compositions, the natural diffusion in concrete is generally too slow. Many procedures are proposed to accelerate it, in particular by imposing an electric field where migration is concerned. Depending on the mechanisms involved, which must be determined from the magnitude obtained, and according to the usage, it is important to choose the appropriate procedure (Rozière, 2007). Obtaining easily measurable quantities by avoiding tedious procedures in fairly short time frames is a solution frequently researched. In this study, migration in non-steady-state conditions under electric fields by the NT BUILD 492 method (Nordtest, 1999) is investigated. The electric field migration test involves applying a potential difference on either side of the concrete sample through electrodes. Under the influence of the electric field created, the movement of the chlorides, as well as that of the other ionic species, is accelerated towards the electrode with the opposite sign. This test was undertaken for different ages for 3, 7, 14, 28, 90 and 180 d of curing in lime-saturated water on slices of 100 × 50 mm from cast cylinders. According to method NT BUILD 492 (Nordtest, 1999), the chloride migration coefficient can be calculated from Equation 5.

$$D_{\mathrm{nSSm}} = \frac{0{\cdot}0239(273+T)L}{(U-2)t} \times \left(X_{\mathrm{d}} - 0{\cdot}0238\sqrt{\frac{(273+T)LX_{\mathrm{d}}}{U-2}} \right) \tag{5}$$

where D_{nSSm} is the migration coefficient (×10–12 m^2/s), U is the applied voltage (V), T is the anolyte solution temperature (°C), L is the thickness of the sample (mm), X_{d} is the penetration depth of chlorides (mm) and t is the duration of the test (hours).

Test for the determination of chloride penetration in marine environments

To evaluate the resistance to chloride penetration of the studied SCCs in a marine environment, two different exposure zones are considered and experimentally simulated: the tidal zone and immersion zone. Immersion conditions were simulated by a total immersion of SCC samples in saline solution. Tidal conditions were simulated by imposing wetting and drying cycles. In one chamber, saline water was pumped and drained from another chamber every 6 h to simulate immersion and tidal conditions by using a water pump and electric valve. The test was carried out on slices of cylinders at 90 d of ageing with dimensions of 100 × 50 mm (three samples per formulation). All surfaces of samples are sealed except one, the upper surface, by which chloride diffusion will ingress in order to ensure a one-dimensional penetration of chlorides (Deby, 2010). The samples sealed by resin are saturated in lime water for almost a week (up to arrival at a constant mass). The concentration of the saline solutions prepared for the tidal and immersion media is identical and equal to 40 g/l. The duration of this test for both immersion and tidal conditions is fixed at 90 d in this study. At the end of the test, the depth of penetration of chloride ions was determined by spraying 0·1 M silver nitrate solution on split surfaces.

Electrical resistivity

In this study, the measurement of the electrical resistivity of SCC was carried out according to the FM 5-578 method (FDOT, 2004) using a surface resistivity measuring device with a four-probe Wenner linear array, with a range of 0–100 K Ω-cm, with a resolution of 0·1 KΩ-cm and an accuracy of ±2% of reading. This test was carried out on cylindrical specimens with diameter 100 mm and 200 mm in height (three specimens per formulation) at various ages of concrete: 3, 7, 14, 28, 90 and 180 d.

Result and discussion

Microstructure investigation

SEM and XRD analysis

Observations by SEM were performed on the studied cementitious material using a JSM 7100F SEM. Figures 6 and 7 present SEM images of limestone filler and sand filler. As indicated, limestone filler is composed of small particles in comparison to sand filler. SEM images of metakaolin powder are illustrated in Figure 8. Metakaolin appears to be composed of dense particles that resemble crystal material. XRD results are shown in Figure 9. XRD analysis of metakaolin showed that it is composed of anorthite and quartz minerals, which are siliceous natural metakaolin.

Fresh state

In order to evaluate the effects of pozzolan on the fresh properties of SCC, slump flow, V-funnel and L-box tests were performed according to the procedure recommended by the EFNARC Committee (EFNARC, 2002); sieve, J-ring and U-tube tests were also performed according to the procedures recommended by AFGC 2000 (AFGC, 2000), ASTM C 1621 (ASTM, 2017) and BS EN 206-1 (BSI, 2001), respectively. The slump flow of all SCC mixes

Figure 6 SEM images and chemical analysis of limestone filler

Figure 7 SEM images and chemical analysis of sand filler

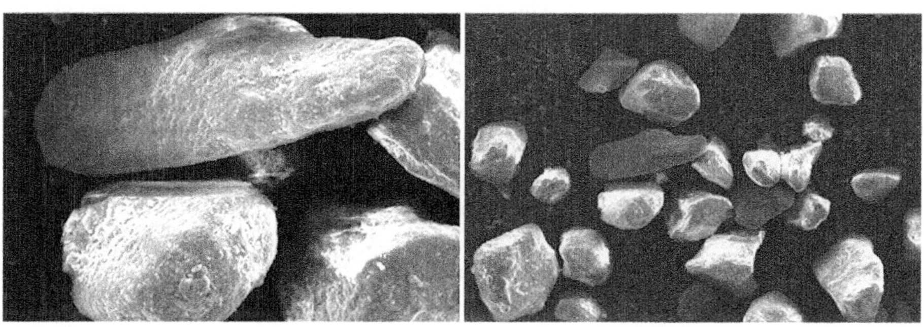

Figure 8 SEM images and chemical analysis of metakaolin powder

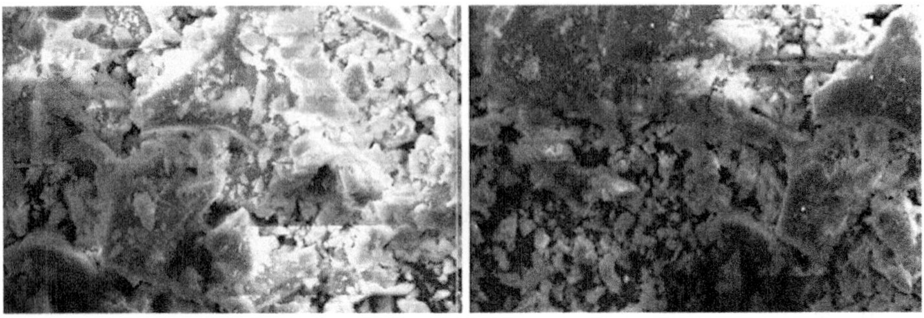

Figure 9 XRD analysis of metakaolin pozzolan. A full-colour version of this figure can be found on the ICE Virtual Library website (www.icevirtuallibrary.com)

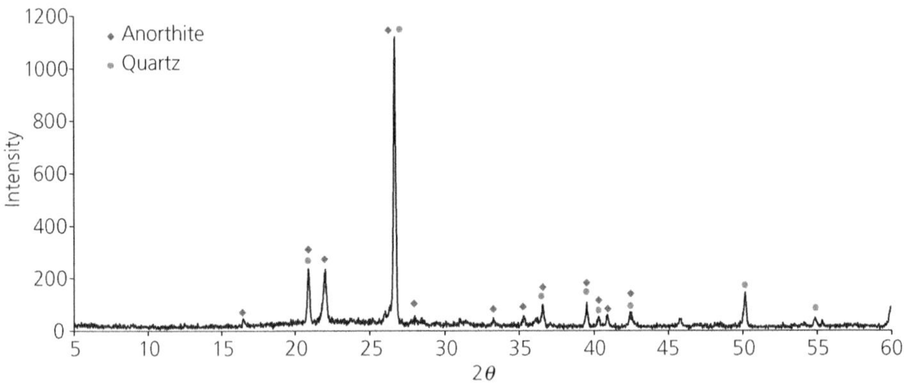

was near to 700±30 mm and no segregation was observed in the mixes (Table 3). The spreading rate of the concrete is also an indication often taken into account (T_{50} – e.g. time to reach a 500-mm spreading diameter; if it takes more than 5 s a large plastic viscosity is concluded, and if the measured time is less than 1 s, lower viscosity is indicated – in these cases, the risk of segregation and bleeding will increase and an aureole may be created). The ability of concrete to flow through a restricted area without blockage and segregation was evaluated using a V-funnel test. The time of flow from the opening of the outlet to the seizure of flow was recorded and the results presented in Table 3. The V-funnel flow time also depends on the type of application, but it is grouped into two classes by EFNARC (2002): the VF1 class has a flow time of less than 10 s and the VF2 class has a flow time of between 7 and 27 s. Although there are other test methods for assessing passing ability, for example the L-box test, the most common test method is the J-ring test. The experimental flows of a J-ring for concrete mixes based on ASTM C 1621 (ASTM, 2017) are indicated in Table 5. An L-box ratio test was carried out on concrete mixes to measure the cohesiveness and ability of SCC to pass through reinforcements without segregation. The L-box ratio is reported to be between 0·7 and 0·9 for

Table 3 Test results of fresh SCCs

Mix name	Slump flow		L-box			V-funnel t: s		Sieve test	J-ring		U-tube
	Dia: mm	T_{50}: s	h_2/h_1	T_{20}: s	T_{40}: s	1 min	5 min	Segregation: %	ΔH: mm	Dia: mm	$h_2–h_1$: mm
SCCL	730	2·0	0·9	1·3	2·0	9·00	8·00	5·00	13	704·6	9
SCCS	719	2·7	0·73	2·2	5·7	6·60	8·40	5·50	5	700·0	26
SCCM10	690	4·8	0·71	2·0	5·7	10·63	12·40	4·24	15	650·0	6
SCCM15	725	2·5	0·8	1·9	5·2	12·37	17·15	7·00	15	630·0	14

normal SCC; however, a range of 0·8 to 1·0 is also proposed by EFNARC guidelines (EFNARC, 2002). Results of h_2/h_1 ratios relative to this study are presented in Table 5. It can be seen that the values range from 0·71 to 0·9, which are within the specified limits for SCC. The U-box test allows the mobility of confined concrete to be characterised and verifies that the installation of the concrete will not be opposed by phenomena such as unacceptable blockages; the value limit is equal to $h_2 - h_1 = 10$ mm. This test was performed for all mixes and the results are presented in Table 3. It can be seen that the range of values is within the specified limits, approximately. The implementation of SCC, under only the effect of gravity, requires a very high fluidity of the material but it is also essential that the concrete maintains a satisfactory stable and perfect homogeneity. Various tests can be used to characterise the resistance to static segregation of SCC to remain homogeneous after its placement until it begins to set. One such test is the so-called 'stability sieve' developed by GTM (AFGC, 2002) to assess the weight percentage milt (the P_{milt} noted later). The acceptable limitations are as follow: (*a*) $0\% < P_{milt} < 15\%$: satisfactory stability, (*b*) $15\% < P_{milt} < 30\%$: critical stability (segregation test necessary on site) and (*c*) $P_{milt} > 30\%$: very poor stability (systematic segregation, unusable concrete).

Compressive strength

The results of the compressive strength of the mixes at the age of 1, 3, 7, 14, 28, 90 and 180 d are plotted as a function of time in Figure 10. From the results shown in Figure 10, the compressive strength of all formulations at all maturities is between 6·3 and 67·5 MPa. According to these results, a higher resistance development is noted when using sand filler. This is true up to 90 d of maturation; beyond, the resistance stabilises for the concrete with sand filler but continues to increase for the SCC with limestone filler. This may be explained by the adhesion between the matrix and the calcareous grains, which becomes better with time. The greatest resistance of 1 to 3 d and 7 to 180 d was obtained for SCCS and SCCM15, respectively. Concrete of 15% metakaolin had a very positive effect on the compressive strength from 7 d of the cure. This result is quite consistent with the results given by Siddique and Klaus (2009). On the other hand, the increase in the compressive strength of the concrete containing 10% metakaolin compared to the control concrete occurred from the age of 90 d. This increase for SCCM10 is more visible at the age of 180 d. The compressive strength in concrete containing 10% metakaolin is 1·02 and 1·1 times higher than that of the control concrete at 90 and 180 d of curing, respectively. In this study, the increase of compressive strength in SCC containing metakaolin is in a good agreement with the experimental results of

Figure 10 Compressive strength of the different formulations studied

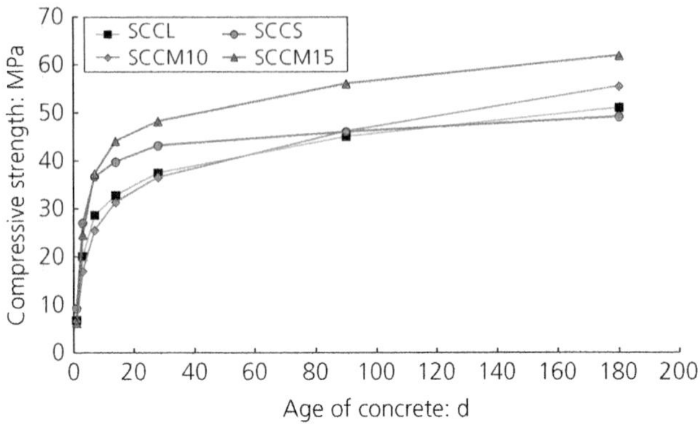

previous studies performed on SCC, normal concrete and geopolymer concrete whose cement has been replaced by metakaolin at various percentages (Alanazi *et al.*, 2017; Badogiannis and Tsivilis, 2009; Dinakar *et al.*, 2013; Güneyisi *et al.*, 2012; Hassan *et al.*, 2012; Khatib and Clay, 2004; Siddique and Klaus, 2009).

The major difference between the metakaolin found in Iran and that found in Europe is the silicon dioxide (SiO_2) and aluminium oxide (Al_2O_3) content. As a matter of fact, silicon dioxide is greater in metakaolin found in Iran. Also, the compressive strength of concrete with metakaolin from Europe, during early age, is less than that of concrete mixtures without additives. On the other hand, the Iranian metakaolin increases the compressive strength from an earlier age. This behavioural difference is well documented in research work conducted by the present authors and Gilan *et al.* (2012) in Iran, as well as in research conducted by San Nicolas *et al.* (2014) in France.

Evolution of evaporable water and non-evaporable water content

The main purpose of conducting thermal analysis is to measure the content of evaporable and non-evaporable water at 105°C of binder pastes. The amount of evaporable water can be linked to the porosity. The non-evaporable water is considered to be an important and reliable indicator of the hydration progress in the modification of the microstructure. In this way, thermal measurement of evaporable and non-evaporable water content can be useful to justify the mechanical behaviour and durability properties of the studied SCCs. The results relating to the evaporable and non-evaporable water at 1, 3, 7, 14, 28, 90 and 180 d of ageing are shown in Figures 11 and 12, respectively. An increase over time of maturation of non-evaporable water content was found, accompanied by a decrease in evaporable water for all formulations. This first observation is predictable due to the continued hydration of the binders. Moreover, a paste with metakaolin, even if it contains less cement and therefore less clinker, has been able to develop a non-evaporable water content close to that of the control, which is seen to be greater in the case of binders with 15%, and this from 7 d of maturation. These results indicate the presence of additional hydration, which may be related to the pozzolanic activity and the formation of hydrates.

Figure 11 Evaporable water content in the studied mixtures at different ages

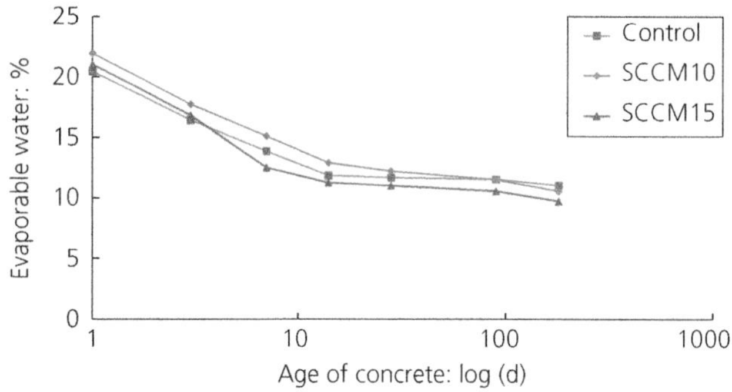

Figure 12 Non-evaporable water content at 105°C in the studied mixtures at different ages

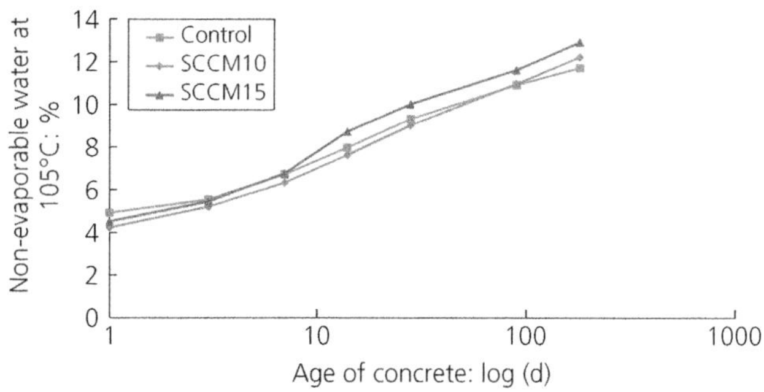

Water absorption and permeable voids

The test results for water absorption and void permeability are presented respectively in Figures 13 and 14.

The results indicated that using metakaolin in place of cement and sand filler instead of limestone filler in SCC can reduce water absorption and the number of permeable pores in comparison to control concrete at all conservation durations. According to the results illustrated in Figure 13, the water absorption rate in SCCM15 is lower than in all other formulations. The water absorption in SCCM15 is 1·14, 1·15, 1·16, 1·15 and 1·16 times lower than that of the control concrete at 7, 14, 28, 90 and 180 d of ageing, respectively. The number of permeable pores of the studied SCCs is presented in Figure 14. According to the results illustrated in this figure, the SCC containing metakaolin shows good behaviour in the reduction of the number of permeable pores compared to the control concrete. This fact is highlighted by the increase in the additive level of metakaolin from 10 to 15% cement replacement. The reason can be found in the conversion of continuous pores to more discontinuous pores due to the pozzolanic reaction of metakaolin. Its pozzolanic reaction can contribute to the formation of additional hydration, which can lead to lower connected

Figure 13 Water absorption (%) of studied SCCs

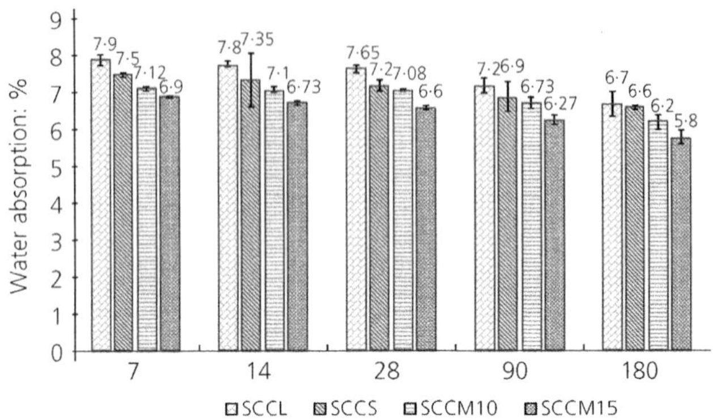

Figure 14 Volume of permeable pores (voids) (%) of studied SCCs

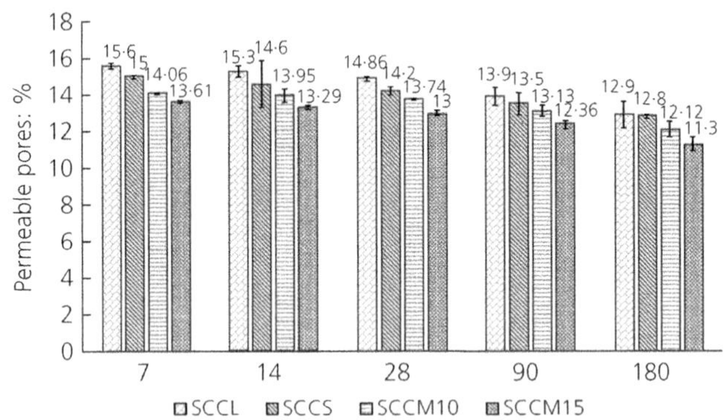

porosity. Indeed, the decreased void content increases the load-carrying capacity of concrete, and therefore the compressive strength of SCCM10 and SCCM15 is increased. Moreover, the amount of non-evaporable water in paste with metakaolin and the compressive strength of SCCM10 and SCCM15 are completely consistent with the results with regard to water absorption and permeable pores. This fact is more visible with long curing times. The different properties studied here can be related to compressive strength, as shown in Figures 15 and 16. A linear equation seems to describe correctly the evolution of water absorption and permeable pores with compressive strength. Compressive strength increases with the decrease of these properties.

Water penetration

The rate of water penetration into the concrete is a parameter that allows the performance of the concretes to be compared with respect to permeability. The penetration of water in concrete can cause physical and chemical damage. Water as a solvent is able to dissolve many cementitious components. Many ions cause damage in concrete by the penetration of

Figure 15 Relationship between water absorption and compressive strength

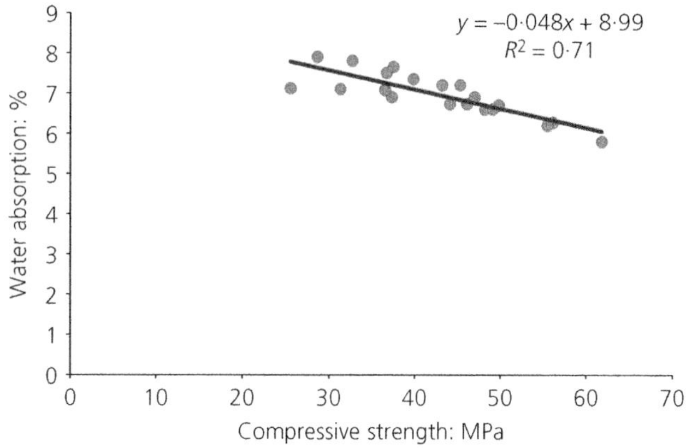

Figure 16 Relationship between the volume of permeable pore and compressive strength

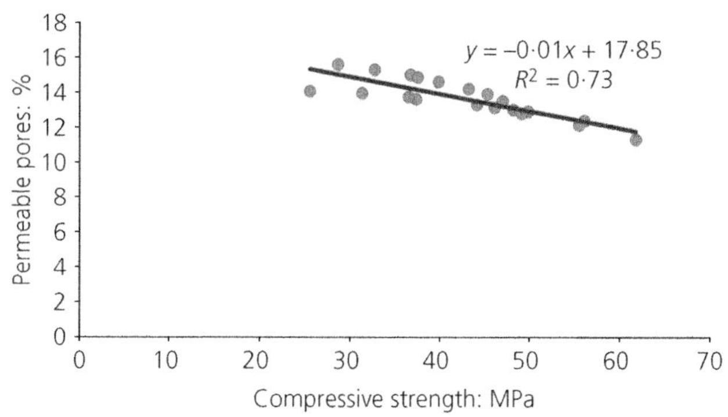

water in concrete. For this reason, the penetration of water in concrete can be used as an indicator in the study of concrete durability. The results of the measurement of water penetration are shown in Figure 17. This figure shows that the substitution of 10 and 15% Portland cement with metakaolin leads to a reduction in the penetration depth of the water and thus to the reduction of the water permeability. The maximum and minimum water penetration depth in the SCCs studied after all ages of hardening was obtained for SCCS and SCCM15, respectively. The depth of water penetration of the control SCC was 1·39, 1·45, 1·47, 1·67 and 2·5 times higher compared to that of the SCCM15 at 7, 14, 28, 90 and 180 d, respectively. By contrast, the water penetration depth of the control concrete is 1·56, 1·66, 1·8, 2·05 and 2·47 times lower than that of the SCCs at 7, 14, 28, 90 and 180 d, respectively. According to the results given in Figure 17, the SCC containing 15% metakaolin, SCCM15, exhibits the best performance. This can be explained by its higher pozzolanic reactions, which have a strong impact on the decreased permeability in SCC.

Figure 17 Water penetration in the studied SCCs after 7, 14, 28, 90 and 180 d of curing

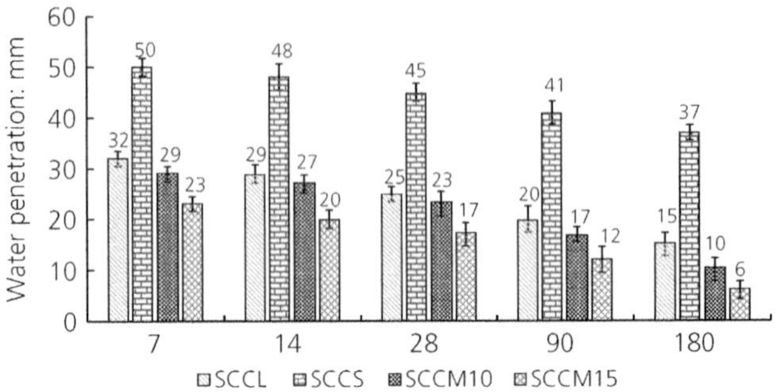

Capillary absorption (sorptivity)

The higher the capillary absorption, the more quickly the material is able to be invaded by the liquid in contact. The sorptivity coefficients (S) after 24 h of testing are presented for the various SCCs at different ages (7, 14, 28, 90 and 180 d) in Figure 18. According to this figure, the absorption coefficient of concrete containing sand filler (SCCS) is slightly higher than that of the control SCC at all ages of curing. From Figure 18, the capillary absorption coefficient of SCCL is 1·08 and 1·19 times higher than that of SCCM10 and SCCM15 aged 180 d, respectively. The results show that the incorporation of metakaolin has a very beneficial effect on the reduction of capillary absorption of SCCs. On the other hand, this reduction is greater when the additional content is increased.

Resistance to chloride penetration

The chloride ions penetrate into concrete and cause, from a certain concentration in the pore solution, at the reinforcements, the local destruction of the passivation film and the initiation of localised corrosion. In general, the chloride in the normal case does not directly effect all the damage of concrete (Dhir *et al.*, 1990; McCarter *et al.*, 1992). The exposure of concrete to

Figure 18 Sorptivity coefficients of the studied SCCs at different curing ages (mm/h$^{0·5}$)

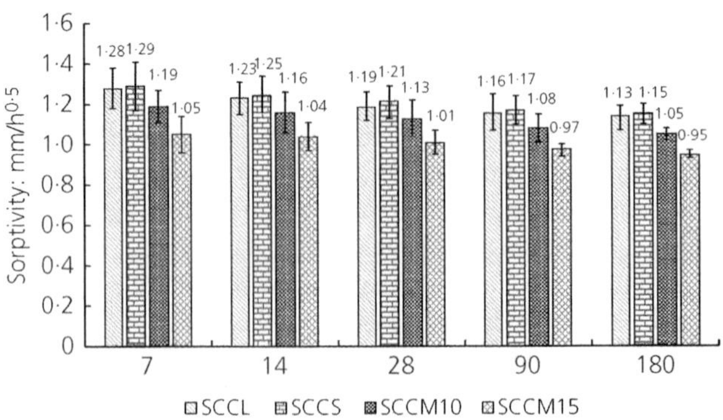

chloride has two main sources: sea water and icing salts. The penetration of chlorides in concrete is established by sophisticated physical and chemical processes from three sources (De Schutter and Audenaert, 2007):

- The location of the structure. In marine environments, the amount of chloride in contact with the concrete depends on the location of the structure, provided that they are completely submerged or placed in the tidal zone or only in contact with the salt fog. However, in situations such as roads, frequent use of icing salts in very cold weather causes difficulties in calculating the amount of pulverised chloride on the structures.
- The properties of chloride penetration of the different materials used in the mixture and the age of the concrete.
- The mechanisms of chloride penetration are not isolated (simple diffusion), but are combined with convection (absorption), chemical and physical connections, and interaction of other coexisting ions. The change in temperature, rain and sun introduces variations which should also be taken into account.

The exposure conditions of the structure in its environment are a key factor for the modelling of structures subjected to chloride corrosion. Depending on the type of exposure, different transport mechanisms are considered for ion chlorides (see Figure 19).

Non-steady-state diffusion. The results for the penetration depth and chlorides diffusion coefficient in non-steady-state conditions for the SCCs studied at 28 d of treatment after 90 and 180 d of immersion in the saline solution are illustrated in Figures 20 and 21, respectively.

Figure 19 Marine environment (De Schutter and Audenaert, 2007)

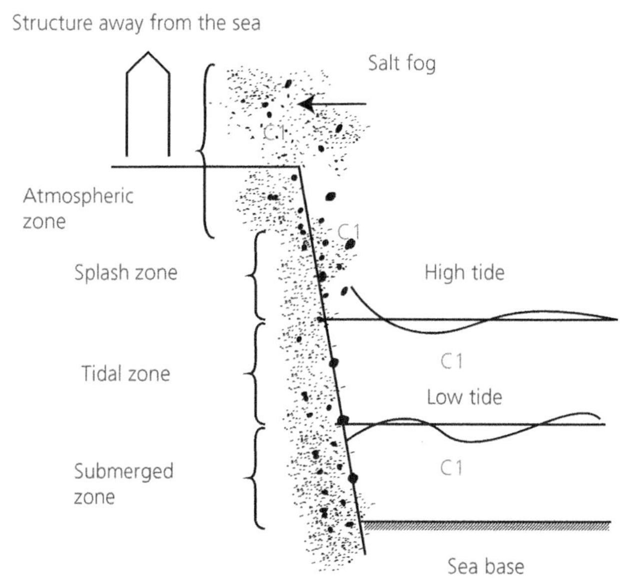

Figure 20 Depth of chloride ions penetration in the studied SCCs at 28 d of ageing after 90 and 180 d of immersion in saline solution

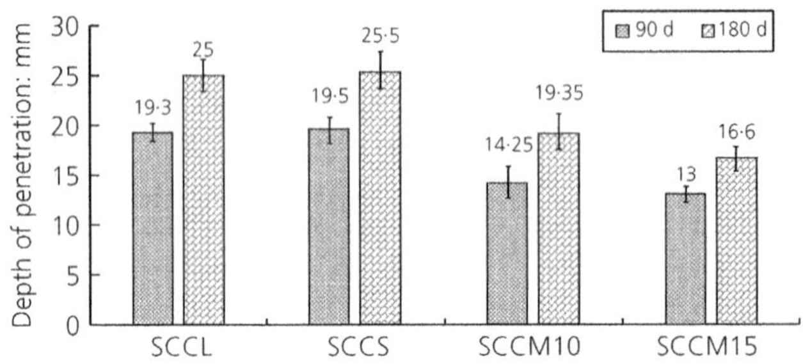

Figure 21 Chloride diffusion coefficient in SCCs at 28 d of ageing after 90 and 180 d of immersion in saline solution

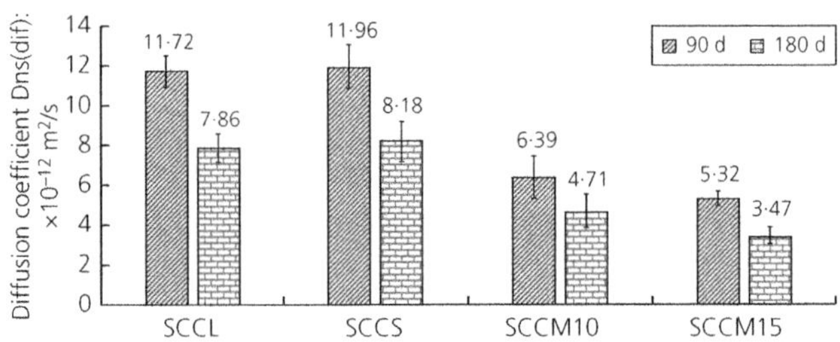

The results illustrated in Figures 20 and 21 show that the penetration depth of the chlorides and consequently the apparent diffusion coefficient of the chlorides in the SCC based on sand filler and the control SCC are almost identical. On the other hand, for SCCs with metakaolin, the penetration depth of the chlorides and therefore the apparent diffusion coefficient are much lower compared to those of the control concrete. The positive effect of incorporating metakaolin is undeniable. Furthermore, SCCs containing 15% metakaolin are relatively more resistant to chloride penetration than SCC based on 10% metakaolin.

Free-chloride concentration profile. Chlorides exist in concrete in two forms: free and bound. The bound chloride is found in the form of a complex salt such as calcium monochloro-aluminate hydrate; it can also be attached to the C–S–H. It is assumed that bound chloride cannot initiate corrosion. However, free chloride ions, which are highly mobile, have a key role in the corrosion of reinforcements. The content of free chlorides is generally determined by measuring the content of water-soluble chlorides. In this study, the content of water-soluble chlorides was measured on concrete powders obtained from the exposed surface using the device explained in the 'Test method' section. The profiles of the free chlorides measured in the concrete are plotted in Figure 22.

Figure 22 Water-soluble chloride penetration profile in the different studied SCCs

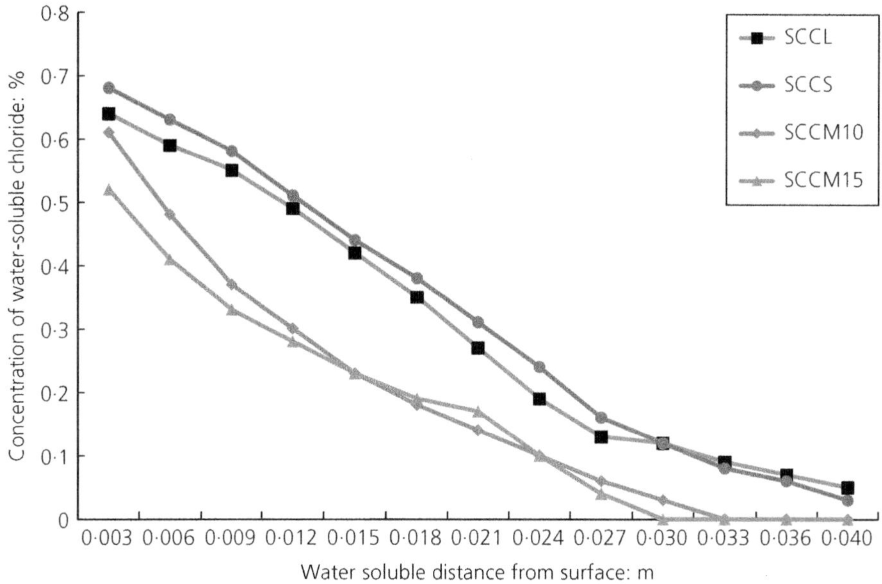

The results presented in Figure 22 show that the SCCs containing metakaolin have lower penetration depths of the chlorides compared to the control concrete, which is in agreement with the colorimetric measurements. The chloride penetration profiles deduced from the water-soluble distance from the surface are higher than those measured using a colorimetric method based on silver nitrate solution for diffusion in non-steady state. This can be explained by the lesser sensitivity of the colorimetric method to the low values of chloride content and probably also to the possible ingress of chloride into the sound zone of the sample during the conservation time between the end of the diffusion test and the measurement of water-soluble chloride. These results are in agreement with the results of the capillary absorption measurements, which were found to be higher for the control concrete compared to the mixtures with pozzolan at all ages of curing. The higher capillary absorption indicates a greater connected porosity and consequently a higher free-chloride content in the concrete.

Non-steady-state migration. The aim of this part of the experimental study is to evaluate the effect of sand filler and metakaolin used on the evolution of the chloride migration coefficient in the four treated SCCs. The cure period effect is also taken into account by considering different conservation ages: 3, 7, 14, 28, 90 and 180 d. The evolution of the migration coefficient of the chlorides during the maturation time for the different SCCs is shown in Figure 23.

According to the results illustrated in Figure 23, the SCC with sand filler behaves in a very similar way to the control concrete. On the other hand, the evolution of this coefficient over time is different in the case of SCCs with metakaolin. On very short terms (3 d), SCCs with metakaolin have higher values compared to control concrete. This is due to their greater

Figure 23 Migration coefficient of different mixes at different ages

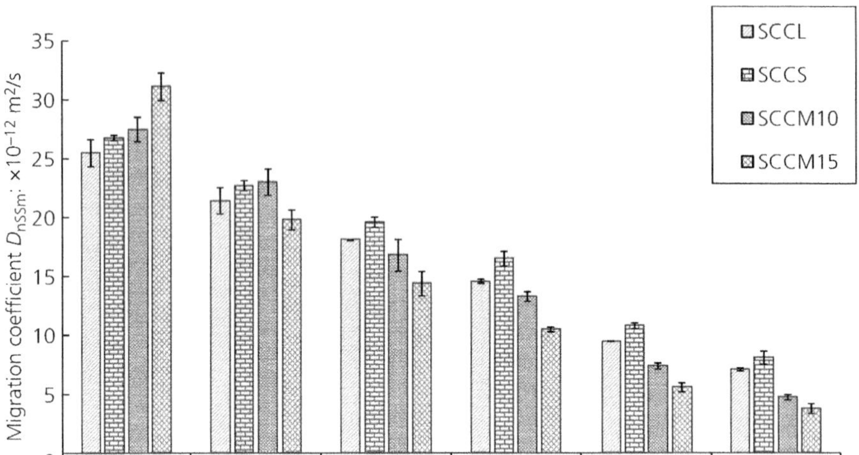

	3	7	14	28	90	180
SCCL	25·47	21·39	18·07	14·57	9·47	7·07
SCCS	26·77	22·67	19·57	16·49	10·77	8·07
SCCM10	27·47	22·97	16·77	13·27	7·37	4·73
SCCM15	31·07	19·77	14·37	10·47	5·56	3·77

porosity. However, the migration coefficient of these SCCs will quickly drop and then be lower than that of the control concrete. At the age of 180 d, the migration coefficient of the SCCM10 and SCCM15 is 1·5 and 1·9 times lower, respectively, than that of the control SCC. Moreover, Figure 23 shows the curing effect on the decrease in the chlorides migration coefficient. In this study, the positive influence of metakaolin in reducing the chloride penetration is in agreement with the results obtained by Kim *et al.* (2007) and Ramezanianpour and Bahrami (2012). A very rapid decrease is observed in the mixtures with metakaolin, whereas this decrease is slower and less important in the case of the control SCC and the SCC containing sand filler. This result can be explained at least partially by the low porosity, as can be expected from the capillary absorption results. Another explanation for the positive effect of metakaolin is related to the reactivity of this pozzolan and consequently the consumption of calcium hydroxide, which most likely results in increasing tortuosity and reduction of hydroxide ions in the interstitial solution. Therefore, the ionic concentration in the interstitial solution decreases. Consequently, the conductivity and penetration of chloride ions will decrease.

The values obtained at 180 d make it possible to classify the control concrete and the SCC containing metakaolin with good resistance against the penetration of chlorides according to the classification given by Tange Jepsen *et al.* (2001) and indicated as follows

- $D<2\times10^{-12}\,m^2/s$: very good resistance against the penetration of chlorides
- $D<8\times10^{-12}\,m^2/s$: good resistance against the penetration of chlorides
- $D<16\times10^{-12}\,m^2/s$: moderate resistance against chloride penetration
- $D>16\times10^{-12}\,m^2/s$: not suitable for an aggressive environment.

The SCC with sand filler is also considered to have almost good resistance to chloride penetration at the same maturity of 180 d.

The quality of the concrete can also be assessed by measuring the current that passes when applying 30 V across the sample. This means that the greater the current, the lower the resistance of the concrete to the penetration of chloride ions. The amount of initial current of different mixes at different ages due to the RCMT apparatus is presented in Figure 24. According to these results, the amount of initial current in the SCC containing sand filler is greater than in the other formulations. From 7 d of cure, the initial current passed in SCCM15 is lower than in the other mixtures. This can be justified because the 15% substitution of metakaolin in SCC using pozzolanic reaction can create a denser microstructure by converting continuous pores to discontinuous pores. Figure 25 shows that there is a direct relationship between the migration coefficient and the initial current. The initial current measurement appears to be a good indicator of sustainability, which can be directly related to the migration coefficient.

Resistance to chloride penetration in immersion and tidal conditions. Numerous other investigations have been carried out on the penetration of chloride ions in the vicinity of seashores and ports (Bermudez and Alaejos, 2010; Conciatori *et al.*, 2008). It is well known that Kuzel's salt ($3CaO.Al_2O_3{\cdot}0{\cdot}5CaSO_4{\cdot}0{\cdot}5CaCl_2{\cdot}10(11)H_2O$) and Friedel's salt ($3CaO.Al_2O_3CaCl_2{\cdot}10H_2O$) can be formed when the chloride ions are present during cement hydration. When the chloride ions react with C3A, Friedel's salt is formed. Also, the reaction of the chloride ion with C4AF leads to the formation of Kuzel's salt. In order to evaluate the studied SCCs' resistance to chloride penetration in the marine environment, two different conditions of exposure were considered and experimentally simulated: tidal and immersion.

Figure 24 Initial current of different mixes at different ages due to RCMT apparatus

	3	7	14	28	90	180
SCCL	132	125·5	110	86	51·33	37
SCCS	142·33	133	118	95	57·33	43
SCCM10	155	125	100·33	74	43	29
SCCM15	196·5	113·5	90	66	34	20·67

Figure 25 Evolution of initial current according to migration coefficient of the studied SCCs

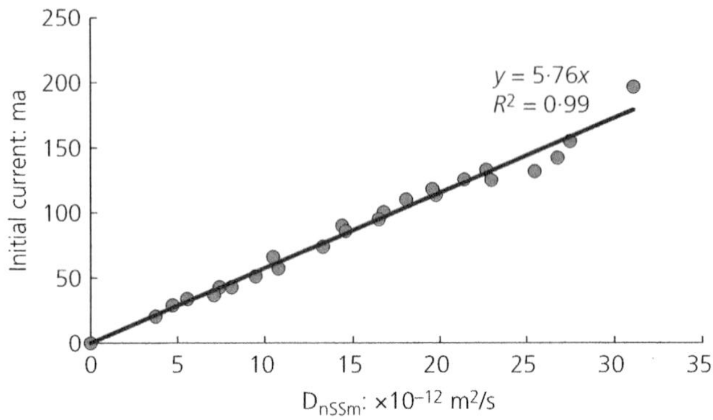

The immersion condition was simulated by the total immersion of the SCC samples in saline solution. The tidal condition was simulated by imposing wetting and drying cycles as explained previously. At the end of the tests, after 90 d of ageing, the penetration depth of the chloride ions in all formulations was determined for both immersion and tidal conditions. The results are shown in Figure 26 and illustrated in Figure 27. The penetration depth of the chlorides in all mixtures is higher under tidal conditions than immersion conditions due to the drying of the surface of the concrete. Its value for the SCCL, SCCS, SCCM10 and SCCM15 is respectively 1·07, 1·08, 1·15 and 1·11 times higher under the wetting–drying conditions compared to the immersion conditions. This increase in chloride penetration should be higher under more severe drying conditions, as in the case of the presence of wind and sun. The mixtures with metakaolin show better resistance to chloride penetration, and this under both tidal and immersion conditions, which shows the positive effect of the use of metakaolin.

Figure 26 Depth of chloride ions penetration in SCCs at age of 90 d after 90 d of immersion, and tidal in saline solution

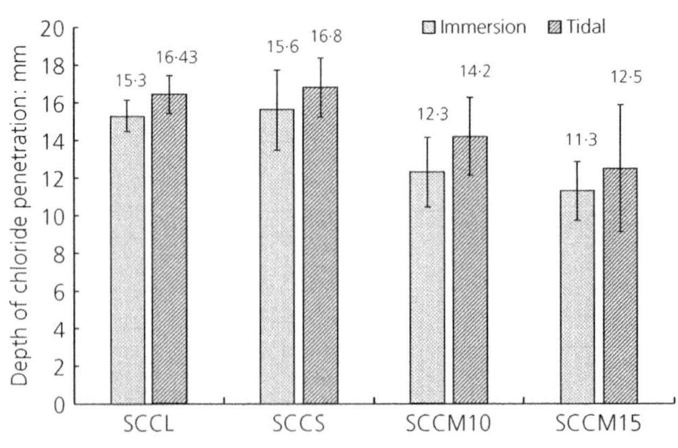

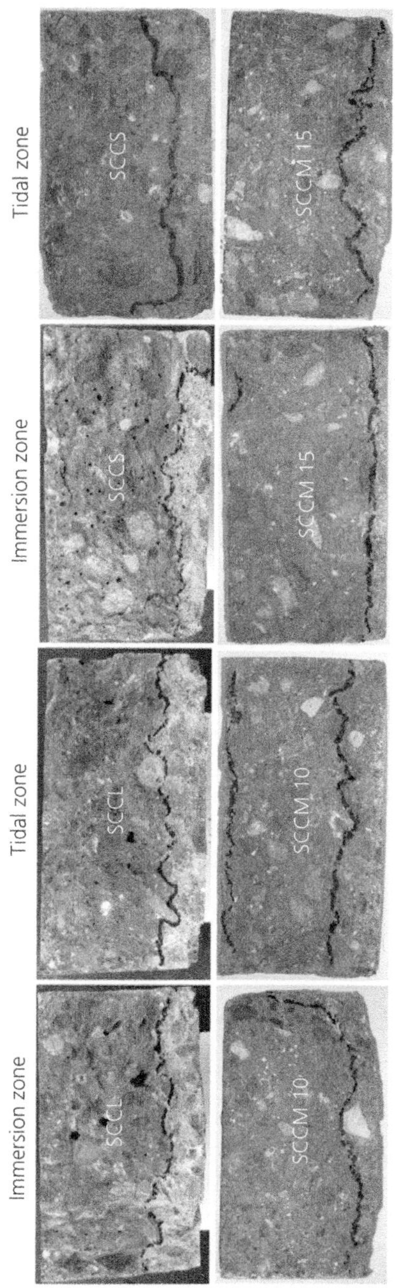

Figure 27 Chloride penetration depth of SCCs samples after 90 d of testing, highlighted using silver nitrate solution: left: in immersion conditions; right: in tidal conditions

In this study, the results showing the increase in resistance of SCCs containing metakaolin in marine environments are in good agreement with the experimental results of research studies performed on concrete blended with metakaolin in a seawater environment, as presented, for example, by Duan *et al.* (2015).

In fact, chloride penetration should be greater in tidal conditions than in immersion conditions, based on the following two points:

- The increase of concrete's diffusion speed rate: in this situation, the saline solution penetrates in concrete through capillary suction by means of the drying and wetting of the concrete. As such, the humidity and chloride ions increase inside the concrete, causing the chloride ion diffusion speed to increase.
- The increase of concrete's demolition rate: in this situation, due to the lack of humidity in tidal zones, saline crystals appear on the surface of concrete samples resulting in increased internal stress and micro-crack. This chloride ion penetration increase can be much greater when the hard drying conditions are magnified by wind and/or sun (Samimi *et al.*, 2017).

Electrical resistivity

Electrical resistivity is one of the main indicators of the durability of cementitious materials. In general, high electrical resistivity is correlated with a high resistance to chloride penetration and consequently a high resistance to corrosion. The results of electrical resistivity for the different mixtures are presented in Figure 28. This figure clearly shows the very significant influence of metakaolin on the increase in electrical resistivity at different ages: 3, 7, 14, 28, 90 and 180 d. This increase is more remarkable when the percentage of metakaolin addition is increased. This may be related to the pozzolanic reaction, which leads to a dense structure and a lower concentration of OH- in the interstitial solution. This fact ultimately increases the electrical resistivity and improves the durability of the concrete against the penetration of chlorides. For SCC containing sand filler, the electrical resistivity is always lower than that of the control concrete. According to Figure 28, the age of the concrete has a positive influence on

Figure 28 Evolution of the electrical resistivity with curing time

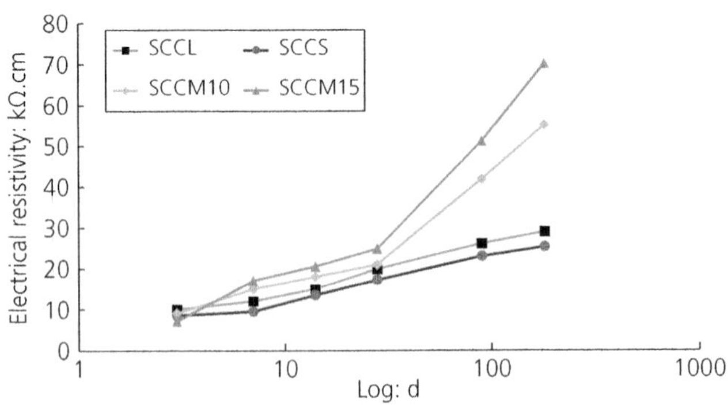

Figure 29 Evolution of the electrical resistivity according to migration coefficient

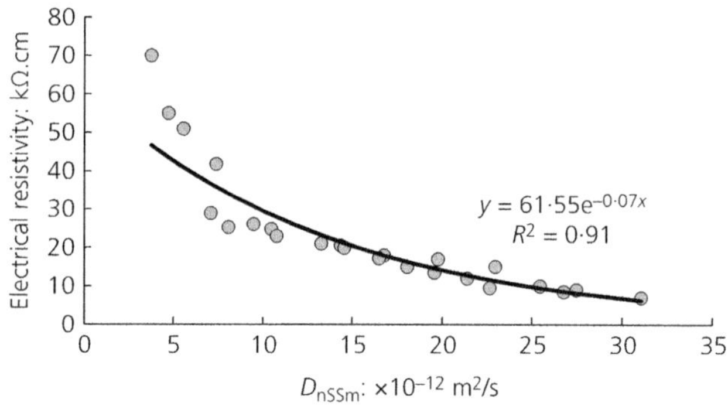

the increase of the electrical resistivity, which is predictable due to the further hydration of the binder. However, this increase is greater when metakaolin is used.

The electrical resistivity in concrete increases when the chloride migration coefficient decreases. This can be clearly observed in Figure 29. According to the relation between electrical resistivity and the chloride migration coefficient, it can be concluded that there is a close relationship between the two last properties. These results are in good agreement with the experimental results of previous studies performed on concrete composites based on various supplementary materials (Ramezanianpour and Bahrami, 2012).

Conclusion

Based on the obtained results from this experimental study performed on a reference SCC with limestone filler (SCCL), two SCC composites with limestone filler and 10% and 15% metakaolin substitution (SCCM10 and SCCM15, respectively), and one other SCC composite based on sand filler (SCCS) instead of limestone, the following conclusions can be drawn:

(*i*) The results of the mineralogical analysis showed that the studied metakaolin is composed of a mixture containing glass phases as well as mineral phases such as quartz and anorthite.

(*ii*) The results of the fresh characterisation showed that the substitution of 10 and 15% Portland cement with metakaolin as well as the total substitution of the limestone filler by sand filler makes it possible to obtain stable SCCs with good capacities of filling without blocking.

(*iii*) The results of the mechanical characterisation of the four SCCs showed that the total substitution of the limestone filler by sand filler makes it possible to obtain an SCC with better mechanical resistances compared to the control SCC. This is true up to 90 d of maturation; beyond this, the resistance stabilises for the concrete with sand filler but continues to increase for the control SCC based on limestone filler.

(*iv*) The results also showed that the substitution of 10% Portland cement by metakaolin reduces the resistance in the very short term (from 1 to 28 d), but from 90 d of maturation the compressive strength is superior to that of the control concrete. On the other hand, the SCC containing 15% of the metakaolin shows a higher compressive strength from 7 d of cure in comparison with the other formulations. This was confirmed by the measurements of evaporable water and non-evaporable water content.

(*v*) The results relating to the compressive strength and porosity are coherent – namely, that the more compact the matrix is, which would be due to the granular skeleton (sand filler) or the pozzolanic reaction (metakaolin), the greater the compressive strength and the greater the absorption after immersion or after immersion/boiling, and permeable pore number decreases.

(*vi*) The results of capillary absorption show that the incorporation of metakaolin has a very beneficial effect on the reduction of capillary absorption. On the other hand, this reduction is greater with the increase in the addition content. The depth of penetration of the water is thus minimal in the case of the presence of metakaolin and maximal in the case of sand filler.

(*vii*) The SCC with sand filler behaves in a way very similar to the control concrete with values of migration coefficient that are almost identical to 90 and 180 d. On the other hand, the evolution of this coefficient over time is different in the case of SCCs with metakaolin. At very short terms (3 d), the SCCs with metakaolin exhibit higher values of migration coefficient compared to the control concrete. However, the migration coefficient of these SCCs will quickly drop and then be lower than that of the control concrete.

(*viii*) The penetration depth of the chlorides in all mixtures is higher under tidal than immersion conditions due to the drying of the surface of the concrete. Mixtures with metakaolin show much better resistance to chloride penetration than control concrete and SCC with sand filler in both tidal and immersion conditions, which shows the positive effect of the use of metakaolin.

Acknowledgements

The authors would like to specifically thank Mr Rahmatolah Hakimi, President of the Imen Rah Consulting Engineers Company for financial support of this research. The authors would like to thank Ms Minoo Shahrokhi for her assistance with some laboratory measurements. The authors acknowledge the physics center of Shiraz University for its help in microstructure observations and the Mana Peysanjesh Consulting Engineering Laboratory for their authorisation to use apparatus.

REFERENCES

AFGC (Association Française de Génie Civil) (2000) Bétons autoplaçants – Recommandations provisoires. *Annales du Bâtiment et des Travaux Publics* **2000(3)**: 15–38 (in French).

AFGC (2002) *Interim Recommendations for Use of Self-Compacting Concrete*. AFGC, Paris, France, pp. 65–128.

Ahmadi B, Sobhani J, Shekarchi M and Najimi M (2014) Transport properties of ternary concrete mixtures containing natural zeolite with silica fume or fly ash. *Magazine of Concrete Research* **66(3)**: 150–158, https://doi.org/10.1680/macr.13.00224.

Alanazi H, Yang M, Zhang D and Gao Z (2017) Early strength and durability of metakaolin-based geopolymer concrete. *Magazine of Concrete Research* **69(1)**: 46–54, https://doi.org/10.1680/jmacr.16.00118.

Assié S (2004) *Durabilité des bétons autoplaçants*. Thèse de doctorat, *INSA, Toulouse, France*.

ASTM (1997) C 642: Standard test method for density, absorption, and voids in hardened concrete. ASTM International, West Conshohocken, PA, USA.

ASTM (2002) C 1543: Standard test method for determining the penetration of chloride ion into concrete by ponding. ASTM International, West Conshohocken, PA, USA.

ASTM (2009) C 150: Standard specification for Portland cement. ASTM International, West Conshohocken, PA, USA.

ASTM (2012) C 31/C 31M-12: Standard practice for making and curing concrete test specimens in the field. ASTM International, West Conshohocken, PA, USA.

ASTM (2013) C 511: Specification for moist cabinets, moist rooms, and water storage tanks used in the testing of hydraulic cements and concretes. ASTM International, West Conshohocken, PA, USA.

ASTM (2017) C 1621/C 1621M: Standard test method for passing ability of self-consolidating concrete by J-ring. ASTM International, West Conshohocken, PA, USA.

Badogiannis E and Tsivilis S (2009) Exploitation of poor Greek kaolins: durability of metakaolin concrete. *Cement and Concrete Composites* **31(2)**: 128–133.

Baroghel-Bouny V, Belin P, Maultzsch M and Henry D (2007) $AgNO_3$ spray tests: advantages, weaknesses, and various applications to quantify chloride ingress into concrete. Part 2: non-steady-state migration tests and chloride diffusion coefficients. *Materials and Structures* **40(8)**: 783–799.

Bermudez MA and Alaejos P (2010) Models for chloride diffusion coefficients of concrete in tidal zone. *ACI Materials Journal* **107(1)**: 3–11.

Bessa-Badreddine A (2004) *Étude de la Contribution des Additions Minérales aux Propriétés Physiques, Mécaniques et de Durabilité des Mortiers*. Doctoral dissertation, Université de Cergy Pontoise, Cergy Pontoise, France (in French).

BSI (1997) BS 8110-1:1997: Structural use of concrete. Code of practice for design and construction. BSI, London, UK.

BSI (2001) BS EN 206–1: European standards for concrete. BSI, London, UK.

BSI (2009) BS EN 12390-8: 2012: Testing hardened concrete. Depth of penetration of water under pressure. BSI, London, UK.

Caldarone MA, Gruber KA and Burg RG (1994) High reactivity metakaolin (HRM): a new generation mineral admixture for high performance concrete. *Concrete International* **16(11)**: 37–41.

Conciatori D, Sadouki H and Brühwiler E (2008) Capillary suction and diffusion model for chloride ingress into concrete. *Cement and Concrete Research* **38(12)**: 1401–1408.

Deby F (2010) *Rapport de synthèse sur les mesures de migration des chlorures*. LMDC/INSA-Univ, Toulouse, France.

De Larrard F (1999) *Concrete Mixture Proportioning: A Scientific Approach*. CRC Press, Boca Raton, FL, USA.

De Schutter G and Audenaert K (2007) *Durability of Self-Compacting Concrete, State-of-the-Art Report of RILEM Technical Committee 205-DSC*. RILEM, Paris, France, RILEM Report 38.

Dhir RK *et al.* (1990) Rapid estimation of chloride diffusion coefficient in concrete. *Magazine of Concrete Research* **152(42)**: 177–185, https://doi.org/10.1680/macr.1990.42.152.177.

Dinakar P, Sahoo PK and Sriram G (2013) Effect of metakaolin content on the properties of high strength concrete. *International Journal of Concrete Structures and Materials* **7(3)**: 215–223.

Duan P, Zhou W and Yan C (2015) Investigation of pore structure and ITZ of concrete blended with mineral admixtures in a seawater environment. *Magazine of Concrete Research* **67(15)**: 812–820, https://doi.org/10.1680/macr.14.00282.

Dubey A and Banthia N (1998) Influence of high-reactivity metakaolin and silica fume on the flexural toughness of high-performance steel fiber reinforced concrete. *Materials Journal* **95(3)**: 284–292.

Durabilité des Bétons (1997) *Méthodes Recommandées pour la Mesure des Grandeurs Associées à la Durabilité, Compte-rendu des Journées Techniques AFPC AFREM*. INSA, Toulouse, France (in French).

EFNARC (2002) *Guidelines for Self-Compacting Concrete. Specification, Production and Use*. EFNARC, London, UK.

FDOT (Florida Department of Transportation) (2004) *Florida Method of Test for Concrete Resistivity as an Electrical Indicator of its Permeability, FM 5-578*. FDOT, Tallahassee, FL, USA.

Gilan SS, Jovein HB and Ramezanianpour AA (2012) Hybrid support vector regression–Particle swarm optimization for prediction of compressive strength and RCPT of concretes containing metakaolin. *Construction and Building Materials* **34**: 321–329.

Güneyisi E, Gesoğlu M, Karaoğlu S and Mermerdaş K (2012) Strength, permeability and shrinkage cracking of silica fume and metakaolin concretes. *Construction and Building Materials* **34**: 120–130.

Hassan AA, Lachemi M and Hossain KM (2012) Effect of metakaolin and silica fume on the durability of self-consolidating concrete. *Cement and Concrete Composites* **34(6)**: 801–807.

Hubertova M and Hela R (2007) The effect of metakaolin and silica fume on the properties of lightweight self-consolidating concrete. *Special Publication* **243**: 35–48.

Kannan V and Ganesan K (2014) Synergic effect of pozzolanic materials on the structural properties of self-compacting concrete. *Arabian Journal for Science and Engineering* **39(4)**: 2601–2609.

Khatib JM and Clay RM (2004) Absorption characteristics of metakaolin concrete. *Cement and Concrete Research* **34(1)**: 19–29.

Kim HS, Lee SH and Moon HY (2007) Strength properties and durability aspects of high strength concrete using Korean metakaolin. *Construction and Building Materials* **21(6)**: 1229–1237.

McCarter WJ *et al.* (1992) Absorption of water and chloride into concrete. *Magazine of Concrete Research* **44(158)**: 31–37, https://doi.org/10.1680/macr.1992.44.158.31.

Najimi M, Sobhani J, Ahmadi B and Shekarchi M (2012) An experimental study on durability properties of concrete containing zeolite as a highly reactive natural pozzolan. *Construction and Building Materials* **35**: 1023–1033.

Nordtest (1999) NT BUILD 492: Concrete, mortar and cement-based repair materials: chloride migration coefficient from non-steady-state migration experiments. Nordtest, Oslo, Norway.

Okamura H and Ouchi M (2003) Self-compacting concrete. *Journal of Advanced Concrete Technology* **1(1)**: 5–15.

Pera J, Husson S and Guilhot B (1999) Influence of finely ground limestone on cement hydration. *Cement and Concrete Composites* **21(2)**: 99–105.

Ramezanianpour A and Bahrami Jovein H (2012) Influence of metakaolin as supplementary cementing material on strength and durability of concretes. *Construction and Building Materials* **30**: 470–479.

Rozière E (2007) *Etude de la durabilité des bétons par une approche performantielle*. Thèsethèse de doctorat, l'Ecole Centrale de Nantes, France (in French).

Sadrmomtazi A, Tahmouresi B and Kohani Khoshkbijari R (2017) Effect of fly ash and silica fume on transition zone, pore structure and permeability of concrete. *Magazine of Concrete Research*, https://doi.org/10.1680/jmacr.16.00537.

Samimi K, Kamali-Bernard S, Maghsoudi AA, Maghsoudi M and Siad H (2017) Influence of pumice and zeolite on compressive strength, transport properties and resistance to chloride penetration of high strength self-compacting concretes. *Construction and Building Materials* **151**: 292–311.

San Nicolas R, Cyr M and Escadeillas G (2014) Performance-based approach to durability of concrete containing flash-calcined metakaolin as cement replacement. *Construction and Building Materials* **55**: 313–322.

Siddique R and Klaus J (2009) Influence of métakaolin on the properties of mortar and concrete: a review. *Appl Clay Science* **43(3)**: 392–400.

Tange Jepsen M, Munch-Petersen C and Bager D (2001) Durability of resource saving 'green' types of concrete. In *Proceedings of the FIB-symposium Concrete and Environment, Berlin, Germany.*

ice Publishing

Dhir and Lye
ISBN 978-0-7277-6461-4
https://doi.org/10.1680/icetscc.64614.301
ICE Publishing:

Chapter 14
Performance and microstructure characteristics of self-curing self-compacting concrete

Madduru Sri Rama Chand
Research Scholar, Department of Civil Engineering, NIT Warangal, Warangal, India
(corresponding author: maddurusriram@gmail.com)

Pancharathi Rathish Kumar
Associate Professor, Department of Civil Engineering, NIT Warangal, Warangal, India

Pallapothu Swamy Naga Ratna Giri
Research Scholar, Department of Civil Engineering, NIT Warangal, Warangal, India

Garje Rajesh Kumar
Professor, Department of Civil Engineering, NIT Warangal, Warangal, India

The role of hydrophilic (polyethylene glycol) and hydrophobic (light liquid paraffin wax) chemicals as self-curing compounds to improve the curing efficiency of self-compacting concrete (SCC) was investigated in terms of performance and microstructure. The parameters considered in the investigation were the concrete grade, dosage of self-curing compound, curing regime, temperature and duration of curing. Performance investigations included water retention tests, compressive strength test and durability studies (porosity, sorptivity, rapid chloride permeability and accelerated corrosion permeability tests). Microstructural analyses studies (X-ray diffraction analysis and scanning electron microscopy) were conducted to confirm the results obtained from the macrostructural studies. Two grades of concretes, with compressive strengths of 70 MPa and 50 MPa, were used. Based on experimentation and microstructural studies, it was concluded that the addition of optimum dosages of hydrophilic and hydrophobic chemicals arrested loss of moisture, resulting in improved water retention and progressive hydration of SCC.

Notation

A	cross-section of specimen in contact with water
b, k, n, c	empirical constants
d	density of water
i	water absorption coefficient
p	porosity
p_0	porosity at zero strength
S	sorptivity coefficient

t	time
W_d	specimen dry weight
W_{ssd}	specimen weight in saturated surface dry condition
W_w	weight of saturated submerged specimen
Δw	amount of water adsorbed
σ	concrete strength
σ_0	concrete strength at zero porosity

Introduction

Concrete is a complex material that contains random features at different length scales (millimetre to nanometre) and also in different physical and mechanical states. These features mainly depend on the constitutive materials added during mixing (Xotta *et al.*, 2015). It is well known that concrete requires sufficient curing during the hydration process, particularly in the first few days, because the rate and heat of hydration will be very high. If curing is not sufficiently addressed during the initial period, the performance of the concrete will be significantly affected (Leach and CAM Sales, Inc., 1992; Meeks and Carino, 1999; Timms *et al.*, 1952; Zhao *et al.*, 2012). In addition, changes in temperature, humidity and vapour pressure lead to the evaporation of water, causing self-desiccation and leading to autogenous shrinkage (Bentz, 2009; Bentz and Jensen, 2004; De la Varga *et al.*, 2012; Geiker *et al.*, 2004). Another major hurdle in acquiring superior concrete is improper compaction during construction. This problem is all the more important in case of high-strength concretes, where vibration is required.

All these hurdles can be overcome by the use of self-compacting concrete (SCC) (Das and Chatterjee, 2012; Domone, 2006; Okamura and Ouchi, 2003; Okamura and Ozawa, 1994; Okamura *et al.*, 2000), which is a sustainable material as it uses mineral admixtures (Bouzoubaa and Lachemi, 2001; Gesoğlu and Özbay, 2007; Uysal and Sumer, 2011). However, like other concretes, SCC requires a sufficient amount of water for hydration of the cementitious particles. In regions where temperatures are of the order of 50°C, the problem of evaporation of water from concrete is significant (Palou *et al.*, 2016). It is also a matter of concern that, due to non-availability of or inaccessibility to water, concrete elements might not receive the amount of water required for correct curing.

Against this background, it is considered that the retention of mixing water added during concreting would be beneficial for the continuation of hydration (Bentz, 2009; Bentz *et al.*, 2005; Kovler and Jensen, 2005) and use of lightweight aggregates (LWA), super-absorbent polymers (SAP) or hydrophilic/hydrophobic chemicals has been found to benefit the hydration process by retaining embedded water (Bentz and Weiss, 2011; Oliveira *et al.*, 2015; Zhutovsky and Kovler, 2012). However, the use of LWA or SAP in concrete reduces its mechanical strength. Given this, it has been found that internal curing can be effectively accomplished using water-sealing materials such as polyethylene glycol (PEG), liquid paraffin wax (LPW), polyacrylic acid and so on (Dhir *et al.*, 1994, 1995, 1996, 1998; El-Dieb, 2007; El-Dieb *et al.*, 2012; Jau, 2011; Kumar, 2011). These polymer-based self-curing compounds reduce the evaporation of water from the concrete surface and thus increase the water retention capacity required for concrete hydration. The necessary and beneficial effect of the use of self-curing chemicals in SCC to restrict self-desiccation and enhance the hydration of cementitious particles has been established by the authors in previous studies (Chand *et al.*, 2014, 2016a, 2016b).

Although concrete at the macro level is considered a uniform and homogenous material with bulk properties such as compressive strength, creep and other aspects, at the micro level, it is an extremely complex system of solid phases, pores and water with a high degree of heterogeneity (Mehta and Monteiro, 2014). Concrete strength depends on the interfacial transition zone (ITZ) and hydrated cement paste as these phases are subjected to change over time and with environmental humidity and temperature. Microstructural images of the formation of calcium silicate hydrate (C–S–H) gel, calcium hydroxide ($Ca(OH)_2$) and ettringite in concretes under different curing conditions are helpful to analyse strength properties.

Research significance

Lack of proper curing and difficulties in vibration gave rise to the development of concretes as such as self-curing concretes and SCCs. At higher temperatures, concreting is difficult due to the evaporation of moisture from the surface of concrete. The hydration process is hindered due to unavailability of the required amount of water. In such cases the use of hydrophilic or hydrophobic materials can be useful for controlling the moisture loss by sealing the pores. This paper reports on an exhaustive experimental investigation on the use of self-curing compounds in SCC at room temperature and elevated temperature (50°C) and offers important findings in situations where high temperatures, curing issues and vibration problems dominate the process of concrete making. The performance aspects of the concretes were investigated at both macro and microstructural levels.

Experimental programme

The experimental work reported in this paper dealt with investigating the strength, durability and microstructural aspects of SCC with the addition of self-curing compounds. The parameters investigated included

- the grade of concrete (70 MPa (mix A) or 50 MPa (mix B))
- the type of curing compound added (hydrophilic PEG 4000 (i.e. PEG with a molecular weight of 4000) or hydrophobic LPW-light)
- dosage of the curing compound (0%, 0·1%, 0·5% or 1·0%)
- type of curing regime (water curing, no curing or self-curing)
- curing temperature (ambient temperature (27±2°C) or elevated temperature (50°C)).

To study the performance of the self-curing SCCs, tests for water retention capacity, compressive strength, porosity and sorptivity were conducted, along with rapid chloride permeability tests (RCPTs) and accelerated chloride permeability tests (ACPTs). Microstructure studies include scanning electron microscopy (SEM) and X-ray diffraction (XRD) analyses in order to understand the internal structure of the concrete. The mix proportions of the two SCC mixes considered in this study are given in Table 1.

Nomenclature of specimens

The specimens were designated according to grade, type of self-curing chemical and dosage. The notation A represents the higher grade concretes, while B represents the lower grade concretes. The hydrophilic chemical (PEG 4000) is designated as H, while V represents the hydrophobic chemical (LPW). N represents no curing (0% dosage of self-curing compound) and W represents

Table 1 Mix proportions of concrete mixes

	Mix A	Mix B
Cement: kg/m^3	500	430
Fine aggregate: kg/m^3	800	847
Coarse aggregate: kg/m^3	775	783
Water: kg/m^3	190	194
Fly ash: kg/m^3	110	180
Microsilica: kg/m^3	40	–

Table 2 Physical properties of aggregates

Properties	Coarse aggregate	Fine aggregate
Bulk density: g/cm^3	1·50	1·45
Specific gravity	2·80	2·65
Fineness modulus	7·15	2·50
Water absorption: %	1·00	2·00

conventional water curing. The dosages of self-curing compound were 0·1%, 0·5% and 1·0% by weight of cement. For example, AH 0·1 represents a specimen of higher grade containing PEG 4000 at a dosage of 0·1%. Similarly, BV 1·0 represents a specimen of lower grade LPW at a dosage of 1·0%.

Materials

The cement used in the investigation conformed to ASTM C 150 (ASTM, 2016a). The specific gravity of the cement was 3·10 and the specific surface area was 225 m^2/g. The initial and final setting times were 50 min and 540 min, respectively. River sand with fineness modulus of 2·50 was used as the fine aggregate. Crushed granite of nominal size 20 mm well-graded aggregate was used as the coarse aggregate. The physical properties of the aggregates are listed in Table 2 and their grading curves are shown in Figure 1. A polycarboxylic-ether-based water-reducing admixture conforming to ASTM C 494 (ASTM, 2016b) was used as a superplasticiser to improve workability. The self-curing compounds PEG 4000 and LPW-light were added with the water during mixing. The physical properties of PEG 4000 and LPW-light are given in Tables 3 and 4, respectively.

Testing

Fresh properties

Selection of the mix design was based on the key properties of SCC (filling ability, passing ability and resistance to segregation) as well as on reproducibility and repeatability. Standard tests for fresh properties (slump flow, L-box, J-ring and V-funnel tests) were conducted on the

Figure 1 Grading curves of coarse and fine aggregates

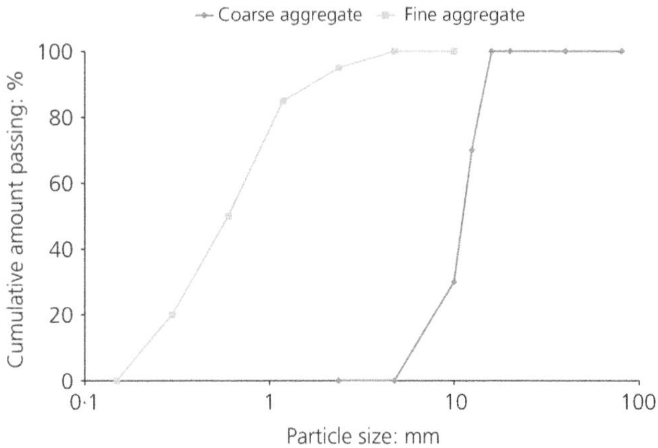

Table 3 Properties of Carbowax PEG 4000

Average molecular weight: g/mol	Average hydroxyl number: mg potassium hydroxide/g	Liquid density: g/cm^3			Melting or freezing range: °C	Water solubility at 20°C: wt%	Viscosity at 100°C
		20°C	60°C	80°C			
3600–4400	25–32	Solid	1·093	1·077	53 to 59	66	140·4

Table 4 Properties of liquid paraffin wax light

Property	
Specific gravity @ 25°C	0·820–0·860
Dynamic viscosity @ 20°C: mPa.s	25–80
Appearance	Clear colourless liquid
Solubility	Passes
Flash point (Pensky–Martens closed-cup): °C	Minimum 150°C

SCCs without and with self-curing compounds according to specifications published by the European Federation of Specialist Construction Chemicals and Concrete Systems (Efnarc, 2005). Table 5 presents details of the fresh properties of SCC mixes A and B. The results show that the dosage of self-curing compound significantly affected the flow properties of the concretes. Concrete flowability improved with an increase in dosage of the hydrophilic compound (PEG 4000) while it decreased with an increase in dosage of the hydrophobic compound (LPW-light). This is due to the highly viscous nature of the hydrophobic chemical compared with the hydrophilic chemical. There was, however, no segregation in any of the mixes.

Table 5 Fresh properties of SCC with self-curing compounds

Property	Efnarc range (Efnarc, 2005)	Mix	Plain SCC	Dosage of PEG 4000: %			Dosage of LPW-light: %		
				0·1	0·5	1·0	0·1	0·5	1·0
H-flow: mm	550–850	A	700	711	727	744	665	655	645
		B	750	755	761	777	690	675	670
T_{50}: s	2–5	A	2·89	2·82	2·76	2·53	3·37	3·54	3·69
		B	2·66	2·46	2·38	2·28	3·26	3·33	3·47
J-ring: mm	0–10	A	6·0	6·0	6·0	6·0	8·0	9·0	10·0
		B	5·0	5·0	4·0	4·0	7·0	8·0	9·0
L-box: h2/h1	0·8–1·0	A	0·93	0·93	0·95	0·96	0·85	0·84	0·83
		B	0·94	0·95	0·96	0·96	0·89	0·88	0·87
V-funnel: s	6–12	A	10·05	9·36	9·02	8·59	11·25	11·46	11·75
		B	9·15	8·60	8·36	8·02	10·60	11·44	11·82
$V_{5\,min}$: s	9–15	A	10·50	10·11	9·75	9·52	13·00	13·77	14·00
		B	10·04	10·00	9·81	9·57	11·42	11·92	12·45

Water retention

Loss of moisture due to water evaporation from the surface of concrete can be indirectly measured by a water retention test; that is, periodically evaluating the mass of concrete cubes of size 150 mm × 150 mm × 150 mm on an electronic balance with an accuracy of 0·1 g. In this study the initial mass of each specimen was taken as the reference and the mass loss was estimated at the end of 1, 3, 7 and 28 d.

Figures 2–5 show the average mass losses of SCC specimens cured without and with the addition of self-curing compounds. Figures 2(a) and 2(b) show the average mass losses of mix A SCC specimens with PEG 4000 self-curing compound at ambient (27°C) and elevated temperature (50°C), respectively. It can be clearly seen that the concrete specimens cured at ambient temperature had less mass loss than those cured at elevated temperature. For both curing temperatures, the specimen with a 0·1% dosage of the hydrophilic compound (PEG 4000) exhibited less mass loss than the specimens with other dosages of PEG 4000. The mass loss was highest in the non-cured specimens at both curing temperatures, and this was true for both mixes A and B, as shown in Figures 2 and 4.

In the case of the addition of LPW-light, the optimum dosage was 0·1% for both mixes A and B (Figures 3 and 5), whereas the optimum dosage of PEG 4000 for mix A was 0·1% and 1·0% for mix B (Figures 2 and 4). At elevated temperature, although the mass losses were high compared with those at ambient temperature, the self-cured specimens formed polymer bonds, which decreased the evaporation of moisture from inside the concrete specimen.

Compressive strength

Cube specimens of size 150 mm × 150 mm × 150 mm were cast and tested for compressive strength after a curing period of 28 d as per BS EN 12390-3 (BSI, 2009). The compressive strength of specimens cured at ambient (27°C) and elevated temperature (50°C) was recorded as the average value of six specimens in each case.

Curing of concrete at higher temperatures speeds up the rate of hydration initially and, with time, moisture loss results in a decrease in the water available for hydrating the cement particles (Magzoub Ahmed, 2015). In tropical climates, high early strength but lower ultimate strength than the same concrete cured at ambient temperature can be expected. Figures 6(a) and 6(b) show the average 28 d compressive strength results of mix A SCC specimens, self-cured by PEG 4000 and LPW-light at ambient and elevated temperatures. Although the maximum compressive strength

Figure 2 Average weight loss of mix A without and with PEG 4000: (a) curing at 27°C; (b) curing at 50°C

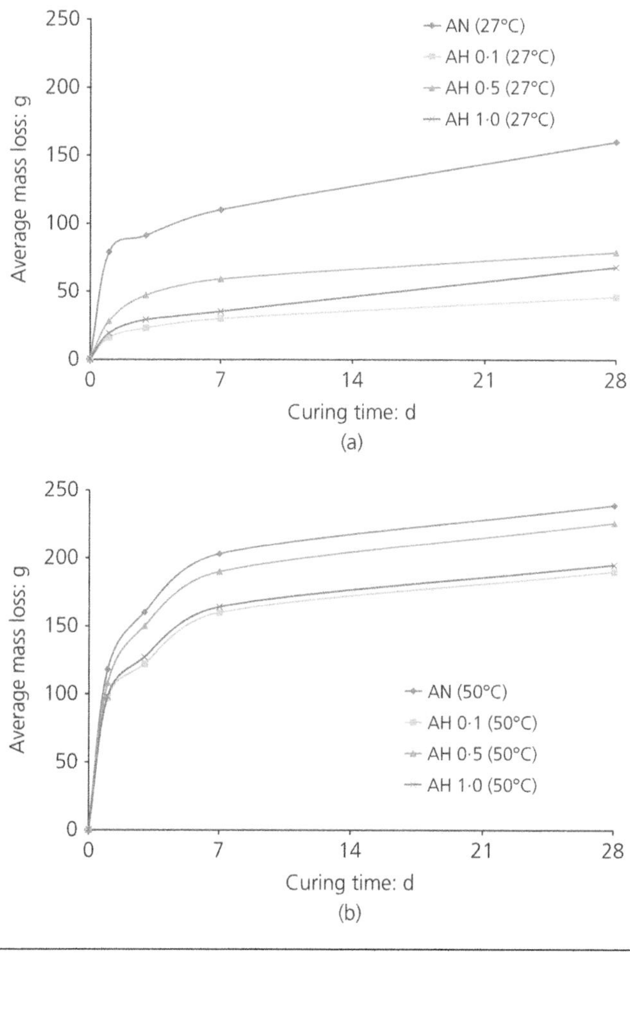

Figure 3 Average weight loss of mix A without and with LPW-light: (a) curing at 27°C; (b) curing at 50°C

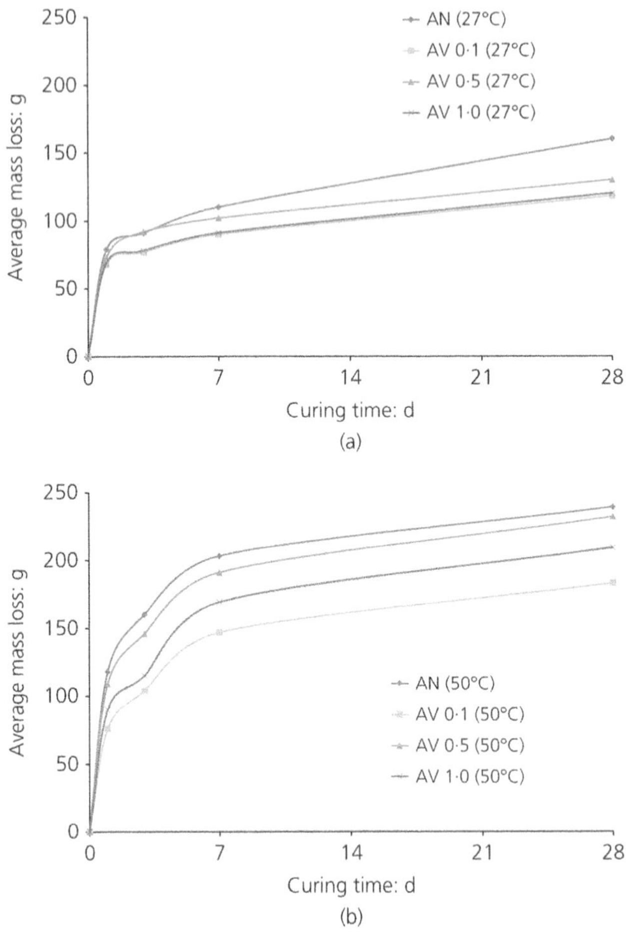

was obtained by the water-cured specimens, the self-cured specimens with optimum dosages (specimens AH 0·1 and AV 0·1) attained strength on par with the strengths obtained at ambient temperature. Although the 28 d compressive strengths decreased at elevated temperature, the self-cured specimens attained higher strengths than the non-cured specimens. The 28 d strength of specimens cured with self-curing compounds (PEG 4000 and LPW-light) at 50°C was about 90% of those cured at 27°C. This indicates that self-curing compounds sealed the internal pores on the concrete, preventing the escape of moisture and resulting in improved strengths even at higher temperatures (Safi *et al.*, 2013). Similarly, Figures 7(a) and 7(b) show the average 28 d compressive strength results for mix B SCC specimens cured with PEG 4000 and LPW-light at the two curing temperatures. The results were similar to those of the mix A specimens. The optimum dosage of self-curing chemicals for mix B was 1·0% for PEG 4000 and 0·1% for LPW-light at both curing temperatures.

Figure 4 Average weight loss of mix B without and with PEG 4000: (a) curing at 27°C; (b) curing at 50°C

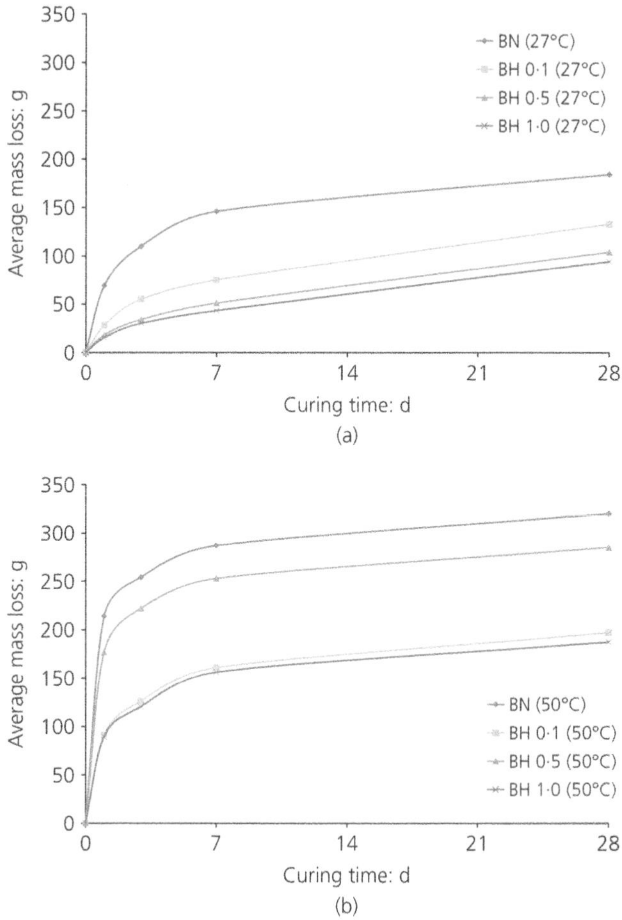

Open porosity

The porosity of the concretes was determined by the oven drying method as per ASTM C 642 (ASTM, 2006). The porosity of concrete can be calculated using

$$p = \frac{(W_{\text{ssd}} - W_{\text{d}})}{(W_{\text{ssd}} - W_{\text{w}})} \times 100 \tag{1}$$

where p is the porosity (in %), W_{ssd} is the specimen weight in the saturated surface dry condition, W_{d} is the specimen dry weight and W_{w} is the weight of the saturated submerged specimen.

The porosity of concrete specimens cured under different curing conditions 28 d is reported in Table 6. The specimens with higher porosity values yielded low compressive strength, as was

Figure 5 Average weight loss of mix B without and with LPW-light: (a) curing at ambient 27°C; (b) curing at 50°C

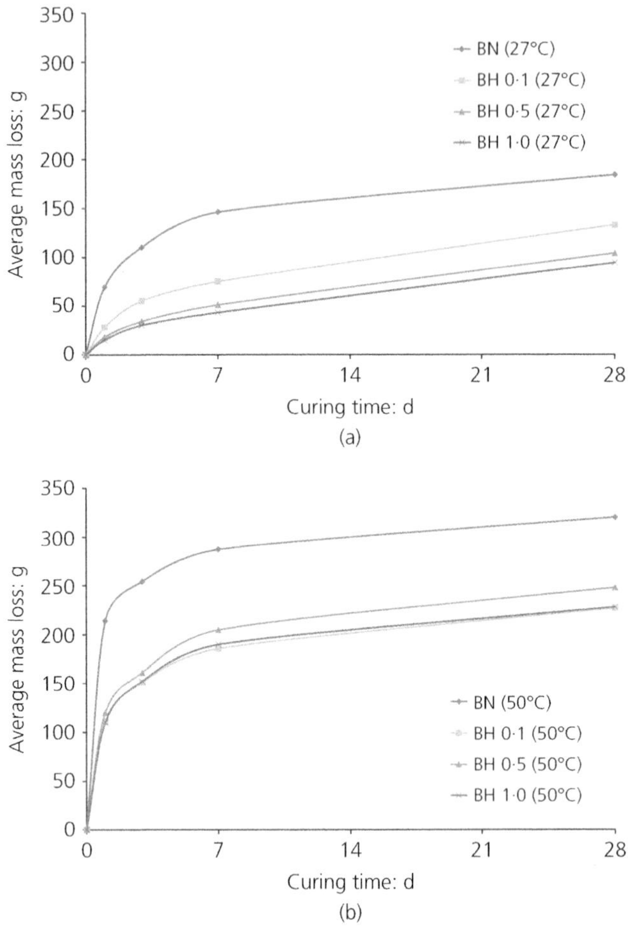

evident in the case of the non-cured specimens. The specimens cured with optimum dosages of self-curing chemicals exhibited less porosity, similar to that of the water-cured specimens. It can also be observed that at elevated-temperature curing, the porosity of concrete specimens was high and there was a decrease in the strength.

The experimentally obtained strength–porosity relationship was compared with various models for cement-based materials in the literature (Kumar and Bhattacharjee, 2003). The strength–porosity relationships considered were

$$\sigma = \sigma_0(1 - p)^b \tag{2}$$

Figure 6 28 d compressive strength results of mix A with self-curing compounds: (a) PEG 4000; (b) LPW-light (W indicates water curing and N indicates no curing)

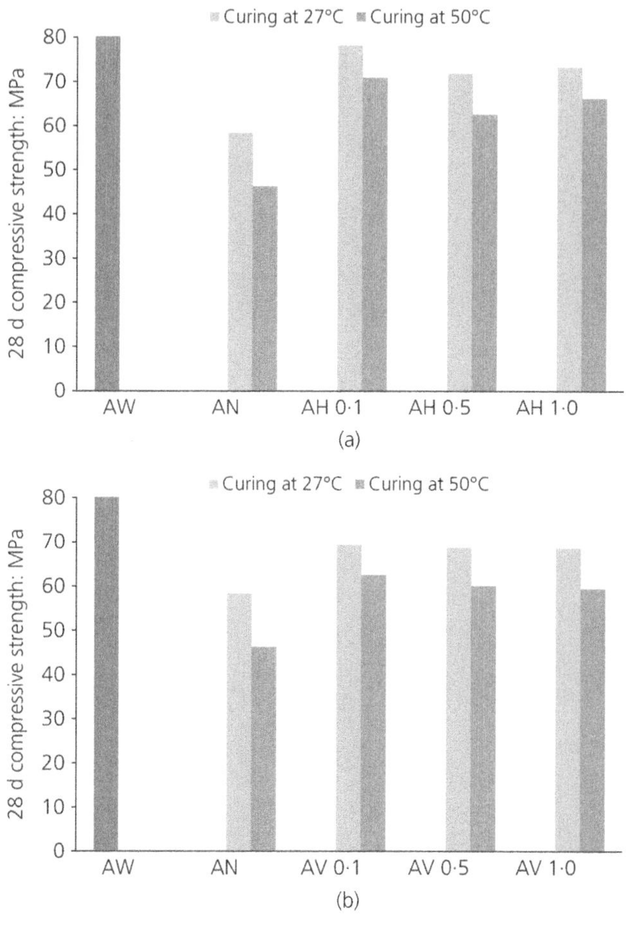

$$\sigma = \sigma_0 e^{-kp} \tag{3}$$

$$\sigma = n \ln(p_0/p) \tag{4}$$

$$\sigma = \sigma_0 - cp \tag{5}$$

For ease of reference, these equations, along with the relevant sources, are listed in Table 7. The experimentally obtained relationship was plotted for both ambient- and elevated-temperature curing and is shown in Figure 8. The experimental values fitted well with the equation

Figure 7 28 d compressive strength results of mix B with self-curing compounds: (a) PEG 4000; (b) LPW-light (W indicates water curing and N indicates no curing)

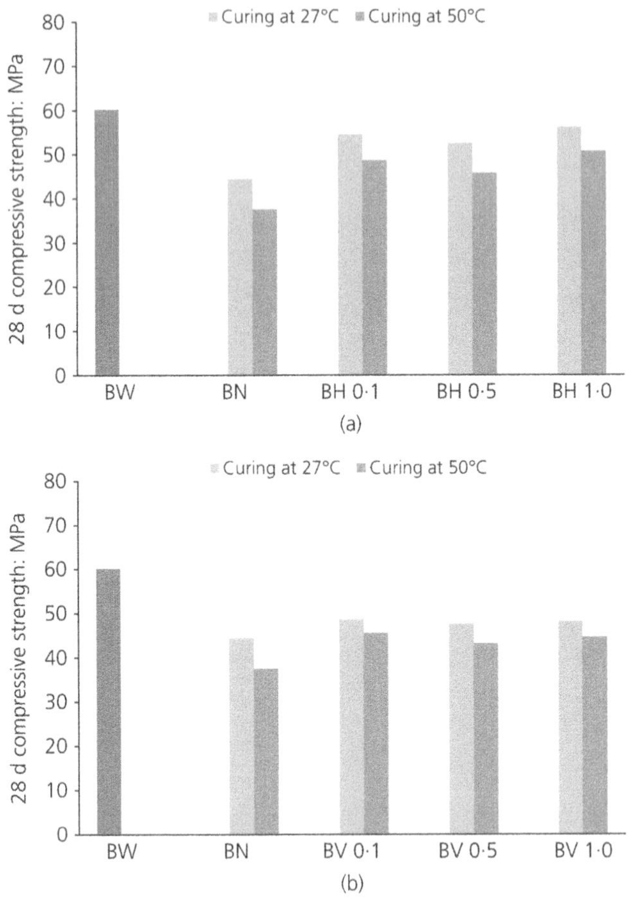

proposed by Hasselman (1969) (Equation 5). These equations were also tested by Kumar and Bhattacharjee (2003).

Sorptivity

The rate of absorption (sorptivity) of water is determined by obtaining the increase in mass of a specimen due to absorption of water as a function of time when only one surface of the specimen is exposed to water. A standard test specimen (100 mm dia. disc of thickness 50 mm) was cast and tested as per ASTM C 1585 (ASTM, 2013). A schematic diagram of the sorptivity test is shown in Figure 9. The sorptivity was determined using the expression

$$\frac{\Delta w}{A.d \times \sqrt{t}} = S \tag{6}$$

Table 6 28 d porosity results

Specimen	Porosity: %			
	Mix A (70 MPa) Ambient curing (27°C)	Mix A (70 MPa) Elevated curing (50°C)	Mix B (50 MPa) Ambient curing (27°C)	Mix B (50 MPa) Elevated curing (50°C)
Water curing (W)	3·55	–	6·37	–
No curing (N)	11·95	17·42	13·99	18·43
0·1% PEG 4000 (H 0·1)	6·08	13·41	8·62	16·42
0·5% PEG 4000 (H 0·5)	10·12	14·93	8·15	17·27
1·0% PEG 4000 (H 1·0)	9·32	15·06	7·57	17·73
0·1% LPW-light (V 0·1)	6·22	12·94	7·57	16·42
0·5% LPW-light (V 0·5)	9·12	13·23	10·27	17·14
1·0% LPW-light (V 1·0)	9·12	13·18	9·50	17·58

Table 7 Strength–porosity relationships

Proposer	Form of equation	Equation
Balshin (1949)	$\sigma = \sigma_0(1 - p)^b$	$\sigma = 87{\cdot}4(1 - p)^{4{\cdot}496}$
Ryshkevitch (1953)	$\sigma = \sigma_0 e^{-kp}$	$\sigma = 90{\cdot}19e^{-5{\cdot}07p}$
Schiller (1971)	$\sigma = n \ln(p_0/p)$	$\sigma = 26{\cdot}55 \ln(0{\cdot}0772/p)$
Hasselman (1969)	$\sigma = \sigma_0 - cp$	$\sigma = 83{\cdot}28 - 280{\cdot}6p$

Figure 8 Compressive strength–porosity relationships

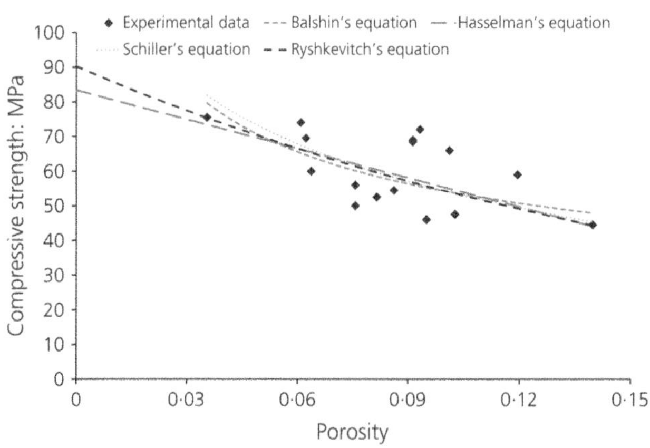

Figure 9 Schematic diagram of sorptivity test setup as per ASTM C 1585 (ASTM, 2013)

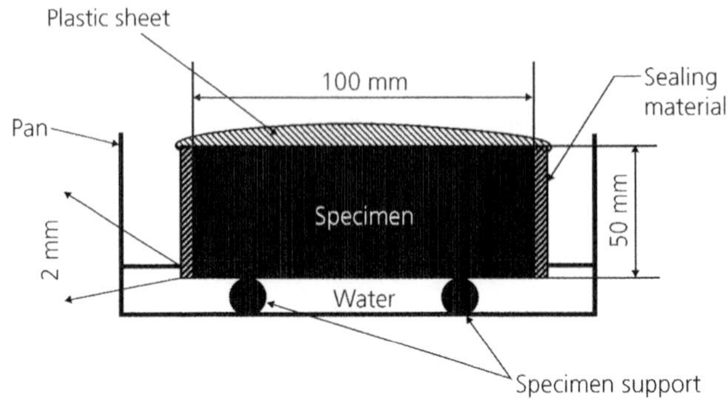

where Δw is the amount of water adsorbed (g), A is the cross-section of the specimen in contact with water (mm^2), d is the density of water (10^{-3} g/mm^3), t is time (min), i is the water absorption coefficient ($= \Delta w/Ad$) and S is the sorptivity coefficient of the specimen (mm/$min^{0\cdot5}$) ($= i/(t)^{1/2}$).

Figures 10 and 11 show the relationship between water absorption coefficient (i) and $time^{0\cdot5}$ ($(t)^{1/2}$) for SCC mixes A and B, respectively, cured under different regimes at ambient temperature. The figures show that addition of PEG or LPW resulted in a decrease in sorptivity at an optimum dosage of 0·1% for the higher grade of concrete and at dosages of 1·0% PEG 4000 and 0·1% LPW for the lower grade of concrete. Figures 12 and 13 show the same relationship for mixes A and B cured under different regimes at elevated temperature. The water absorption coefficient was high for specimens cured at elevated temperature, irrespective of the type of curing or concrete grade.

Figure 14 depicts the variations in sorptivity coefficient for different grades of concrete cured under different regimes. It can be noted from this figure that the sorptivity decreased with an increase in concrete grade at optimum dosages of the self-curing compounds.

RCPT

The RCPT indicates the permeability of chloride ions into concrete and is based on the charge passed through the sample: the higher the amount of charge passed, the higher is the permeability of the concrete. According to ASTM C 1202 (ASTM, 2012), a water-saturated, 50 mm thick, 100 mm dia. concrete specimen is subjected to a 60 V applied DC voltage for 6 h in RCPT apparatus. The total charge passed is determined and used to rate the concrete according to the criteria listed in Table 8.

Figures 15 and 16 depict the RCPT values of concrete mixes A and B cured by using PEG 4000 and LPW-light. It can be seen from Figure 15 that the chloride ion penetration was low for mix A concrete specimens cured at ambient temperature compared with those cured at elevated temperature. The specimens cured with PEG 4000 and LPW-light performed better than non-cured

Figure 10 Water absorption coefficient (i) plotted against $t^{1/2}$ for SCC cured at ambient temperature: (a) mix A SCC with PEG 4000; (b) mix A SCC with LPW-light

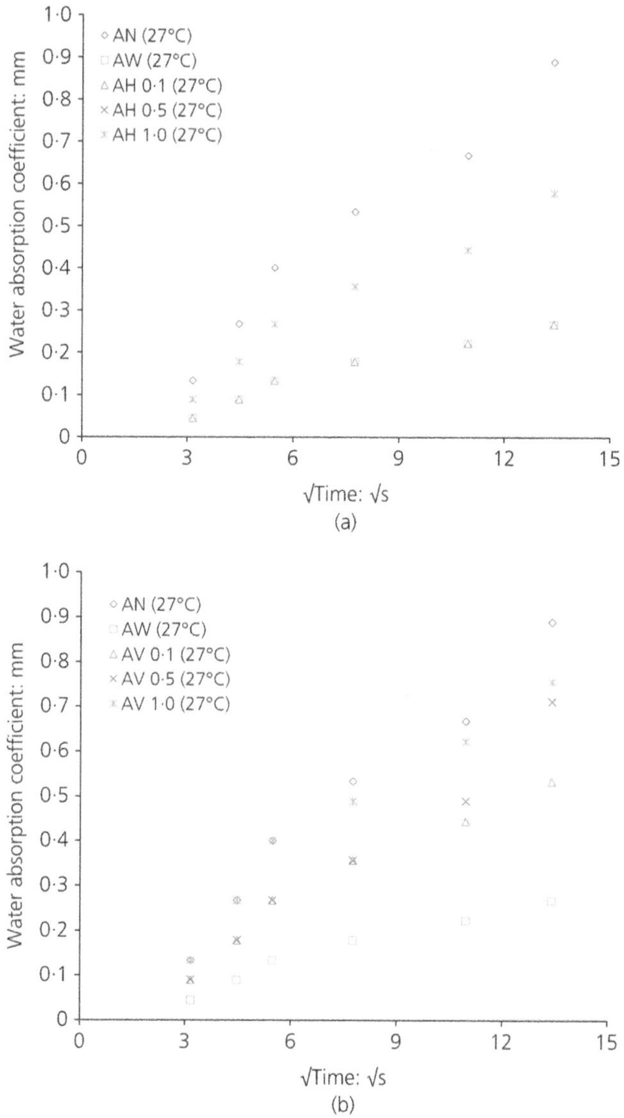

specimens even at elevated temperature. The charge passed was high for non-cured specimens at both curing temperatures, but maximum values were obtained at elevated temperature. The concrete subjected to conventional water curing attained the least charge of all the specimens. Figure 16 shows the RCPT values for mix B concrete specimens. The figure shows that the total charge passed in mix B specimens was moderate for the water-cured and self-cured specimens and

Figure 11 Water absorption coefficient (i) plotted against $t^{1/2}$ for SCC cured at ambient temperature: (a) mix B SCC with PEG 4000; (b) mix B SCC with LPW-light

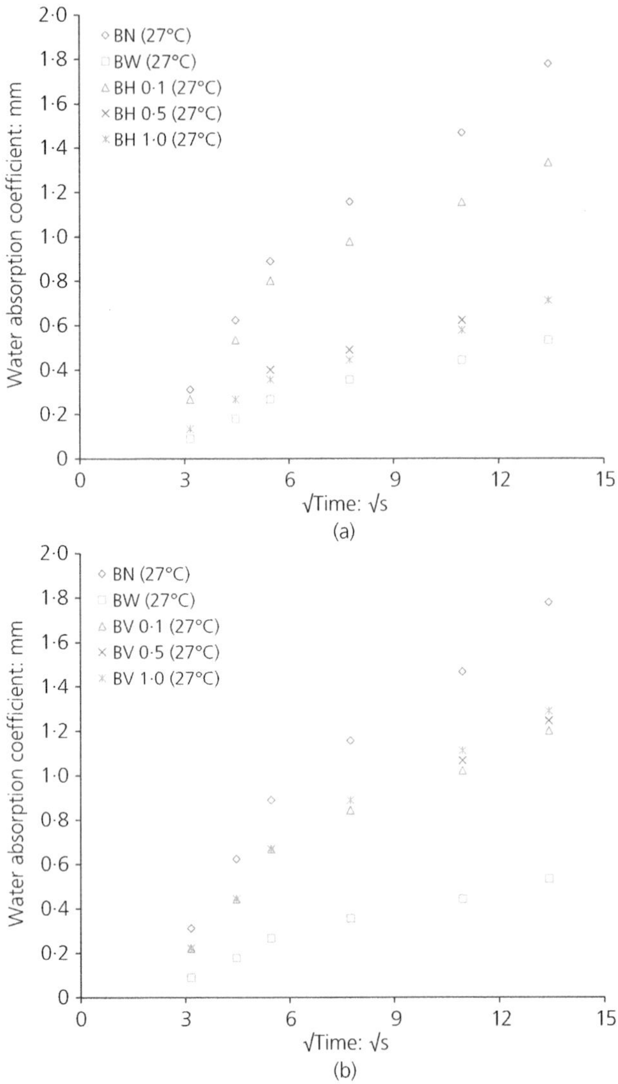

high for the non-cured specimens. The addition of optimum dosages of self-curing compounds minimised the chloride ion permeability, even at elevated temperature. The self-cured SCC specimens always exhibited lower permeability values at ambient and elevated curing temperatures, indicating a lower degree of susceptibility to steel corrosion.

Figures 15 and 16 show that, after the water-cured samples, the RCPT values were the lowest for the samples dosed with 0·1% PEG 4000. The optimum dosage of LPW-light was also 0·1%.

Figure 12 Water absorption coefficient (i) plotted against $t^{1/2}$ for SCC cured at elevated temperature: (a) mix A SCC with PEG 4000; (b) mix A SCC with LPW-light

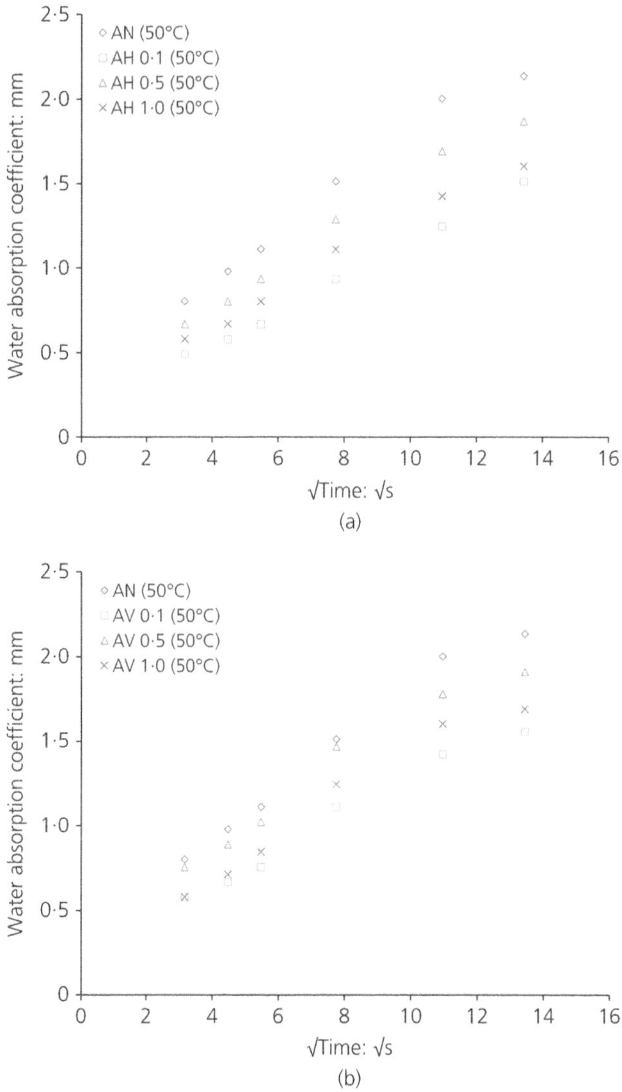

ACPT

The test setup for the accelerated corrosion test (also known as the impressed voltage test) includes a DC power supply, two stainless steel plates, a data logger, test specimen and a container with the required dosage of sodium chloride (NaCl) solution (3%). Concrete specimens of 100 mm dia. and 200 mm height embedded with a 10 mm dia. steel bar were cast and tested at the required age. The steel bar (anode/working electrode) of the beam specimen was connected to the positive terminal and the stainless steel plates (cathode/counter electrode)

Figure 13 Water absorption coefficient (i) plotted against $t^{1/2}$ for SCC cured at elevated temperature: (a) mix B SCC with PEG 4000; (b) mix B SCC with LPW-light

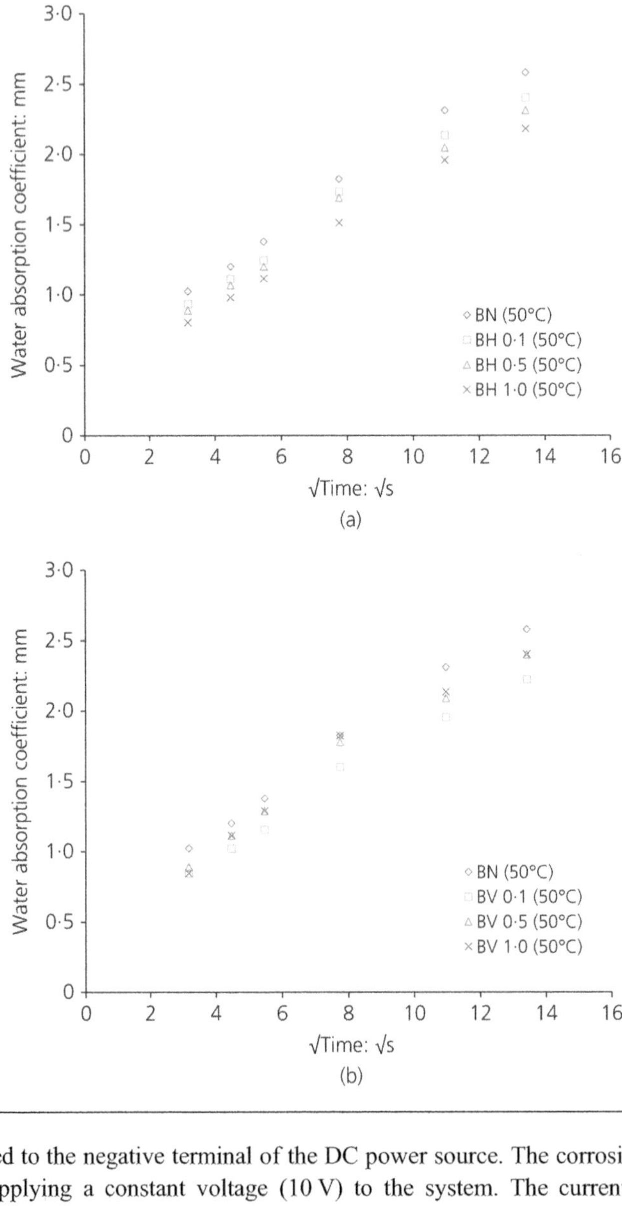

were connected to the negative terminal of the DC power source. The corrosion process was initiated by applying a constant voltage (10 V) to the system. The current response was continuously monitored and recorded by data logger. In addition to this, the specimens were periodically inspected for the onset of cracking. The data logger was set at a sampling frequency of 10 min for recording the corrosion current of the circuit. The ACPT was terminated after cracking of the specimen; this corresponds to the condition where the rate of increase in corrosion current with time is negligible.

Figure 14 Sorptivity coefficients of SCCs cured under different curing regimes

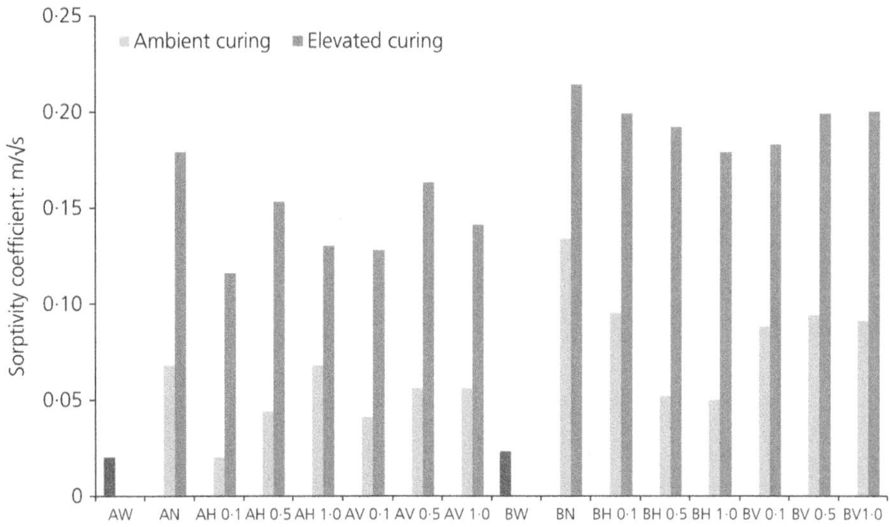

Table 8 RCPT ratings as per ASTM C 1202 (ASTM, 2012)

Charge passed: C	Chloride ion permeability
4000	High
2000–4000	Moderate
1000–2000	Low
100–1000	Very low
<100	Negligible

Figure 17 illustrates the generalised nature of the current variations in mix A SCC specimens treated under ambient temperature. The final critical corrosion currents and post-depassivation crack times for all the self-cured specimens are presented in Table 9. It can be noticed from the table and Figure 17 that, initially, some relatively small-amplitude variations were encountered; at a later stage, a larger increase in current occurred. This corresponds to the growth of cracks within the concrete and is attributed to a decrease in electrical resistance. With incremental increases in crack growth, the tensile stresses are relieved, leading to the arrest of concrete cracking and giving rise to the accumulation of additional solid products. For the conventional water-cured specimens, initial failure was observed 11 d after the start of the experiment and cracks were observed 14 d after the start of the experiment. The post-depassivation crack time of these specimens was 14 d. In case of no curing, specimen failure was observed on the seventh day and a failure crack was observed after 10 d (the post-depassivation crack time was 10·2 d). In the case of specimens with an optimum dosage of PEG 4000 (0·1%), the failure time was noted as 11 d and cracking was observed 13·3 d after the start of the experiment. In case of LPW-light,

Figure 15 RCPT values of mix A with PEG 4000 and LPW-light

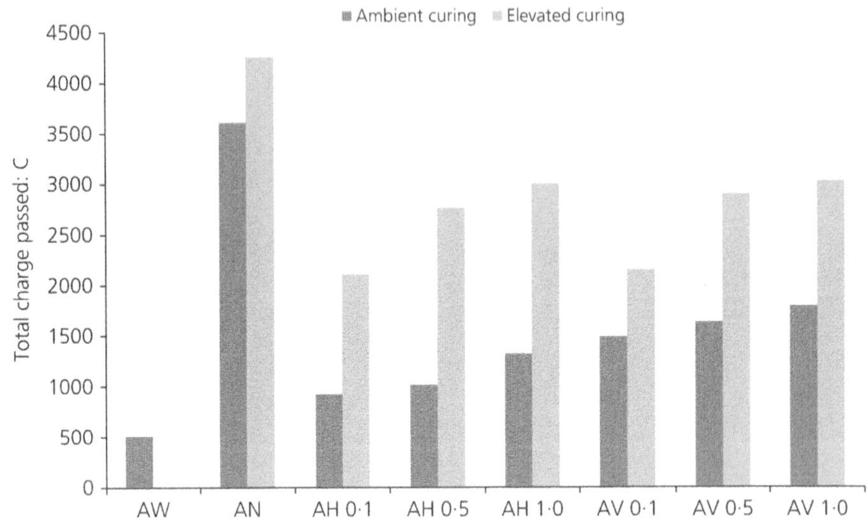

Figure 16 RCPT values of mix B with PEG 4000 and LPW-light

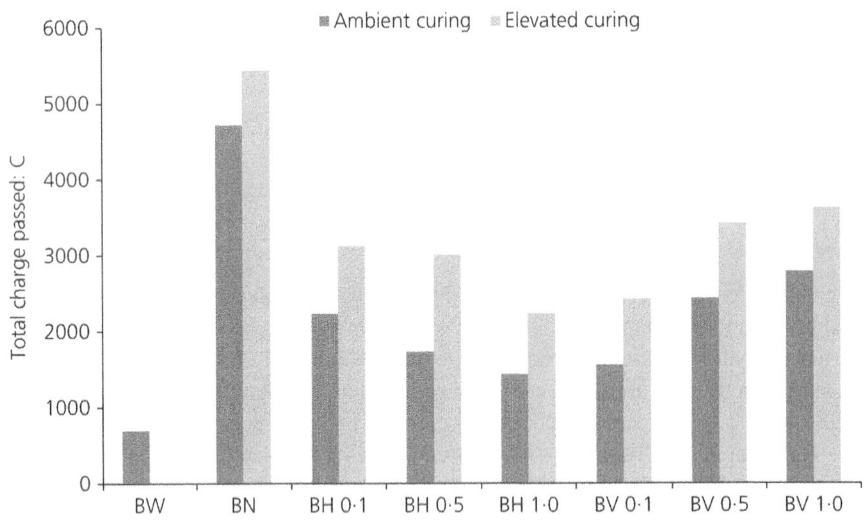

at an optimum dosage of 0·1%, the failure time and crack times were 11 d and 14 d, respectively, similar to the conventional water-cured specimens.

Failure after longer times (11 d and 14 d) is an indication of improved durability in terms of corrosion and progressive hydration, unlike in the case of the non-cured specimens. In the

Figure 17 Generalised nature of current variations in SCC with variation of age

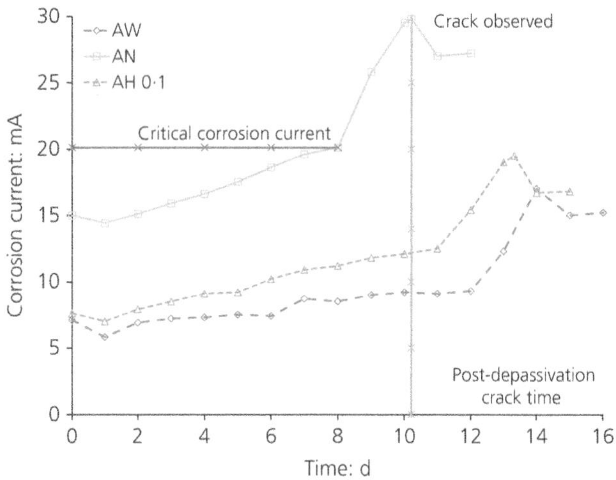

self-cured specimens with optimum dosages of PEG 4000 and LPW-light, the delayed failure due to corrosion was due to better hydration of the concrete (Table 9). The similar advantage of the addition of self-curing chemicals was also observed in mix B: the critical corrosion current was least for the conventional water-cured and self-cured specimens while it was high for the non-cured specimens. The post-depassivation time period was less in the non-cured specimens, indicating early failure after the initial failure, while better resistance to failure to corrosion was observed in the self-cured and conventional water-cured specimens. A delay in initial cracking was observed.

Table 9 Accelerated corrosion test results

Specimen	Mix A		Mix B	
	Critical corrosion current: mA	Post-depassivation crack time: d	Critical corrosion current: mA	Post-depassivation crack time: d
Water curing	9·3	14·0	15·7	12·0
No curing	20·1	10·2	30·8	8·8
0·1% PEG 4000	12·5	13·3	28·8	10·4
0·5% PEG 4000	16·3	12·4	25·9	10·9
1·0% PEG 4000	16·8	12·8	25·4	11·3
0·1% LPW-light	12·3	13·2	20·2	11·6
0·5% LPW-light	14·7	12·6	24·5	11·0
1·0% LPW-light	16·4	12·1	21·8	11·2

Microstructural studies

XRD analysis

The mineral composition of the concretes after 28 d of curing was investigated using a PANalytical XRD instrument equipped with an X'Celerator ultrafast detector. The sample preparation technique described by Buhrke *et al.* (1998) was adopted and the sample was scanned using Cu-Kα radiation at 60 kV, 55 mA, counts/s = 1000, width 2·5 mm, speed 2°/min and scanned with an angle of 2θ from 6–70° with a step size of 0·02°. The scans were analysed using X-Pert High Score software and the compounds present in the sample were determined by matching the peak intensities with the PCPDFWIN database.

XRD analysis of SCC specimens cured at ambient temperature

Figure 18 shows typical XRD patterns for mix A SCC specimens cured for 28 d under different curing regimes at ambient temperature. XRD analysis of the water-cured, non-cured and self-cured samples indicated quantitative estimates of the different mineral phases and the results are summarised in Table 10. Figure 18 shows that the SCC cured with conventional water curing contained 21% quartz (SiO_2), 11% portlandite ($Ca(OH)_2$), 35% synthetic hatrurite (Ca_3SiO_5), 30% larnite (Ca_2SiO_4) and 3% gismondine ($CaAl_2Si_2O_8$). The major peaks of quartz appeared at $2\theta \approx 20{\cdot}83°$ ($d \approx 0{\cdot}421$ nm) and $2\theta \approx 26{\cdot}63°$ ($d \approx 0{\cdot}334$ nm). The major peaks of portlandite appeared at $2\theta \approx 18{\cdot}01°$ ($d \approx 0{\cdot}492$ nm) and $2\theta \approx 34{\cdot}10°$ ($d \approx 0{\cdot}263$ nm). The gismondine mineral peak was observed at $2\theta \approx 12{\cdot}06°$ ($d \approx 0{\cdot}732$ nm). The major peaks of synthetic hatrurite minerals were identified at $2\theta \approx 29{\cdot}43°$ ($d \approx 0{\cdot}304$ nm) and $2\theta \approx 32{\cdot}56°$ ($d \approx 0{\cdot}275$ nm). The larnite mineral major peaks were observed at $2\theta \approx 37{\cdot}05°$ ($d \approx 0{\cdot}242$ nm) and $2\theta \approx 39{\cdot}27°$ ($d \approx 0{\cdot}229$ nm). The larger quantities of hatrurite and larnite minerals and the lesser quantities of gismondine and portlandite minerals indicate rich formation of C–S–H gel and low amounts of portlandite and aluminate–ferrite compounds. These results match with the higher values of compressive strength (Figure 6).

Figure 18 XRD peaks of SCC subjected to different curing regimes at 27 ± 2°C. A full-colour version of this figure can be found on the ICE Virtual Library (www.icevirtuallibrary.com)

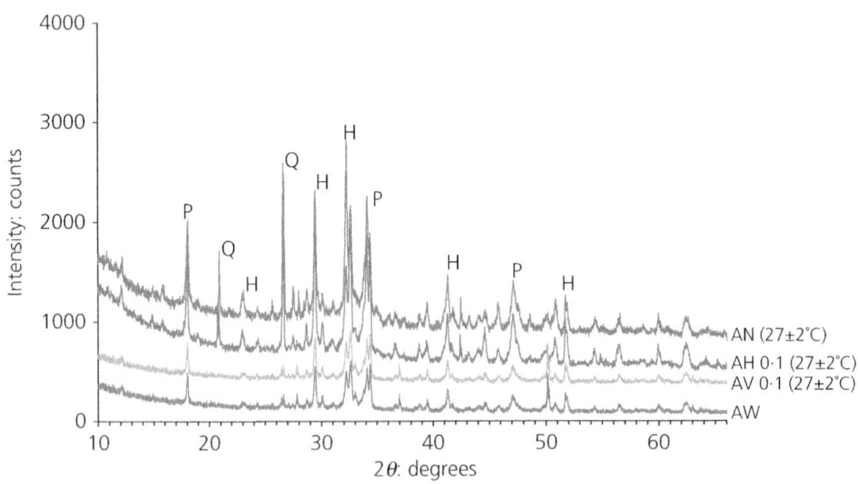

Table 10 Estimates of the mineral phases in SCC specimens cured under different curing conditions

	AW	Ambient curing (27°C)			Elevated curing (50°C)		
		AN	AH 0·1	AV 0·1	AN	AH 0·1	AV 0·1
Quartz (Q): wt%	21	29	20·2	17	38·6	12·1	30
Portlandite, synthetic (P): wt%	11	18	8·1	9	23·8	10·1	8·4
Hatrurite, synthetic (H): wt%	35	26	29·3	33	13·8	27·3	31
Larnite (L): wt%	30	9	32·3	32	19·3	28·3	23
Gismondine (G): wt%	3	–	10·1	–	4·5	10	7·6
Vesuvianite (V): wt%	–	–	–	–	–	12·2	–
Brownmillerite, synthetic (B): wt%	–	–	–	9	–	–	–
Ettringite (E): wt%	–	18	–	–	–	–	–

In the SCC treated with 0·1% PEG 4000, the estimated mineral quantities present were 20·2% quartz, 8·1% portlandite, 29·3% synthetic hatrurite, 32·3% larnite and 10·1% gismondine. For the specimen dosed with 0·1% LPW-light, the estimated mineral quantities were 17% quartz, 9% portlandite, 33% synthetic hatrurite, 32% larnite and 9% synthetic brownmillerite ($Ca_2(Al,Fe)_2O_5$). The mineral quantities in the non-cured specimens were estimated to be 29% quartz, 18% portlandite, 26% synthetic hatrurite, 9% larnite and 18% synthetic ettringite ($Ca_6A_{12}(SO_4)_3(OH)_{12}$ $26H_2O$). From a comparison of the X-ray difractograms of specimens subjected to different curing regimes (Figure 18) and based on the summary given in Table 10, the following points are evident.

- The presence of ettringite was observed only in large quantities in the non-cured specimens. In other specimens, either gismondine or brownmillerite minerals were mainly identified. Ettringite formation can be identified from visible displacement and cracking. Ettringite increases the risk of freeze/thaw attack or reinforcement corrosion (secondary forms of distress).
- Compared with the self-cured and water-cured specimens, large amounts of portlandite deposits were observed in the non-cured samples. Portlandite crystals in concrete are converted to minute crystals of calcite ($CaCO_3$), leading to carbonation and lime leaching effects in concrete. The observation of excessive carbonation in concrete, in association with microcracking extending inwards from the surface, is important in diagnosing the cause of corrosion of steel in concrete.
- Gismondine is a zeolite-type phase or hydrated alumino silicate that has an orthorhombic and dipyramidal structure. The presence of gismondine increases the strength of concrete, but not as much as hatrurite or larnite minerals.
- Brownmillerite was present in the SCC specimen cured with 0·1% LPW-light. The presence of brownmillerite does not influence concrete strength.

XRD analysis of SCC specimens cured at elevated temperature

Figure 19 shows the XRD patterns of mix A SCC specimens subjected to no curing and self-curing (0·1% PEG 4000 or 0·1% LPW-light for 28 d) at elevated temperature (50°C).

Figure 19 XRD peaks of SCC subjected to different curing regimes at 50°C. A full-colour version of this figure can be found on the ICE Virtual Library (www.icevirtuallibrary.com)

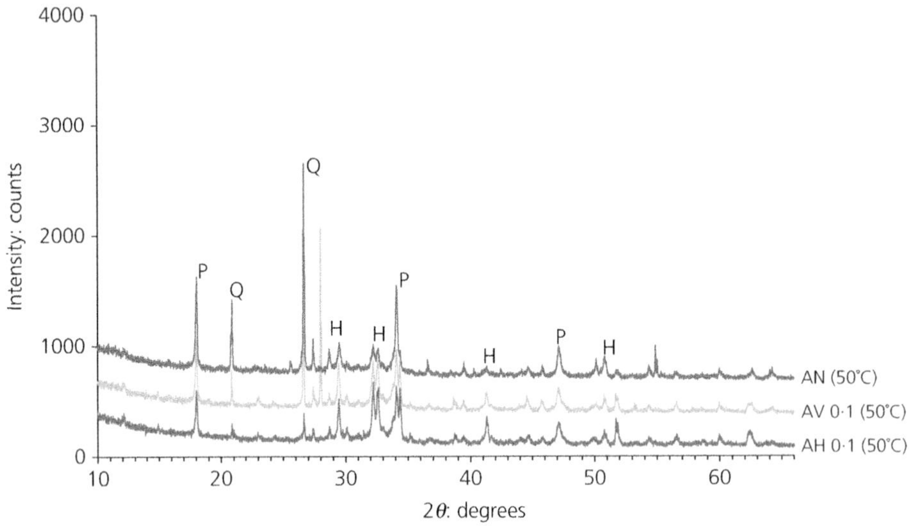

The mineral quantities in the specimens dosed with 0·1% PEG 4000 were estimated to be 12·5% quartz, 10·1% portlandite, 27·3% synthetic hatrurite, 28·3% larnite, 12·2% vesuvianite ($Ca_{19}Fe(Al_{6\cdot48}Fe_{1\cdot52})Al_4(SiO_4)_{10}(Si_2O_7)_4(OH)_8F_2$) and 10% gismondine. The mineral quantities in the specimens dosed with 0·1% LPW-light were estimated to be 30% quartz, 8·4% portlandite, 31% synthetic hatrurite, 23% larnite and 7·6% gismondine. The non-cured specimen contained 38·6% quartz, 23·8% portlandite, 13·8% synthetic hatrurite, 19·3% larnite and 4·5% gismondine.

The concrete cured at elevated temperature without self-curing compounds resulted in maximum moisture loss, which resulted in reduced compressive strength (Figure 6). The same was reflected in the XRD analysis as the compounds responsible for C–S–H gel formation were less in the non-cured specimens than in the self-cured specimens.

Comparison of the specimens cured at ambient and elevated temperature reveals that the estimated quantities of hatrurite and larnite compounds at 28 d were higher in the specimens cured at ambient temperature than those cured at elevated temperature. The presence of these minerals in concrete enriches the formation of C–S–H gel, leading to improved strength, as was reflected in the compressive strength results shown in Figure 6.

SEM analysis

Microstructural investigations were carried out by SEM (Tescan Vega 3 LMU) coupled with energy dispersive X-ray analysis (EDXA), (Penta FET precision, Oxford Instruments X-act). A representative sample of size $1{\cdot}0 \times 1{\cdot}0 \times 0{\cdot}5$ cm containing aggregate, cement mortar and ITZ was taken from the core of the cube after identifying the relevant portion by means of an optical microscope. Using a bench grinder, the sample was ground on all surfaces except the surface to be tested. To make the concrete sample conductive, gold sputtering was applied prior to the SEM examination.

Along with an assessment of physical and mechanical properties, microstructure analysis by means of SEM is important to identify the type, amount, size, shape and distribution of phases present in solid concrete (Bentz and Stutzman, 1994; Stutzman, 1994). The ITZ of concrete samples was analysed to determine the distribution of atomic ratios in the aggregates and hydrated phases of the ITZ. In order to detect the main composition of hydrated phases in the ITZ region and bulk cement paste (C–S–H, calcium hydroxide and AFm), the criteria used were (Rossignolo, 2009)

- C–S–H: $0{\cdot}8 \leq$ calcium/silicon $\leq 2{\cdot}5$; (aluminium + iron)/calcium $\leq 0{\cdot}2$
- calcium hydroxide: calcium/silicon ≥ 10; (aluminium + iron) calcium $\leq 0{\cdot}4$; Sulphur/calcium $\leq 0{\cdot}04$
- AFm: calcium/silicon $\geq 4{\cdot}0$; (aluminium + iron)/calcium $> 0{\cdot}40$; silicon/calcium $> 0{\cdot}15$

SEM analysis of SCC specimens cured at ambient temperature

Figures 20–23 show backscattered electron (BSE) images of SCC (mix A) cured with conventional water curing, no curing, self-cured with 0·1% PEG 4000 and self-cured with 0·1% LPW-light at ambient temperature (27°C), respectively. The elemental composition obtained from EDXA spectra of the concrete samples subjected to different curing regimes is given in Table 11.

Figure 20 BSE image of conventional water-cured SCC

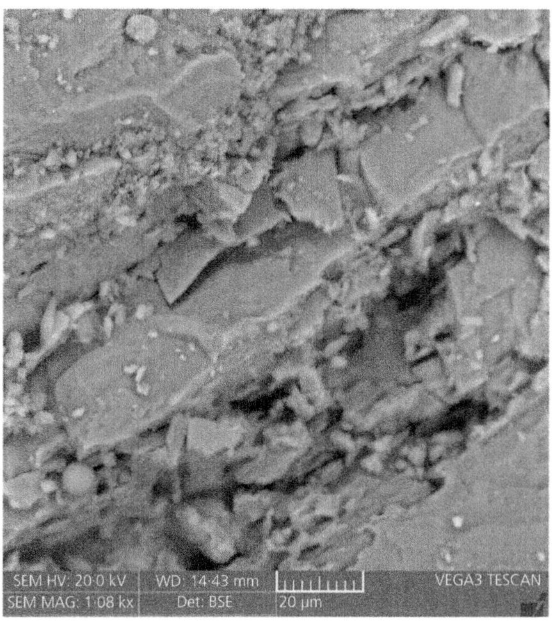

The BSE image of the water-cured sample (Figure 20) shows dense gel formation with few pores. The calcium/silicon and (aluminium + iron)/calcium ratios were obtained as 0·93 and 0·53, which indicate large amounts of C–S–H gel and small amounts of aluminate–ferrite phases. The C–S–H observed was reticular in structure and formed a dense microstructure, resulting in few pores.

Figure 21 shows the BSE image of non-cured specimens at ambient temperature. The structure was poorly crystalline. It is evident that, for the sample not subjected to any curing, water was lost due to evaporation and therefore self-desiccation of the concrete occurred, resulting in unhydrated cement particles. As the specimen was not water-cured but kept open at ambient temperature, the alite (C_3S) and belite (C_2S) remained unhydrated. The calcium/silicon ratio was high compared with the conventional water-cured specimens, indicating the presence of calcium hydroxide crystals rather than C–S–H gel. It is thus evident that the required amount of hydration had not taken place, resulting in a more porous structure (i.e. a less dense concrete matrix).

Figure 22 shows the BSE image of an identical specimen self-cured with 0·1% PEG 4000 at ambient temperature. It is clear from the image that, compared with the non-cured specimen, the ITZ was more dense. The low calcium/silicon and (aluminium + iron)/silicon ratios indicate the presence of C–S–H and hence better hydration. The concrete specimen cured with 0·1% LPW-light (AV 0·1) showed similar behaviour (Figure 23). Adding self-curing compounds to SCC has thus been shown to be advantageous, with a reduction in pores and hence

Figure 21 BSE image of non-cured concrete at 27°C

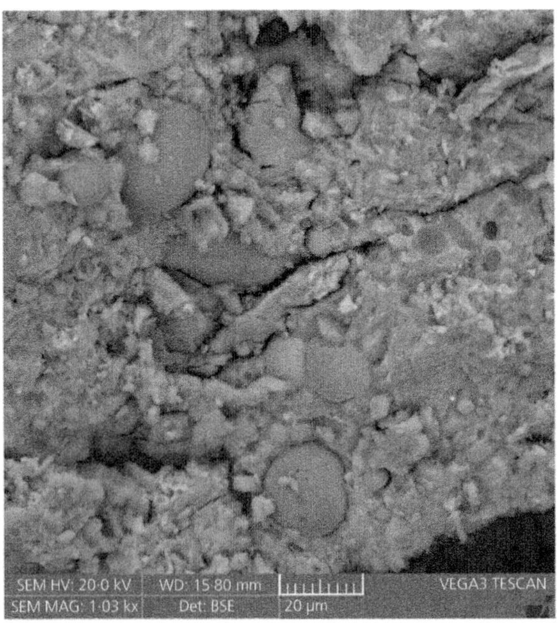

Figure 22 BSE image of self-cured SCC (0·1% PEG 4000) at 27°C

Figure 23 BSE image of self-cured SCC (0·1% LPW-light) at 27°C

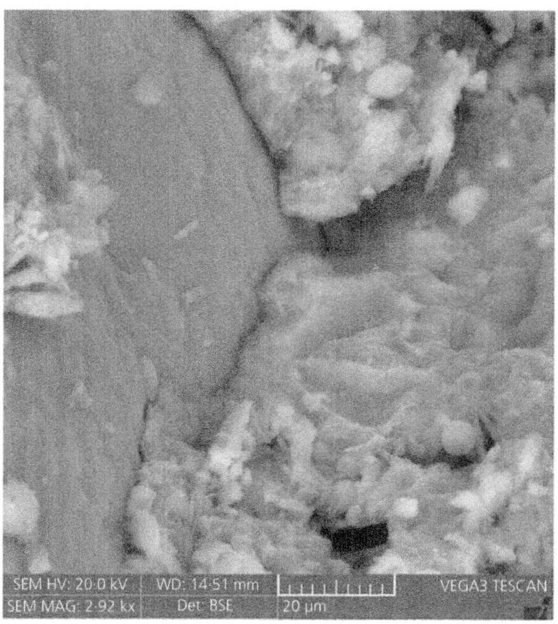

Table 11 Elemental composition of SCC subjected to different curing regimes

Element: atomic%	AW	Ambient curing (27°C)			Elevated curing (50°C)		
		AN	AH 0·1	AV 0·1	AN	AH 0·1	AV 0·1
Oxygen (O)	66·56	62·06	66·57	68·77	68·92	71·60	66·58
Aluminium (Al)	4·97	4·11	4·10	4·60	8·41	3·92	5·24
Silicon (Si)	13·85	9·74	14·29	11·88	6·08	9·28	10·20
Calcium (Ca)	12·82	22·25	13·76	13·04	14·70	14·44	16·80
Iron (Fe)	1·80	1·84	1·28	0·90	1·89	0·76	1·18
Potassium (K)	–	–	–	0·81	–	–	–
Total	100·00	100·00	100·00	100·00	100·00	100·00	100·00

densification of the concrete matrix. The hydration process was advanced due to sealing of the pores, meaning that more water was available for curing.

SEM analysis of SCC specimens cured at elevated temperature

Figures 24–26 show BSE images of SCC specimens cured at elevated temperature under different curing regimes. The EDXA spectrum indicated the calcium/silicon and (aluminium + iron)/silicon ratios of the non-cured SCC to be 2·42 and 0·70 respectively. These ratios

Figure 24 BSE image of non-cured SCC at 50°C

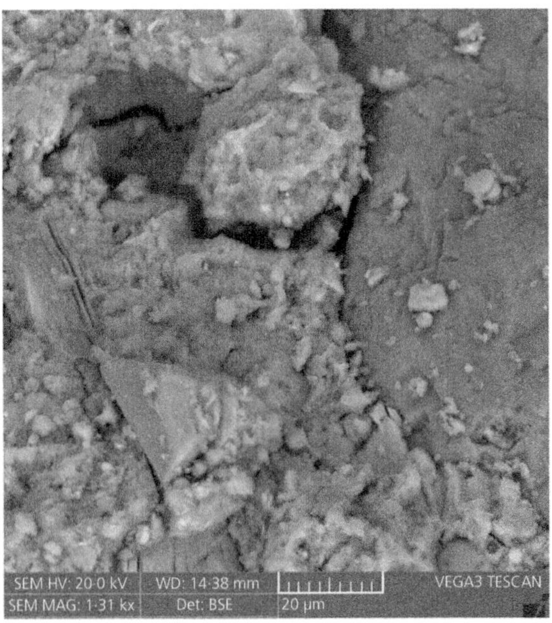

Figure 25 BSE image of self-cured SCC (0·1% PEG 4000) at 50°C

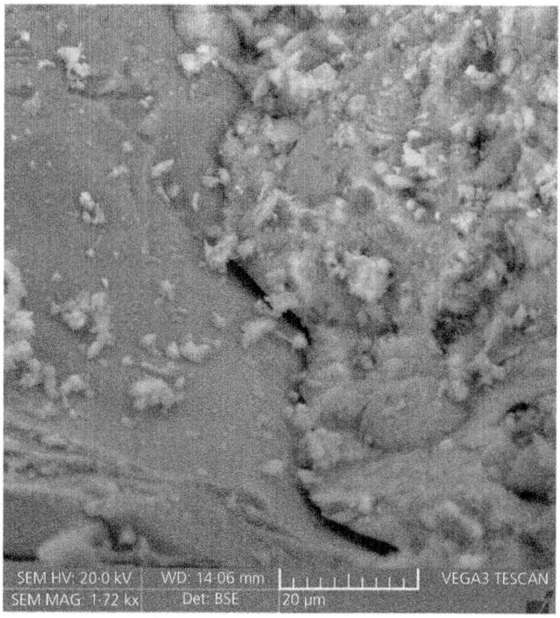

Figure 26 BSE image of self-cured SCC (0·1% LPW-light) at 50°C

indicate the presence of large quantities of aluminate–ferrite hydrate phases, portlandite crystals and a minimum quantity of C–S–H phases. Curing at elevated temperature led to the evaporation of water, resulting in a porous microstructure, as shown in Figure 24. The interlinking of microcracks along with pores was observed in the ITZ of the SCC specimens without the addition of a curing compound cured at elevated temperature.

As shown in Figures 25 and 26, the self-cured specimens showed a better microstructure with dense C–S–H gel formation. The calcium/silicon and (aluminium + iron)/silicon ratios for the SCC specimen cured with PEG 4000 were 1·56 and 0·32, while these ratios were 1·65 and 0·38 for the SCC specimen cured with LPW-light. These ratios and the SEM images of specimens cured at elevated temperature indicate inferior performance compared with the specimens cured at ambient temperature. However, it can be concluded that concrete with the addition of self-curing compounds is a very good alternative in water-scarce areas and high-temperature regions.

Conclusions

Based on the series of experiments conducted on SCC specimens subjected to various curing conditions, the following important conclusions were drawn.

- Optimum dosages of hydrophilic (PEG 4000) and hydrophobic (LPW-light) chemicals can be used as self-curing compounds in SCC, particularly in hot arid regions where the evaporation of water from the surface of concrete is a major issue.
- Use of 0·1% PEG 4000 for mix A (70 MPa), 1·0% PEG 4000 for mix B (50 MPa) and 0·1% LPW-light for both mixes resulted in improved water retention capacity.
- In the case of SCC specimens with a low water/cement ratio cured at ambient temperature, there was a 27% decrease in compressive strength for the non-cured specimens compared with conventional water-cured specimens. With the addition of the optimum amounts of PEG 4000 (0·1%) and LPW (0·1%), the strength loss was only 2·2% and 13·1%, respectively.
- The 28 d compressive strengths of SCC specimens cured with PEG 4000 and LPW-light at 50°C were about 90% of the strengths of specimens cured at 27°C.
- Self-cured SCC specimens with minimum moisture loss exhibited superior performance in terms of compressive strength, open porosity and sorptivity. Better water retention decreases concrete porosity, leading to better gel formation.
- Use of optimum dosages of the hydrophilic (0·1% PEG 4000) and hydrophobic (0·1% of LPW-light) chemicals increased the post-depassivation crack time compared with non-cured specimens. This indicates delayed initiation of corrosion.
- The addition of hydrophilic and hydrophobic chemicals resulted in a dense microstructure with a calcium/silicon ratio of 0·96 and 1·10, respectively. This is an indication of stable C–S–H gel formation compared. The non-cured specimens exhibited poor microstructure with interlinking of microcracks and ettringite formation.

REFERENCES

ASTM (2006) C 642-06: Standard test method for density, absorption and voids in hardened concrete. ASTM International, West Conshohocken, PA, USA.

ASTM (2012) C 1202-12: Standard test method for electrical indication of concrete's ability to resist chloride ion penetration. ASTM International, West Conshohocken, PA, USA.

ASTM (2013) C 1585-13: Standard test method for measurement of rate of absorption of water by hydraulic-cement concretes. ASTM International, West Conshohocken, PA, USA.

ASTM (2016a) C 150/C 150M-16e1: Standard specification for Portland cement. ASTM International, West Conshohocken, PA, USA.

ASTM (2016b) C 494/C 494M: Standard specification for chemical admixtures for concrete. ASTM International, West Conshohocken, PA, USA.

Balshin MY (1949) Relation of mechanical properties of powder metals and their porosity and the ultimate properties of porous-metal ceramic materials. *DoklAskd SSSR* **67(5)**: 831–834.

Bentz DP (2009) Influence of internal curing using lightweight aggregates on interfacial transition zone percolation and chloride ingress in mortars. *Cement & Concrete Composites* **31(5)**: 285–289.

Bentz DP and Jensen OM (2004) Mitigation strategies for autogenous shrinkage cracking. *Cement & Concrete Composites* **26(6)**: 677–685.

Bentz DP and Stutzman PE (1994) SEM analysis and computer modeling of hydration of Portland cement particles. In *Petrography of Cementitious Materials* (DeHayes SM and Stark D (eds)). ASTM International, West Conshohocken, PA, USA, p. 60.

Bentz DP and Weiss WJ (2011) *Internal Curing: A 2010 State-of-the-Art Review*. National Institute of Standards and Technology, Gaithersburg, MD, USA.

Bentz DP, Lura P and Roberts JW (2005) Mixture proportioning for internal curing. *Concrete International* **27(2)**: 35–40.

Bouzoubaa N and Lachemi M (2001) Self-compacting concrete incorporating high volumes of class F fly ash: preliminary results. *Cement and Concrete Research* **31(3)**: 413–420.

BSI (2009) BS 12390-3: 2009: Testing hardened concrete. Compressive strength of test specimens. *BSI, London, UK.*

Buhrke VE, Jenkins R and Smith DK (eds) (1998) *Practical Guide for the Preparation of Specimens for X-ray Fluorescence and X-ray Diffraction Analysis*. Wiley-VCH, New York, NY, USA.

Chand MSR, Giri PSNR, Kumar GR and Kumar PR (2014) Paraffin wax as an internal curing compound in ordinary concrete. *Magazine of Concrete Research* **67(2)**: 82–88, http://dx.doi.org/10.1680/macr.14.00192.

Chand MSR, Kumar PR, Giri PSNR, Kumar GR and Rao MVK (2016a) Influence of paraffin wax as a self-curing compound in self-compacting concretes. *Advances in Cement Research* **28(2)**: 110–120, http://dx.doi.org/10.1680/jadcr.15.00062.

Chand MSR, Giri PSNR, Kumar PR, Kumar GR and Raveena C (2016b) Effect of self-curing chemicals in self compacting mortars. *Construction and Building Materials* **107(6)**: 356–364.

Das D and Chatterjee A (2012) A comparison of hardened properties of fly-ash-based self-compacting concrete and normally compacted concrete under different curing conditions. *Magazine of Concrete Research* **64(2)**: 129–141, http://dx.doi.org/10.1680/macr.10.00118.

De la Varga I, Castro J, Bentz D and Weiss J (2012) Application of internal curing for mixtures containing high volumes of fly ash. *Cement & Concrete Composites* **34(9)**: 1001–1008.

Dhir RK, Hewlett PC, Lota JS and Dyer TD (1994) An investigation into the feasibility of formulating 'self-cure' concrete. *Materials and Structures* **27(10)**: 606–615.

Dhir RK, Hewlett PC and Dyer TD (1995) Durability of 'self-cure' concrete. *Cement and Concrete Research* **25(6)**: 1153–1158.

Dhir RK, Hewlett PC and Dyer TD (1996) Influence of microstructure on the physical properties of self-curing concrete. *Materials Journal* **93(5)**: 465–471.

Dhir RK, Hewlett PC and Dyer TD (1998) Mechanisms of water retention in cement pastes containing a self-curing agent. *Magazine of Concrete Research* **50(1)**: 85–90, http://dx.doi.org/10.1680/macr.1998.50.1.85.

Domone PL (2006) Self-compacting concrete: an analysis of 11 years of case studies. *Cement & Concrete Composites* **28(2)**: 197–208.

Efnarc (European Federation of Specialist Construction Chemicals and Concrete Systems) (2005) *The European Guidelines for Self-Compacting Concrete – Specification, Production and Use.* Efnarc, Farnham, UK.

El-Dieb AS (2007) Self-curing concrete: water retention, hydration and moisture transport. *Construction and Building Materials* **21(6)**: 1282–1287.

El-Dieb AS, El-Maaddawy TA and Mahmoud AA (2012) Water-soluble polymers as self-curing compounds in cement mixes. *Advances in Cement Research* **24(5)**: 291–299, http://dx.doi.org/10.1680/adcr.11.00030.

Geiker MR, Bentz DP and Jensen OM (2004) Mitigating autogenous shrinkage by internal curing. American Concrete Institute, Farmington Hills, MI, USA, ACI SP: 218–9, pp. 143–154.

Gesoğlu M and Özbay E (2007) Effects of mineral admixtures on fresh and hardened properties of self-compacting concretes: binary, ternary and quaternary systems. *Materials and Structures* **40(9)**: 923–937.

Hasselman DPH (1969) Griffith flaws and the effect of porosity on tensile strength of brittle ceramics. *Journal of the American Ceramics Society* **52(8)**: 457.

Jau WC (2011) *Self-Curing Concrete.* US Patent 8 016 939, Sept.

Kovler K and Jensen OM (2005) Novel techniques for concrete curing. *Concrete International* **27(9)**: 39–42.

Kumar GR (2011) Self-curing concrete. In *International Rilem Conference on Advances in Construction Materials through Science and Engineering* (Leung C and Wan K (eds)). Rilem Publications, Paris, France, pp. 116–123.

Kumar R and Bhattacharjee B (2003) Porosity, pore size distribution and in situ strength of concrete. *Cement and Concrete Research* **33(1)**: 155–164.

Leach CB and CAM Sales, Inc. (1992) *Method for Curing Concrete Articles.* US Patent 5 089 198, Feb.

Magzoub Ahmed F (2015) *Effect of Curing on Concrete Strength.* Doctoral dissertation, University of Khartoum, Khartoum, Sudan.

Meeks KW and Carino NJ (1999) *Curing of High-Performance Concrete: Report of the State-of-the-Art.* National Institute of Standards and Technology, Gaithersburg, MD, USA.

Mehta PK and Monteiro PJM (2014) *Concrete: Microstructure, Properties and Materials.* McGraw-Hill Education, New York, NY, USA.

Okamura H and Ouchi M (2003) Self-compacting concrete. *Journal of Advanced Concrete Technology* **1(1)**: 5–15.

Okamura H and Ozawa K (1994) Mix design method for self-compactable concrete. *Doboku Gakkai Ronbunshu* **1994(496)**: 1–8.

Okamura H, Ozawa K and Ouchi M (2000) Self-compacting concrete. *Structural Concrete* **1(1)**: 3–17.

Oliveira MJ, Ribeiro AB and Branco FG (2015) Curing effect in the shrinkage of a lower strength self-compacting concrete. *Construction and Building Materials* **93(18)**: 1206–1215.

Palou MT, Kuzielová E, Žemlička M, Boháč M and Novotný R (2016) The effect of curing temperature on the hydration of binary Portland cement. *Journal of Thermal Analysis and Calorimetry* **125(3)**: 1301–1310.

Rossignolo JA (2009) Interfacial interactions in concretes with silica fume and SBR latex. *Construction and Building Materials* **23(2)**: 817–821.

Ryshkevitch R (1953) Compression strength of porous sintered alumina and zirconia. *Journal of the American Ceramics Society* **36(2)**: 65–68.

Safi B, Ghernouti Y, Rabehi B and Aboutaleb D (2013) Effect of the heat curing on strength development of self-compacting mortars containing calcined silt of dams and Ground Brick Waste. *Materials Research* **16(5)**: 1058–1064.

Schiller KK (1971) Strength of porous materials. *Cement and Concrete Research* **1(4)**: 419–422.

Stutzman PE (1994) Applications of scanning electron microscopy in cement and concrete petrography. In *Petrography of Cementitious Materials* (DeHayesSM (ed.)). ASTM International, West Conshohocken, PA, USA, pp. 74–90.

Timms AG, Robinson DL, Gilkey HJ *et al.* (1952) Curing of concrete. *Journal of the American Concrete Institute* **48(5)**: 701–724.

Uysal M and Sumer M (2011) Performance of self-compacting concrete containing different mineral admixtures. *Construction and Building Materials* **25(11)**: 4112–4120.

Xotta G, Mazzucco G, Salomoni VA, Majorana CE and Willam KJ (2015) Composite behavior of concrete materials under high temperatures. *International Journal of Solids and Structures* **64(10)**: 86–99.

Zhao H, Sun W, Wu X and Gao B (2012) Effect of initial water-curing period and curing condition on the properties of self-compacting concrete. *Materials & Design* **35(3)**: 194–200.

Zhutovsky S and Kovler K (2012) Effect of internal curing on durability-related properties of high performance concrete. *Cement and Concrete Research* **42(1)**: 20–26.

Section 5

Use of self-compacting concrete in structural concrete

ice Publishing

Dhir and Lye
ISBN 978-0-7277-6461-4
https://doi.org/10.1680/icetscc.64614.337

Chapter 15
Steel–concrete bond behaviour of self-compacting concrete with recycled aggregates

Huai-liang Wang
Associate Professor, Civil and Architectural Engineering College, Dalian University, Dalian, PR China;
Visiting Scholar, Joint Research Center of Urban Resource Recycling Technology of the Graduate School at Shenzhen, Tsinghua University and Shenzhen Green Eco-Manufacturer High-Tech, Shenzhen, PR China (whuailiang@163.com)

Tests were conducted to examine the bond of reinforcement in self-compacting concrete (SCC) produced using recycled coarse aggregate (RCA). The study considered one natural coarse aggregate, two sources of RCA (concrete tested in the laboratory and concrete from demolished old structures), two types of RCA (pre-treated and untreated), three RCA replacement percentages (25%, 50% and 100%) and two types of steel rebar (plain and deformed). The results indicate that the compressive and tensile strengths of the recycled aggregate self-compacting concretes (RA-SCCs) were greatly influenced by the RCA replacement ratio, and the RCA source and type. Using pre-treated RCA, the RCA replacement ratio had little effect on the bond strength of RA-SCC. There was an obvious reduction of the RA-SCC's bond capacities when the pre-treated RCA was replaced with the low-quality untreated RCA obtained from demolished old structures. Based on regression analysis of the experimental data, an empirical bond stress–slip relationship between RA-SCC and steel rebar was established and it was also proved that the bond strength predictions based on the provisions of existing codes (ACI 318 and GB 50010) were compatible with the experimental results.

Notation

d	nominal rebar diameter
f_c'	compressive strength of concrete
$f_{cu,150}$	cubic compressive strength of concrete
f_{sp}	splitting tensile strength of concrete
f_t	direct tensile strength of concrete
l	bonded length of steel bar
P_u	peak load
s_0	slip corresponding to τ_b
s_a	relevant water absorption ratio of each coarse aggregate
s_t	equivalent water absorption of coarse aggregates
W_a	each coarse aggregate weight in mix design
W_t	total weight of coarse aggregate in mix design

α, β	regression parameters in the bond–slip relationship
τ, s	bond stress and corresponding slip at any stage of loading
$\bar{\tau}, \bar{s}$	dimensionless bond stress and slip, respectively
τ_0	normalised bond strength
τ_b	peak bond stress between concrete and steel rebar

Introduction

One of the main directions of concrete development is research and applications of high-performance concrete (HPC). Self-compacting concrete (SCC) is a type of environmentally friendly concrete that has high flowability and so can pass through congested reinforcement areas and fill formwork completely under its own gravity without the need for any vibration. Compared with normal concrete, SCC has a higher fine particle content and improved flowing, filling, passing and segregation-resistance ability, which offers great advantages in special structures that involve difficulties in concrete casting and vibration, such as large thin-walled structures and beam–column joints with densely arranged steel bars. The aggregates used in traditional concrete, such as gravel, crushed rock, river and mining sand, and even some lightweight aggregates, can also be used in SCC. In general, aggregate usually accounts for 55–60% of the SCC volume and plays a substantial role in determining the workability, strength and durability of the concrete (Choi *et al.*, 2006; Su *et al.*, 2001). The types of aggregate used also affect the cost of the concrete cast, so low-priced aggregate in SCC can also result in better economic benefits.

The construction boom in developing countries and high levels of reconstruction in developed countries in many regions of the world have led to a serious shortage of natural coarse aggregate (NCA), and there is therefore a growing demand for alternative aggregates for concrete construction. More importantly, the large amounts of waste concrete arising from old demolished buildings, experiments and new construction are a problem in that the removal and disposal of such wastes brings significant environmental problems. In recent decades, the lack of NCA and increased landfill costs have promoted the application of recycled coarse aggregate (RCA) in the production of new concrete. In this context, the use of RCA in SCC is also very important because it not only finds a solution to the disposal problem of concrete wastes and industry wastes such as fly ash or ground granulated blast-furnace slag, which are also usually used in the production of SCC to reduce the material cost (Bingol and Tohumcu, 2013), but also helps to reduce the demand for NCA. Recycled aggregate self-compacting concrete (RA-SCC) – a new ‘green’ HPC – thus offers the advantages of using construction waste as recycled aggregate and self-consolidating without vibration.

The bond behaviour between concrete and steel bars is an important index to measure the collaborative workability for both kinds of materials, and many scholars have carried out research on the bond mechanism between concrete and steel bars. Xiao and Falkner (2007) investigated the bond behaviour between recycled aggregate concrete (RAC) and steel rebars. They found that the bond strength between RAC and plain rebars decreased with an increase in RCA replacement percentage, whereas the bond strength between RAC and deformed rebars had no obvious relation with the RCA replacement percentage. Prince and Singh (2013, 2014a, 2014b) concluded that relative bond strengths increased with RCA replacement levels and the highest values were obtained for 100% replacement of NCA with RCA. However, different conclusions have also been reported in the literature: some authors noted reductions of 9–19% (Butler *et al.*, 2011, 2014) whereas others (Sindy *et al.*, 2014) obtained decreases of 12% and 13% for concretes with 100% recycled aggregate content compared to conventional aggregate. These different results can be explained by the fact that the sources, production conditions and properties of the recycled aggregates used in the experiments were different. A limited amount of research has been conducted on the use of RCA in SCC, but it has been shown that RCA could be a reliable alternative to NCA in SCC. Murat *et al.* (2014) and Zoran *et al.* (2010) produced SCCs using RCA as partial and full replacements of NCA. However, these studies placed more emphasis on the fresh and hardened properties, and no comprehensive work investigating the effects of RCA on the bond properties of RA-SCC was found by the author. More systemic and comprehensive studies and comparisons with ordinary concrete should thus be carried out.

In this study, RCAs derived from tested concrete and old concrete buildings were used to produce SCC, substituting 0%, 25%, 50% and 100% NCA by weight. The effects of RCA replacement percentage and quality on the bonding properties of RA-SCC were investigated and the results compared with ordinary SCC and vibrated RAC. The findings presented will add to the current state-of-the-art on RA-SCC research, provide an extensive new dataset on the bond performance of RA-SCC concrete members, and help to promote the use of RCA in sustainable structural HPC applications.

Experimental programme

Materials

Crushed limestone (diameter 5–20 mm) was used as NCA and natural river sand was used as the fine aggregate. RCA was obtained by crushing waste concrete using a jaw crusher (Figure 1). Considering the large differences in the properties of RCAs obtained from different waste

Figure 1 Laboratory crushing machine used to produce RCA

concrete sources, which notably influence the material properties of concrete such as compressive and tensile strengths and elastic modulus, two sources of RCA were used in this study

- RCA1, from concrete tested in the laboratory (3 months old, strength grade C40 to C80)
- RCA2, from a demolished concrete structure in Dalian, China (concrete age more than 20 years, strength grade below C30).

The two types of recycled aggregate are shown in Figure 2. The appearance of RCA1 was close to natural crushed aggregate, with rough surfaces. RCA2 had more smooth and edgeless

Figure 2 Appearance of (a) RCA1 and (b) RCA2

(a)

(b)

surfaces and contained many harmful impurities such as sawdust, asphalt, metal material, and plastic and ceramic tile fragments.

Both RCAs were separated into two parts in order to investigate the effect of RCA replacement percentage and RCA properties on the mechanical properties of RA-SCC. The first part of each RCA, called pre-treated RCA, was submerged in hydrogen chloride (HCl) solution at 10% molarity for 24 h and then submerged in distilled water for another 24 h to remove the acidic solution. As shown in Figure 3, the pre-treated RCA samples were then wet sieved over a 5 mm sieve to ensure that part of the loose old adhered mortar on the RCA surface was removed following the submersions in hydrogen chloride solution and water. This method of acid treatment, which was adopted from Tam *et al.* (2007), has been shown to be effective at improving concrete quality in terms of compressive strength, flexural strength and elastic modulus compared with concretes containing untreated RCA. The second part of the RCAs and the NCA were pre-soaked in water for 24 h. Before casting, all the coarse aggregates were spread out in air for 24 h to ensure an air-dried moisture state was reached. The typical physical characteristics of the pre-treated RCAs, the untreated RCAs and the NCA used in this study are listed in Table 1.

Ordinary Portland cement type 52·5, conforming to Chinese standard GB 175-2007, was used for the test specimens. Class I fly ash, 39% of the total binder by mass, was incorporated as a viscous admixture during the concrete mixing process. The viscosity and flowability of the RA-SCCs were enhanced by adding a commercial high-range water reducer (HRWR). The designed mixture ratios of all the mixtures and the workability of the fresh concretes are listed in Table 2. The percentages of pre-treated RCA used were 25%, 50% and 100%. The percentage of untreated RCA was 100%. The content of HRWR in each mix had been adjusted to achieve the target workability without having to add extra water. The necessary tests to check

Figure 3 Wet screening of pre-treated RCA

Table 1 Basic properties of the coarse aggregates used

Coarse aggregate	Apparent density: kg/m^3	24 h water absorption: %	ACV: %	Abrasion loss: %, by mass
NCA	2640	1·2	8·5	14
Pre-treated RCA1	2524	3·7	10·5	20
Untreated RCA1	2480	6·6	12·2	35
Pre-treated RCA2	2500	4·8	10·6	29
Untreated RCA2	2410	7·2	18·7	39

Table 2 Mixture proportions and fresh properties of SCC and RA-SCC mixtures

	Mixture[a]				
	SCC0	SCC*i*–25	SCC*i*–50	SCC*i*-100	RAC*i*–100
Water/binder ratio	0·27	0·27	0·27	0·27	0·27
Water: kg/m^3	178	178	178	178	178
Cement: kg/m^3	410	410	410	410	410
Fly ash: kg/m^3	260	260	260	260	260
Natural sand: kg/m^3	676	676	676	676	676
NCA: kg/m^3	860	645	430	–	–
Pre-treated RCA: kg/m^3	–	215	430	860	–
Untreated RCA: kg/m^3	–	–	–	–	860
HRWR: kg/m^3	3·1	4·8	5·6	7·4	9·1
Slump flow: mm	750	710–740			
T_{500} slump flow time: s	1·6	2·0–4·1			
V-funnel flow time, T_V: s	7·4	9·6–21·8			
L-box passing ratio, H_2/H_1	0·96	0·81–0·95			
Sieve segregation ratio: %	9·4	5·2–8·6			

[a] SCC0 is concrete with 0% RCA. SCC*i*–25, SCC*i*–50 and SCC*i*–100 represent concretes made with 25%, 50% and 100% of pre-treated RCA*i*, respectively. RAC*i*–100 represents concrete made with untreated RCA*i*. *i* = 1 or 2 represents the RCA source.

the self-consolidating properties of the mixes, such as slump flow test, J-ring test, L-box passing test, V-funnel test and sieve segregation test, were carried out on the fresh concretes (Figure 4). Cubic specimens of dimensions 150 × 150 × 150 mm were cast to determine the 28 d compressive and splitting strengths of the mixed concretes.

Figure 4 Examples of workability tests

(a)

(b)

Plain steel bars measuring 10 mm in diameter and deformed steel bars measuring 16 mm in diameter were adopted for the investigation. Their average yield strengths were 296 MPa and 410 MPa, respectively.

Preparation of specimens

Cubic specimens with a side length of 150 mm, shown in Figure 5, were used in the pull-out test. The embedded length was five times the bar diameter for all the specimens, which is reasonable for defining the bond stress–slip relationship (Xiao and Falkner, 2007). A bar was fixed through a hole in a steel mould, as shown in Figure 6. To avoid additional restraints,

Figure 5 Dimensions (in mm) of pull-out specimens

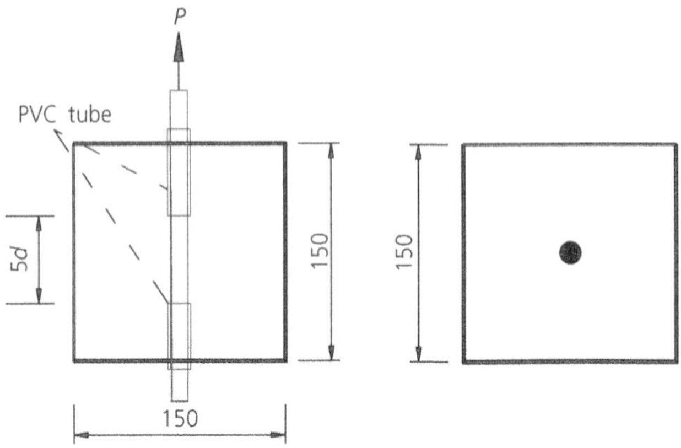

Figure 6 Steel moulds

the unbounded part of the bar was encased in a polyvinyl chloride (PVC) tube. Fresh SCC was poured into the mould from one side, allowing it to flow to the other side. Sufficient workability of SCC was maintained so that no segregation was observed. For each batch of concrete, six cubes of side 150 mm were prepared for the experiment. After 24 h of casting, the specimens were demoulded and transferred to a standard curing room. Before testing, they were allowed to air dry in the laboratory.

Test equipment

All the specimens were tested in a stiff electrohydraulic test frame using a specially fabricated mild steel rig rigidly connected to the testing machine. The test setup is shown in Figure 7, and a schematic illustration of the test instrumentation used in the pull-out tests is presented in Figure 8. The specimen was fixed tightly by a loading frame fastened to the base of the testing machine for the purpose of self-aligning. During testing, the pull-out load was applied

Figure 7 Pull-out specimens at test

Figure 8 Pull-out test setup

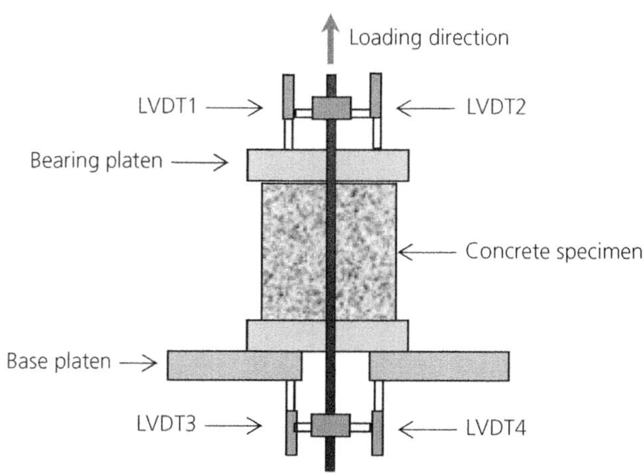

orthogonally through a vertical piston at a constant rate of 0·02 mm/s. The loaded-end slip was measured using linear variable differential transformers (LVDT1 and LVDT2 in Figure 8) and the net slip at the free end was measured using LVDT3 and LVDT4. The output of the LVDTs was recorded by a data acquisition system and the test was stopped when a slip of at least 10 mm was recorded.

Results and discussion

For ease of comparison, the cubic compressive strength ($f_{cu,150}$), splitting strength (f_{sp}) and bond strength (τ_b) of the different series of RA-SCCs are presented in Table 3. It can be noted from Table 3 that the compressive and tensile strengths decreased with increasing RCA

Table 3 Experimental results of pull-out tests

Specimen[a]	$f_{cu,150}$: MPa	SV: MPa[b]	f_{sp}: MPa	SV: MPa	τ_b: MPa	SV: MPa	Slip, s_0: mm	SV: mm
SCC0-D	69·2	1·6	4·84	0·27	20·82	0·62	0·747	0·033
SCC0-P					9·02	0·50	0·118	0·009
SCC1-25-D	67·0	1·8	4·73	0·35	20·59	0·75	0·680	0·025
SCC1-25-P					8·64	0·82	0·112	0·006
SCC1-50-D	63·9	1·7	4·48	0·66	21·69	0·64	0·837	0·023
SCC1-50-P					8·53	0·93	0·147	0·014
SCC1-100-D	58·6	2·0	4·22	0·24	20·92	1·22	0·944	0·012
SCC1-100-P					8·26	0·43	0·185	0·005
RAC1-100-D	55·2	2·2	3·75	0·87	20·04	1·05	0·919	0·013
RAC1-100-P					7·94	1·12	0·164	0·008
SCC2-25-D	63·6	1·4	4·51	0·18	20·21	0·91	0·878	0·042
SCC2-25-P					8·42	0·70	0·144	0·007
SCC2-50-D	62·5	1·5	4·24	0·34	21·45	0·35	0·726	0·018
SCC2-50-P					8·85	1·35	0·126	0·004
SCC2-100-D	56·7	1·3	4·08	0·26	20·75	0·45	0·935	0·035
SCC2-100-P					7·92	1·10	0·173	0·012
RAC2-100-D	52·9	2·3	3·17	0·43	19·97	0·94	0·904	0·032
RAC2-100-P					7·66	0·52	0·152	0·018

[a] D represents deformed bar and P represents plain bar.
[b] SV is standard variation, calculated from $(\sum(x-\bar{x})^2/(n-1))^{0.5}$, where $n=3$ represents the number of specimens tested for each loading scenario.

replacement percentage. When the pre-treated RCA replacement percentage was 100%, compared with SCC0, the compressive strength was approximately 15% lower for RCA1 and 18% lower for RCA2, and the tensile strength was 12% lower for RCA1 and 15% lower for RCA2. When the NCA was completely replaced with untreated RCA, the compressive strength decreased by more than 20% for both RCAs and the tensile strength decreased by more than 22% for RCA1 and 34% for RCA2.

Failure mode

Figure 9 shows the typical pull-out failures of the specimens with different replacement ratios of pre-treated RCA. After testing, the concrete block of the pull-out specimen was split into two pieces to allow closer investigation of the bond failure patterns. As shown in Figure 10, concrete crushing was found around the bar ribs in both SCC0 and SCC100 specimens, and it was especially evident in the SCC containing untreated RCA.

On the fracture surface of SCC0 in Figure 10(a), one can see that almost half the NCA was broken away from the mortar at the interface while the other half of the NCA fractured directly. This indicates that the interface strength was almost as strong as the strength of the natural aggregate for the NA-SCC in this investigation. For SCC100 containing 100% pre-treated RCA (Figure 10(b)), very little of the original natural aggregate was fractured, but most of the RCA was broken away from old adhered mortar on the surface of the RCA. For RAC100 containing 100% untreated RCA (Figure 10(c)), no fracture phenomenon of the original natural aggregate could be observed by the naked eye. Other researchers (Kim and Yun, 2013; Prince and Singh, 2013, 2014a, 2014b) have concluded that the bond behaviour between concrete and rebars is not only influenced by the compressive strength of the concrete but also by the interlocking of coarse aggregates around the rebar. The interlocking effect of coarse aggregates in RAC100 containing 100% untreated RCA was much weaker than that of SCC0 due to the large amount of loose mortar present in untreated RCA. Thus, the bond strength of RAC100 was lower than that of SCC0, as shown in Table 3, irrespective of the use of a deformed bar or a plain bar.

Figure 9 Pull-out failure modes

Figure 10 Cross-sections of pull-out specimens with different types of coarse aggregate: (a) SCC0-D, (b) SCC100-D, (c) RAC100-D

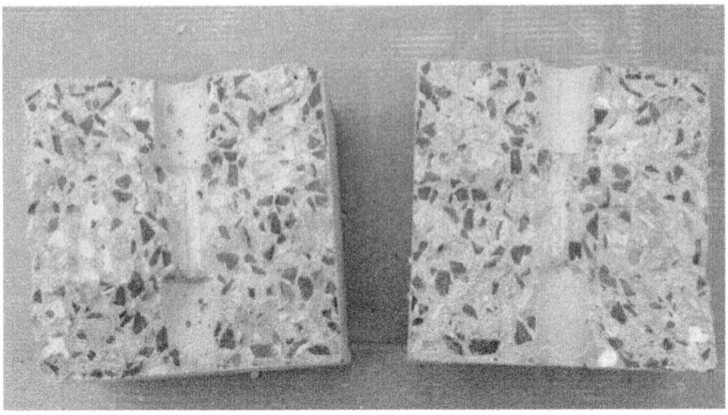

(a)

(b)

(c)

Bond strength

The tests results on the bond strength of SCC mixes were evaluated by dividing the ultimate load (P_u) by the bonded area (πdl). Given that the bonded length (l) was five times the bar diameter (d) in the current work, the ultimate bond strength can be expressed as

$$\tau_b = \frac{P_u}{5\pi d^2} \tag{1}$$

The bond strengths of SCC mixtures with different RCA replacement percentages and different RCA types obtained from the test results are summarised in Figure 11 and Figure 12. It can be seen that the bond strengths for the RA-SCCs incorporating 100% untreated RCA were lower than the value of the conventional mix for both deformed and plain rebars, mainly due to their lower compressive strengths. The bond strength was reduced by 3·7% and 12·0% for RAC1-100 with deformed and plain rebars, respectively, and by 4·1% and 15·1% for RAC2-100 with deformed and plain rebars, respectively. With the deformed rebars, the bond strengths of the RA-SCCs incorporating different dosages of pre-treated RCA were almost the same as that of the conventional mix. However, for the plain rebars, the bond strengths of the RA-SCCs incorporating higher dosages of pre-treated RCA tended to be lower than that of the conventional mix. This conclusion is the same as that reported by Xiao and Falkner (2007), although vibrated RAC was used in the experiment of Xiao and Falkner (2007).

Figure 11 Comparison of bond strength for concretes with different percentages of RCA1

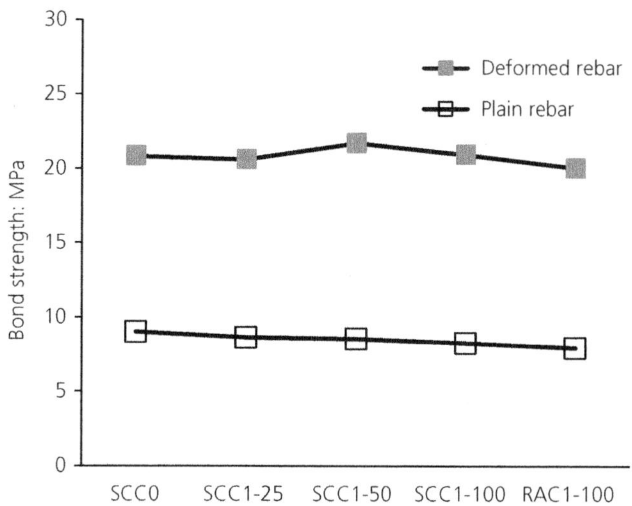

Figure 12 Comparison of bond strength for concretes with different percentages of RCA2

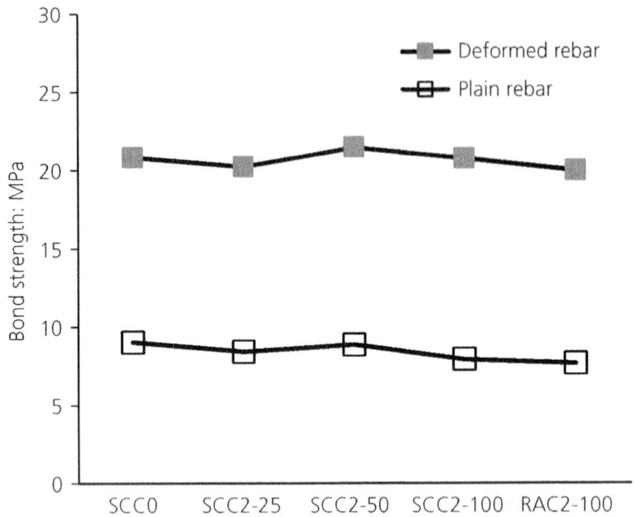

As stated by several researchers (Kim and Yun, 2013; Prince and Singh, 2013, 2014a, 2014b; Sindy *et al.*, 2014; Xiao and Falkner, 2007), the concrete compressive strength is a main factor affecting the bond strength between steel reinforcement and concrete. In order to eliminate the variation of concrete compressive strength, the bond strengths were re-evaluated in terms of normalised bond strength (τ_0)

$$\tau_0 = \frac{\tau_b}{(f_{cu})^{0.5}} \tag{2}$$

Figure 13 and Figure 14 show the effect of RCA replacement ratio and type on the mean values of normalised bond strength of the RA-SCCs. It can be seen from the figures that, for a plain round steel rebar, the normalised bond strength was almost the same despite an increase in RCA replacement ratio and regardless of RCA type. However, for deformed rebars, the normalised bond strength increased with a higher replacement percentage. When the coarse aggregate was replaced with 100% untreated RCA, the normalised bond strength of the RA-SCC was slightly lower than that of SCC containing 100% pre-treated RCA.

Influence of aggregate properties

It is generally accepted that the basic properties of vibrated RAC and RA-SCC vary proportionally with the amount of old mortar present in RCA (Güneyisi *et al.*, 2014; Tam *et al.*, 2007). However, it is very difficult to measure the precise amount of old mortar present in RCA. The aggregate crushing value (ACV) and water absorption capacities are usually used to measure the quality of RCA, and these are listed for each aggregate type in Table 1. The correlation between the splitting strength and the ACVs of the NCA, untreated RCA and

Figure 13 Comparison of normalised bond strength for concretes with different percentages of RCA1

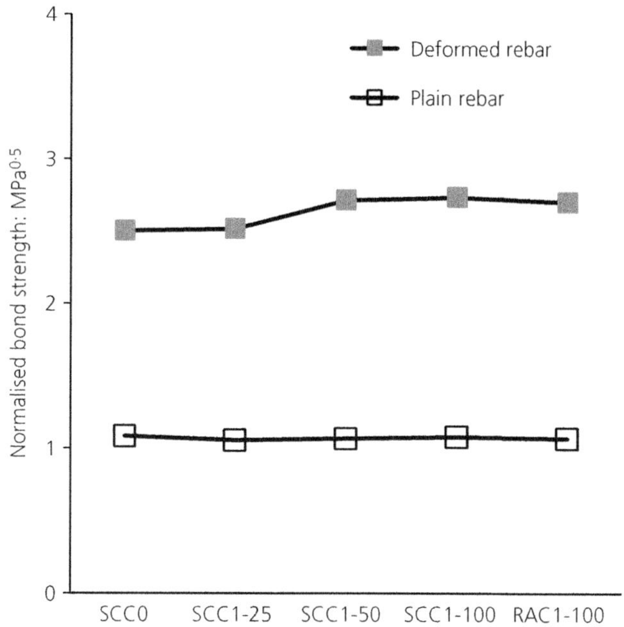

Figure 14 Comparison of normalised bond strength for concretes with different percentages of RCA2

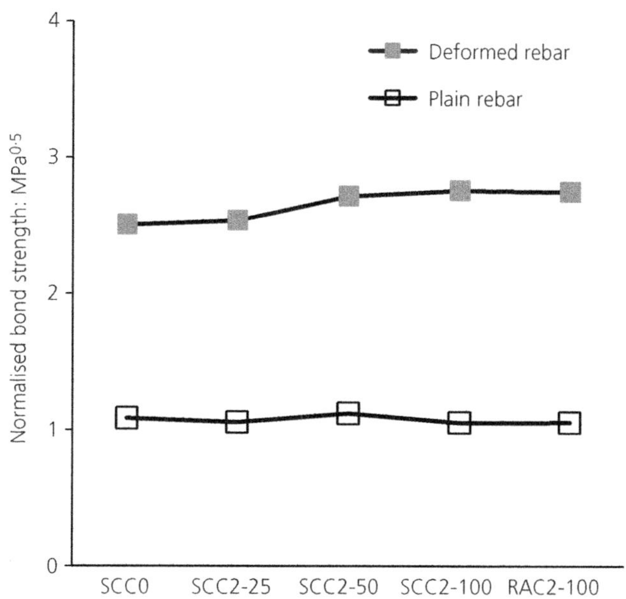

pre-treated RCA used in this research is shown in Figure 15. It is apparent that the splitting strength of RC-SCCs generally decreased with an increase in the ACV of the coarse aggregate. For SCCs incorporating 100% NCA, untreated RCA and pre-treated RCA, the relationship between normalised bond strength and ACV is shown in Figure 16. The normalised bond

Figure 15 Relationship between ACV and splitting strength

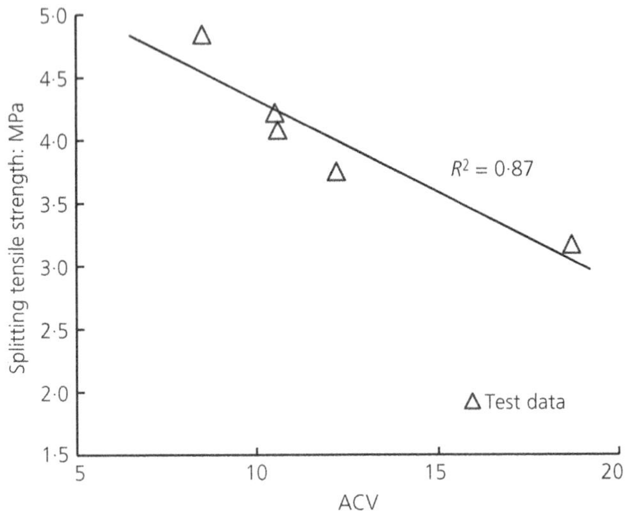

Figure 16 Relationship between ACV and normalised bond strength

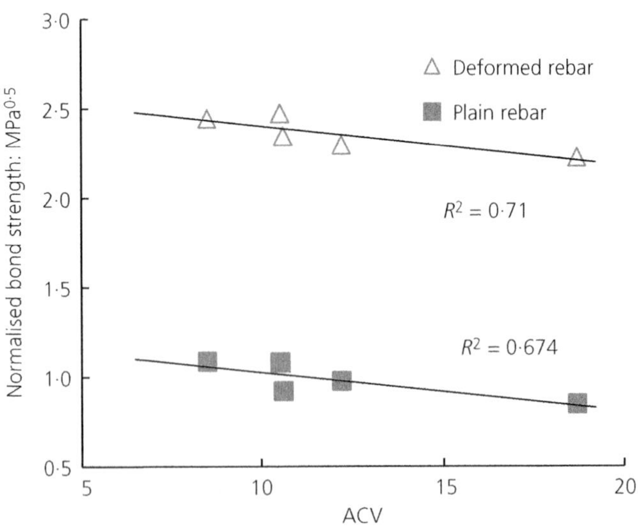

strength of specimens with both deformed and plain rebars decreased only slightly with an increase in ACV, demonstrating that the normalised bond strength was less sensitive to the ACV of the RCA. Regarding the RCA sources used in the SCC mixtures, specimens SCC1-100-D, SCC1-100-P, RAC1-100-D and RAC1-100-P showed slightly higher values of normalised bond strength than specimens SCC2-100-D, SCC2-100-P, RAC2-100-D and RAC2-100-P specimens, respectively. This indicates that the source and surface treatment of RCAs have a great effect on the mechanical strength of RA-SCCs but only a limited effect on the bond capacity between the rebar and concrete. In conclusion, the relationship between the normalised bond strength of RA-SCC and the ACV obtained in this work showed the same trend as vibrated RAC test results that were obtained using beam-end RAC specimens (Butler *et al.*, 2011, 2014).

Considering that the same fine aggregate was used in all the SCC mixtures, the equivalent water absorption (EWA) of aggregates was computed and then divided by the total weight of the coarse aggregates as

$$s_t = \frac{\sum W_a \times s_a}{W_t} \tag{3}$$

where s_t is the EWA of coarse aggregates (%), W_a is each coarse aggregate weight, s_a is the relevant water absorption ratio (%) of each coarse aggregate and W_t is the total weight of coarse aggregate in the mix design.

The relationships between the splitting tensile strengths of the various concrete specimens and the EWA of coarse aggregates are shown in Figure 17. It can be seen that the splitting tensile

Figure 17 Relationship between splitting tensile strength and EWA of the aggregate

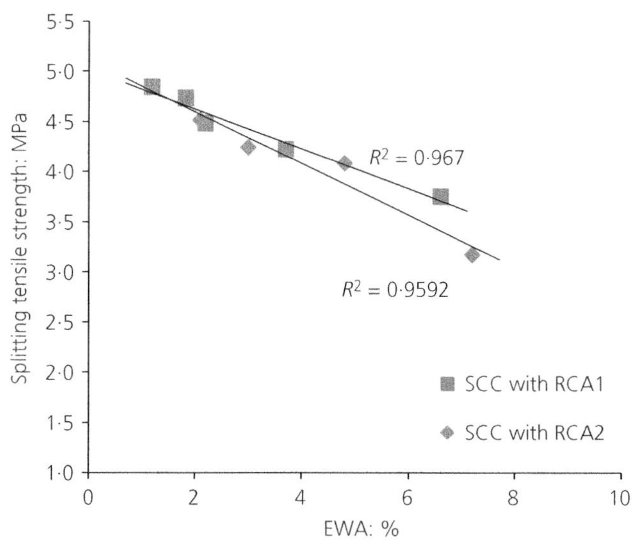

strength of the SCC decreased with an increase in the EWA of coarse aggregate (i.e. an increase in RCA content) for all concrete mixes. Moreover, the splitting tensile strengths of mixes with RCA1 were higher than those of the concretes with RCA2 as the quality of RCA2 was inferior to that of RCA1.

Figure 18 shows the normalised bond strength as a function of EWA of the aggregate. The test results of Xiao and Falkner (2007) and Prince and Singh (2013, 2014a) are also plotted in Figure 18. As shown in the figure, the normalised bond strength of RA-SCC with deformed bars showed a tendency to increase in proportion to the EWA (or the RCA replacement ratio), which is similar to the vibrated RAC test results. As for plain bars, the normalised bond strength of RA-SCC was almost the same with an increase in the EWA of coarse aggregate. The almost-even slope shows that the bond strength for a plain rebar is less sensitive to aggregate EWA. It should be noted that the concretes produced using RCA in the other studies plotted in Figure 18 were all vibrated concrete. Nonetheless, visually, Figure 18 seems to indicate that the normalised bond strengths of the RA-SCCs were as high as or higher than those of the vibrated RACs in other studies for the same steel rebar type and diameter. This enhanced bond behaviour could be explained by the larger amount of fine aggregates in RA-SCC mixtures and the higher fluidity of SCC, resulting in better containment of the reinforcement bars. However, more tests need to be carried out to verify this point.

Figure 18 Relationship between normalised bond strength and EWA of the aggregate

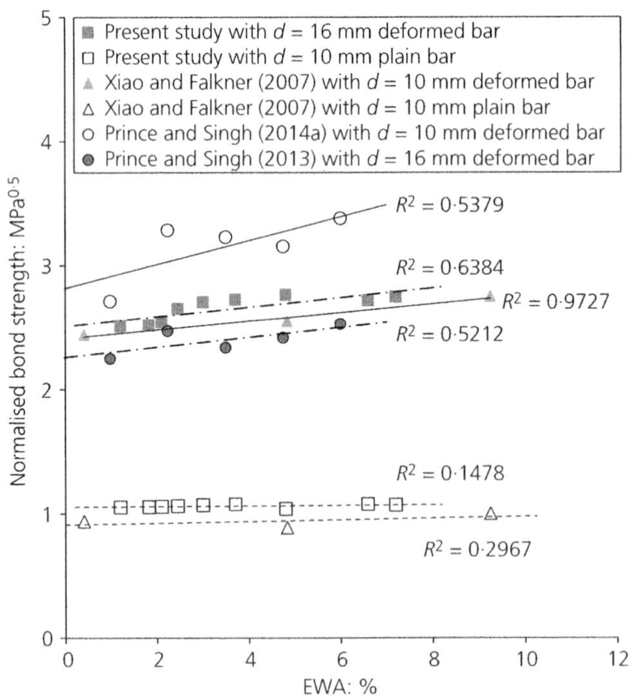

Bond stress–slip relationship

The bond test results summarised in Table 3 include the average maximum bond stress (τ_b) and the corresponding average peak slip (s_0) during the tests. The average peak slip was calculated from the loaded-end slip (s_l) and free-end slip (s_f), $s_0=(s_l+s_f)/2$, at the maximum bond stress. Table 3 shows that the average peak slips of deformed bars in different SCC mixtures varied from 0·680 mm to 0·944 mm, which are in the range of peak slips reported elsewhere for such rebars (Kim and Yun, 2013; Prince and Singh, 2013, 2014a, 2014b; Sindy *et al.*, 2014; Xiao and Falkner, 2007). The average peak slips of plain bars in different SCC mixtures varied from 0·112 mm to 0·185 mm, which again are in the narrow range for plain bars reported in other work (Xiao and Falkner, 2007).

In order to investigate the effect of RCA on the shape of the bond–slip relationship curves, every bond–slip relationship was normalised in terms of the dimensionless parameters $\bar{\tau}$ and $\bar{s}$

$$\bar{\tau} = \frac{\tau}{\tau_b}, \quad \bar{s} = \frac{s}{s_0} \tag{4}$$

where τ and s are, respectively, the bond stress and the corresponding slip at any stage of loading while τ_b is the peak bond stress (i.e. bond strength) and s_0 is the slip corresponding to τ_b.

Figure 19 shows the normalised experimental bond–slip curves in order to compare the effects of both RCA fraction and RCA type. In this investigation, the form of the normalised bond stress–slip relationship did not show significant differences between RA-SCC and SCC mixes for deformed and plain rebars. Each curve reflects the bond behaviour at the different stages (i.e. micro-slip, internal cracking, pull-out, descending and residual). Figure 19 shows that the bond stress–slip relationships of the control specimens (SCC0-D and SCC0-P) had a steeper and more upward linear branch than those of the SCCs containing pre-treated RCA and untreated RCA, indicating SCC0's higher ability to resist slip between the steel bar and concrete. This could be related to the old adhered mortar in RCA, which obstructs the chemical adhesion between concrete and steel. This result is in agreement with the work of Xiao and Falkner (2007). Furthermore, the treatment of RCA minimised the loose mortar attached to the surface of the RCA particles and decreased the water absorption and the ACV of the RCA, so the bond stress–slip relationships of SCCs containing pre-treated RCA show a steeper initial tangent inclination and a more linear part than the curves for SCC containing untreated RCA for the same steel bar type and RCA source.

For numerical analysis of reinforced RA-SCC members, modelling of the bond behaviour at the steel–concrete interface was necessary. To this end, the normalised bond–slip curves of all the mixtures (Figure 19) were compared with analytical bond stress–slip curves under monotonic loading proposed by Harajli (1994) and by Xiao and Falkner (2007). The model describes the ascending and descending branches of the normalised bond–slip relationship as

$$\bar{\tau} = \begin{cases} (\bar{s})^{\alpha} & \bar{s} \le 1 \\ \dfrac{\bar{s}}{\beta(\bar{s}-1)^2+\bar{s}} & \bar{s} > 1 \end{cases} \tag{5}$$

Figure 19 Typical normalised bond stress–slip relationships of specimens

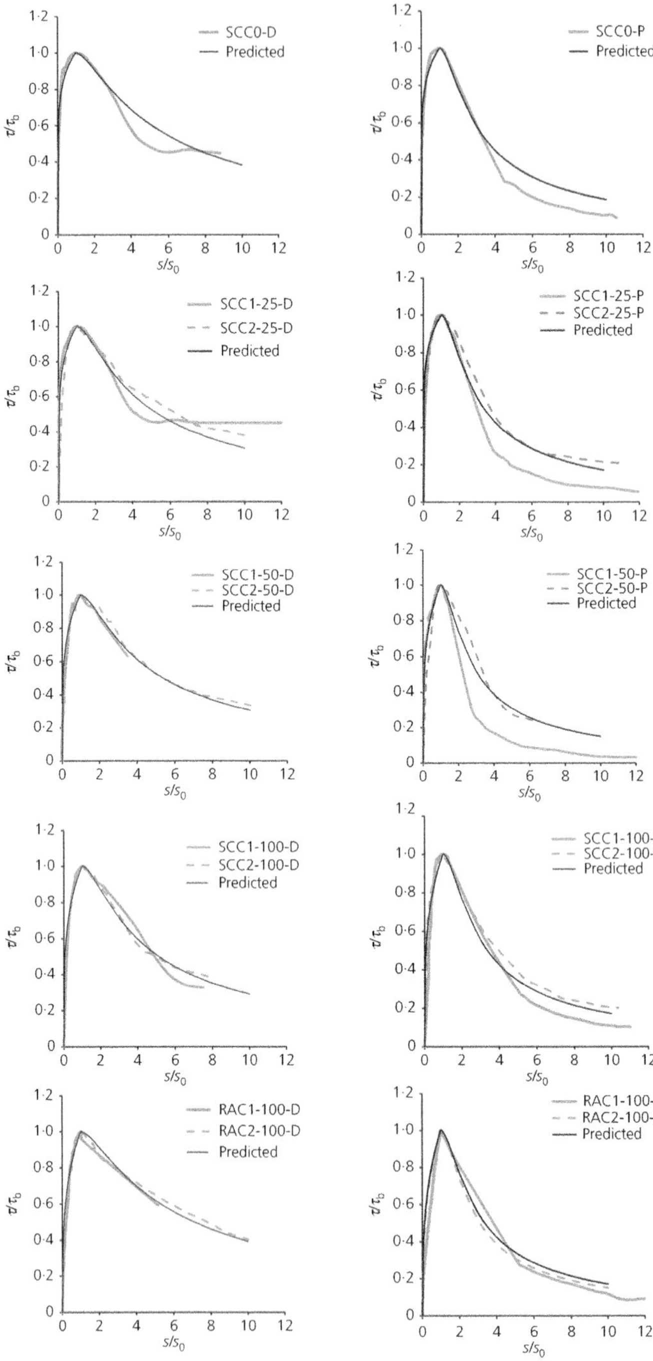

Table 4 Regression parameters α and β

Rebar types	RCA fraction: %	α	β
Deformed bar	0	0·18	0·20
	25	0·18	0·15
	50	0·20	0·30
	100	0·25	0·30
	100 (untreated)	0·25	0·20
Plain bar	0	0·15	0·54
	25	0·15	0·54
	50	0·24	0·60
	100	0·40	0·60
	100 (untreated)	0·40	0·60

in which α is a curve-fitting parameter calibrated for deformed and plain bars from the current bond test results and β is a parameter based on curve-fitting of the experimental data. The values of α and β are shown in Table 4: α and β were affected by the RCA replacement ratio and rebar type, but had almost no relationship with the RCA source. The predicted curves fitted well with the test curves, demonstrating that Equation 5 could be used to simulate the whole process of the bond–slip relationship of RA-SCC with plain and deformed bars. Xiao and Falkner (2007) and Prince and Singh (2013, 2014a, 2014b) also reported accurate predictions of the measured bond–slip relationships of their vibrated RAC concrete mixtures with the help of Equation 5, which therefore lends support to the validity of this equation for the predictive assessment of the bond–slip behaviour of vibrated RAC.

Bond strength prediction model

Various bond strength calculation models have been developed for predicting the bond strength of steel bars in concrete. Bond prediction models can either be expressed in terms of the predicted bar stress or they can be rearranged in terms of a required development/splice length. For direct comparison with test results, it is convenient to express bond prediction equations in terms of bond strength.

The tension splice provisions for deformed steel rebars of ACI 318 (ACI, 2011) are based on a study by Orangun (1977), which suggested a basic form for the bond strength equation as in Equation 6. It is also noted that Darwin *et al.* (1992) and Harajli (1994) adopted almost the same format as that suggested by Orangun (1977).

$$\tau_b = \left(0{\cdot}1 + 0{\cdot}25\frac{c_{min}}{d} + 4{\cdot}15\frac{d}{l}\right)(f'_c)^{0{\cdot}5} \qquad (6)$$

where τ_b is the bond strength (MPa), f_c' is the compressive strength of concrete (MPa), d is the bar diameter (mm), c_{min} is the minimum concrete cover (mm) and l is the embedded bar length (mm).

Xu *et al.* (1994) proposed a modified expression (in SI units) for the bond strength of a plain round bar

$$\tau_b = \left(\frac{1 + 2{\cdot}1d}{l}\right) f_t \tag{7}$$

where f_t is the direct tensile strength of concrete (MPa), taken as $0{\cdot}395 f_{cu}^{0{\cdot}55}$ or $0{\cdot}9 f_{sp}$. The model proposed by Xu *et al.* (1994) was the basis of anchorage length provisions for plain steel rebar of GB 50010 (China National Standards, 2010).

The experimental bond strengths were compared to those predicted by Equations 6 and 7, and the results are summarised in Figure 20. Overall, Equations 6 and 7 considerably underestimate all

Figure 20 Comparative performance of predicted and experimental bond strengths for deformed steel bar Orangun (1977) and plain steel bar Xu *et al.* (1994)

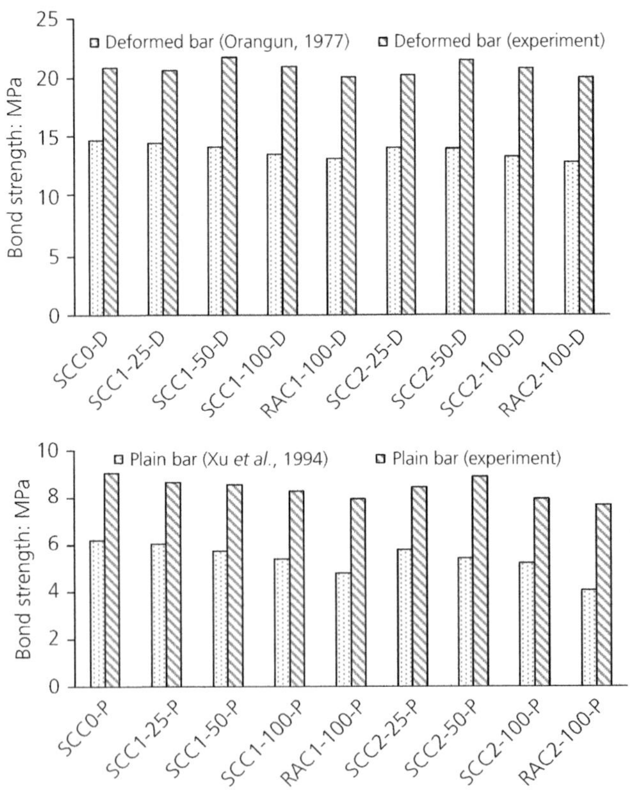

the bond strengths of the RA-SCC specimens for both deformed and plain rebars. Based on Figure 20, the bond strengths of the deformed rebars were 42–55% higher than the predicted values, while the bond strengths of the plain rebars were 42–89% higher than those predicted.

The results of the tests on SCC carried out in this work, as also observed in vibrated RAC test results in the literature, confirm that the normalised bond strength of RA-SCC was only slightly affected by the EWA of the aggregate (RCA replacement level) or the ACV (RCA type and source). Additionally, it seems that the normalised bond strengths of the RA-SCCs are as high as or higher than the values for vibrated RAC. This indicates that the anchorage lengths of steel bars in RA-SCC could be smaller than those in vibrated RAC. As a conservative measure, it is suggested that, for both plain and deformed rebars, similar anchorage lengths could be adopted for conventional vibrated concrete and SCC containing recycled concrete aggregate. Thus, the anchorage length format based on the code equations of GB 50010 (China National Standards, 2010) and ACI 318 (ACI, 2011) can also be used in RA-SCC under the assumptions of no transverse reinforcement, bottom-cast bars and normal-density concrete. However, as regards SCC containing untreated RCA of inferior quality, both the bond strength and normalised bond strength were lower than those of SCC containing the higher quality pre-treated RCA. This unfavourable factor should be carefully considered in structural design, particularly in anchorage and lapping zones.

Conclusions

The following conclusions can be drawn from the results of this experimental work.

(*a*) The general shape of the bond stress–slip curve between recycled aggregate self-compacting concrete (RA-SCC) and steel rebar was similar to that of normal concrete, which includes micro-slip, internal cracking, pull-out, descending and residual stages. The only differences were that at the beginning of loading, the bond stress–slip relationship of the control SCC had a steeper and more upward linear branch than those of the RA-SCCs. The normalised bond–slip relationship for RA-SCC could be fitted using Equation 5. Both the fraction and type of recycled coarse aggregate (RCA) have some effect on the steel–concrete bond behaviour of RA-SCC.

(*b*) Compared with the natural aggregate SCC specimens, the RA-SCC specimens containing pre-treated (acid treatment) RCA had almost the same bond strength even at higher RCA replacement levels. However, the bond strengths of RA-SCCs containing 100% untreated RCA were reduced for both deformed and plain rebars due to their lower compressive strengths, especially for the low-quality crushed aggregate RCA2.

(*c*) Compressive strength and splitting tensile strength were strongly affected by RCA replacement ratio, source and type. It was also found that there was a slight increase in normalised bond strength for deformed bars with higher RCA replacement ratios, whereas the normalised bond strength between RA-SCC and a plain rebar has no obvious relation with RCA replacement percentage. The pre-treatment of RCA can improve the tensile and compressive strengths of RA-SCC, and enhance the bond strength to some extent.

(*d*) For all RCA types and sources, as the aggregate crushing value increased (indicating lower strength aggregate), the tensile strength of the RA-SCC decreased obviously, while the decrease in normalised bond strength was very small. For both recycled aggregate concrete (RAC) specimens with different RCA replacement ratios, as the equivalent water absorption (EWA) of mixed aggregates increased, the tensile strength showed a tendency to decrease proportionally, while the normalised bond strength increased slightly or remained unchanged. It seems that the normalised bond strengths of steel bars in RA-SCC were higher than that those in vibrated RAC for the same EWA and the same steel bar. This is due to the self-compacting character of SCC compared with vibrated RAC, which reduces the amount of water in the interface between steel and concrete.

(*e*) The bond strengths of the RA-SCCs were higher than the predicted bond strengths for both deformed and plain rebars. The anchorage length of a steel rebar embedded in RA-SCC could thus be chosen as that used for normal concrete under the condition of the same compressive strength of concrete. However, the fact that the use of untreated RCA, with more old adhered mortar, decreased the bond strength of RA-SCC should be taken into consideration.

Acknowledgements

This work was sponsored by the Research Foundation of the Joint Research Center of Urban Resource Recycling Technology of the Graduate School at Shenzhen, Tsinghua University and Shenzhen Green Eco-Manufacturer High-Tech (no. URRT2014006), the Natural Science Foundation of Liaoning Province (no. 2015020593) and the Open Foundation of State Key Laboratory of Hydrology-Water Resources and Hydraulic Engineering of Hohai University (no. 2013491811). The author also thanks Mr Zhang Ji and Miss Zhang Nan for their contributions in realising the tests.

REFERENCES

ACI (American Concrete Institute) (2011) ACI 318-11: Building code requirements for structural concrete. American Concrete Institute, Farmington Hills, MI, USA.

Bingol AF and Tohumcu I (2013) Effects of different curing regimes on the compressive strength properties of self-compacting concrete incorporating fly ash and silica fume. *Materials & Design* **51(10)**: 12–18.

Butler L, West J and Tighe S (2011) The effect of recycled concrete aggregate properties on the bond strength between RCA concrete and steel reinforcement. *Cement and Concrete Research* **41(10)**: 1037–1049.

Butler L, West J and Tighe S (2014) Bond of reinforcement in concrete incorporating recycled concrete aggregates. *Journal of Structural Engineering* **141(3)**: 1–12.

Choi YW, Kim YJ, Shin HC and Moon HY (2006) An experimental research on the fluidity and mechanical properties of high-strength lightweight self-compacting concrete. *Cement and Concrete Research* **36(9)**: 1595–1602.

China National Standard (2010) GB50010-2010: Code for design of concrete structures. China Building Industry Press, Beijing, China.

Darwin D, McCabe SL, Idun EK and Schoenekase SP (1992) Development length criteria: bars not confined by transverse reinforcement. *ACI Structural Journal* **89(6)**: 709–720.

Güneyisi E, Mehmet G, Zeynep A and Yazıcı H (2014) Effect of surface treatment methods on the properties of self-compacting concrete with recycled aggregates. *Construction and Building Materials* **64(14)**: 172–183.

Harajli MH (1994) Development/splice strength of reinforcing bars embedded in plain and fibre reinforced concrete. *ACI Structural Journal* **91(5)**: 511–520.

Kim SW and Yun HD (2013) Influence of recycled coarse aggregates on the bond behavior of deformed bars in concrete. *Engineering Structures* **48(3)**: 133–143.

Murat T, Ali MA and Kambiz R (2014) Freeze–thaw resistance, mechanical and transport properties of self-consolidating concrete incorporating coarse recycled concrete aggregate. *Materials & Design* **53(1)**: 983–991.

Orangun CO (1977) A revaluation of test data on development length and splices. *ACI Structural Journal* **74(3)**: 114–122.

Prince MJR and Singh B (2013) Bond behaviour of deformed steel bars embedded in recycled aggregate concrete. *Construction and Building Materials* **49(12)**: 852–862.

Prince MJR and Singh B (2014a) Bond behaviour between recycled aggregate concrete and deformed steel bars. *Materials and Structures* **47(3)**: 503–516.

Prince MJR and Singh B (2014b) Bond strength of deformed steel bars in high-strength recycled aggregate concrete. *Materials and Structures* **48(12)**: 3913–3928.

Sindy SP, Belen GF, Javier EL and Manuel FH (2014) Bond behavior between steel reinforcement and recycled concrete. *Materials and Structures* **47(1)**: 323–334.

Su N, Hsu KC and Chai HW (2001) A simple mix design method for self-compacting concrete. *Cement and Concrete Research* **31(12)**: 1799–1807.

Tam VWY, Tam CM and Le KN (2007) Removal of cement mortar remains from recycled aggregate using pre-soaking approaches. *Resources, Conservation and Recycling* **50(1)**: 82–101.

Xiao J and Falkner H (2007) Bond behaviour between recycled aggregate concrete and steel rebars. *Construction and Building Materials* **21(2)**: 395–401.

Xu Y, Shen W and Wang H (1994) An experimental study of bond-anchorage properties of bars in concrete. *Journal of Building Structures* **15(3)**: 26–37.

Zoran JG, Gordana AT, Iva MD and Nenad SR (2010) Properties of self-compacting concrete prepared with coarse recycled concrete aggregate. *Construction and Building Materials* **24(7)**: 1129–1133.

ice Publishing

Dhir and Lye
ISBN 978-0-7277-6461-4
https://doi.org/10.1680/icetscc.64614.363

Chapter 16
Behaviour of reinforced self-consolidating concrete frames

A. Said PhD
Assistant Professor, University of Nevada Las Vegas, USA

M. Nehdi PhD, PEng
Associate Professor, University of West Ontario Canada

Multi-storey reinforced concrete (RC) structural frames represent some of the most congested structural elements. Placing and consolidating concrete in such structural frames imposes substantial challenges. This offers a unique area of application for self-consolidating concrete (SCC) because of its inherent ability to flow under its own weight and fill congested sections, complicated formwork and hard-to-reach areas. Research is, however, needed to demonstrate the ability of SCC structural frames adequately to resist vertical and lateral loads. In the present study, full-scale 3 m high beam-column joints reinforced as per the Canadian Standards CSA A23·3-94 and ACI-352R-02 were made with normal concrete (NC) and SCC. They were tested under reversed cyclic loading applied at the beam tip and at a constant axial load applied on the column. The beam–column joint specimens were instrumented with linear variable displacement transducers and strain gauges to determine load–displacement traces, cumulative dissipated energy and secant stiffness. The current paper compares the performance of reinforced NC and SCC structural frames and discusses the potential use of SCC in such structural elements. Results indicate that reducing the coarse aggregate content in SCC mixtures can reduce the contribution of the aggregate interlock mechanism to total shear resistance, which leads to more rapid deterioration under cyclic loading. Further research is needed to ensure the safety of using low coarse aggregate content in SCC in highly seismic areas and assess the safety of already existing buildings cast using SCC.

1. Introduction

Self-consolidating concrete (SCC) is a relatively recent development in concrete technology; its inception was in Japan in the late 1980s.[1] It has, however, been predicted that within the next decade, SCC would replace a significant portion of normal concrete (NC),[2,3] especially in developed countries. SCC has been generating significant interest and is gaining wider use in various projects worldwide. Its advantages over conventionally vibrated NC include: reduced labour; less noise and equipment on construction sites; and faster placement. Moreover, SCC ensures improved finish, hence reducing surface remedial costs and minimising wear and tear on formwork. It is a highly flowable yet stable concrete that can easily flow and consolidate, even in congested sections or complicated formwork, with little or no vibration and without undergoing significant segregation or bleeding.

SCC is usually produced using readily available conventional concrete materials. Its mixture proportions are based on creating high flowability while maintaining a relatively low water/cementitious materials ratio. This can be achieved through the use of high-range water reducing admixtures (HRWR) often in conjunction with rheology-modifying admixtures to ensure the stability and homogeneity of the mixture. The coarse aggregate content and its particle size gradation must be carefully controlled to enhance the flow of SCC, prevent blocking in densely reinforced sections and avoid segregation problems. The effects of the reduced coarse aggregate content, often used in formulating SCC, on its structural behaviour remain, however, largely unexplored.

A considerable portion of research performed on SCC was dedicated to its mixture proportions and rheological properties, while only a few studies focused on its structural applications. For instance, Nagai *et al.*[4] performed a research on the use of super-workable concrete in thin-walled prestressed precast concrete members. Their study showed an exceptional capacity of SCC to fill voids in heavily reinforced sections as thin as 60 mm with no significant segregation and no deleterious effects on durability. Research was also conducted on the compatibility of SCC with NC in sandwiched construction in order to reduce cost.[5] It was found that when casting structural members with SCC in congested areas and NC elsewhere in a sandwiched manner, such hybrid members behave satisfactorily.

Very limited studies were concerned with the structural performance of SCC in congested members compared with that of NC[6,7] and no study dedicated to the seismic performance of SCC was accessible in the open literature. Persson[3] performed a comparative study on NC and SCC to conclude that, for similar concrete strength, both concrete types behave similarly in terms of modulus of elasticity, creep and shrinkage. The relatively lower coarse aggregate content in SCC may, however, result in lower contribution to the shear resistance produced by aggregate interlock. The study performed by Schiessl and Zilch[9] confirmed such a behaviour through a monotonic test. Using roughness measurements, they found that crack surfaces of SCC were smoother than those of NC and that at a similar normal stress across the crack, SCC specimens exhibited lower shear stress resistance. Studies on the shear strength of high-strength concrete have yielded similar conclusions. The improvement in concrete strength was achieved through enhancement of the transition zone between aggregates and cement paste. Thus, microcracks tend to propagate through aggregates rather than around them, and fracture surfaces become therefore smoother than those in NS concrete.[10] Accordingly, the shear capacity of SCC could be a concern, especially in the case of seismic loading.

For instance, moment-resisting frames (MRFs) are usually congested areas of reinforcement in which casting and consolidating concrete represents a substantial challenge. Such frames would be among the applications most benefiting from SCC since this material needs little or no vibration to be consolidated. The nature of reversed cyclic loading of MRFs in the event of an earthquake and the resulting plastic hinging would, however, impose cautiousness when SCC is used in such applications. Although SCC has been used in several buildings such as the Millennium Point Building in Birmingham, UK[1] without reported problems, investigations on the behaviour of SCC under cyclic loading are needed to ensure that a wide implementation of this material in earthquake-resistant structures is safe.

Fig. 1 Beam–column joint failure in the Kocaeli 1999 earthquake (O'Rourke,[11] with permission)

Several recent earthquakes demonstrated that beam–column joints are vital elements in keeping structural integrity. Fig. 1[11] shows a close-up of a non-ductile failure of a beam–column joint owing to the Kocaeli earthquake and illustrates that adequate performance of such joints during earthquakes is key for ensuring structural safety.

In the present paper, the behaviour of SCC beam–column joints under reversed cyclic loading is investigated and compared with that of NC beam–column joints in terms of load–displacement envelope, cumulative dissipated energy, steel and concrete contributions to beam and joint shear resistance and secant stiffness and the use of SCC in structural frames is discussed.

2. Experimental programme

Beam–column joints can be isolated from plane frames at the points of contraflexure. The beam of the current test unit is taken to the mid-span of the bay, while the column is taken from the mid-height of one storey to the mid-height of the next storey. Two standard beam–column joints (J1 and J3) were designed as per the current CSA A23·3-94[12,13] requirements with adequate shear reinforcement in the joint area and in the hinging areas of the column and beam. The column is 3000 mm high with cross-section dimensions of 250 × 400 mm. The beam's length was 1750 mm from the face of the column to its free end with a cross-section of 250 × 400 mm. The longitudinal reinforcement used in the column was 14 M15 bars corresponding to a 2·8% reinforcement ratio (M15 is equivalent to a 16·0 mm diameter bar). The transverse reinforcement in the column was two M10 closed rectangular ties (M10 is equivalent to a 10·0 mm diameter bar).

The column ties are spaced at 80 mm inside the joint and for 500 mm above and below it (one-sixth of the floor's height) and then spaced at 125 mm for the rest of the column's height. The top and bottom longitudinal reinforcements of the beam are six M15 bars each corresponding to a 1·2% reinforcement ratio. The transverse reinforcement of the beam is M10 rectangular ties starting at 50 mm from the face of the column. The ties were spaced at 80 mm for the 800 mm adjacent to the column (equivalent to twice the beam's depth) and then spaced at 120 mm for the remaining 840 mm ending at 60 mm from the free end of the beam. The longitudinal rebar size and transverse reinforcement for the joint and hinging zones confinement satisfy current code requirements.

Reinforcement details for the tested specimens are shown in Fig. 2. NC and SCC were used to cast specimens J1 and J3, respectively. The concrete mixture proportions for both specimens are shown in Table 1. No vibration was used for casting the SCC specimen. Upon the release of the formwork, it was clear that the specimen constructed with SCC had less surface irregularities in

Fig. 2 Reinforcement details for the tested specimens (dimensions in mm)

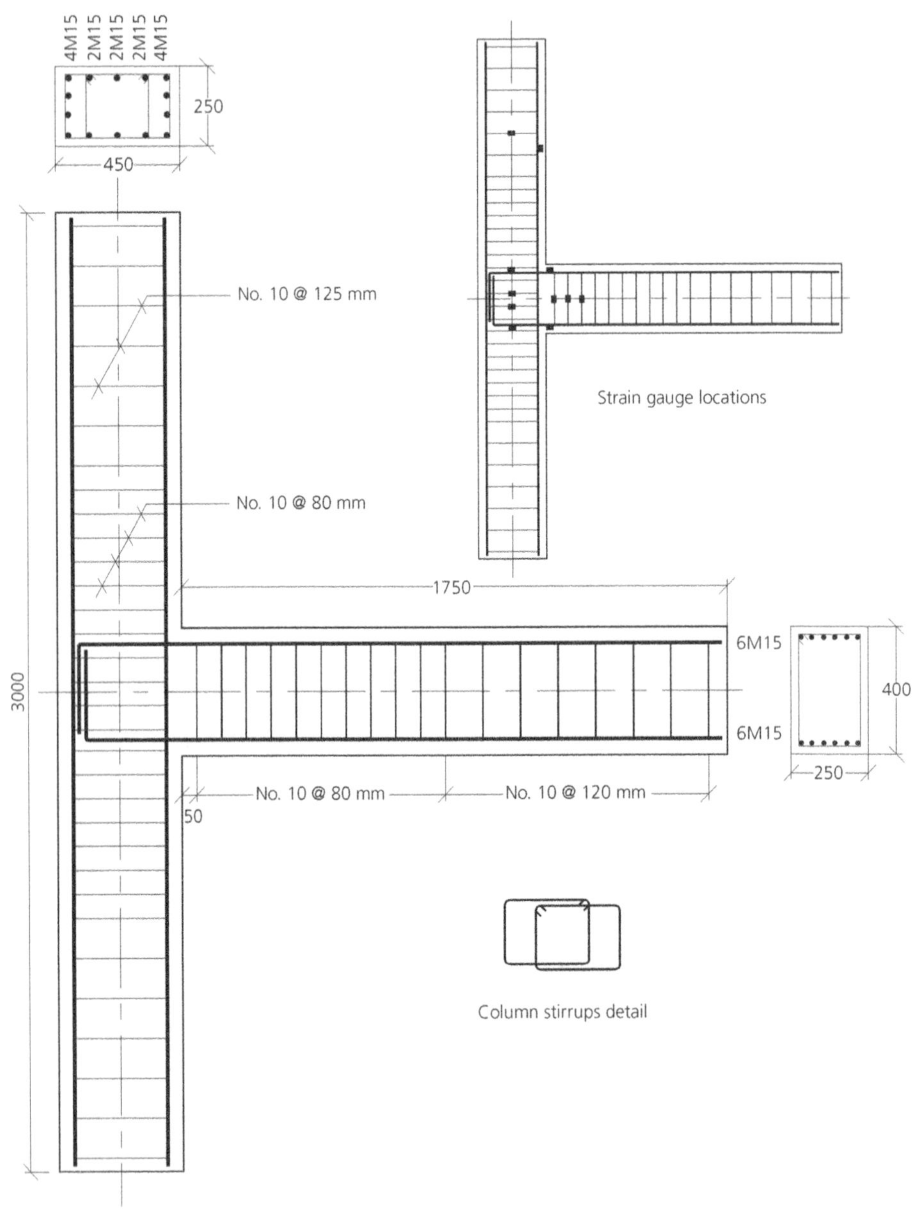

Table 1 Concrete mixture proportions for the tested specimens

	NC–J1	SCC–J3
Cement: kg/m^3	330	400
Sand: kg/m^3	790	850
Gravel: kg/m^3	1130	850
W/C	0·34	0·40
HRWR: l/m^3	–	4·0
WRA: ml/m^3	6·6	–
VMA: g/m^3	–	120
7 df'_c : MPa	37·2	33·3
28 df'_c : MPa	50·9	50·4
Slump: mm	50	–
Slump flow: mm	–	600

Fig. 3 Surface irregularities in NC specimen (J1)

comparison with the one made with NC. Fig. 3 illustrates exposed steel reinforcement which took place at several locations of specimen J1 despite the use of vibration. Specimen J3 cast without vibration did not exhibit such features.

2.1 Test set-up and procedure

The beam–column joint specimens were tested under a constant axial load of 600 kN applied on the column and reversed quasi-static cyclic loading applied at the beam tip. The selected loading pattern is intended to cause forces that simulate high levels of inelastic deformations that may be experienced by the frame during a severe earthquake. The selected load history consisted of two phases. The first phase was load-controlled followed by a displacement-controlled loading phase as shown in Fig. 4.

In the first phase of loading, two load cycles at approximately 10% of the estimated strength of the specimen were applied to check the test set-up and ensure that all data acquisition channels were functioning properly. This was followed by two load cycles reaching the concrete flexural cracking load in the beam at the column face. These in turn were followed by two cycles at the load causing initial yield in the beam measured through the load–displacement trace on the data acquisition monitor. The displacement at initial yield of the beam section at the column face, δ_y, was recorded and used in the subsequent displacement-controlled phase of loading.

The second phase of loading after first steel yield was displacement-controlled and consisted of applying the yield displacement for a second time then subsequently adding incremental multiples of the yield displacement δ_y (previously recorded at initial yield). Two load cycles were applied at each ductility level to verify the stability of the specimen. The ductility level is expressed in terms of a ductility factor, μ, which is defined as the ratio of the beam tip displacement, δ, to the displacement at first yield of the principal steel reinforcement, δ_y. The test was stopped when the load-carrying capacity of the subassemblage dropped to about 50% of its maximum value. The effect of cyclic loading on the load-carrying capacity, strength degradation, energy dissipation and joint deformation was examined.

The effect of cyclic loading on the load-carrying capacity, energy dissipation, concrete contribution to shear resistance in the beam and the joint as well as strength degradation were examined.

The specimens were placed in the test rig as shown in Fig. 5 to mimic a hinge support at the base of the column and a roller support at the top part of the column. The roller support was created using a 2 cm vertical slot, which allowed vertical deformation in the column as well as

Fig. 4 Load history for the reversed cyclic load test used in the present study

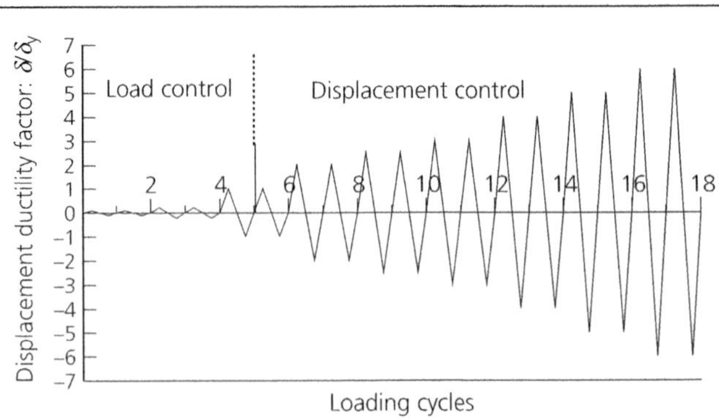

Fig. 5 View of the test rig

the transmission of the column's axial load from the hydraulic jack to the lower hinge support. The cyclic load was applied at the beam tip using a loading ram through a greased pin connection located at 1670 mm from the column face. A view and a schematic of the test set-up are shown in Figs. 5 and 6, respectively.

2.2 Instrumentation and data acquisition

Data from different monitoring devices were separated into analogue (load cells, displacement transducers) and digital (strain gauges) channels connected to the data acquisition unit, which was attached to a portable computer to record readings at a constant time interval of 5 s. The measured loads were the column's axial load and the beam's tip load. Two load cells were used for this purpose. Five linear variable displacement transducers (LVDTs) were used to measure the displacement at various locations on the specimen as shown in Fig. 6. One principal LVDT measured the beam's tip displacement and was fixed to an independent frame. Two half-inch LVDTs were used in the joint area to measure the joint's distortion. These LVDTs were fixed to previously installed bolts in the joint panel of the specimen. Two other one-inch LVDTs were

Fig. 6 Test set-up

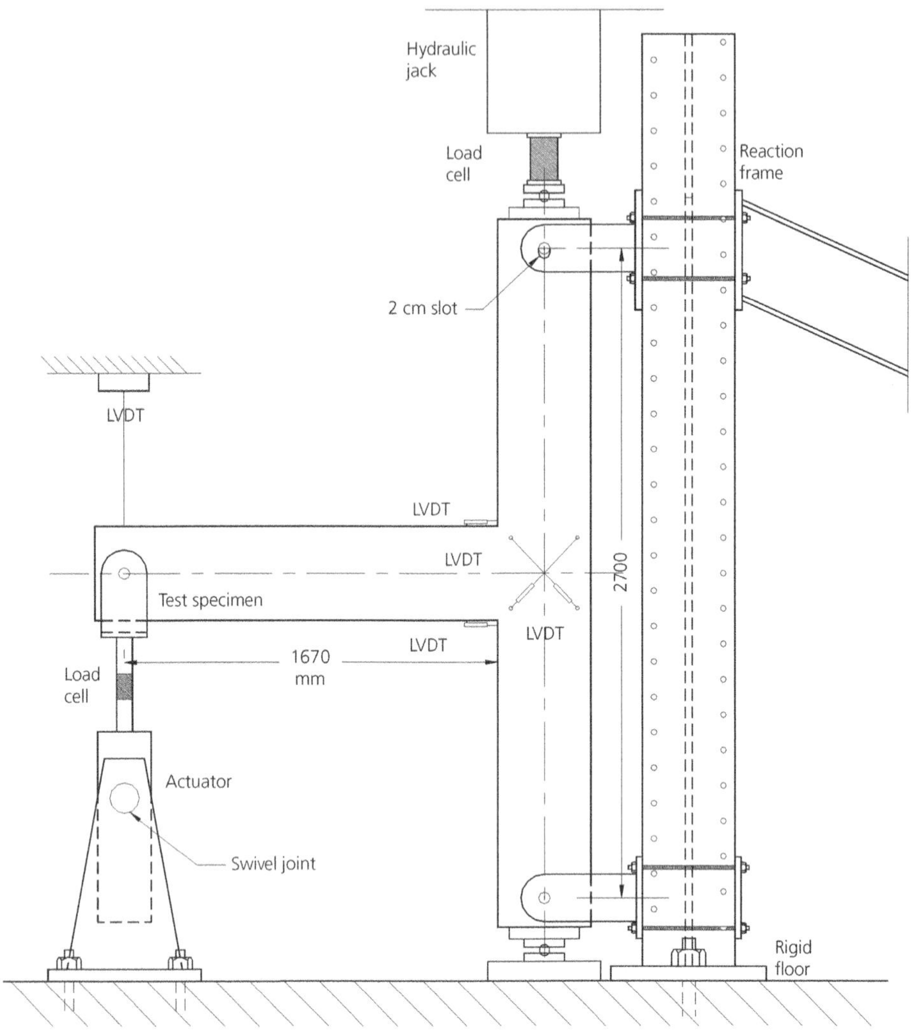

fixed on the top and bottom of the beam to measure the slippage of bars in it. Small aluminium plates were fixed for these LVDTs stems to butt against.

Strains in the reinforcing steel were monitored using electrical resistance strain gauges, which were fixed to the deformed bars after being ground to expose straight clean flat surface. The wiring was bundled to come out of the concrete at specific spots. The location of strain gauges on the longitudinal and transverse reinforcement bars in the beam and column are shown in Fig. 2.

3. Results and discussion

3.1 Behaviour of specimens

The load–storey drift plots for the NC (J1) and SCC (J3) specimens are shown in Figs. 7 and 8, respectively. First flexural cracking of the beam section subjected to maximum bending moment appeared at a beam tip load of 15 and 17 kN corresponding to a drift of 0·17% and 0·18% for specimens J1 and J3, respectively. The onset of diagonal cracks in the joint area took place at a beam tip load of 60 and 65 kN corresponding to a drift of 0·6% and 0·5% for specimens J1 and J3, respectively. Additional cracks appeared thereafter as loading progressed at a uniform spacing, but remained within a very fine width throughout the test.

For the NC specimen, the yield of the beam's longitudinal steel was reached at an average beam tip load of 107 kN and the corresponding average yield displacement was 28 mm (corresponding to a drift of 1·5%, based on push up and pull down values), whereas for the SCC specimen, the yield load was 104 kN at a displacement of 27 mm (corresponding to a drift of 1·4%).

Fig. 7 Load–storey drift relationship for NC specimen J1

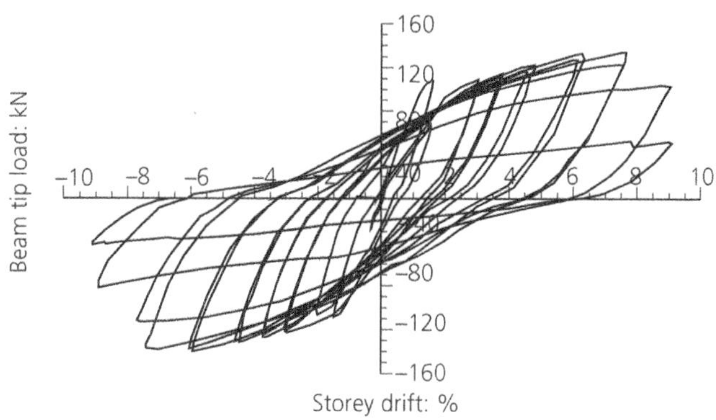

Fig. 8 Load–storey drift relationship for SCC specimen J3

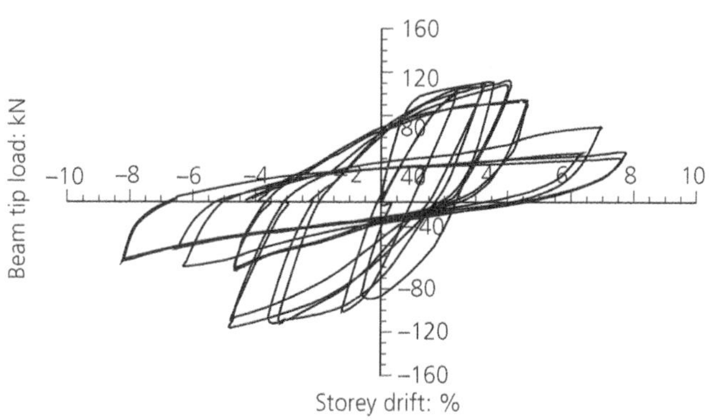

Fig. 9 Final crack pattern for NC specimen J1

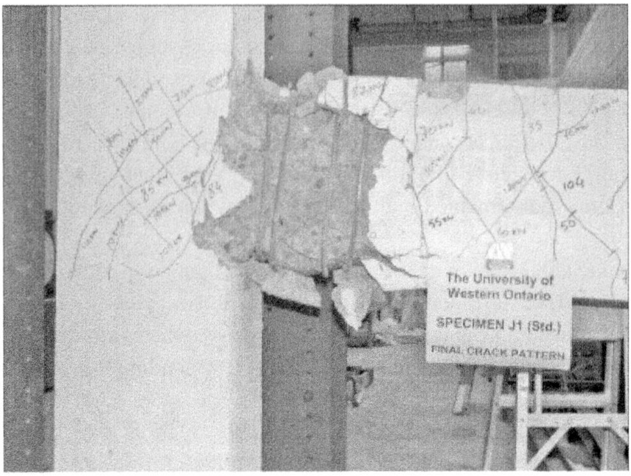

Fig. 10 Final crack pattern for SCC specimen J3

At a ductility factor of 2 (corresponding to a drift of 3·0%), both specimens developed extensive cracking in the beam along a distance equal to its depth from the face of the column. At a ductility factor of 3 (corresponding to a drift of 4·5%), the SCC specimen started exhibiting lower load-carrying capacity compared with that of the NC specimen. This became even clearer in subsequent load cycles. For both specimens the column's axial load was maintained and the joint areas were still intact, except the presence of fine diagonal cracks.

Fig. 11 Load–storey drift envelopes for the NC and SCC specimens

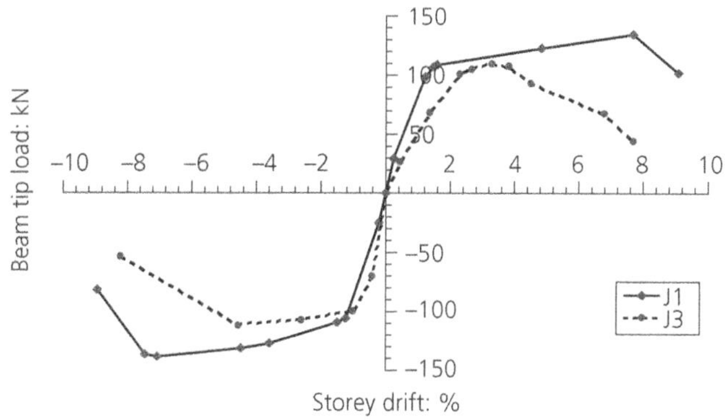

The faster decline of the load-carrying capacity of the SCC specimen could be attributed to the fact that its lower coarse aggregate reduced the contribution of friction due to aggregate interlock to the total shear resistance mechanisms, especially at high levels of displacement. Final crack patterns for the NC (J1) and the SCC (J3) beam–column joint specimens are shown in Figs. 9 and 10, respectively.

3.2 Load–storey drift envelope relationship

For each of the beam–column joint specimens, the envelope of the beam tip load–displacement relationship is plotted in Fig. 11. The SCC specimen (J3) had a comparable capacity to that of the NC (J1) specimen up to a displacement level of about 75 mm (corresponding to a 4·5% drift), which could be considered as structurally adequate. Subsequently, the reserve strength of the SCC specimen was lower and a plastic hinge formed in the beam. The maximum displacement ductility achieved by the NC specimen was 6 compared with 5 for the SCC specimen.

3.3 Cumulative dissipated energy

The capability of a structure to survive an earthquake depends on its ability to dissipate the energy input from ground motion. Despite the fact that energy input during a ground movement event is difficult to estimate, a satisfactory design should ensure a larger energy dissipation capability of the structure than the demand. The cumulative energy dissipated by the beam–column joint specimens during the reversed cyclic load tests was calculated by summing up the energy dissipated in consecutive load–displacement loops throughout the test. The energy dissipated in a cycle is calculated as the area that the hysteretic loop encloses in the corresponding beam tip load–displacement plot.

Figure 12 shows a plot of the cumulative energy dissipation versus the storey drift for the NC specimen (J1) and the SCC specimen (J3). Results show that the SCC joint had higher energy dissipation up to a ductility level of 3 corresponding to about 6% storey drift. Afterwards, the NC joint specimen showed higher energy dissipation capacity with an overall 38% superiority.

Fig. 12 Cumulative energy dissipated for the NC and SCC specimens

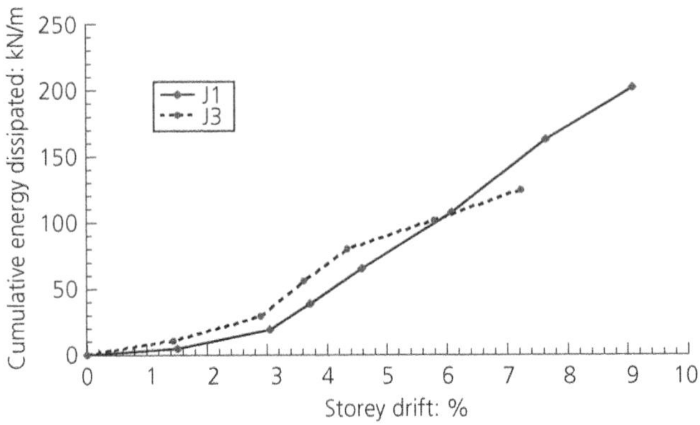

3.4 Shear in beams

The shear force applied to the beam section is resisted by the steel and concrete. Ever since Ritter[14] followed by Mörsch[15] formulated the truss analogy approach, several methods emerged to calculate the shear capacity of reinforced concrete (RC) beams based on their approach. Analytical methods in the literature such as the modified compression field theory[16] can more precisely estimate the shear capacity of RC beams and calculate the individual contribution of concrete and steel to shear strength. Despite their scientific appeal, however, such methods still remain complicated and involve intensive calculations. Accordingly, the 45° truss analogy approach remains quite widely adopted.

Typically, the shear capacity of the beam is assumed to be simultaneously supplied by concrete and the transverse steel reinforcement according to the expression:

$$V_t = V_c + V_s \tag{1}$$

where, V_t is the total shear force in the beam, V_c is the shear force resisted by concrete and V_s is the shear force resisted by the transverse steel reinforcement. Adopting the 45° truss analogy, the shear resistance provided by the transverse reinforcement can be expressed as:

$$V_s = \frac{A_v f_s d}{s} \tag{2}$$

where s is the spacing of transverse steel reinforcement, A_v is the area of transverse reinforcement over a distance s, f_s is the stress in the transverse reinforcement and d is the distance from the extreme compression fibre to tension reinforcement.

The strains measured in the stirrups were converted into stresses then used to evaluate the shear resistance of the transverse reinforcement according to equation (2). The stress in the steel, f_s, was based on the average strain obtained from three strain gauge measurements mounted on the first three stirrups in the beam from the column face.

The Canadian Standard Association concrete design provisions[12] evaluate the concrete shear strength, V_c, for nominal ductility members as

$$V_c = 0{\cdot}2\sqrt{f_c'}\,b_w d \tag{3}$$

where f_c' is the compressive strength of concrete cylinder, b_w is the beam's breadth and d is the effective beam's depth. Figs. 13 and 14 illustrate the individual contribution of concrete and steel to the shear resisted by concrete and transverse steel reinforcement for specimens J1 and J3, respectively. Fig. 13 shows that for specimen J1, the maximum concrete contribution to the beam shear occurred at a ductility level of 2 (corresponding to a drift of 3·0%) with a value of 110 kN (corresponding to $0{\cdot}18\sqrt{f_c'}\,b_w d$). At a displacement ductility factor of 6 (corresponding to a drift of 9·0%), the concrete was still resisting some of the beam shear load with a

Fig. 13 Beam shear force resisted by concrete and transverse steel reinforcement plotted against storey drift envelopes for the NC specimen J1

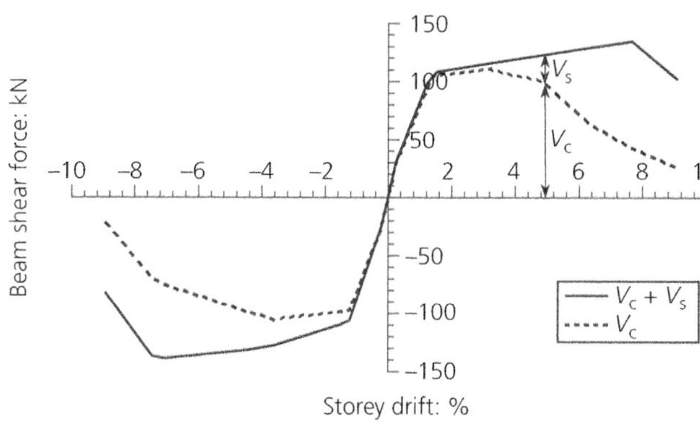

Fig. 14 Beam shear force resisted by concrete and transverse steel reinforcement plotted against storey drift envelopes for the SCC specimen J3

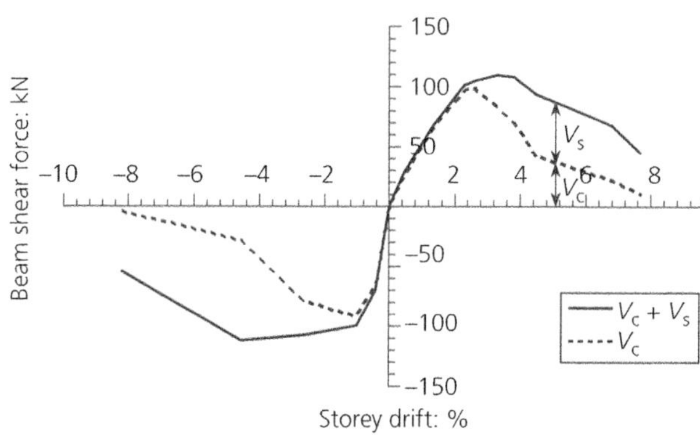

value of 26 kN (corresponding to $0 \cdot 05\sqrt{f_c'}\,b_w d$). Fig. 14 shows that for specimen J3 the maximum concrete contribution to the beam shear resistance occurred at displacement ductility of 2·0 (corresponding to a drift of 2·8%) with a value of 98 kN (corresponding to $0{\cdot}16\sqrt{f_c'}\,b_w d$). At a displacement ductility of 5·0, the contribution of concrete was minor and the transverse reinforcement resisted almost all the shear load. An examination of Figs. 13 and 14 reveals that the concrete in the NC specimen carried a higher percentage and magnitude of the shear force applied on the beam compared with that in the SCC specimen. Fig. 15, which shows the beam-tip load plotted against strain in the second stirrup in the beam from the face of the column, indicates higher levels of strain in the stirrups of specimen J3 compared with that in specimen J1 at corresponding loading levels. An analogy to the case of the smooth crack surface in SCC is high-strength concrete (HSC). A study performed by Walraven[17] indicated that the shear friction in HSC beams is up to 35% lower than that in NSC beams. This is attributed to the fact that the shear fracture surface in HSC members is usually transgranular (propagates across coarse aggregates) and is therefore smoother than that in NSC members.[11] It is worth noting that the beams' plastic hinges formed at equal distances from the column face for both specimens.

Fig. 15 Beam tip load plotted against second steel stirrup strain for specimen: (a) J1 and (b) J3

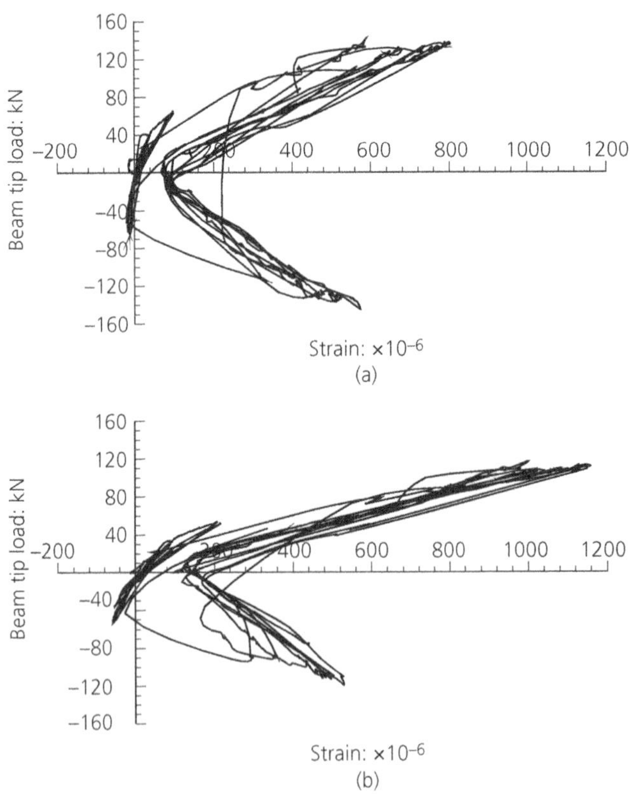

3.5 Joint behaviour

The shear force input to the joint from the beam is expressed as

$$V_{jt} = A_s \, f_s - V_{col} \tag{4}$$

where V_{jt} is the total joint shear force, A_s is the cross-sectional area of the tensile reinforcement in the beam, f_s is the stress in the longitudinal tensile reinforcement in the beam, and V_{col} is the column shear force that can be obtained by static equilibrium of the subassemblage (Fig. 16(b)) as follows

$$V_{col} = \frac{P(L/2)}{H} \tag{5}$$

The joint shear is resisted by both the concrete and the transverse reinforcement. The shear force resisted by the transverse reinforcement can be calculated as

$$V_{js} = A_v \, f_{st} \tag{6}$$

where A_v is the total area of the transverse reinforcement in the joint, and f_{st} is the stress in the joint transverse reinforcement calculated using the averaged strains measured by the strain gauges on stirrups in the joint.

Fig. 16 Exterior beam–column subassemblages in (a) displaced frame and (b) test rig

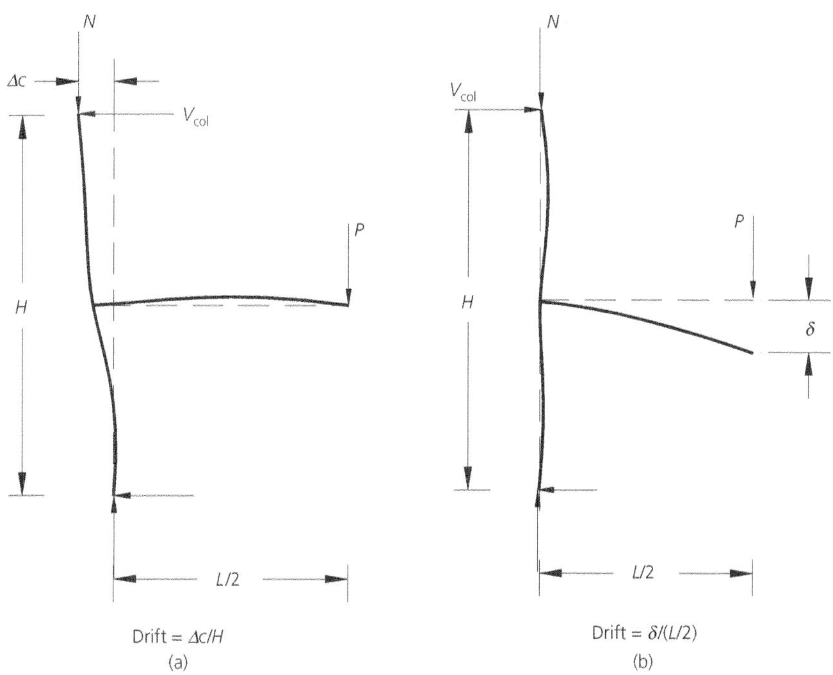

The joint shear is plotted against storey drift for tested specimens and is presented in Figs. 17 and 18. Strain gauge readings revealed that the first values of strain were recorded at the onset of cracks in the joint area for both specimens. The results given in Fig. 17 for specimen J1 show that concrete carried the majority of joint shear up to approximately 1·5% drift, corresponding to the yielding of steel in the beam section. As the test progressed, the contribution of steel slightly increased, but the concrete contribution remained significant. Fig. 18 similarly shows that concrete was the major contributor to joint shear resistance. As the test progressed, the beam capacity declined resulting in a decrease in joint shear.

It was noted that the steel contribution to shear resistance was lower in the joint panel area than that at the beam hinging area. This is mainly attributed to the extent of cracking and the associated slippage along the cracks, which is significantly higher in the case of the hinging

Fig. 17 Joint shear force resisted by concrete and transverse steel reinforcement plotted against storey drift envelopes for the NC specimen J1

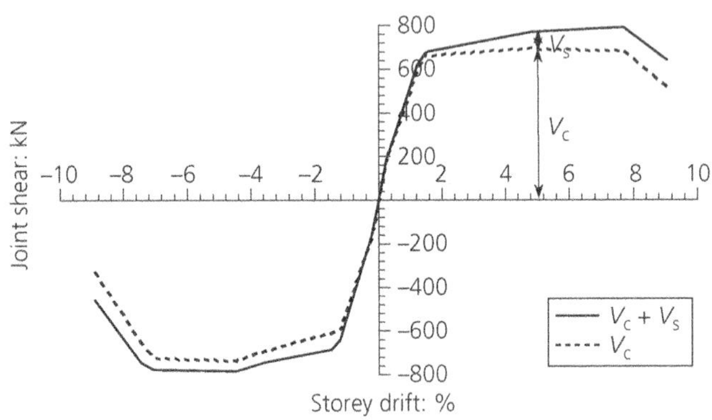

Fig. 18 Joint shear force resisted by concrete and transverse steel reinforcement plotted against storey drift envelopes for the SCC specimen J3

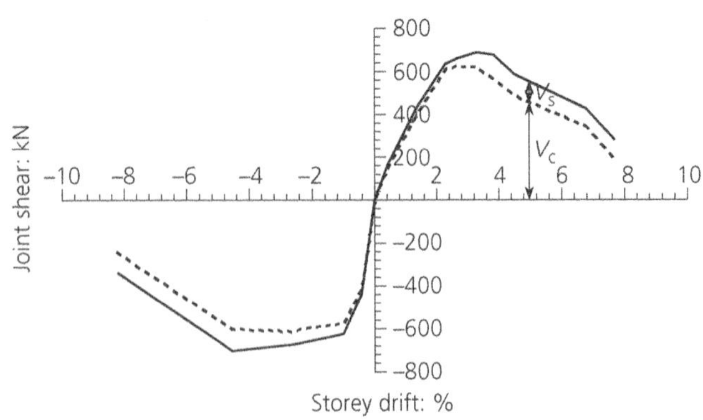

Fig. 19 Secant stiffness–storey drift for the NC and SCC specimens

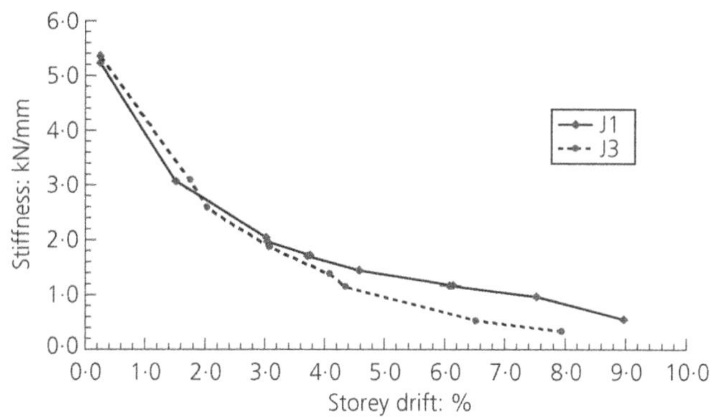

zone of the beam than that in the slightly cracked joint panel. It is worth mentioning that for both specimens, at the same levels of joint shear input, the calculated joint deformations were comparable.

3.6 Secant stiffness

Secant stiffness is evaluated as the peak-to-peak stiffness of the beam tip load–displacement relationship. It is calculated as the slope of the line joining the peak of positive and negative loads at each given cycle. The secant stiffness is an index of the response of the specimen during a cycle and its strength degradation from one cycle to the following cycle.

Figure 19 shows plots of the secant stiffness for the NC and SCC beam–column joint specimens plotted against the storey drift. The storey drift is calculated as shown in Fig. 16 by relating the subassemblage deformation in the test rig to the actual displaced frame case. An examination of the plots indicates that the SCC specimen (J3) had higher initial stiffness. After a drift angle of 2%, the NC standard specimen (J1) had higher stiffness up to the end of the test. Nonetheless, the SCC specimen (J3) exhibited stable strength degradation up to failure. The maximum drift achieved was 9·0% and 7·9% for specimens J1 and J3, respectively.

4. Conclusions

Reversed cyclic loading tests were performed on full-scale beam–column joint specimens to compare the performance of NC and SCC in moment-resisting frames. Based on experimental observations and analysis of test results, the following conclusions can be drawn.

1. SCC beam–column joints have comparable load capacity to that of NC joints up to a certain ductility level. At high ductility levels, SCC specimens may not maintain the same load-carrying capacity as NC specimens. While this could be attributed to the fact that the lower coarse aggregate content in SCC reduced the contribution of the aggregate interlock to the total shear resistance mechanism, further research is required fully to understand this behaviour.

2. The performance of SCC under shear stress in the joint panel was comparable to that of NC in terms of cracking and deformations.
3. The SCC beam–column joint specimen performed adequately in terms of the mode of failure and ductility requirements, assuming that the expected minimum drift requirement is 3%, as recommended in the literature for ductile frame buildings.[18]
4. The contribution of concrete to the shear resistance in the beam of the NC beam–column specimen was higher than that of the SCC specimen, especially at high drifts. In the case of the joint, the concrete contribution to shear resistance remained significant until the end of the test for both specimens. The difference in both cases was attributed to the smaller slippage along cracks in the joint area compared with that in the beam hinging zone.
5. Further studies are needed to investigate the behaviour of SCC under cyclic loading in hinging zones and to quantify aggregate interlock contribution mechanisms for different coarse aggregate contents and maximum aggregate size along with the effect of other mixture design parameters.

Acknowledgements

The continuous support of the Natural Science and Engineering Research Council of Canada (NSERC), the Ontario Centres of Excellence (OCE), the Canada Foundation for Innovation (CFI) and the Ontario Innovation Trust (OIT) to the research programmes of Professor Moncef Nehdi has been essential for the development of this research.

REFERENCES

1. Nagamoto N. and Ozawa K. Mixture properties of self-compacting, high-performance concrete. *Proceedings of the Third CANMET/ACI International Conference on Design and Materials and Recent Advances in Concrete Technology*, ACI SP 172, Kuala Lumpur, Malaysia, 1997, pp. 627–637.
2. Campion M. J. and Jost P. Self-compacting concrete—expanding the possibilities of concrete design and placement. *Concrete International*, 2000, **22**, No. 4, 31–34.
3. Henderson N. Self compacting concrete at millennium point. In *Concrete for the Construction Industry*, The Concrete Society, UK, 2000, pp. 26–27.
4. Nagai T., Kojima T. and Miura T. Application of high strength/superworkable concrete to thin-wall prestressed concrete products. *Magazine of Concrete Research*, **51**, No. 3, 153–162.
5. Ho D. W. S., Sheinn A. M. M. and Tam C. T.. The sandwich concept of construction with SCC. *Cement and Concrete Research*, 2001, **31**, No. 9, 1377–1381.
6. Khayat K. H., Paultre P. and Tremblay S.. Structural performance and in-place properties of self-consolidating concrete used for casting highly reinforced columns. *ACI Materials Journal*, 2001, **98**, No. 5, 371–378.
7. Sonebi M., Bartos P. and Tamimi A. Flexural response of reinforced beams cast with self-compacting concrete. *Proceedings of the Second International Symposium on Self-Compacting Concrete*, Tokyo, Japan. RILEM Publications S.A.R.L., 2001, pp. 517–526.
8. Persson B. A comparison between mechanical properties of self-compacting concrete and the corresponding properties of normal concrete. *Cement and Concrete Research*, 2001, **31**, No. 2, 193–198.

9. Schiessl A. and Zilch K. The effect of the modified composition of SCC on shear and bond behaviour. *Proceedings of the Second International Symposium on Self-Compacting Concrete*, Tokyo, Japan. RILEM Publications S.A.R.L., 2001, pp. 501–506.
10. El Chabib H., Nehdi M. and Said A. Predicting shear capacity of NSC and HSC slender beams without stirrups using artificial intelligence. *Computers and Concrete*, 2005, **2**, No. 1, 79–96.
11. O'rourke T. D. *The Turkey, Taiwan, and Mexico City Earthquakes: Lessons Learned.* Hearing of the Subcommittee on Basic Research U.S. House of Representatives Committee on Science, October 20, 1999. See http://mceer.buffalo.edu/outreach/Congress/TRtestimony.asp.
12. Canadian Standards Association. *Design of Concrete Structures. CSA A23·3-94*, CSA, Rexdale, Ontario, Canada, 1994, 209 p.
13. American Concrete Institute/American Society of Civil Engineers. *Recommendations for Design of Beam-Column Connections in Monolithic Reinforced Concrete Structures.* Joint ACI/ASCE Committee 352, Technical Committee Document 352R-02, 37 pp.
14. Ritter W. Die Bauweise Hennebique. *Schweizerische Bauzeitung*, 1899, **33**, No. 7, 59–61.
15. Mörsch E. *Concrete Steel Construction*, 3rd edn (1st edn in 1902). Engineering News Publishing Co, New York, 1909, 368 p. Translated into English by E. P. Goodrich of Der Eisenbetonau.
16. Vecchio F. J. and Collins M. P. The modified compression field theory for reinforced concrete elements subjected to shear. *ACI Journal Proceedings*, 1986, **83**, No. 2, 219–231.
17. Walraven J. C. Shear friction in high strength concrete. In *Progress in Concrete Research*, Delft University of Technology, The Netherlands, 1995, Vol. 4, pp. 57–65.
18. Corley W. G. Ductility of columns, walls, and beams—how much is enough? *Proceedings of the Thomas Paulay Symposium on Recent Developments in Lateral Force Transfer in Buildings*, ACI SP-157, La Jolla, California, 1995, pp. 331–350.

ice Publishing

Dhir and Lye
ISBN 978-0-7277-6461-4
https://doi.org/10.1680/icetscc.64614.383

Chapter 17
Structural behaviour of reinforced concrete beams containing crumb rubber and steel fibres

Mohamed K. Ismail
Faculty of Engineering and Applied Science, Memorial University of Newfoundland, St. John's, NL, Canada

Assem A. A. Hassan
Faculty of Engineering and Applied Science, Memorial University of Newfoundland, St. John's, NL, Canada (corresponding author: ahassan@mun.ca)

Amgad A. Hussein
Faculty of Engineering and Applied Science, Memorial University of Newfoundland, St. John's, NL, Canada

This paper presents experimental work to investigate the strength and cracking characteristics of optimised self-consolidating and vibrated rubberised concrete mixtures with/without steel fibres (SFs) using large-scale reinforced concrete beams. The test beams were cast with varying percentages of crumb rubber (CR) (0 to 35%), SF volume fractions (0, 0·35 and 1%) and SF lengths (35 and 60 mm). The performance of some design codes and published empirical equations was evaluated in predicting the shear capacity and first cracking moment of the tested beams. The results showed that the inclusion of SFs could alleviate the reduction in the shear capacity and first cracking moment that resulted from the addition of CR. In addition, combining CR and SFs contributed to developing sustainable concrete beams with high deformability, reduced self-weight and improved shear capacity. The composite effect of CR and SFs also helped to narrow the developed cracks and change the failure mode from brittle shear failure into ductile flexural failure, particularly for the SF volume of 1% (35 mm length). Comparisons of the predicted and experimental results indicate that most of the proposed equations can satisfactorily estimate the shear strength, but overestimate the first cracking moment.

Notation

b	width of beam (mm)
d	effective depth of beam (mm)
f_{ck}	characteristic compressive cylinder strength of concrete at 28 d
f_{ctm}	mean value of axial tensile strength of concrete
f'_c	28 d compressive strength

f'_{cf}	characteristic flexural tensile strength of concrete
h	height of cross-section of beam
I_g	second moment of area of gross section
I_u	second moment of area of uncracked transformed section
M_{cr}^{exp}	experimental first cracking moment
M_{cr}^{pred}	predicted first cracking moment
V_{exp}	experimental shear capacity
V_{pred}	predicted shear capacity
V_u	ultimate shear load
v_{nz}	normalised shear strength (v_{nz})
x_u	distance from neutral axis of the section to the extreme top fibre
y_t	distance from the centroidal axis of the gross section to the extreme tension fibre
Z	section modulus ($=I/y$)

Introduction

Reutilisation of scrap rubber in concrete is being increasingly adopted as a possible safe technique to exploit huge volumes of this problematic waste material and produce a new type of sustainable concrete. In the last two decades, extensive research has been conducted to evaluate the impact of using rubber on the properties of concrete. From the data available in the literature, adding rubber to small-scale samples (i.e. cubes, cylinders and prisms) appeared to boost their impact resistance, damping properties, energy dissipation and ductility, and also reduced the self-weight of the concrete (Ismail and Hassan, 2016a; Najim and Hall, 2012; Zheng *et al.*, 2008). However, all the studies confirmed that the compressive and tensile strengths of these concrete composites were negatively affected by the inclusion of rubber. This was related to (*a*) the poor strength of the rubber–mortar interface and (*b*) the significant difference between the modulus of elasticity of rubber particles and that of cement mortar (Najim and Hall, 2010).

Some studies involved the use of rubber in large-scale structural members. For example, Ganesan *et al.* (2013) investigated the influence of shredded rubber on the performance of self-consolidating rubberised concrete (SCRC) beam–column joints under monotonic and cyclic loading. They reported that the inclusion of rubber increased the energy absorption capacity, crack resistance and ductility of the concrete. Youssf *et al.* (2015) observed similar effects when rubber was added; the hysteretic damping ratio and energy dissipation of concrete columns subjected to cyclic loading were found to be improved when crumb rubber (CR) was added, but the ultimate lateral load slightly decreased. Other researchers (Ismail and Hassan, 2015a; Najim and Hall, 2014) have reported that reinforced beams made with SCRC and vibrated rubberised concrete (VRC) showed lower flexural capacity and stiffness compared with beams made without rubber, while the deformation capacity and energy absorption increased with an increase in rubber content. Sadek and El-Attar (2014) reported that the inclusion of rubber in masonry walls increased their toughness, deformation capacity and capability to withstand post-failure loads. None of the studies conducted have reported on the shear behaviour of rubberised concrete, although cracking behaviour and the cracking mechanism, which affect the shear behaviour of concrete (Taylor, 1974), may be influenced by the addition of CR.

Reinforcing rubberised concrete with steel fibres (SFs) can effectively contribute to the development of composites with higher tensile strength, toughness and impact strength (Abadel *et al.*, 2015; Altun and Aktas, 2013; Ismail and Hassan, 2016a; Khaloo *et al.*, 2014; Olivito and Zuccarello, 2010). SFs also play a significant role in transferring stress along cracked sections, which in turn provides a residual strength to concrete, and hence increases the shear capacity of concrete beams, as reported by other researchers (Ding *et al.*, 2011, 2012; Tahenni *et al.*, 2016). In addition, the fibres' bridging mechanism can help to restrain the development of cracks and limit crack openings (Ganesan *et al.*, 2014; Lee *et al.*, 2016).

The beneficial effects that come from using SFs in rubberised concrete can be maximised when self-consolidating concrete (SCC) is used. Coupling rubber and SFs in SCC can offer a novel type of sustainable concrete that enjoys the desirable properties of SCC in the fresh state and also has higher ductility, toughness, impact resistance and reduced self-weight in the hardened state. However, optimising the fresh properties of SCC incorporating rubber and SFs is a big challenge. As shown in past studies, the addition of rubber has negative effects on the flowability, passing ability and stability of SCC mixtures (Güneyisi, 2010; Ismail and Hassan, 2015b; Topçu and Bilir, 2009). Similarly, SCC with the addition of SFs appears to suffer from high interference between the SFs and coarse aggregates, which leads to a significant decrease in the flowability and passing ability of mixtures (Ding *et al.*, 2008; Iqbal *et al.*, 2015).

The universal design codes provide several models to estimate the shear strength of concrete. Some of these models were proposed based purely on empirical procedures, such as ACI 318 (ACI, 2008) and Eurocode 2 (EC2) (BSI, 2005), while other models were derived from fundamentals presented in the modified compression field theory, such as the Aashto LRFD code (Aashto, 2007). With much easier assumptions, the shear design model of CSA A23.3 (CSA, 2004) was developed based on simplified modified compression field theory. However, these code models do not include the effect of SFs, which has encouraged many investigators to adapt the code equations or propose new empirical models in attempts to estimate the contribution of SFs to the shear capacity of reinforced concrete beams. For example, Ashour *et al.* (1992) and Khuntia *et al.* (1999) adjusted the ACI 318 equation (ACI, 2008) to account for the contribution of SFs and incorporated new terms such as the SF fraction volume, aspect ratio and geometry. On the other hand, other researchers have developed shear models based on their own experiments and results available in the literature. Sharma (1986) proposed a simple model derived from experiments conducted in his investigation and data obtained from Baston *et al.* (1972). Although this model was validated by a significant number of studies, it lacks some factors that are known to greatly affect the shear strength, such as fibre volume, fibre aspect ratio and flexural reinforcement ratio. These factors, in addition to others such as the effect of the fibres' bond stresses, were covered in a model proposed by Narayanan and Darwish (1987). Imam *et al.* (1994) modified the equation proposed by Bazant and Sun (1987), which was suggested to predict the shear strength of conventional concrete beams in order to consider the contribution of fibres. The modified model appeared to be different from most of the other available models in that the effect of maximum aggregate size on the shear strength of concrete beams was considered. An equation proposed by Zsutty (1971) was adapted by Kwak *et al.* (2002) to account for the effect of tensile strength on arching action, with a new term added to consider the effect of fibres on the shear strength of beams.

Despite active research aiming to evaluate the performance of rubberised concrete with/without SFs, no study has covered the effect of CR with/without SFs on the shear behaviour of large-scale beams, especially when SCC is used. In addition, evaluating the capability of the proposed equations to predict the performance of rubberised concrete beams with/without SFs in terms of shear capacity and first cracking moment is lacking, because this type of concrete is a novel material. The study reported here was thus carried out to highlight the effect of CR with/without SFs on the shear behaviour and cracking characteristics of large-scale reinforced concrete beams with no stirrups. The performance of some code design equations and different proposed models on the shear behaviour and cracking moment was also evaluated. The tested beams were fabricated with varying percentages of CR (0 to 35%), different SF volume fractions (0, 0·35 and 1%) and different SF lengths (35 and 60 mm).

Research significance

Developing structural concrete with low-density waste rubber aggregate and reinforced with SFs can contribute to the development of sustainable concrete with a high-impact resistance and deformability and reduced self-weight. This research was conducted to evaluate the influence of CR with/without SFs on the shear capacity and cracking characteristics of reinforced concrete beams made with SCC and VRC. The research also compared the experimental results with the performance of the code design equations commonly used to predict the shear strength and cracking moment of concrete beams. A review of the literature indicated that no study has covered the shear behaviour and cracking characteristics of large-scale rubberised concrete beams with/without SFs, especially when made with SCC. This investigation is therefore intended to help further the understanding of the shear behaviour and cracking characteristics of SCRC, VRC, steel-fibre-reinforced SCRC (SFSCRC) and steel-fibre-reinforced VRC (SFVRC) in order to extend the possible applications of CR in the construction industry.

Experimental programme

Material properties

Type GU Canadian Portland cement, similar to type 1 ASTM C150 cement (ASTM, 2012b), metakaolin (MK) conforming to ASTM C618 class N (ASTM, 2012a) and fly ash similar to ASTM C618 type F (ASTM, 2012a) were used as binders for all the developed mixtures. Natural crushed stone of 10 mm maximum size and natural sand were used for the coarse and fine aggregates, respectively, with a specific gravity of 2·6 and water absorption of 1%. In this study, the fine aggregate was partially replaced by CR aggregate, which had a maximum size of 4·75 mm, a specific gravity of 0·95 and negligible water absorption. Two types of SFs with hooked ends (Dramix 3D) were used: the first type were 35 mm long and 0·55 mm in diameter with an aspect ratio of 65; the second type were 60 mm long and 0·9 mm in diameter, again with an aspect ratio of 65. Each type of SF had a tensile strength of 1050 MPa, Young's modulus of 210 GPa and density of 7·85 kg/m^3. A polycarboxylate-based high-range water-reducing admixture (HRWRA) similar to ASTM C494 type F (ASTM, 2013), with a specific gravity of 1·2, volatile weight of 62% and pH of 9·5, was used to achieve the required workability of the test mixtures. The steel bars used in the constructed beams had an average yield stress of 417 MPa.

Concrete mixtures

In total, 12 concrete mixtures were developed to cast 12 reinforced concrete beams. These mixtures were derived from a comprehensive investigation conducted by the authors on optimising the fresh and mechanical properties of SCRC, VRC, SFSCRC and SFVRC mixtures. The results of this study indicated that a total binder content of at least 550 kg/m^3 with a minimum water/binder (w/b) ratio of 0·4 should be used to obtain acceptable slump flow and no visual sign of segregation. Replacement of cement by 30% fly ash and 20% MK was found to be an optimum choice for adjusting the mixtures' viscosity, in which a good particle suspension was achieved with reasonable flowability. In addition, 20% MK contributed to the development of mixtures with enhanced mechanical properties, suitable for structural applications (Ismail and Hassan, 2016b). The coarse to fine aggregate ratio was kept constant at 0·7 for all the tested mixtures.

The chosen mixtures were used in two stages of work and details of all the mixtures are given in Table 1.

Table 1 Mixture designs for tested beams

Beam	Mixture	Cement: kg/m^3	MK: kg/m^3	Fly ash: kg/m^3	CA: kg/m^3	FA: kg/m^3	CR: kg/m^3	SF: kg/m^3	Density: kg/m^3
SCRC/SFSCRC mixtures									
B1	SCC-0CR	275	110	165	620·3	886·1	0·0	–	2246
B2	SCC-5CR	275	110	165	620·3	841·8	16·2	–	2207
B3	SCC-15CR	275	110	165	620·3	753·2	48·6	–	2128
B4	SCC-25CR	275	110	165	620·3	664·6	80·9	–	2041
B5	SCC-5CR-0·35SF	275	110	165	616·5	836·7	16·1	27·48	2217
B6	SCC-15CR-0·35SF	275	110	165	616·5	748·6	48·3	27·48	2138
VRC/SFVRC mixtures									
B7	VC-25CR	275	110	165	620·3	664·6	80·9	–	2048
B8	VC-35CR	275	110	165	620·3	576·0	113·3	–	2014
B9	VC-35CR-0·35SF	275	110	165	616·5	572·5	112·6	27·48	2040
B10	VC-35CR-1SF	275	110	165	609·6	566·0	111·4	78·50	2073
B11	VC-35CR-0·35LSF	275	110	165	616·5	572·5	112·6	27·48	2040
B12	VC-35CR-1LSF	275	110	165	609·6	566·0	111·4	78·50	2073

Note: All mixtures had a 0·4 w/b ratio and a 0·7 coarse/fine aggregate ratio.
CA, coarse aggregate; FA, fine aggregate; SF, steel fibre; LSF, long steel fibre.

The first stage included (*a*) four SCRC mixtures with CR percentage varying from 0 to 25% (mixtures 1–4), developed to evaluate the effect of CR on the shear behaviour and cracking of SCRC beams, and (*b*) two SFSCRC mixtures with 0·35% SFs (35 mm) having 5% and 15% CR (mixtures 5 and 6), developed to investigate the effect of combining CR and SFs on the shear behaviour and cracking of SCRC beams. Mixtures 1–6 were developed with the maximum percentage of CR (with/without SFs) that could be used in SCC mixtures to successfully fulfil the self-compactability criteria, as per the European guidelines for self-compacting concrete (Efnarc, 2005). Further increasing the percentage of CR and/or SFs in these mixtures (1–6) led to a significant reduction in the passing ability below the acceptable limits (H_2/H_1 of L-box $\geq$ 0·75) (Efnarc, 2005). Similarly, in this stage, fibres longer than 35 mm could not be used because high blockage occurred in the L-box, which made it difficult for mixtures to pass through the rebars.

For these reasons, stage 2 was constructed to evaluate the effect of using higher percentages of CR (with/without SFs) and longer fibres on the strength and cracking of the tested beam. Unlike SCC mixtures, vibrated concrete does not require high flowability and passing ability. Therefore, it was possible to combine a maximum of 35% CR and 1% SF in the VRC mixtures in stage 2. Different SF lengths (35 and 60 mm) could also be tested in this stage. The second stage thus included (*a*) two VRC mixtures with 25% and 35% CR (mixtures 7 and 8), (*b*) two SFVRC mixtures with 35% CR having 0·35% and 1% SFs (35 mm length) and (*c*) two SFVRC mixtures with 35% CR having 0·35% and 1% SFs (60 mm length). The maximum percentage of CR (35%) in the second stage was chosen based on the 28 d compressive strength: a further increase in CR content resulted in a significant reduction in compressive strength. However, a VRC mixture with 25% CR was included in this study to investigate the influence of concrete type (i.e. SCRC compared to VRC).

The tested beams/mixtures were designated by concrete type (whether SCC or vibrated concrete (VC)), percentage of CR, and the volume and size of SFs used (see Table 1). For example, a beam/mixture using SCC, 15% CR and 0·35% short SFs (35 mm length) is labelled as SCC-15CR-0·35SF and a beam/mixture using VC, 35% CR and 1% long SFs (60 mm length) is labelled as VC-35CR-1LSF.

The behaviour of the SCC/SCRC/SFSCRC mixtures in the fresh state was evaluated using slump flow, V-funnel, L-box and sieve segregation tests according to Efnarc (2005). A slump test was used to assess the workability of VRC/SFVRC according to ASTM C143 (ASTM, 2015). The percentage of air content in all the developed mixtures was measured in the fresh state according to ASTM C231 (ASTM, 2014). Tests for 28 d compressive strength and splitting tensile strength (STS) were conducted using 100 mm diameter × 200 mm height concrete cylinders according to ASTM C39 (ASTM, 2011a) and C496 (ASTM, 2011b), respectively. The compressive strength and STS tests were implemented after the sample had been exposed to a curing condition similar to that used

Table 2 Fresh and mechanical properties of tested mixtures

Beam	Mixture	T_{50}: s	V-funnel T_0: s	L-box ratio: H_2/H_1	SR: %	Slump: mm	Air: %	HRWRA: l/m^3	f_c': MPa	STS: MPa
SCRC/SFSCRC										
B1	SCC-0CR	2	7·0	0·91	2·1	–	1·5	3·43	65·6	4·0
B2	SCC-5CR	2·4	8·5	0·88	2·7	–	2·0	3·43	58·4	3·7
B3	SCC-15CR	3	10·6	0·82	5·8	–	3·1	3·75	48·4	3·3
B4	SCC-25CR	3·4	14·3	0·77	8·3	–	4·6	3·75	38·4	2·8
B5	SCC-5CR-0·35SF	2·6	9·8	0·8	2·9	–	2·4	4·63	59·2	4·4
B6	SCC-15CR-0·35SF	3·3	12·1	0·75	6·0	–	3·5	4·63	49·5	3·9
VRC/SFVRC										
B7	VC-25CR		–			180	3·0	3·18	40·3	2·8
B8	VC-35CR		–			145	3·3	3·18	29·7	2·5
B9	VC-35CR-0·35SF		–			185	3·0	3·64	31·1	3·2
B10	VC-35CR-1SF		–			85	3·1	3·64	32·4	4·4
B11	VC-35CR-0·35LSF		–			170	3·2	3·64	30·7	3·4
B12	VC-35CR-1LSF		–			80	3·4	3·64	31·5	4·7

SR, segregation resistance.

for the test beams. The results of the fresh properties, 28 d compressive strength and STS of the tested mixtures are given in Table 2.

Shear test setup, instrumentation and loading procedure

All beams were cast without shear reinforcement and tested under a four-point symmetrical vertical loading condition until shear failure (Figure 1). A single load was applied through a hydraulic actuator (with a capacity of 500 kN) and then distributed into two-point loads acting on the top surface of the beam. The shear span-to-depth ratio was kept constant at 2·5 for all tested beams. The mid-span deflection of the tested beams was measured using a linear variable differential transformer. The beams were loaded gradually, with a constant loading rate up to the first cracking load and the failure load. After each stage of loading, the cracks were marked and a crack detection microscope (60× magnification with 0·02 mm least count) was used to accurately measure their widths. The overall behaviour of the beams, including the cracking characteristics (patterns, widths, heights and angles) and failure modes, was observed

Figure 1 Typical test setup, dimensions and reinforcement of tested beams

and sketched for all beams (see Figure 2). The results obtained from the shear testing of the 12 beams are presented in Table 3.

Discussion of test results

Experimental load–deflection curve

Figure 3 shows the load–mid-span deflection curves of the tested beams. Figure 3(a) shows that the beams made with CR (B2, B3 and B4) appeared to behave similarly to the beam B1 without CR. In general, up to the first cracking load, all the tested beams showed elastic behaviour where the deflection was linearly proportional to the applied load. Beyond the first crack, the curves were almost linear but with a slightly lower slope as the beams' stiffness decreased due to the formation of macrocracks. With a further increase in applied load, the curves deviated from linearity and a higher rate of deflection was exhibited until the occurrence of failure. From Figure 3(a), it can also be seen that increasing the percentage of CR from 0 to

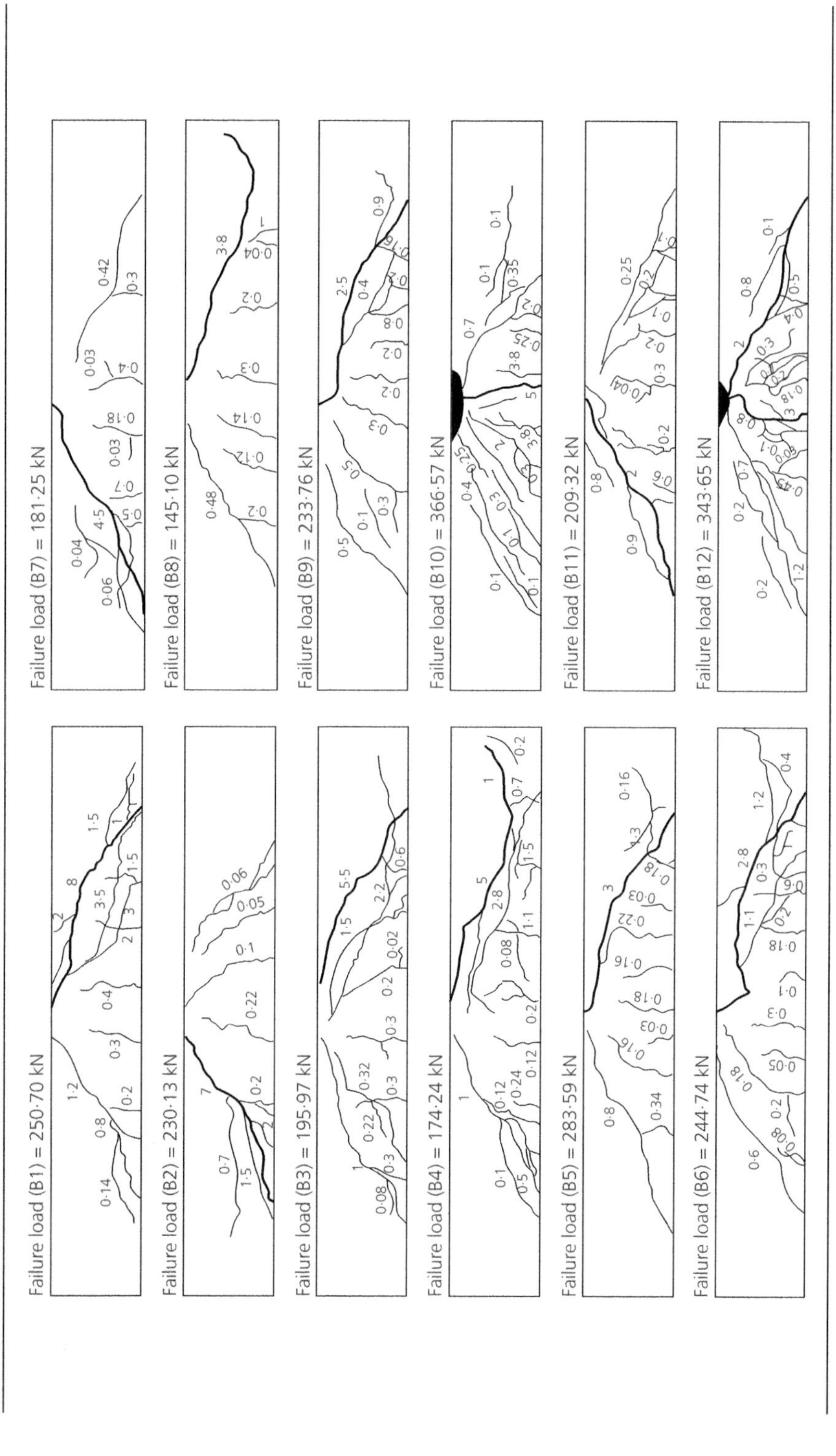

Figure 2 Crack patterns of tested beams at failure (crack widths in mm)

Table 3 Results of beam shear tests

Beam	Experimental first cracking moment: kN.m	Ultimate shear load, V_u: kN	Normalised shear strength, v_{nz}	Failure mode	At failure		
					Number of cracks	Maximum crack width: mm	Failure angle: degrees
SCRC/SFSCRC							
B1	10·9	125·35	0·31	Shear	9	8	37
B2	9·7	115·07	0·30	Shear	10	7	28
B3	8·0	97·99	0·29	Shear	11	5·5	33
B4	6·5	87·12	0·28	Shear	11	5	30
B5	11·0	141·80	0·37	Shear	10	3	31
B6	9·1	122·37	0·35	Shear	12	2·8	32
VRC/SFVRC							
B7	6·7	90·63	0·29	Shear	8	4·5	27
B8	5·3	72·55	0·27	Shear	9	3·8	21
B9	6·2	116·88	0·42	Shear	9	2·5	28
B10	7·7	183·29	0·65	Flexure	12	5	–
B11	5·8	104·66	0·38	Shear	11	2	26
B12	7·2	171·83	0·62	Flexure-shear	13	3	–

25% (B1 to B4) appeared to increase the deformability of the SCRC beams at a given load, which indicates a reduction in beam stiffness with increasing CR content. Similar behaviour was observed in the VRC beams when the percentage of CR was increased from 25 to 35% (B7 to B8), as shown in Figure 3(b). Such a decrease in stiffness was due to replacing the conventional sand with flexible rubber particles, which in turn decreased the overall stiffness of the tested SCRC and VRC beams. Comparing the load–deflection curve of the SCRC beam (B4) with its counterpart VRC beam (B7) in Figure 3(b) reveals that, up to almost 70% of the failure load, both beams showed comparable behaviour but beyond this load level the SCRC beam (B4) experienced slightly higher deformation than the VRC beam (B7).

Assessing the effect of including 0·35% SFs, Figures 3(c) and 3(d) show that the SFSCRC (B5 and B6) and SFVRC (B9) beams with 0·35% SFs presented a slightly increased stiffness and maximum deflection at ultimate load compared with their counterpart beams without SFs (B2 and B3, and B8, respectively). However, adding a higher volume of fibres (1%) increased the beams' stiffness and allowed the beams to sustain higher ultimate load accompanied with larger corresponding deflections (B10 and B12 compared with B8). This indicates a less brittle behaviour and an obvious improvement in the energy absorption of the tested beams (the area under the load–mid-span deflection curve up to failure load).

Figure 3 Experimental load–mid-span deflection responses

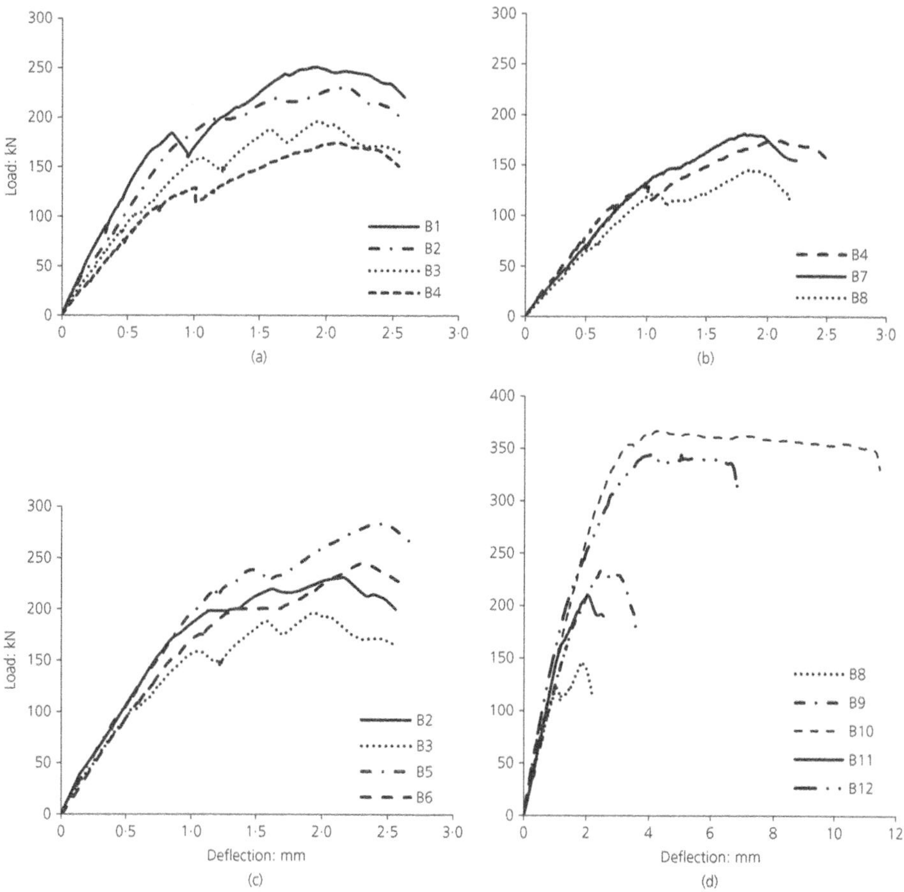

Cracking behaviour

Effect of CR

As shown in Figure 2 and Table 3, all the SCC/SCRC beams (B1–B4) exhibited a brittle shear failure mode that occurred upon the formation of a major shear diagonal crack starting at the middle height of the beam near one of the supporting points and extending towards the loading zone (compression zone). The failure pattern of beams B2–B4 was characterised by a relatively higher number of cracks and a smaller maximum crack width compared to the beam with no CR (B1). For example, increasing the percentage of CR from 0 to 25% raised the number of cracks at failure from 9 to 11, while the maximum diagonal crack width decreased from 8 mm to 5 mm. Such observations may be related to the fact that the increased deformability of the beams due to inclusion of CR encourages the development of higher levels of cracking in terms of the number of cracks and crack widths. Since the tensile strength of concrete is negatively affected by the inclusion of CR, a larger number of cracks is most likely to develop instead of continuous widening of one localised crack, thus lowering the overall crack widths. Comparing the SCRC and VRC beams (B4 and B7, respectively), it can be seen that the SCRC beam showed larger crack widths and a higher number of cracks at the failure stage, as shown in Table 3 and Figure 2. This finding may be attributed to the higher ultimate deflection exhibited by the SCRC beam compared with its VRC beam counterpart, as illustrated earlier.

Effect of combining SFs with CR

Figure 2 and Table 3 show that the inclusion of 0·35% SFs (35 or 60 mm) did not change the failure mode of the beams: both SFSCRC beams (B5 and B6) and SFVRC beams (B9 and B11) failed in a shear mode, similar to their SCRC and VRC counterpart beams without SFs (B2 and B3 and B8, respectively). However, the SFs played a significant role in stitching the cracks and developing bridging mechanisms along the crack faces, which in turn limited the crack openings and delayed their propagation. For example, in the SFSCRC beams with 0·35% SFs and 5% and 15% CR (B5 and B6, respectively) the maximum diagonal crack width at failure appeared to decrease by 57·1% and 49·1%, respectively, compared with beams with no SFs (B2 and B3). A similar effect for SFs was confirmed in the SFVRC beams (B9 or B11 compared with B8). Increasing the volume of fibres to 1% further limited the diagonal cracks, in addition to changing the beam failure mode from brittle shear failure to flexural and flexural-secondary shear mode, as shown in B10 and B12, respectively, compared with B8.

Shear strength

Since the beam mixtures had different 28 d compressive strengths (as shown in Table 2) due to the inclusion of CR with/without SFs, the ultimate shear load (V_u) was normalised to account for the effect of different compressive strengths. Knowing that the shear strength is proportional to the square root of the compressive strength of the concrete (f'_c), the normalised shear strength (v_{nz}) was calculated using

$$v_{nz} = \frac{V_u}{bd\sqrt{f'_c}} \tag{1}$$

Effect of CR

The normalised shear strengths of all the tested beams are shown in Table 3: the normalised shear strengths of the SCC beams decreased as the percentage of CR increased. Increasing the percentage of CR from 0 to 25% (B1–B4) reduced the normalised shear strength up to 9·1%. Similarly, in the VRC beams (B7 and B8) the normalised shear strength appeared to decline with increases in the percentage of CR. This behaviour may highlight the contribution of CR to decaying the tensile strength of concrete within the region of shear, which in turn allowed diagonal cracks to develop in the beam at relatively lower loads, eventually causing failure. However, the inclusion of CR beneficially contributed to producing semi-lightweight concrete (CSA, 2004); for example, varying the percentage of CR from 15 to 35% achieved mixtures with a density ranging from 2128 to 2014 kg/m^3. By comparing the performances of VRC and SCRC (B7 compared with B4), the results show that both VRC and SCRC beams had a comparable normalised shear strength.

Effect of combining SFs with CR

Table 3 shows that including SFs increased the normalised shear strength. For instance, adding 0·35% SFs to SCRC mixtures improved the normalised shear strength by an average of 23%, as shown by comparing B5 and B6 with B2 and B3, respectively. The reasons for such an increase could be related to (*a*) the contribution of fibres to restricting the widening and propagation of diagonal cracks, thus increasing the aggregate interlock and preserving a bigger uncracked compression zone, and increasing the shear capacity of the beams (Ding *et al.*, 2011) and (*b*) fibres perpendicular to the diagonal cracks may act as aggregate particles with a very elongated shape (Tahenni *et al.*, 2016), which can develop a special type of aggregate–mortar interlock, thus improving the shear transfer capacity. The beneficial impact of SFs on enhancing the normalised shear strength of beams was also confirmed in the VRC beams. The effect of SFs appeared to be more pronounced in beams with high percentages of CR (B9) than those with low CR content (B5 and B6). Adding 0·35% SFs to beams with 35% CR raised the normalised shear strength by 57·5% (B9 compared with B8), and this was more than twice the improvement observed in B5 and B6 with 5% CR and 15% CR, respectively. This may be attributed to the fact that at a higher CR content (35%) the rubber–concrete composite became weaker and more susceptible to the fibres' mechanism in arresting crack growth and improving the aggregate interlock, which then greatly increased the shear capacity of the beams. A similar effect was observed in the STS results: at low CR contents (5% and 15%), adding 0·35% SFs (35 mm) increased the tensile strength by 15·6–17·2%, while with higher CR content (35%) this increase reached 29·1% (Table 2). Increasing the volume of SFs to 1% led to further improvement in the normalised shear strength, reaching up to 242% as much as beams with no SFs (B8). The results also indicate that adding 60 mm long SFs increased the normalised shear strength of beams, but this increase was less than that observed with use of the 35 mm SFs. Using 0·35% and 1% of 60 mm SFs (B11 and B12, respectively) resulted in increases in the normalised shear strength of 42% and 230%, compared with 57·5% and 242% increases in beams with 35 mm SFs (B9 and B10, respectively). These results may be attributed to the fact that, for a given volume of fibres, the number of single fibres increases as the diameter and/or length of fibres decrease, which may raise the probability of single fibres being oriented perpendicularly to the diagonal cracks. This could lead to greater bridging/stitching actions along cracks, which boosts the contribution of fibres to the shear capacity of beams. It is worth noting that combining 15% CR and 0·35% SFs (35 mm) in B6 resulted in a normalised shear strength 12·4% higher than that of the control beam, while a combination of 35% CR and 1% SFs (VRC beams B10 or B12) helped to produce semi-lightweight concrete beams with a normalised shear strength almost twice the value obtained in the beam with no CR (B1).

Theoretical predictions of shear strength

SCRC and VRC beams

Four design codes that have gained great approbation worldwide – namely ACI 318 (ACI, 2008), EC2 (BSI, 2005), Aashto LRFD (Aashto, 2007) and CSA A23.3 (CSA, 2004) – were used to predict the ultimate shear capacity of the tested SCC, SCRC and VRC beams. It should be noted that since these code models do not include the effect of SFs in their calculations, only beams without SFs were compared with the predictions from these codes. Details of the equations used are shown in Table 4.

Table 4 Theoretical models of the shear strength for beams with/without SFs and with no stirrups

Code	Model
ACI (2008)	$V_c = \left(0{\cdot}16\lambda\sqrt{f'_c} + 17\rho\frac{d}{a}\right) bd \leq 0{\cdot}29\sqrt{f'_c}\,bd$ f'_c = cylinder compressive strength (MPa) λ = lightweight concrete modification factor (1·0 for normal-weight concrete, 0·85 for sand–lightweight concrete, and 0·75 for all lightweight concrete) ρ = tensile reinforcement ratio d = effective depth of beam (mm) a = shear span (mm) b = width of beam (mm)
EC2	$V_c = [0{\cdot}18\,\eta\,K(100\rho fc')^{1/3}]bd$ η = factor to account for lightweight concrete ($\eta = 0{\cdot}4 + 0{\cdot}6\rho/2200$) K = size effect factor ($K = 1 + \sqrt{200/d} \leq 2{\cdot}0$)
CSA (2004)	$V_c = \beta\lambda\sqrt{f'_c}\,bd_v$ β = factor accounting for shear resistance of cracked concrete d_v = effective shear depth (the greater of 0·9 times the beam depth or 0·72 times the beam height)
Aashto (2007)	$V_c = 0{\cdot}083\beta\lambda\sqrt{f'_c}\,bd_v$
Investigator	
Narayanan and Darwish (1987)	$v_u = \left[2{\cdot}8\frac{d}{a}\left(0{\cdot}24\left(\frac{f'_c}{20-\sqrt{F}} + 0{\cdot}7 + \sqrt{F}\right) + 80\rho\frac{d}{a}\right) + 0{\cdot}41\,\tau F\right]$ For $a/d \leq 2{\cdot}8$ F = fibre factor = $[V_f(l_f/d_f)\,D_f]$ V_f = fibre volume l_f/d_f = fibre aspect ratio D_f = the bond factor dependent on the shape of the SFs (0·5 for circular section plain fibre, 0·75 for crimped fibre or hooked fibre, and 1 for indented fibre) τ = fibre–matrix interfacial bond strength, taken as 4·15 MPa based on recommendations of Swamy *et al.* (1974)
Khuntia *et al.* (1999)	$v_u = \left[\left(0{\cdot}167\left(2{\cdot}5\frac{d}{a}\right) + 0{\cdot}25\,F\right)\sqrt{f'_c}\right]$ For $a/d \leq 2{\cdot}5$ (ACI Code Modification) D_f = 2/3 for plain and round, 1·0 for hooked or crimped fibres

Table 4 (*Continued*)

Code	Model
Ashour *et al.* (1992)	$v_u = \left[\left(0.7\sqrt{f'_c} + 7F\right)\frac{d}{a} + 17.2\rho\frac{d}{a}\right]$ ACI Code Modification
Sharma (1986)	$v_u = \left[\frac{1}{2}f'_t\left(\frac{d}{a}\right)^{0.25}\right]$ f'_t = splitting tensile strength of concrete (MPa)
Imam *et al.* (1994)	$v_u = \left[0.6\Psi\sqrt[3]{\omega}\left((f'_t)^{0.44} + 275\sqrt{\frac{\omega}{(a/d)^{0.44}}}\right)\right]$ where, $\Psi = \frac{1+\sqrt{5.08/d_a}}{\sqrt{1+d/(25\,d_a)}}$ = size effect d_a is the maximum aggregate size mm; ω = reinforcement factor = $\rho(1+4F)$; the bond coefficient (D_f) for the F factor is taken as 1·0 for hooked fibres = 0·9 for deformed fibres = 0·5 for smooth fibres
Kwak *et al.* (2002)	$v_u = \left[3.7\left(3.4\left(\frac{d}{a}\right)\left(\frac{f'_c}{20-\sqrt{F}} + 0.7 + \sqrt{F}\right)^{2/3}\left(\rho\frac{d}{a}\right)^{1/3}\right) + 0.8\left(0.41\,\tau F\right)\right]$ For $a/d \leq 3.4$

Figure 4(a) shows the ratio of experimental to predicted shear capacity (V_{exp}/V_{pred}) for each tested beam. The figure shows that all the code equations highly underestimated the shear strength of SCC, SCRC and VRC beams. Looking closely at the four models, it can be seen that EC2 (BSI, 2005) had the lowest V_{exp}/V_{pred} ratio (i.e. the most accurate predictions). These ratios ranged from 1·10 to 1·39 with a mean and standard deviation (SD) of 1·25 and 0·10, respectively. CSA A23.3 (CSA, 2004) exhibited lower V_{exp}/V_{pred} ratios for B1 and B2 compared with those obtained by the ACI 318 (ACI, 2008) and Aashto LRFD (Aashto, 2007) methods, while the CSA code appeared to be more conservative than ACI 318 and Aashto LRFD for B3 and B4, and B7 and B8. This is attributed to the fact that the predictions of CSA A23.3 for B3, B4, B7 and B8 were subjected to a reduction by a factor of $\lambda = 0.85$ as the density of concrete used in those beams fell within the range 1850–2150 kg/m^3 (which is classified as a semi-lightweight concrete in CSA A23.3 (CSA, 2004)). Unlike the CSA code, for the case of replacing the total volume of fine aggregate by lightweight sand, ACI 318 (ACI, 2008) and Aashto LRFD (Aashto, 2007) recommend that the reduction factor of $\lambda = 0.85$) be applied. Since the rubber partially replaced the fine aggregate, the reduction factor λ was linearly interpolated between 0·85 and 1. The results also indicated that increasing the CR content appeared to decrease the V_{exp}/V_{pred} ratios for all the design codes. This is due to the fact that the equations do not take the impact of CR into account, which may indicate a need for further investigations to highlight the rubber–concrete composite's contribution to the shear strength of beams.

Figure 4 Ratio of experimental shear strength to shear strength predicted by (a) code design equations and (b) researchers' models

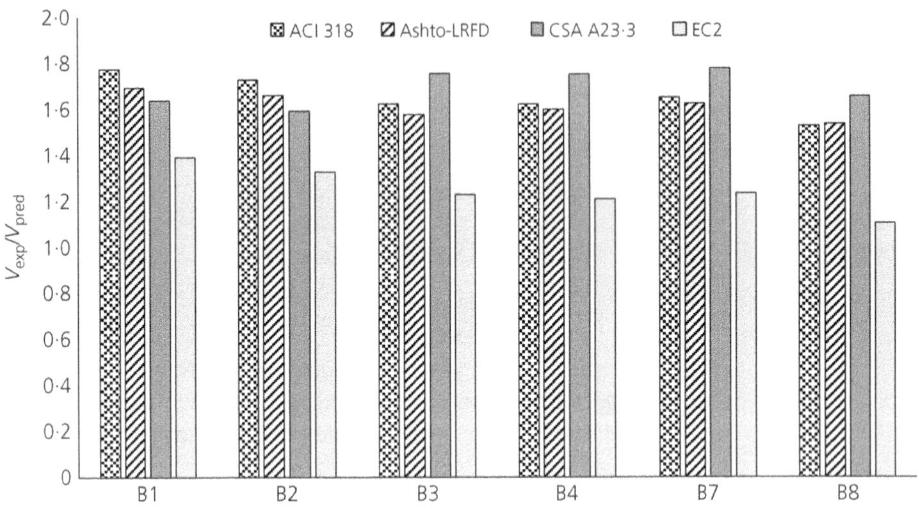

(a)

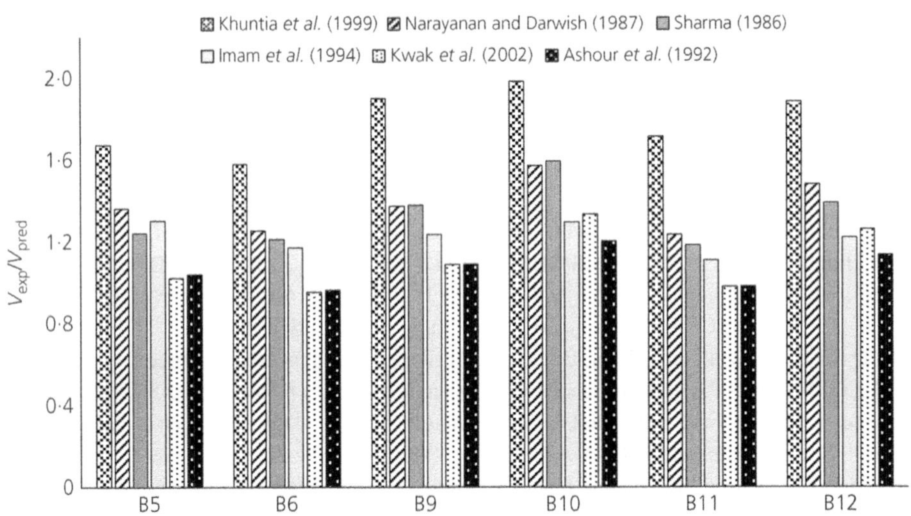

(b)

SFSCRC and SFVRC beams

The ultimate shear strengths of SFSCRC and SFVRC beams obtained from the conducted experiments were compared with those predicted by six of the existing models available in the literature; these models include the effect of SFs in their calculations. Details of the equations used are shown in Table 4.

The V_{exp}/V_{pred} ratio for each beam is shown in Figure 4(b). The model developed by Khuntia *et al.* (1999) appears to be the most conservative model of those used here, showing V_{exp}/V_{pred} ratios ranging from 1·58 to 1·98. The models developed by Narayanan and Darwish (1987), Imam *et al.* (1994), and Kwak *et al.* (2002) provided better predictions, with V_{exp}/V_{pred} in the range 1·23–1·57, 1·11–1·30 and 0·95–1·33, respectively. The closest predictions were from the model proposed by Ashour *et al.* (1992), with V_{exp}/V_{pred} in the range 0·96–1·20. Despite the simplicity of the model proposed by Sharma (1986), which does not consider the fibre volume, fibre aspect ratio and/or reinforcement ratio, the model was able to conservatively predict the shear strength of beams with V_{exp}/V_{pred} in the range 1·18–1·59. Examining the predicted values of each model, it can be clearly seen that increasing the percentage of CR resulted in a reduction in V_{exp}/V_{pred} ratio. Further research is thus required to include the effect of CR in these equations in order to safely predict the contribution of CR to the shear strength of fibre–concrete composites.

Cracking moment

Table 3 lists the cracking moments (M_{cr}^{exp}) associated with the first flexural crack. It should be noted that the first flexural crack load was visually observed and then compared/verified with values associated with the change in slope of the load–deflection obtained from the test. The observed cracking moment was also compared with predicted values (M_{cr}^{pred}), calculated based on various codes as follows.

According to ACI 318 (ACI, 2008)

$$M_{cr}^{pred} = f_r \frac{I_g}{y_t} \tag{2}$$

where $f_r = 0{\cdot}62\lambda\sqrt{f'_c}$ for normal-weight concrete (λ is taken as equal to 1 for normal-density concrete and 0·85 for lightweight sand), y_t is the distance from the centroidal axis of the gross section to the extreme tension fibre and I_g is the second moment of area of the gross section (the steel bars are not considered).

As per CSA A23.3 (CSA, 2004)

$$M_{cr}^{pred} = f_r \frac{I_g}{y_t} \tag{3}$$

where $f_r = 0{\cdot}6\lambda\sqrt{f'_c}$ for normal-weight concrete (λ is taken as equal to 1 for normal-density concrete and 0·85 for semi-low-density concrete (density in the range 1850–2150 kg/m^3).

According to AS 3600 (SA, 1988)

$$M_{cr}^{pred} = Zf'_{cf} \tag{4}$$

where f'_{cf} is the characteristic flexural tensile strength of the concrete ($=0{\cdot}6\sqrt{f'_c}$) and Z ($=I/y$) is the section modulus, which is calculated based on the uncracked transformed section, referring to the extreme fibre at which cracking occurs.

EC2 (BSI, 2005) gives

$$M_{cr}^{pred} = f_{ctm}\frac{I_u}{(h - x_u)} \tag{5}$$

where f_{ctm} is the mean value of axial tensile strength of concrete ($=0{\cdot}3f_{ck}^{0{\cdot}67}$, f_{ck} is the characteristic compressive cylinder strength of the concrete at 28 d), I_u is the second moment of area of the uncracked transformed section, x_u is the distance from the neutral axis of the section to the extreme top fibre and h is the height of the cross section of the beam.

SCC, SCRC and VRC beams

As shown in Table 3, the first cracking moment was negatively affected by the inclusion of CR. Varying the CR content from 0 to 25% reduced M_{cr}^{exp} by 40·4% (B1–B4). Such a reduction is attributed to the significant decay in the tensile strength of the concrete when the CR content increased, as shown in the STS test results (Table 2). A similar trend was observed in the SFSCRC and VRC beams (B5–B8), in which M_{cr}^{exp} decreased as the CR content increased.

Figure 5 presents the experimental to predicted cracking moment ratio ($M_{cr}^{exp}/M_{cr}^{pred}$) for each tested beam. The predicted cracking moments in this figure were calculated using the code equations based on either a tensile strength derived proportionally from the compressive strength (as proposed by each code design) or using the experimental STS. Figure 5(a) shows $M_{cr}^{exp}/M_{cr}^{pred}$ in which M_{cr}^{pred} was calculated based on compressive strength. For the beam with no CR (B1), using the equations proposed by ACI 318, CSA A23.3, AS 3600 and EC2 (Equations 2–5) appeared to overestimate the cracking moment, showing $M_{cr}^{exp}/M_{cr}^{pred}$ ratios ranging from 0·75 to 0·86. Similar observations were reported by other researchers (Fathifazl *et al.*, 2009) when the ACI 318 equation was used to predict the cracking moment of conventional reinforced concrete beams. Figure 5(a) also shows that the inclusion of CR seemed to heighten the error of the codes' predictions. For example, the VRC beam with 35% CR (B8) showed $M_{cr}^{exp}/M_{cr}^{pred}$ in the range 0·52–0·73 (for all code equations). This finding highlights the influence of CR on reducing the tensile strength of concrete.

Using the experimental STS in the code equations to calculate M_{cr}^{pred} (instead of that derived based on compressive strength) generally appeared to exhibit better predictions, as shown in Figure 5(b). For instance, the $M_{cr}^{exp}/M_{cr}^{pred}$ ratios of B1 ranged from 0·93 to 1·05, while B8 (with 35% CR) exhibited $M_{cr}^{exp}/M_{cr}^{pred}$ ratios of 0·68–0·96. The CSA code showed the most conservative predictions, while EC2 and AS 3600 gave higher estimations compared to those of ACI 318 and the CSA model.

SFSCRC and SFVRC beams

It can be clearly seen from Table 3 that the moment corresponding to the first crack increased with the inclusion of SFs. The SFSCRC beams (B5 and B6) with 0·35% SFs (35 mm) showed an increase in M_{cr}^{exp} (reaching an average of 13·6%) compared with the beams with no SFs (B2 and B3). Similar to the trend of STS and shear capacity results, adding the same amount of SFs to VRC beams with a higher percentage of CR (35%) resulted in a greater improvement in M_{cr}^{exp}, which reached 17%, as shown by comparing B9 and B8. Increasing the volume of SFs (35 mm) to 1% (as in B10) raised the M_{cr}^{exp} up to 45·3% higher than the M_{cr}^{exp} of beams with no

Figure 5 Ratio of the experimental first cracking moment to that predicted by code design equations: (a) based on compressive strength; (b) based on STS

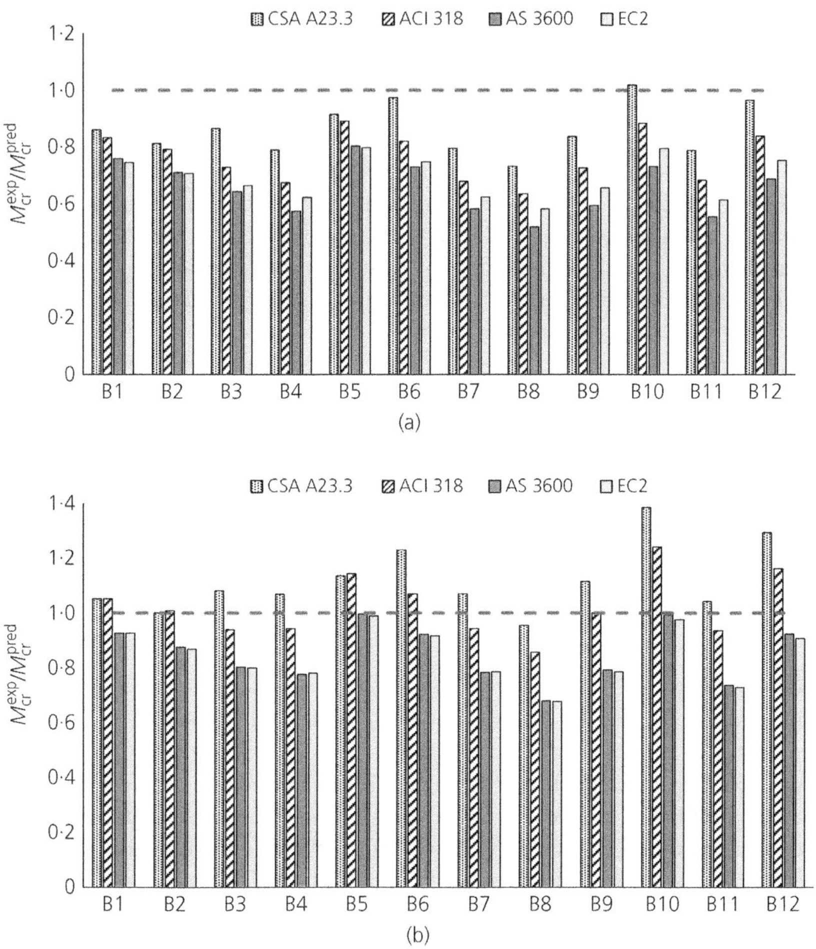

SFs (B8). These results are attributed to the fibres' mechanism in controlling the development of microcracks and delaying the formation of macrocracks (Ding *et al.*, 2011; Ganesan *et al.*, 2013; Tahenni *et al.*, 2016). Using longer SFs (60 mm) in B11 and B12 also increased the M_{cr}^{exp}, but with values 5·9% and 7·2% less than the values exhibited by B9 and B10 with 35 mm SFs, respectively. This may be related to the fact that, at a given fibre volume, using smaller SFs increases the number of fibres that may align in the tensile direction, which effectively helps to delay the formation of macrocracks, as explained earlier.

Examining the predictions of the code equations based on compressive strength shown in Figure 5(a), it can be seen that the inclusion of SFs seemed to decrease the percentage of error in M_{cr}^{pred}. For instance, adding 0·35% SFs (35 mm) to SFSCRC beams (B5 and B6) resulted in

$M_{cr}^{exp}/M_{cr}^{pred}$ ranging from 0·78 to 0·97, and these values were higher than those shown by B2 and B3 (SFSCRC counterparts). Further increases in the amount of SFs led to more discrepancy between M_{cr}^{pred} and M_{cr}^{exp}. With 1% 35 mm SFs (B10), the $M_{cr}^{exp}/M_{cr}^{pred}$ ratios ranged from 0·73 to 1·02. Such results are attributed to the fact that the code equations used do not take into account the effect of fibres in the calculation of M_{cr}^{pred}, but only consider the 28 d compressive strength, which was unaffected by the inclusion of SFs, resulting in a misleading prediction. Using 60 mm SFs showed a similar effect to the 35 mm SFs. In the second method of calculating the M_{cr}^{pred} based on STS, it was found that using the STS values of SFSCRC or SFVRC in the code equations highly overestimated the cracking moment. It should be noted that, although the tensile strength associated with the first cracking load may have been improved by the inclusion of SFs (as shown by the improved M_{cr}^{exp} of B9–B12 compared with B8), it might not be reliable to use ultimate STS test results to predict the cracking moment. This is because, as observed in the STS tests, after the tested cylinders experienced first cracking, the fibres' bridging mechanism allowed the samples to sustain more loading until pullout of the fibres. Since it is not easy to detect the contribution of fibres to increasing the tensile strength of the first crack in STS tests, it may be possible to use the STS of the counterpart SFSCRC and SFVRC mixtures with no SFs in order to predict the cracking moment. For example, as seen from Figure 5(b), using the STS values of B2 and B3 mixtures to calculate the M_{cr}^{pred} of B5 and B6 showed acceptable estimations, in which the $M_{cr}^{exp}/M_{cr}^{pred}$ ratio ranged from 0·92 to 1·23. Similar results were observed when the STS of mixture B8 was used to predict the M_{cr}^{pred} of the SFVRC beams (B9–B12). However, as explained earlier, similar to the results on SCRC and VRC, EC2 and AS 3600 exhibited lower $M_{cr}^{exp}/M_{cr}^{pred}$ ratios than those of ACI 318 and CSA A23.3.

Conclusions

The following conclusions can be drawn from the results of this study.

(*a*) Using lightweight crumb rubber (CR) aggregate with low stiffness contributed to increasing the deformability of the tested beams at a given load and also reduced their self-weight. However, increasing the percentage of CR (0 to 25%) in self-consolidating rubberised concrete (SCRC) beams decreased their normalised shear strength by 9·1%. This reduction seemed to be higher for larger percentages of CR (more than 25%).

(*b*) The addition of 0·35% steel fibres (SFs) (35 mm length) to SCRC beams with up to 15% CR increased their normalised shear strength by 23% (on average), while this increase reached 57·5% in vibrated rubberised concrete (VRC) beams with a higher percentage of CR (35%). Increasing the volume of SFs to 1% (in VRC beams with 35% CR) showed a 242% improvement in the normalised shear strength. Also, using longer SFs (60 mm) (in VRC beams with 35% CR) increased the normalised shear strength, but to less than that obtained by the short SFs (35 mm).

(*c*) Beams with up to 35% CR showed a failure mode similar to the beam with no CR (shear failure), but which was characterised by a larger number of cracks and narrower crack widths. The inclusion of 0·35% SFs (35 or 60 mm) and up to 35% CR in steel-fibre-reinforced SCRC and VRC (SFSCRC and SFVRC) beams continued to reduce the crack widths but with no change in the failure mode. Increasing the SF volume to 1% further narrowed the crack widths and changed the failure mode from shear to shear-flexure failure (in beams with 60 mm long fibres) or to flexure failure (in beams with 35 mm long fibres).

(*d*) The investigated design codes (ACI 318 (ACI, 2008), EC2 (BSI, 2005), Aashto LRFD (Aashto, 2007) and CSA A23.3 (CSA, 2004)) were found to be conservative in predicting the ultimate shear strength of beams without SFs (SCC, SCRC and VRC). EC2 showed the most reasonable predictions for the shear strength compared with the other design codes. Comparing the CSA, ACI and Aashto-LRFD results, the CSA code gives a better prediction for normal-weight concrete. However, for semi-lightweight concrete, the values predicted by the Aashto-LRFD appeared to be closer to the values obtained in the experiments.

(*e*) In the shear strength prediction of beams with SFs (SFSCRC and SFVRC), the equation developed by Khuntia *et al.* (1999) showed the highest V_{exp}/V_{pred} values of all the equations used. The accuracy of the estimation was improved using the models proposed by Sharma (1986), Narayanan and Darwish (1987) and Imam *et al.* (1994). On the other hand, although the most reasonable prediction was shown by the equations of Ashour *et al.* (1992) and Kwak *et al.* (2002), these models slightly overestimated some strengths with a percentage of less than 5%, which may lead to unsatisfactory design.

(*f*) Although most of the code equations and shear design models satisfactorily predicted the shear strength of beams containing CR with/without SFs, increasing the CR content negatively affected their conservatism. This finding may indicate that further investigations are required to take the influence of CR into account in the prediction of the shear capacity of beams with/without SFs.

(*g*) Use of the ACI 318, CSA, AS 3600 (SA, 1988) and EC2 equations (based on tensile strength derived from compressive strength) obviously overestimated the first cracking moment values compared with those obtained from experiments, even for the beam with no CR. Increasing the CR content appeared to further reduce the experimental first cracking moment compared with the predicted values in all the tested code-based equations. On the other hand, using the investigated code equations based on the experimental splitting tensile strength (STS) (instead of the value derived from compressive strength) generally gave better predictions, especially when the CSA and ACI 318 equations were used. Similarly, estimating the cracking moment of SFSCRC and SFVRC beams using the STS of their counterpart mixtures with no SFs provided more reliable and satisfactory predictions.

Acknowledgements

The authors would like to acknowledge the Research & Development Corporation of Newfoundland and Labrador (RDC) for sponsoring this work as part of a larger research project.

REFERENCES

Aashto (Association of State Highway and Transportation Officials) LRFD (2007) *Bridge Design Specifications and Commentary, SI Units*. Aashto, Washington, DC, USA.

Abadel A, Abbas H, Almusallam T, Al-Salloum Y and Siddiqui N (2015) Mechanical properties of hybrid fibre-reinforced concrete – analytical modelling and experimental behaviour. *Magazine of Concrete Research* **68(16)**: 1–21, http://dx.doi.org/10.1680/jmacr.15.00276/.

ACI (American Concrete Institute) (2008) ACI 318R-08: Building code requirements for structural concrete and commentary. ACI, Farmington Hills, MI, USA.

Altun F and Aktas B (2013) Investigation of reinforced concrete beams behavior of steel fibre added lightweight concrete. *Construction and Building Materials* **38**: 575–581.

Ashour SA, Hasanain GS and Wafa FF (1992) Shear behavior of high-strength fiber-reinforced concrete beams. *ACI Structural Journal* **89(2)**: 176–184.

ASTM (2011a) ASTM C39: Standard test method for compressive strength of cylindrical concrete specimens. ASTM International, West Conshohocken, PA, USA.

ASTM (2011b) C496: Standard test method for splitting tensile strength of cylindrical concrete specimens. ASTM International, West Conshohocken, PA, USA.

ASTM (2012a) C618: Standard specification for coal fly ash and raw or calcined natural pozzolan for use in concrete. ASTM International, West Conshohocken, PA, USA.

ASTM (2012b) C150/C150M: Standard specification for Portland cement. ASTM International, West Conshohocken, PA, USA.

ASTM (2013) C494/C494M: Standard specification for chemical admixtures for concrete. ASTM International, West Conshohocken, PA, USA.

ASTM (2014) C231/C231M: Standard test method for air content of freshly mixed concrete by the pressure method. ASTM International, West Conshohocken, PA, USA.

ASTM (2015) C143/C143M: Standard test method for slump of hydraulic-cement concrete. ASTM International, West Conshohocken, PA, USA.

Baston G, Jenkins E and Spatney R (1972) Steel fibers as shear reinforcement in beams. *ACI Journal Proceedings* **69(10)**: 640–644.

Bazant Z and Sun H (1987) Size effect in diagonal shear failure: influence of aggregate size and stirrups. *ACI Materials Journal Proceedings* **84(4)**: 259–272.

BSI (2005) *BS EN 1992-1-1: Eurocode 2 – Design of Concrete Structures – Part 1-1: General rules and rules for buildings*. BSI, London, UK.

CSA (Canadian Standards Association) (2004) *CSA A23.3-04: Design of concrete structures*. CSA, Rexdale, ON, Canada.

Ding Y, Liu S, Zhang Y and Thomas A (2008) The investigation of workability of fibre cocktail reinforced self-compacting high performance concrete. *Construction and Building Materials* **22(7)**: 1462–1470.

Ding Y, You Z and Jalali S (2011) The composite effect of steel fibres and stirrups on the shear behaviour of beams using self-consolidating concrete. *Engineering Structures* **33(1)**: 107–117.

Ding Y, Zhang F, Torgal F and Zhang Y (2012) Shear behaviour of steel fibre reinforced self-consolidating concrete beams based on the modified compression field theory. *Composite Structures* **94(8)**: 2440–2449.

Efnarc (European Federation for Specialist Construction Chemicals and Concrete Systems) (2005) *The European Guidelines for Self-Compacting Concrete: Specification, Production and Use*. Efnarc, Norfolk, UK.

Fathifazl G, Razaqpur AG, Isgor OB *et al.* (2009) Flexural performance of steel-reinforced recycled concrete beams. *ACI Structural Journal* **106(6)**: 858–867.

Ganesan N, Bharati RJ and Shashikala AP (2013) Behavior of self-consolidating rubberized concrete beam–column joints. *ACI Materials Journal* **110(6)**: 697–704.

Ganesan N, Indira PV and Santhakumar A (2014) Influence of steel fibres on tension stiffening and cracking of reinforced geopolymer concrete. *Magazine of Concrete Research* **66(6)**: 268–276, http://dx.doi.org/10.1680/macr.13.00273.

Güneyisi E (2010) Fresh properties of self-compacting rubberized concrete incorporated with fly ash. *Materials and Structures* **43(8)**: 1037–1048.

Imam M, Vandewalle L and Mortelmans F (1994) *Shear Capacity of Steel Fiber High Strength Concrete Beams*. American Concrete Institute, Farmington Hills, MI, USA, ACI SP 149, pp. 227–242.

Iqbal S, Ali A, Holschemacher K and Thomas AB (2015) Mechanical properties of steel fibre reinforced high strength lightweight self-compacting concrete (SHLSCC). *Construction and Building Materials* **98**: 325–333.

Ismail MK and Hassan AAA (2015a) Performance of full-scale self-consolidating rubberized concrete beams in flexure. *ACI Materials Journal* **113(2)**: 207–218.

Ismail MK and Hassan AAA (2015b) Influence of mixture composition and type of cementitious materials on enhancing the fresh properties and stability of self-consolidating rubberized concrete. *Journal of Materials in Civil Engineering*.

Ismail MK and Hassan AAA (2016a) Impact resistance and mechanical properties of self-consolidating rubberized concrete reinforced with steel fibers. *Journal of Materials in Civil Engineering* **28(1)**, http://dx.doi.org/10.1061/(ASCE)MT.1943-5533.0001338.

Ismail MK and Hassan AAA (2016b) Use of metakaolin on enhancing the mechanical properties of self-consolidating concrete containing high percentages of crumb rubber. *Journal of Cleaner Production* **125**: 282–295.

Khaloo A, Raisi EM, Hosseini P and Tahsiri H (2014) Mechanical performance of self-compacting concrete reinforced with steel fibers. *Construction and Building Materials* **51**: 179–186.

Khuntia M, Stojadinovic B and Goel S (1999) Shear strength of normal and high-strength fiber-reinforced concrete beams without stirrups. *ACI Structural Journal* **96(2)**: 282–289.

Kwak YK, Eberhard M, Kim WS and Kim J (2002) Shear strength of steel fibre-reinforced concrete beams without stirrups. *ACI Structural Journal* **99(4)**: 530–538.

Lee JH, Yoo DY and Yoon YS (2016) Mechanical behaviour of concrete with amorphous metallic and steel fibres. *Magazine of Concrete Research* **68(24)**: 1253–1264, http://dx.doi.org/10.1680/jmacr.15.00493.

Najim KB and Hall MR (2010) A review of the fresh/hardened properties and applications for plain-(PRC) and self-compacting rubberised concrete (SCRC). *Construction and Building Materials* **24(11)**: 2043–2051.

Najim KB and Hall MR (2012) Mechanical and dynamic properties of self-compacting crumb rubber modified concrete. *Construction and Building Materials* **27(1)**: 521–530.

Najim KB and Hall MR (2014) Structural behaviour and durability of steel-reinforced structural plain/self-compacting rubberised concrete (PRC/SCRC). *Construction and Building Materials* **73**: 490–497.

Narayanan R and Darwish IYS (1987) Use of steel fibres as shear reinforcement. *ACI Structural Journal* **84(3)**: 216–227.

Olivito RS and Zuccarello FA (2010) An experimental study on the tensile strength of steel fibre reinforced concrete. *Composites Part B: Engineering* **41(3)**: 246–255.

SA (Standards Australia) (1988) *AS 3600: Concrete structures*. SA, Sydney, Australia.

Sadek DM and El-Attar MM (2014) Structural behaviour of rubberized masonry walls. *Journal of Cleaner Production* **89**: 174–186.

Sharma AK (1986) Shear strength of steel fiber reinforced concrete beams. *ACI Journal Proceedings* **83(4)**: 624–628.

Swamy RN, Mangat PS and Rao CVSK (1974) The mechanics of fiber reinforcement of cement matrices. *Fiber Reinforced Concrete*. American Concrete Institute, Farmington Hills, MI, USA, ACI SP 44, pp. 1–28.

Tahenni T, Chemrouk M and Lecompte T (2016) Effect of steel fibres on the shear behavior of high strength concrete beams. *Construction and Building Materials* **105**: 14–28.

Taylor HPJ (1974) *The Fundamental Behavior of Reinforced Concrete Beams in Bending and Shear*. American Concrete Institute, Farmington Hills, MI, USA, ACI SP 42, pp. 43–77.

Topçu IB and Bilir T (2009) Experimental investigation of some fresh and hardened properties of rubberized self-compacting concrete. *Materials and Design* **30(8)**: 3056–3065.

Youssf O, El-Gawady MA and Mills JE (2015) Experimental investigation of crumb rubber concrete columns under seismic loading. *Structures* **3**: 13–27.

Zheng L, Sharon HX and Yuan Y (2008) Experimental investigation on dynamic properties of rubberized concrete. *Construction and Building Materials* **22(5)**: 939–947.

Zsutty T (1971) Shear strength prediction for separate categories of simple beam tests. *ACI Journal Proceedings* **68(2)**: 138–143.

ice Publishing

Dhir and Lye
ISBN 978-0-7277-6461-4
https://doi.org/10.1680/icetscc.64614.407

Chapter 18
Fatigue lives of self-compacting concrete and self-compacting fibre-reinforced concrete beams

Sanjay Goel ME
Assistant Professor, Department of Civil Engineering, DAV Institute of Engineering and Technology, Jalandhar, India

Surinder Pal Singh PhD
Professor, Department of Civil Engineering, Dr B R Ambedkar National Institute of Technology, Jalandhar, India

Partap Singh PhD
Professor, Department of Civil Engineering, Dr B R Ambedkar National Institute of Technology, Jalandhar, India

This paper presents results of an investigation conducted to estimate the mean and design fatigue lives of self-compacting concrete and self-compacting fibre-reinforced concrete beams containing 0·5% volume fraction of steel fibres. Approximately 19 flexural fatigue tests and 90 complementary static flexural strength tests were conducted on self-compacting concrete and self-compacting fibre-reinforced concrete beam specimens of 100 × 100 × 500 mm size. The distribution of fatigue life of self-compacting concrete and self-compacting fibre-reinforced concrete at different stress levels has been examined and Weibull distribution parameters have been obtained by different methods of analysis, which are subsequently used to estimate the mean and design fatigue lives of self-compacting concrete and self-compacting fibre-reinforced concrete. The performance of self-compacting concrete and self-compacting fibre-reinforced concrete have been found to be better, in terms of reduced variability in the distribution of fatigue lives and higher mean and design fatigue lives, compared to normally vibrated concrete and normally vibrated fibre reinforced concrete.

Notation

$E[N]$	mean fatigue life
$f_{\max}$	maximum fatigue stress
$f_{\min}$	minimum fatigue stress
f_r	static flexural strength
k	number of fatigue data points at given stress level S

L_R	survival probability
N	number of cycles to failure
N_D	design fatigue life
P_f	probability of failure
R	stress ratio $= f_{min}/f_{max}$
S	stress level $= f_{max}/f_r$
u	characteristic life or scale parameter
α	shape parameter
$\Gamma(\,)$	gamma function

1. Introduction

Concrete structures such as bridges, pavements and off-shore structures are normally subjected to millions of cycles of repetitive fatigue load during their service life, making fatigue strength a critical parameter in the design of these structures. The fatigue behaviour of concrete may generally be characterised by the so-called *S–N* curve, which allows the mean fatigue life of concrete under a given stress level to be predicted (Oh, 1991). Hsu (1981) categorised compression and flexure fatigue of normally vibrated concrete (NVC) into three categories, namely low cycle fatigue, high cycle fatigue and super high cycle fatigue. To meet the challenges of the growing travel industry, airport pavements and bridges had to be built. These structures are subjected to several hundred thousands of cycles of repetitive loading. Understanding the fatigue resistance of concrete in this range of applications is essential (Hsu, 1981).

In the past, considerable interest has developed in investigating the fatigue behaviour of NVC, mainly focused on determination of its fatigue strength (Hilsdorf and Kesler, 1966; Hsu, 1981; Klaiber and Lee, 1982; Raithby, 1979; Shi *et al.*, 1993; Tepfers, 1979; Tepfers and Kutti, 1979). In the mid-1970s, enormous interest arose in the study of the mechanical behaviour of normally vibrated, fibre-reinforced concrete (NVFRC), particularly steel-fibre-reinforced concrete (SFRC), owing to its superior crack arrest properties. Extensive investigations on the fatigue behaviour of NVFRC have been conducted; these have mainly been confined to determination of flexural fatigue strength and endurance limits for different types/volume fractions/aspect ratios of steel fibres (Batson *et al.*, 1992; Johnston and Zemp, 1991; Ramakrishnan *et al.*, 1987, 1989, 1996; Tatro, 1987). Since fatigue test data usually show considerable variability, even at a given stress level, under carefully controlled test procedures, the concepts of probabilistic analysis were implemented to model this variability and models have been proposed to predict the flexural fatigue strength of NVC and NVFRC. Oh (1986, 1991), Singh and Kaushik (2000, 2001) and Mohammadi and Kaushik (2005) studied the flexural fatigue life distributions of NVC and SFRC at different stress levels and examined the two-parameter Weibull distribution in this context. Oh (1986), Singh and Kaushik (2001) and Singh *et al.* (2007) estimated the mean and design fatigue lives of NVC and SFRC containing different volume fractions of steel fibres.

Recent years have seen considerable interest in self-compacting concrete (SCC), which has the ability to fill even densely reinforced formworks without segregation and compact under its own weight without the use of mechanical vibrations. Nowadays, owing to its benefits over NVC, SCC has become very popular within the construction industry, dealing with the work of mass concreting in pavements and especially in concrete structures with heavily congested reinforcement, such as bridge decks and piers, high-rise buildings, rapid transportation systems and dams (Domone, 2006). Self-compacting concrete has interested researchers all over the world, who have looked into its fresh and hardened mechanical properties. The majority of the research on SCC and self-compacting fibre-reinforced concrete (SCFRC) has focused attention on the rheological and hardened mechanical properties under statically applied loads. The workability properties of SCC and SCFRC, such as passing ability, filling ability and resistance to segregation, have been extensively investigated (Grunewald and Walraven, 2001; Sahmaran *et al.*, 2005; Sonebi *et al.*, 2007). The mechanical properties of SCC and SCFRC under statically applied loads such as compressive strength, flexural strength, splitting tensile strength, elastic modulus, creep and shrinkage, shear behaviour and so on, have also been a subject of interest to many researchers (Brouwers and Radix, 2005; Cunha *et al.*, 2010; Dhonde *et al.*, 2007; Greenough and Nehdi, 2008; Persson, 2000; Torrijos *et al.*, 2008, Turkel and Kandemir, 2010). The effects of different types of fillers such as fly ash, limestone powder, blast furnace slag, silica fume, metakaolin, quarry dust, chalk powders, and so on, on the rheological and mechanical properties of SCC and SCFRC have also been investigated (Gesoglu, 2009; Sahmaran and Yaman, 2007; Singh *et al.*, 2009; Zhu and Gibbs, 2004). Su *et al.* (2001), Chowdhury and Basu (2010) and Ferrara *et al.* (2007) suggested the mix proportioning methods for SCC and SCFRC, whereas Hans and Michael (2006), Persson (2003a, 2003b) and Corinaldesi and Moriconi (2004) studied their durability aspects under different environmental conditions.

2. Research significance

It can be observed from the review of literature presented in the preceding section that, in spite of a large number of studies on the mechanical properties of SCC and SCFRC under statically applied loads, there has been practically no study on its fatigue strength, which otherwise is an important parameter in the design of concrete structures subjected to such loads. Therefore, the present investigation was planned to estimate the mean and design fatigue lives of SCC and SCFRC. As a prerequisite, the fatigue life distributions of SCC and SCFRC were first examined using two-parameter Weibull distribution; its distribution parameters were obtained using different methods of analysis. Using the average values of the distribution parameters, the mean and design fatigue lives for SCC and SCFRC were estimated and compared with those of NVC and NVFRC containing steel fibres of comparable shape and size, the results of which are available in the literature. It has been observed that both SCC and SCFRC exhibited better performance in terms of mean and design fatigue lives compared to NVC and NVFRC.

3. Experimental programme

Ordinary Portland cement (OPC) and Class F fly ash were used as binder materials, while aggregates consisted of crushed stone (below 12·5 mm) and locally available coarse sand; all materials conformed to the relevant Indian standard specifications. Round, corrugated steel fibres 1 mm in diameter and 30 mm long were used for the SCFRC mix at a volume fraction of 0·5%. Appropriate dosage of a polycarboxylate-based super-plasticiser and viscosity modifying agent were used to obtain the required workability. The mix proportions of SCC and SCFRC designed for the present investigation are listed in Table 1. The workability tests were conducted on SCC and SCFRC mixes following the EFNARC (2005)

Table 1 Mix proportions for SCC and SCFRC (V_f = 0.5%)

Mix type	Water/ binder ratio	Sand/ cement ratio	Aggregate/ binder ratio	Fly ash/ cement ratio	Steel fibres volume fraction, V_f	Super-plasticiser (by weight of cement)	Viscosity modifying agent (by weight of cement)
SCC	0·40	2·06	2·35	0·50	–	1·7%	–
SCFRC	0·40	2·06	2·35	0·50	0·5%	1·9%	0·25%

guidelines. The results of the various workability tests performed for SCC and SCFRC mixes are presented in Table 2. The concrete mix was placed into the moulds without any mechanical vibration or tamping. Beam specimens were cast of dimensions 100 × 100 × 500 mm for static flexural strength and flexural fatigue strength tests and cube specimens were cast of dimensions 150 × 150 × 150 mm for compressive strength tests. The specimens were cast in different batches, each batch consisting of seven standard beam specimens for static flexural and flexural fatigue tests and three cube specimens for compressive strength tests. The specimens were demoulded after 36 h of casting. The quality of each batch of concrete was checked by obtaining its 28-day compressive strength. The average 28-day compressive strengths of the SCC and SCFRC mixes were 35·90 MPa and 38·10 MPa respectively. The beam specimens were water cured under laboratory conditions for 75 days in order to avoid possible strength increase during the fatigue tests.

The estimation of static flexural strength of a particular batch was a prerequisite for the selection of the maximum and minimum load limits for the fatigue tests. The static flexural strength tests on a particular batch of SCC or SCFRC were conducted just before the fatigue testing was carried out. Three specimens selected randomly from a particular batch were tested in static flexure and the average static flexural strength for the same batch was obtained. In the present investigation, the average static flexural strengths for SCC and

Table 2 Workability test results on SCC and SCFRC (V_f = 0.5%)

Workability test	Parameter	Result obtained (SCC)	Result obtained (SCFRC)	EFNARC guidelines
Slump flow test	T_{500}: s	3·12	3·29	2–5
	Slump flow spread: mm	750	705	650–800
J-ring test	T_{500J}: s	3·26	4·53	3–6
	Flow spread: mm	725	710	600–750
	Blocking step Bj: mm	6·5	7·5	0–10
V-funnel test	V-funnel time: s	7·28	7·8	6–12
L-box test	L-box passing ability	0·91	0·90	0·8–1·0

SCFRC mixes were 4·85 MPa and 6·05 MPa respectively. After the static flexural strength of a particular batch was established, the remaining beam specimens (generally four in number) were tested in flexural fatigue. All the static flexural and flexural fatigue tests were conducted on a 100 kN MTS servo-controlled actuator. A four-point load arrangement was adopted and the specimens were simply supported over a span of 450 mm during both static and fatigue tests. Constant-amplitude, sinusoidal, non-reversed flexural loads were applied at a frequency of 10 Hz. The flexural fatigue tests were conducted at different stress levels S ($S = f_{max}/f_r$), ranging from 0·90 to 0·65, maintaining a constant fatigue stress ratio R ($R = f_{min}/f_{max}$) of 0·10 throughout the investigation. The load applied to the specimen was sensed by a load cell and used as a feedback signal to control the load applied by the actuator. The test was terminated as and when the specimen failed or a maximum limit of two million cycles of load applications was reached, whichever was earlier. The number of cycles to failure (N) of each specimen at a particular stress level was recorded as its fatigue life. In total, 119 flexural fatigue tests and 90 complementary static flexural tests were conducted to obtain the data in this investigation.

4. Analysis of fatigue life data

The fatigue life data obtained for SCC and SCFRC, corresponding to different stress levels in this investigation, are listed in Tables 3 and 4 respectively. The main objectives of this paper were to estimate the mean and design fatigue lives of SCC and SCFRC and to compare their

Table 3 Flexural fatigue life data (number of cycles to failure, written in ascending order) for SCC at different stress levels

Specimen no.	Stress level, S				
	0·85	0·80	0·75	0·70	0·65
1	32[a]	8243	29 684	24 862[a]	682 365
2	1292	9268	39 652	89 080	795 624
3	1387	10 724	42 562	97 842	867 475
4	1453	11 892	46 881	129 324	964 688
5	1768	12 984	54 925	139 576	1 135 746
6	1892	13 821	59 493	150 836	1 385 172
7	1918	14 352	63 358	166 709	1 634 125
8	2116	15 257	70 312	182 892	1 771 589
9	2220	16 906	78 237	204 170	1 889 577
10	2328	17 532	83 111	211 587	2 000 000[b]
11	2822	18 112	88 165	244 276	
12		22 464	178 704[a]	270 671	
13		150 187[a]			

[a] Outliers not included in analysis.
[b] Treated as run out, not included in analysis.

Table 4 Flexural fatigue life data (number of cycles to failure, written in ascending order) for SCFRC (V_f = 0.5%) at different stress levels

Specimen no.	Stress level, S				
	0·90	0·85	0·80	0·75	0·70
1	77[a]	972[a]	15 986	75 668	282 014
2	1114	4316	17 414	96 251	380 132
3	1240	5685	19 650	119 562	472 329
4	1327	6106	23 237	131 198	518 928
5	1522	6832	24 363	150 947	635 492
6	1617	7229	31 927	178 217	724 120
7	1726	7987	34 256	191 326	812 596
8	1810	8921	37 236	211 319	893 415
9	1918	9542	42 035	235 282	1 022 520
10	2103	10 146	51 089	346 603	1 698 642
11	2685	13 240	63 549	394 352	1 882 857
12	2948	14 538	72 632	468 581	2 000 000[b]
13		15 831			

[a] Outliers not included in analysis.
[b] Treated as run out, not included in analysis.

performance with NVC and NVFRC. It is well known that relatively large variability usually occurs in the resulting fatigue lives of concrete, even in carefully controlled experimental procedures. The inclusion of fibres in concrete is further expected to enhance the variability in the fatigue test data of SCFRC. Thus it is proposed first to examine the probability distributions for the fatigue life data of SCC and SCFRC using the Weibull distribution. The estimated parameters of the Weibull distribution for the fatigue life of SCC and SCFRC will be used subsequently to estimate their mean and design fatigue lives.

Some fatigue test data points in Tables 3 and 4 may deserve consideration for rejection as outliers. The Chauvenet's criterion was applied to the fatigue tests data at all the stress levels tested in this investigation and points meeting this criterion for rejection were identified and excluded from further analysis (Kennedy and Neville, 1986; Singh and Kaushik, 2001).

4.1 Fatigue life distributions of SCC and SCFRC

As already stated, the primary objective of this paper was to determine the mean and design fatigue lives of SCC and SCFRC. To do so, it is necessary first to establish the probability distributions and estimate the parameters of the Weibull distribution, as shown later. It has been well established that the distribution of fatigue life of NVC (Oh, 1986, 1991; Shi *et al.*, 1993) and

NVFRC (Mohammadi and Kaushik, 2005; Singh and Kaushik, 2000, 2001) at a particular stress level S, approximately follows the two-parameter Weibull distribution. Since the observed fatigue lives of SCC as well as SCFRC specimens of this investigation indicate variability in its distribution, it is proposed in this paper to model this variability by the two-parameter Weibull distribution and to obtain its distribution parameters by different methods of analysis. First, the probability distributions of SCC and SCFRC were examined using the graphical method and subsequently, the parameters of the Weibull distribution were obtained by other methods, such as the method of moments and the method of maximum likelihood estimate.

The graphical method is used here to show that the distribution of fatigue life of SCC and SCFRC at a given stress level S can be approximately modelled by the two-parameter Weibull distribution. The parameters of the Weibull distribution were then estimated by the graphical method, the method of moments and the method of maximum likelihood.

The survivorship function $L_N(n)$ of two-parameter Weibull distribution may be written as follows (Mohammadi and Kaushik, 2005; Oh, 1991)

$$L_N = \exp\left[-\left(\frac{n}{u}\right)^{\alpha}\right] \tag{1}$$

where n is specific value of random variable N, α is the shape parameter at stress level S, and u is the characteristic extreme value of the two-parameter Weibull distribution.

Taking the natural logarithm twice on both sides of Equation 1

$$\ln\left[\ln\left(\frac{1}{L_N}\right)\right] = \alpha\ln(n) - \alpha\ln(u) \tag{2}$$

Equation 2 represents a linear relationship between $\ln[\ln(1/L_N)]$ and $\ln(n)$ and can be used to verify the suitability of the two-parameter Weibull distribution for the fatigue test data at a particular stress level. To obtain a graph between $\ln[\ln(1/L_N)]$ and $\ln(n)$, the fatigue test data at a given stress level are first arranged in ascending order. The empirical survivorship function L_N can be calculated from the following equation (Oh, 1986; Singh and Kaushik, 2000)

$$L_N = 1 - \frac{i}{k+1} \tag{3}$$

where i is the failure order number and k is the number of fatigue data points at given stress level S. A graph is plotted between $\ln[\ln(1/L_N)]$ and $\ln(n)$, and if a linear trend is observed for the fatigue life data at a given stress level S, it can be assumed that the two-parameter Weibull distribution is reasonably valid for the statistical description of fatigue life data at that stress level. The parameters of the Weibull distribution can be obtained from the regression analysis. Typical plots for the graphical analysis of the fatigue life data of SCC and SCFRC at stress level $S = 0{\cdot}85$ are shown in Figures 1 and 2 respectively. It can be observed that the data points fall approximately along a straight line, which indicates that the two-parameter Weibull distribution is reasonably valid for the distribution of fatigue life of SCC as well as SCFRC at stress level $S = 0{\cdot}85$. The calculated parameters for this stress level are also listed in these figures.

Figure 1 Graphical analysis of fatigue life data of SCC at $S = 0{\cdot}85$

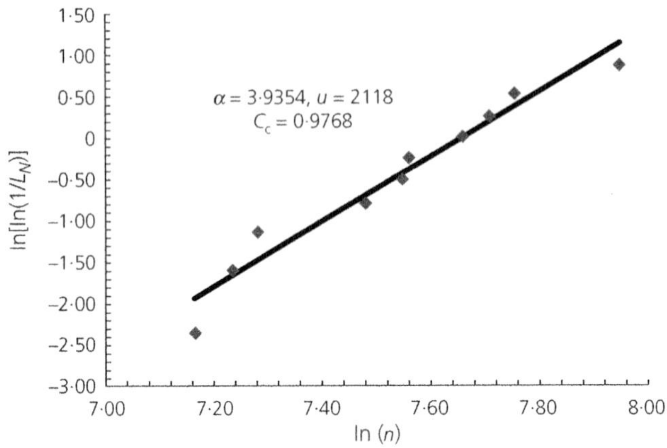

Figure 2 Graphical analysis of fatigue life data of SCFRC, $V_f = 0{\cdot}5\%$ at $S = 0{\cdot}85$

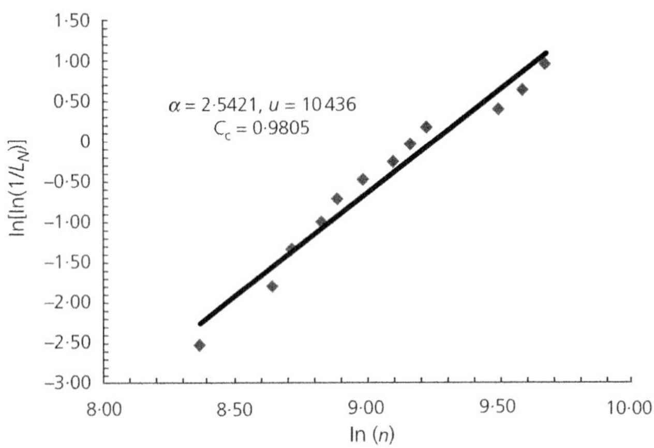

Similarly, the fatigue life data of SCC and SCFRC were analysed at all the other stress levels tested in this investigation and it was observed that the Weibull distribution can approximately be used to describe the fatigue life data of SCC and SCFRC at all the stress levels. The plots for graphical analysis of the fatigue life data at other stress levels are not given.

The distribution parameters for the fatigue life data of SCC and SCFRC have also been obtained by the method of moments and the method of maximum likelihood estimate. The values of Weibull parameters for the fatigue life data of SCC and SCFRC, at different stress levels, obtained by all the three methods of analysis, along with their average values, are listed in Tables 5 and 6 respectively. Figure 3 shows the plot of the shape parameter for the fatigue

Figure 3 Shape parameter for fatigue life of SCC, NVC, SCFRC and NVFRC with V_f = 0·5%

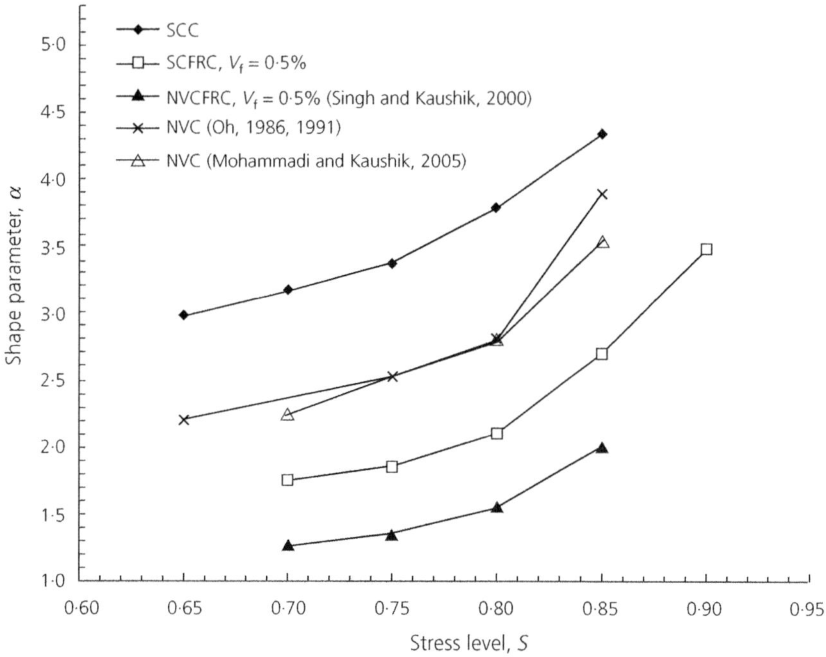

life of SCC and SCFRC against the fatigue stress levels tested in this investigation. The results of the shape parameter for the fatigue life of NVC and NVFRC containing steel fibres of comparable size and shape obtained in some of the previous studies (Mohammadi and Kaushik, 2005; Oh, 1986, 1991; Singh and Kaushik, 2000) have also been plotted for comparison. Relatively higher values of the shape parameter for the fatigue life data of SCC and SCFRC, as compared to that of NVC and NVFRC respectively, indicate that there is reduction in the variability in the distribution of fatigue life of SCC as well as SCFRC when compared to NVC and NVFRC. Detailed results in this respect for SCC are available elsewhere (Goel *et al.*, 2012) and for SCFRC containing 0·5%, 1·0% and 1·5% volume fractions of fibres, the results will be published in a separate paper. As the estimation of mean and design fatigue lives of SCC is the main objective in this paper, this is the focus of the following section.

4.2 Estimation of mean and design fatigue lives of SCC and SCFRC

It has been shown in the previous section that the distribution of fatigue life data of SCC and SCFRC at a particular stress level can be described by the two-parameter Weibull distribution and the parameters of the Weibull distribution have been estimated using different methods of analysis. The mean and design fatigue lives for SCC and SCFRC can now be calculated using the distribution parameters as described in the following paragraphs.

The mean fatigue life $E[N]$ can be obtained by the following relation (Oh, 1986; Singh and Kaushik, 2001)

$$E[N] = C(S)^{-m}\exp\left[\frac{0{\cdot}5772}{\alpha}\right]\Gamma\left(1+\frac{1}{\alpha}\right) \tag{4}$$

where, $S = f_{max}/f_r$ is the stress level, $E[N]$ is mean fatigue life, m and C are empirical constants, Γ is the gamma function, α is the shape parameter of the Weibull distribution as obtained in the previous section and N is the number of cycles to failure. The values of m and C are obtained using the following S–N relationship (Oh, 1986; Singh and Kaushik, 2001)

$$N(S)^m = C \tag{5}$$

Equation 5 is similar to the one used for metals and is more reasonable in the sense that the stress term is treated in the non-dimensional form and thus may have wider applicability (Oh, 1986). Taking the logarithm on both sides of Equation 5

$$\log_{10}(N) = \log_{10}C - m\log_{10}(S) \tag{6}$$

Writing Equation 6 in the following form

$$Y = a + bX \tag{7}$$

where $Y = \log_{10}(N)$, $X = \log_{10}(S)$, $a = \log_{10}C$ and $b = -m$.

The coefficients a and b for SCC and SCFRC can be obtained by fitting Equation 7 into the respective fatigue life data obtained in this investigation. The analysis of the fatigue life data of SCC and SCFRC using Equations 5, 6 and 7 is presented in Figure 4. The values of the empirical constants m and C are obtained from the regression analysis. The S–N relationship depicted by Equation 5 can be represented in terms of the following relations

Figure 4 Estimation of coefficients m and C of Equation 5 for fatigue life data of SCC and SCFRC

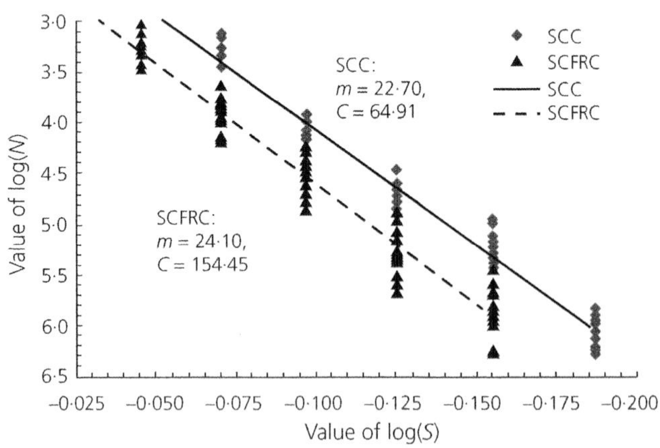

for SCC

$$N(S)^{22 \cdot 70} = 64 \cdot 91 \tag{8}$$

for SCFRC containing 0·5% steel fibres

$$N(S)^{24 \cdot 10} = 154 \cdot 45 \tag{9}$$

Using the average values of the Weibull parameters listed in Tables 5 and 6, and the values of the coefficients m and C as given in Equations 8 and 9 for SCC and SCFRC respectively, Equation 4 can be used to calculate the mean fatigue lives for SCC and SCFRC. The calculated values of mean fatigue lives for SCC and SCFRC corresponding to different stress levels are presented in Table 7. Table 7 also presents the mean fatigue lives obtained in previous investigations for NVC (Oh, 1986) and NVFRC (Singh and Kaushik, 2001) containing 0·5% volume fraction of fibres of comparable shape and size, for the purpose of comparison. The work of Oh (1986) on NVC and Singh and Kaushik (2001) on NVFRC has been selected for comparison with SCC and SCFRC respectively, as the static flexural strengths of the concretes

Table 5 Values of the Weibull parameters for fatigue life of SCC

Stress level, S	Graphical method		Method of moments		Method of maximum likelihood		Average	
	α	u	α	u	α	u	α	u
0·85	3·9354	2118	4·5210	2102	4·5850	2099	4·3471	2106
0·80	3·4942	15 895	3·8877	15 807	3·9754	15 774	3·7858	15 825
0·75	2·9514	67 198	3·4408	66 372	3·7320	66 285	3·3747	66 618
0·70	2·8350	193 667	3·2381	191 453	3·4490	191 237	3·1740	192 119
0·65	2·7279	1 475 243	2·9839	1 385 603	3·2485	1 384 905	2·9867	1 415 250

Table 6 Values of the Weibull parameters for fatigue life of SCFRC (V_f = 0.5%)

Stress level, S	Graphical method		Method of moments		Method of maximum likelihood		Average	
	α	u	α	u	α	u	α	u
0·90	3·2259	2035	3·4572	2023	3·4800	2022	3·4676	2122
0·85	2·5421	10 436	2·7126	10 343	2·8428	10 355	2·6992	10 378
0·80	1·9985	41 439	2·0796	40 773	2·2175	41 006	2·0985	41 073
0·75	1·7834	248 426	1·8224	243 690	1·9610	245 969	1·8556	246 028
0·70	1·6893	973 401	1·7039	949 710	1·8540	961 391	1·7491	961 500

Table 7 Mean fatigue lives for SCC and SCFRC at various stress levels

Stress level, S	Mean fatigue life, $E[N]$			
	SCC	NVC (Oh, 1986)	SCFRC (V_f = 0·5%)	NVFRC (V_f = 0·5%) (Singh and Kaushik, 2001)
0·90	–	–	2273	–
0·85	2932	–	9384	4150
0·80	11 808	1197	42 670	22 213
0·75	51 908	–	208 832	132 481
0·70	250 858	27 573	1 120 849	893 794
0·65	1 362 651	–	–	–
0·60	–	1 030 097	–	–

used in those studies are comparable with those of SCC and SCFRC used in the present investigation. For example, the static flexural strengths of NVC tested by Oh (1986) and SCFRC tested by Singh and Kaushik (2001) are 5·30 MPa and 5·93 MPa respectively, whereas the values are 4·85 MPa for SCC and 6·05 MPa for SCFRC used in the present investigation. Figure 5 shows the comparison of the mean fatigue lives computed in this investigation for SCC and SCFRC with the NVC and NVFRC. It can be seen from Table 7 and Figure 5 that, in general, SCC and SCFRC exhibit higher mean fatigue lives than NVC and NVFRC.

Figure 5 Comparision of mean fatigue lives for SCC, NVC, SCFRC and NVFRC with 0·5% steel fibres

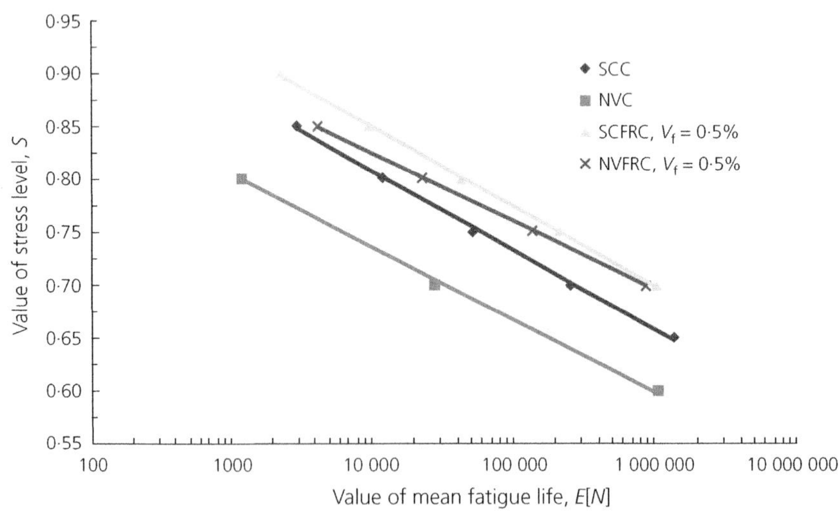

The design fatigue life N_D should be selected such that there is only a small probability that a fatigue failure will occur. Once the distribution function is determined, the design fatigue life may be selected corresponding to an acceptable reliability. Noting that P_f (probability of failure) = $P[N < N_D]$ (design reliability), the design fatigue life corresponding to a permissible probability P_f is determined from the following relation (Oh, 1986; Singh *et al.*, 2007)

$$N_D = u\left(\ln\frac{1}{1-P_f}\right)^{1/\alpha} \tag{10}$$

Equation 10 is used to calculate the design fatigue lives corresponding to various acceptable probabilities of failure, namely 0·01, 0·05, 0·10, 0·15 and 0·25, for SCC and SCFRC. The calculated values of design fatigue lives for SCC and SCFRC with 0·5% steel fibres are listed in Tables 8 and 9 respectively. Figure 6 shows the plot of calculated design fatigue lives for SCC and SCFRC for a probability of failure P_f = 0·10. The design fatigue lives for NVC (Oh, 1986) and NVFRC with 0·5% of corrugated rectangular steel fibres of size 0·6 mm × 2 mm × 30 mm (Singh and Kaushik, 2001) for the same probability of failure obtained in a previous study are also plotted for comparison. It can be observed from the results that, in general, SCC and SCFRC show higher design fatigue lives when compared to NVC and NVFRC.

Table 8 Design fatigue lives for SCC at different probabilities of failure

Stress level, S	Design fatigue life, N_D				
	Probability of failure, P_f				
	0·01	0·05	0·10	0·15	0·25
0·85	731	1063	1255	1387	1581
0·80	4295	7221	8734	9793	11 387
0·75	17 045	27 628	34 197	38 883	46 053
0·70	45 096	75 364	94 549	108 382	129 746
0·65	303 339	523 526	666 206	770 247	932 540

Table 9 Design fatigue lives for SCFRC (V_f = 0.5%) at different probabilities of failure

Stress level, S	Design fatigue life, N_D				
	Probability of failure, P_f				
	0·01	0·05	0·10	0·15	0·25
0·90	563	901	1109	1257	1482
0·85	1888	3453	4509	5294	6541
0·80	4587	9974	14 055	17 279	22 684
0·75	20 622	49 639	73 164	92 413	125 715
0·70	69 303	175 979	265 578	340 256	471 626

Figure 6 Comparision of design fatigue lives for SCC, NVC, SCFRC and NVFRC with 0·5% steel fibres at probability of failure, P_f = 0·10

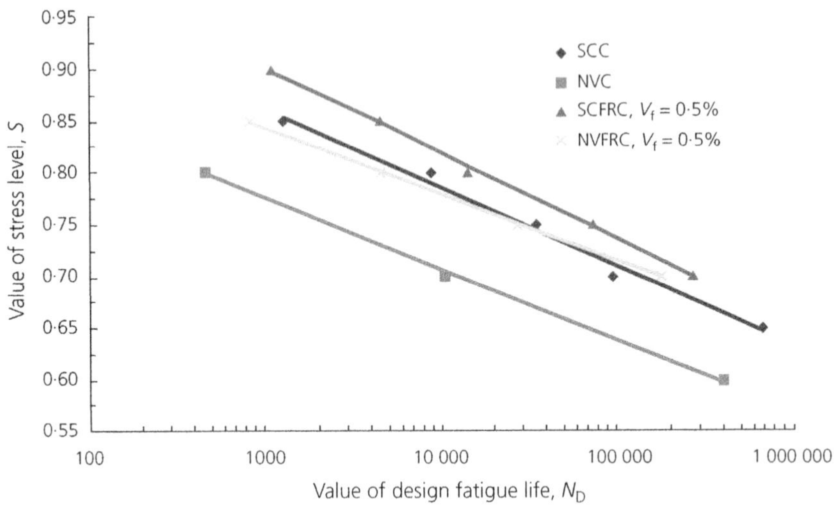

It may be noted that the results obtained in this investigation for SCFRC are applicable to the type, size and volume fraction of the fibres employed and additional research work is required to estimate mean and design fatigue lives for other types, sizes and volume fractions of fibres.

5. Conclusion

The increased use of SCC in structures such as bridges, pavement slabs and rapid transport systems makes the study of its flexural fatigue characteristics essential. In this investigation, both experimental and theoretical studies have been conducted to investigate the fatigue characteristics of SCC and SCFRC. Experiments have been conducted to obtain the flexural fatigue lives for SCC and SCFRC with 0·5% of steel fibres at different stress levels. A probabilistic approach has been used to establish the probability distributions of SCC and SCFRC using the two-parameter Weibull distribution. The parameters of the Weibull distribution have been obtained for the fatigue life data of SCC and SCFRC by different methods of analysis. Subsequently, the mean fatigue lives for SCC and SCFRC corresponding to different stress levels have been estimated. The design fatigue lives for SCC and SCFRC have also been determined corresponding to acceptable probabilities of failure. Comparison of shape parameters, mean and design fatigue lives of SCC and SCFRC with those of NVC and NVFRC indicate that SCC and SCFRC exhibit better performance.

Acknowledgements

The authors gratefully acknowledge the support of the staff of Concrete Structures Laboratories at the DAV Institute of Engineering and Technology, Jalandhar, India and Dr B. R. Ambedkar, National Institute of Technology, Jalandhar, India during the experimental work reported in this paper.

REFERENCES

Batson G, Ball C, Bailey L, Lenders E and Hooks J (1992) Flexural fatigue strength of steel fiber reinforced concrete beams. *ACI Journal* **69(11)**: 673–677.

Brouwers HJH and Radix HJ (2005) Self compacting concrete: theoretical and experimental study. *Cement and Concrete Research* **35(11)**: 2116–2136.

Chowdhury S and Basu PC (2010) New methodology to proportion self consolidating concrete with high volume fly ash. *ACI Materials Journal* **107(3)**: 222–230.

Corinaldesi V and Moriconi G (2004) Durable fiber reinforced self-compacting concrete. *Cement and Concrete Research* **34(2)**: 249–254.

Cunha VMCF, Barros JAO and Sena-Cruz JM (2010) Pullout behaviour of steel fibers in self compacting concrete. *Journal of Materials in Civil Engineering, ASCE* **22(1)**: 1–9.

Dhonde HB, Mo YO, Hsu TCC and Vogel J (2007) Fresh and hardened properties of self consolidating fiber reinforced concrete. *ACI Materials Journal* **104(5)**: 491–500.

Domone PL (2006) Self compacting concrete: an analysis of case studies. *Cement and Concrete Composites* **28(2)**: 197–208.

EFNARC (Experts for Specialised Construction and Concrete Systems) (2005) *The European Guidelines for Self Compacting Concrete, Specification, Production and Use.* EFNARC, p. 68.

Ferrara L, Park Y-D and Shah SP (2007) A method for mix-design of fiber reinforced self-compacting concrete. *Cement and Concrete Research* **37(6)**: 957–971.

Gesoglu M, Guneyisi E and Ozbay E (2009) Properties of self compacting concretes with binary, ternary and quaternary cementitious blends of fly ash, blast furnace slag and silica fume. *Construction and Building Materials* **23(5)**: 1847–1854.

Goel S, Singh SP and Singh P (2012) Flexural fatigue analysis of self-compacting concrete beams. *Proceedings of the Institution of Civil Engineers – Construction Materials* **165(6)**: 367–376.

Greengough T and Nehdi M (2008) Shear behaviour of fiber reinforced self consolidating concrete slender beams. *ACI Material Journal* **105(5)**: 468–477.

Grunewald S and Walraven JC (2001) Parameter study on the influence of steel fibers and coarse aggregate content on the fresh properties of self-compacting concrete. *Cement and Concrete Research* **31(12)**: 1793–1798.

Hans WR and Michael S (2006) Self consolidating concrete in fire. *ACI Materials Journal* **103(2)**: 130–135.

Hilsdorf HK and Kesler CE (1966) Fatigue strength of concrete under varying flexural stresses. *ACI Materials Journal* **63(50)**: 1059–1076.

Hsu TTC (1981) Fatigue of plain concrete. *ACI Materials Journal* **78(4)**: 292–305.

Johnston CD and Zemp RW (1991) Flexural fatigue performance of steel fiber reinforced concrete – Influence of fiber content, aspect ratio type. *ACI Materials Journal* **88(4)**: 374–383.

Kennedy JB and Neville AM (eds) (1986) *Basic Statistical Methods for Engineers and Scientists..* A Dun-Donnelly Publishers, New York, NY, USA, pp. 128–128.

Klaiber FW and Lee DY (1982) Effects of air content, water cement ratio and aggregate type on the flexural fatigue strength of plain concrete. *ACI Materials Journal Special Publication* **75(5)**: 111–132.

Mohammadi Y and Kaushik SK (2005) Flexural fatigue-life distributions of plain and fibrous concrete at various stress levels. *Journal of Materials in Civil Engineering, ASCE* **17(6)**: 650–658.

Oh BH (1986) Fatigue analysis of plain concrete in flexure. *Journal of Structural Engineering, ASCE* **112(2)**: 273–288.

Oh BH (1991) Fatigue-life distributions of concrete for various stress levels. *ACI Materials Journal* **88(2)**: 122–128.

Persson B (2000) A comparison between mechanical properties of self compacting concrete and the corresponding properties of normal concrete. *Cement and Concrete Research* **31(2)**: 193–198.

Persson B (2003a) Internal frost resistance and salt frost scaling of self compacting concrete. *Cement and Concrete Research* **33(3)**: 373–379.

Persson B (2003) Sulphate resistance of self compacting concrete. *Cement and Concrete Research* **33(12)**: 1933–1938.

Raithby KD (1979) Flexural fatigue behaviour of plain concrete. *Fatigue Engineering Materials, Structural* **2(3)**: 269–278.

Ramakrishnan V, Oberling G and Tatnall P (1987) *Flexural Fatigue Strength of Steel Fibre Reinforced Concrete.*. American Concrete Institute, Farmington Hills, MI, USA. ACI Special Publication SP-105-13, pp. 225–245.

Ramakrishnan V, Wu GY and Hosalli G (1989) Flexural fatigue strength, endurance limit and impact strength of fiber reinforced concretes. *Transportation Research Record* **1226**: 17–24.

Ramakrishnan V, Mayer C and Naaman AE (1996) Cyclic behaviour, fatigue strength, endurance limit and models for fatigue behaviour of FRC. In *High Performance Fibre Reinforced Cement Composites 2* (Namaan AE and Reinhardt HW (eds)). E and FN Spon, New York, NY, USA, pp. 101–148.

Sahmaran M and Yaman IO (2007) Hybrid fiber reinforced self compacting concrete with a high volume coarse fly-ash. *Construction and Building Materials* **21(1)**: 150–156.

Sahmaran M, Yurtseven A and Yaman IO (2005) Workability of hybrid fiber reinforced concrete. *Building and Environment* **40(12)**: 1672–1677.

Shi XP, Fwa TF and Tan SA (1993) Flexural fatigue strength of plain concrete. *ACI Materials Journal* **90(5)**: 435–440.

Singh SP and Kaushik SK (2000) Flexural fatigue life distributions and failure probability of steel fibrous concrete. *ACI Materials Journal* **97(6)**: 658–667.

Singh SP and Kaushik SK (2001) Flexural fatigue analysis of steel fiber reinforced concrete. *ACI Materials Journal* **98(4)**: 306–312.

Singh SP, Mahammadi Y, Goel S and Kaushik SK (2007) Prediction of mean and design fatigue lives of steel fibrous concrete beams. *Journal of Advances in Structural Engineering* **10(1)**: 25–36.

Singh B, Kumar R and Mukati AS (2009) SCC mixes with limestone and chalk powder – the Indian experience. *Indian Concrete Journal* **83(12)**: 9–17.

Sonebi M, Grünewald S and Walraven J (2007) Passing-ability and filling-ability of self consolidating concrete. *ACI Materials Journal* **104(2)**: 162–170.

Su N, Hsu KC and Chai HW (2001) A simple mix design method for self compacting concrete. *Cement and Concrete Research* **31(12)**: 1799–1807.

Tatro SB (1987) Performance of steel fiber reinforced concrete using large aggregates. *Transportation Research Record* **1110**: 129–137.

Tepfers R (1979) Tensile fatigue strength of plain concrete. *ACI Journal Proceedings* **76(8)**: 919–933.

Tepfers R and Kutti T (1979) Fatigue strength of plain, ordinary and light weight concrete. *ACI Material Journal* **76(5)**: 635–652.

Torrijos MC, Barragan BE and Zerbino RL (2008) Physical–mechanical properties and mesostructure of plain and fiber reinforced self-compacting concrete. *Construction and Building Materials* **22(8)**: 1780–1788.

Turkel S and Kandemir A (2010) Fresh and hardened properties of SCC made with different aggregate and mineral admixtures. *Journal of Materials in Civil Engineering, ASCE* **22(10)**: 1025–1032.

Zhu W and Gibbs JC (2004) Use of different limestone and chalk powder in self compacting concrete. *Cement and Concrete Research* **35(8)**: 1457–1462.

Section 6
Handling of self-compacting concrete

ice Publishing

Dhir and Lye
ISBN 978-0-7277-6461-4
https://doi.org/10.1680/icetscc.64614.427

Chapter 19
Influence of self-compacting concrete on the lateral pressure on formwork

Michael J. McCarthy BSc, PhD
Senior Lecturer, Concrete Technology Unit, Division of Civil Engineering, University of Dundee, Scotland, UK

Ravinda K. Dhir OBE, PhD, CEng, MIMM, Hon. FICT, FGS
Emeritus Professor, Concrete Technology Unit, Division of Civil Engineering, University of Dundee, Scotland, UK

Sinan Caliskan BSc, MPhil, PhD
Research Fellow, Concrete Technology Unit, Division of Civil Engineering, University of Dundee, Scotland, UK

M. Kashif Ashraf BEng, PhD, CEng, MICE
Civil Engineer, Historic Scotland, Edinburgh, Scotland, UK

The paper describes a study carried out to examine the effect of self-compacting concrete (SCC) on the lateral pressure on formwork. This comprised two series: (*a*) control tests, with SCC and comparative concretes (Portland cement (PC) and superplasticised, high-consistence (SP-HC)) cast in 8·0 m columns and walls, and measurements made of concrete and pore water pressures (ten elements) and (*b*) site tests, with similar properties considered on actual reinforced SCC members during their construction (seven elements). In the control tests, concrete was poured from above to the rising surface at slow (3·0 m/h) and fast (80·0 m/h) rates of rise (with compaction used in the PC and SP-HC concretes only). The results from these indicate that maximum concrete pressures with SCC were slightly lower than those of the comparative concretes at both rates of rise and in columns and walls. Pore water pressures were less than concrete pressures, except with the fast rate of rise for SCC and SP-HC concrete, where the reverse occurred (with the concretes having similar values or SCC slightly higher), which seemed to relate to the concrete properties/pour details and differences in concrete/pore water pressure measurement methods. Maximum pressures for SCC in the control tests were less than hydrostatic and those calculated using the CIRIA report 108 guidance method (Clear and Harrison, 1985). For the site tests, where base-to-top pouring was mainly used (i.e. concrete was passed by way of a hopper at the top of the formwork through a pipe running to its base and kept in place throughout the pour) maximum pressures for SCC were found to exceed hydrostatic in some cases, indicating that this method may give higher pressures. To investigate this further, data from the literature covering SCC pressure measurements on formwork were examined. These showed general agreement with the results of the current study using base-to-top pours and where concrete pumping from below had been adopted, which tended to be at the upper end of the maximum pressure range for SCC. Consideration is given to how SCC may fit within CIRIA 108 for estimating concrete pressure on formwork.

1. Introduction

One of the principal loads that formwork for vertical members must support is the pressure from concrete during construction. In most situations, concrete introduced to these elements is fluidified by compaction in layers, enabling it to pass around reinforcement and spread into formwork (Banfill *et al.*, 1999). As pouring progresses and the formwork fills, so the vertical load of concrete increases. At the same time, following compaction, shear strength develops in the concrete through the formation of a stable particle structure, which is influenced by physical interlock of solids and the setting of cement (Harrison, 1983). The magnitude of the lateral pressure on formwork depends on the combination of these effects and therefore on factors relating to the construction (e.g. concrete rate of rise) and concrete details (e.g. cement type). Indeed, once the stable particle structure has formed, further application of vertical load from the concrete above should have minimal effect on lateral pressure (Clear and Harrison, 1985).

The last 10–15 years have seen several changes to the constituent materials used in concrete, including a broader range of cements and additions (e.g. metakaolin and silica fume) covered in BS EN 197-1 (BSI, 2000) and developments with superplasticising admixtures (Edmeades and Hewlett, 1998). These can affect the fresh properties and period of fluidity retention compared to traditional materials and may influence concrete pressure on formwork. A research project was carried out by the present authors (Dhir *et al.*, 2009) to investigate this and included both laboratory and field experiments. It was found that the new concretes had only a minor effect on the pressure on formwork and therefore were unlikely to influence temporary works design significantly.

In examining this issue, it was noted that the other main innovation with concrete materials over the period was the concept of self-compacting concrete (SCC), which provides further options for engineers in meeting specific construction requirements (Bartos, 2005; Holton, 2004). SCCs contain high powder contents, superplasticising admixture and frequently viscosity-modifying admixture (VMA) to control stability. They normally have high flowability, enabling them to move into and fill formwork and to consolidate without the need for compaction (Concrete Society, 2005). Given the composition and rheology of SCCs, including their tendency to solidify when at rest (thixotropic nature) and that compaction is not used, the influence of these concretes on the pressure on formwork is uncertain. Furthermore, studies examining this aspect of SCC (e.g. Bernabeu, 2000; Billberg, 2003; Brameshuber and Uebachs, 2003; Galeota *et al.*, 2007; Khayat *et al.*, 2005; Leemann and Hoffmann, 2003; Proske and Graubner, 2002) have supported a range of pressures for various test conditions.

Guidance for estimating concrete pressure in practice exists, for example CIRIA report 108 (Clear and Harrison, 1985); however, this does not cover SCC, given its recent development. While hydrostatic behaviour of SCC on formwork can be adopted, it is unclear whether or not this is appropriate. A study was, therefore, set up at the University of Dundee, funded by the

Department of Trade and Industry, and industrial partners, with the aim of determining the influence of SCC on formwork pressure in full-scale elements and its impact on temporary works design.

2. Experimental programme

The approach followed in the experimental programme was similar to that used in the related study carried out by the authors (Dhir *et al.*, 2009). The first part examined elements cast for the project at a specially established site, to enable concretes and construction parameters of interest to be investigated (control tests). Tests were made on SCC containing fly ash (FA) (SCC-I) and ground granulated blast furnace slag (GGBS) (SCC-II), both with VMA, cast in 8·0 m (unreinforced) column and wall elements using formwork instrumented to measure concrete and pore water pressures developing during construction. The latter, which is the pressure of the liquid phase of fresh concrete, has typically been recorded at levels of around 90% of concrete pressure (Assaad and Khayat, 2004; Harrison, 1983). Measurements of changing concrete head with time in formwork and its temperature during placing, which are important factors with regard to pressure development, were also made. In addition to SCC, Portland cement (PC) and superplasticised, high-consistence (SP-HC) concretes were tested for comparative purposes. The other variable considered was rate of rise of concrete in formwork, with slow (3·0 m/h) and fast (80·0 m/h) rates examined.

In the second part of the study, tests were carried out on SCCs used in seven elements at three construction sites. This part of the research followed a similar approach to the control tests and enabled different concretes and construction conditions to be examined. Given the practical nature of the work, it was also intended to explore whether SCC could fit within the CIRIA 108 guidance method, which is currently used to estimate concrete pressure on formwork.

3. Pressure measurements and test procedures

The main test equipment for the pressure measurements was based on that used in earlier work and has been described in detail previously (Dhir *et al.*, 2004; Habgood, 1982). This comprised stainless steel load cells (to measure concrete pressure) positioned across the formwork height (flush with the formwork surface in contact with concrete) and piezometers (pore water pressure transducers (PWPTs)) at lower cell positions. The measuring devices were calibrated in the laboratory with a hydrostatic pressure vessel to establish voltage–pressure relationships. These were checked between pours to ensure their reliability, but indicated little variation during the test period.

All concrete used in both control and site tests was ready-mixed. Before placing, the fresh concrete properties, that is plastic density and slump/slump flow, were measured. As concrete was introduced to the formwork, pressure (voltage) readings (from the load cells and PWPTs) were recorded by data logger (every minute, or more frequently at faster rates of rise) and these were continued for several hours thereafter. The increasing head of concrete was measured with respect to time (manually from the top of the formwork) and concrete and ambient temperatures were also recorded during the test period. Using these techniques, measurements were made without influencing the way concrete was placed.

4. Control tests

4.1 Materials

The PC for the control tests was of strength class 42·5 N (CEM I) to BS EN 197-1 (BSI, 2000). In addition, a Type S FA to BS EN 450-1 (BSI, 2005) and a GGBS to BS 6699 (BSI, 1992) were used.

The coarse aggregate was a natural gravel of designation 4/20 for PC and SP-HC concretes and 4/10 for SCC to BS EN 12620 (BSI, 2002). The natural sand used was of medium grade.

A superplasticising admixture, based on modified polycarboxylic ether, to BS EN 934-2 (BSI, 2001) and a VMA, based on a high-molecular-weight hydroxylated polymer, were also used during the control tests.

4.2 Concrete mixes and production

The concrete mixes were developed based on information and guidance in the literature (EFNARC, 2005; Teychenne *et al.*, 1997), discussions with the concrete producer and trial mixing in the laboratory. The PC and SP-HC concretes (with target slumps of 100 mm and 180 mm) had cement contents of 350 kg/m^3, with water contents adjusted at the site, where necessary, to achieve the required consistence (Dhir *et al.*, 2009). The fine to total aggregate ratio of these was 0·43 and the SP-HC concrete contained superplasticiser, used in accordance with the suppliers' technical information. The SCC mixes, SCC-I (PC/30% FA) and SCC-II (PC/40% GGBS), had cement contents of 530 and 540 kg/m^3, respectively, with water content again adjusted at the site, to give the required fresh properties (target slump flow of 700 mm). Their fine to total aggregate ratio was 0·52 and both contained superplasticiser and VMA, also used in accordance with the suppliers' technical information. Typical fresh properties of the test concretes, before placing, are given in Table 1. Visually all mixes appeared stable following production and could be readily pumped.

4.3 Formwork arrangements

Preliminary tests on similar mixes to those in this study (Dhir *et al.*, 2004) using 3·0 m columns at a rate of rise of 3·0 m/h (with concrete temperatures of approximately 10°C) indicated that maximum pressures for these were in a narrow range (around 60·0 kN/m^2) and within the

Table 1 Typical fresh properties and temperatures of control concretes

Concrete type	Plastic density: kg/m^3	Slump/slump flow: mm	T_{500}: s	Concrete temperature: °C
PC	2395	100	NA	21·0
SP-HC	2430	200	NA	20·5
SCC-I	2345	640	3·1	21·0
SCC-II	2370	680	2·8	20·0

NA: not applicable.

variability of the test equipment (Dhir *et al.*, 2009). In the current work, a greater height was adopted (8·0 m) to provide full-scale elements and conditions likely to give higher pressures at which the material comparisons could be made.

The formwork for the 8·0 m column (500 mm × 500 mm section) and wall (3·0 m long × 500 mm wide) elements were proprietary systems of plywood with aluminium supports. The formwork was slightly higher than 8·0 m and any minor deviations in pour height from this, where noted, were taken into account in calculations of hydrostatic and CIRIA 108 pressures. Details of the element dimensions and load cell positions are given in Figure 1.

Figure 1 Details of column and wall elements and positions of load cells (dimensions in mm)

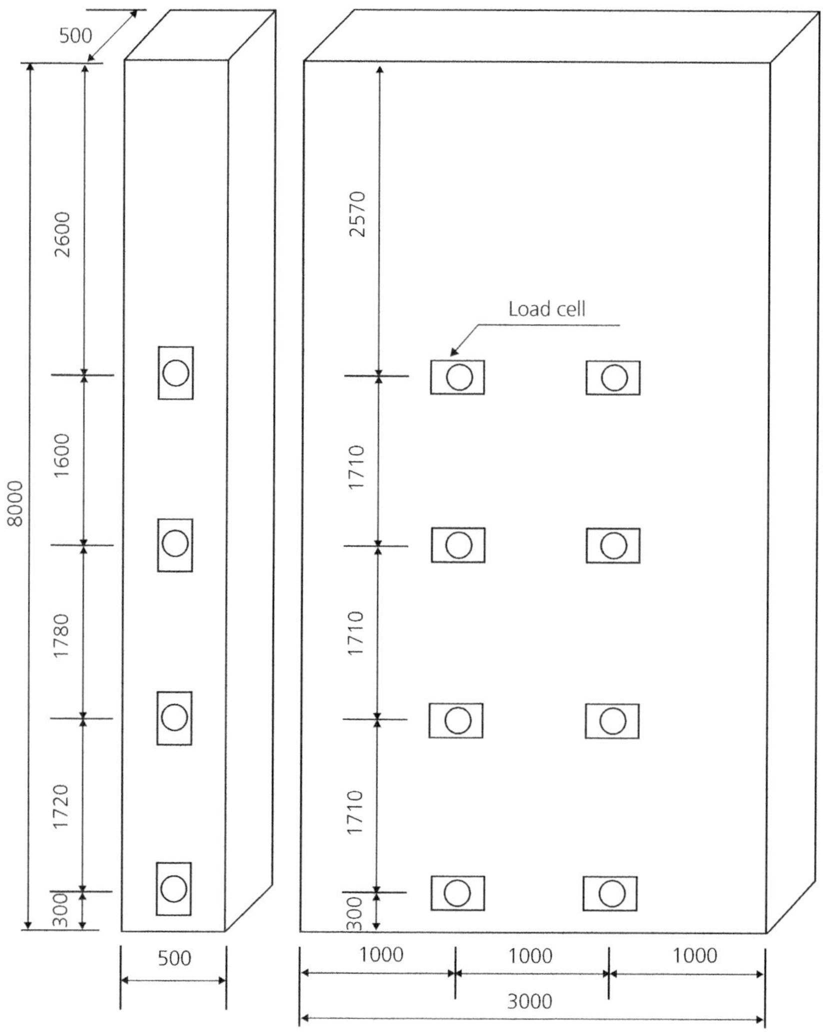

4.4 Placing of concrete

Concrete was transported from the transit mixer, using a pump, through a flexible pipe initially extending to the formwork base. As concrete was discharged, the end of the pipe was raised to maintain a distance from the surface of the rising material of about 0·5 m to limit impact effects. For the walls, the pipe was moved along their length as they were poured.

For the columns and walls tested at 3·0 m/h rate of rise, two batches of concrete were used and material was discharged in increments around 1·0 m high, followed by a rest period before the next layer was placed, to give the required rate of rise. Compaction (not used for SCC) of concrete was carried out internally using a poker until completion of each 1·0 m layer placed. For columns tested at the fast rate of rise, concrete was discharged continuously (from a single batch) into the formwork, which was filled in about 6 min (approximately 80·0 m/h). In this case, compaction (used for the SP-HC concrete only, with PC concrete not tested) was applied throughout until the pour was complete. The concrete temperatures during construction of the elements were mainly around 20°C.

4.5 Results and discussion

As an example of the typical form of data recorded, the development of concrete and pore water pressures at the lowest load cell position of PC and SCC-II concrete columns (3·0 m/h rate of rise) are given in Figure 2. Vertical pressures (above the load cells), that is the product of

Figure 2 Pressure development with time for (a) PC concrete and (b) SCC-II (columns, 3·0 m/h rate of rise)

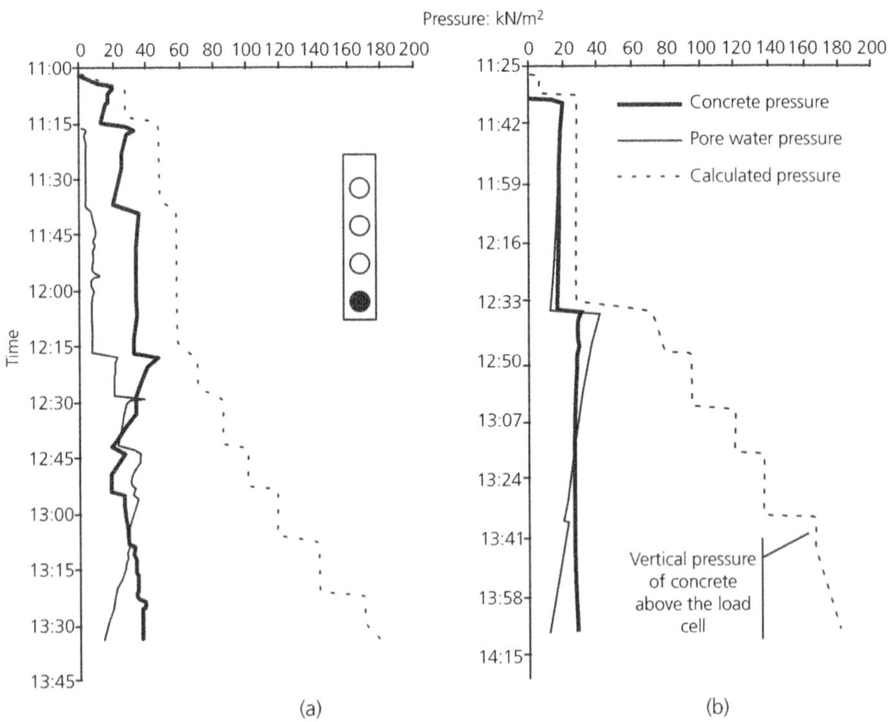

measured concrete pour height with respect to the load cell position, plastic density of concrete and the gravity constant ($9{\cdot}81\ m/s^2$), are shown in the figure (broken lines).

These data demonstrate the stages of placing concrete (step increases in pressure indicate concrete addition to formwork and compaction when used) and that both concrete and pore water pressures followed the vertical pressure. Beyond about 65 to 75 min, the concrete and pore water pressures gave little or no increase, suggesting that the stable particle structure was sufficiently developed at the load cell position and continued load application (from concrete poured above) was transferred vertically. For PC concrete, the minor steps noted in concrete and pore water pressures beyond 12·15 h are likely to reflect periodic vibration. Maximum pressures from these types of plot for each load cell/PWPT were used to examine the effects of SCC.

A comparison of maximum concrete and pore water pressures for the test concretes (3·0 m/h rate of rise) in column elements is given in Figure 3. The calculated maximum pressures (P_{max}), based on CIRIA 108, using Equation 1, given below, are also shown in the figure.

$$P_{max} = D\left[C_1\sqrt{R} + C_2K\sqrt{\left(H - C_1\sqrt{R}\right)}\right] \text{ or} \tag{1a}$$

$$Dh\left(\mathrm{kN/m^2}\right), \text{whichever is smaller} \tag{1b}$$

where D is weight density of concrete (25 kN/m^3 for normal-weight concrete); C_1 is formwork size and shape coefficient (1·0 for walls, 1·5 for columns) ($\sqrt{mh}$); C_2 is concrete constituent material coefficient (0·3 to 0·6) ($\sqrt{m}$); R is vertical rate of concrete rise up the form (m/h); H is vertical form height (m); h is vertical pour height (m); T is temperature of concrete at placing (°C) and K is temperature coefficient = $[36/(T+16)]^2$. When $C_1\sqrt{R} > H$, the fluid pressure (Dh) is taken as the design pressure.

For the comparisons made in the paper h was used rather than H in Equation 1a (although in most cases they were the same) and was the distance from the top of the pour to the lowest pressure measurement point.

Figure 3 Comparison of maximum pressures at 3·0 m/h rate of rise (columns)

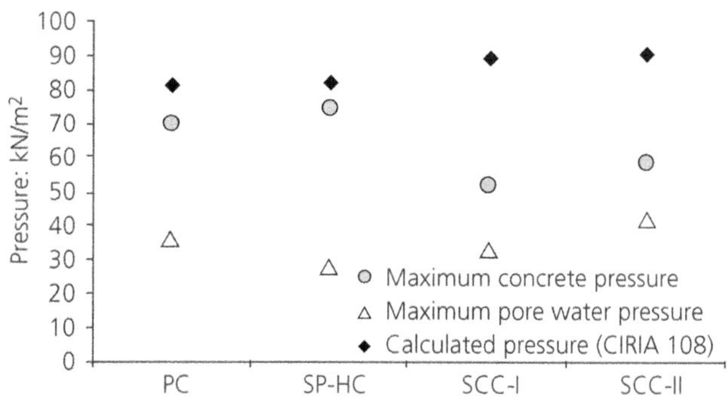

The SP-HC concrete gave maximum concrete pressure around 5 kN/m^2 higher than that of PC, which is within the range of the test equipment variability (Dhir *et al.*, 2009). Maximum pore water pressures were about 40–50% of the concrete pressure for these concretes. The effect of superplasticiser corresponds to its coverage in CIRIA 108, which includes these concretes in the general grouping. Both concretes also gave lower maximum pressures than those estimated by the method.

For the concrete mixes, test conditions and methods used, the results indicate that SCC gave slight reductions in maximum concrete pressure of about 10–20 kN/m^2 compared to those of PC and SP-HC (with SCC-I giving the lower of the two SCCs). This effect may correspond to (*a*) the use of compaction for PC and SP-HC concretes, which may influence the period to stable particle structure formation (Harrison, 1983) and transmit energy to the formwork (Clear and Harrison, 1985) and (*b*) the compositional differences and thixotropic nature of SCC compared to the other concretes (likely to promote structure formation).

The magnitudes of maximum pressure for the concretes in these tests are also close to those obtained during the preliminary work on 3·0 m columns (Dhir *et al.*, 2004). Given:

(*a*) the similarity in pressures, despite the 3·0 and 8·0 m pour heights (and allowing for the different temperatures between these tests)
(*b*) the behaviour in Figure 2 in terms of the time at which no further pressure increases occurred
(*c*) that small differences in maximum pressures were noted between load cells at the lower positions of the 8·0 m columns

it appears that the stable particle structure formed, during the tests, when the concretes were around 3·0 to 4·0 m above the load cells.

While the concrete pressures in the 8·0 m tests were lower for SCCs, their pore water pressures were similar to, or slightly higher than the other concretes. The pore water pressures were also less than the 90% level of the concrete pressure referred to in the literature. These suggest possible differences in particle structure and pore water pressure contributions to the pressure on formwork for the test concretes (Harrison, 1983), than noted elsewhere. However, given these observations, this represents an area requiring further investigation.

The effect of using a fast rate of concrete rise in formwork (80·0 m/h) is shown in Figure 4. This indicates that maximum pressures increased for the concretes tested by about 50–70 kN/m^2 compared to those at the 3·0 m/h rate of rise, and highlights the role of this parameter on pressure development, as noted previously (e.g. Billberg, 2003; Gardner, 1986). Comparing SP-HC and SCCs, reductions in maximum concrete pressure for the latter of between about 15 and 25 kN/m^2 (with SCC-II the lower of these) were noted. For the test conditions used, the maximum pore water pressures were higher than the maximum concrete pressures, with SCC tending to be the slightly higher of these. This seems to relate to the rapid compression

Figure 4 Comparison of maximum pressures at 80·0 m/h rate of rise (columns)

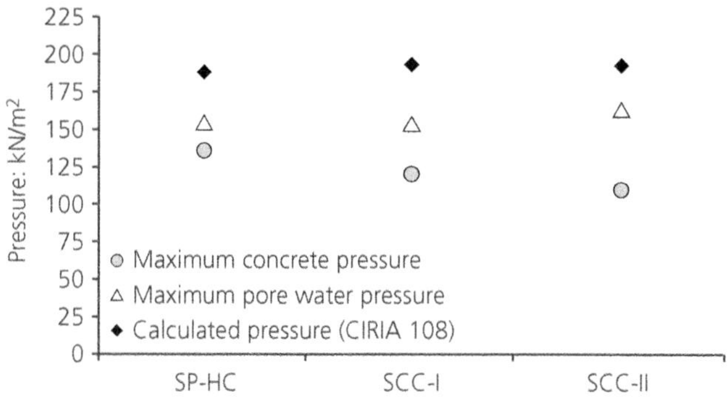

(due to the rate of rise) and high concrete fluidity (Dhir *et al.*, 2009) and to the differences in pressure measurement methods (c.f. load cells and PWPTs). The maximum pressures (pore water or concrete) were, however, less than those calculated by CIRIA 108, which were Dh given the fast rate of rise (i.e. $C_1\sqrt{R} > h$).

The effect of concrete element type is shown in Figure 5. This indicates that the ranking of maximum pressure for the different concretes placed in walls followed that of columns, but with lower values recorded. For normally compacted concretes, similar type differences in maximum pressures between columns and walls reflect greater rest periods between placing of material as the discharge pipe moves along the formwork in the latter. In addition not all concrete in a layer is necessarily mobilised in these elements at the same time and less energy is transmitted to formwork by compaction (Clear and Harrison, 1985). It appears that the periods at rest and thixotropic nature of SCC (Bartos, 2005) lead to a similar effect with the material between element types.

Figure 5 Comparison of maximum pressures at 3·0 m/h rate of rise for columns and walls

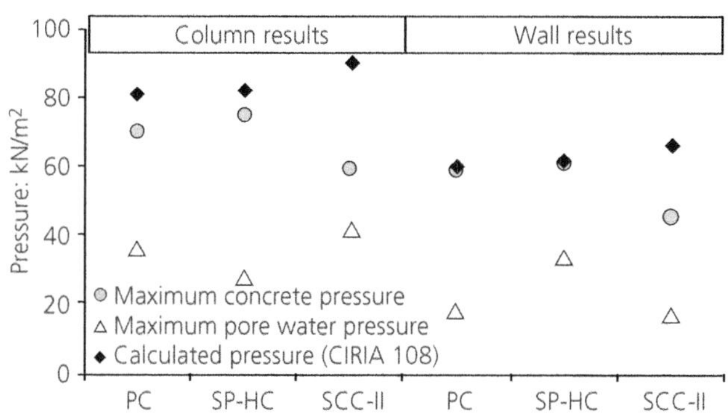

5. Construction site tests

Having investigated the effect of SCC on the pressure on formwork in the control tests, the next stage examined this during casting of elements at actual construction sites. Measurements were made on seven reinforced concrete elements, (a column and six walls) at three sites, with load cells located across the height and PWPTs at the lower test positions of their formwork, Table 2. The concretes at all sites were believed to contain limestone (LS) as an addition/filler. Compaction was not used during the construction of the elements.

At site 1, an L-shaped column (element 1) and three walls (elements 2–4) were considered. The former was located on the corner of two previously cast walls and was 4·8 m high, with outer legs of the 'L' about 1000 mm. The walls of similar height were approximately 200 and 500 mm wide and cast in 7·0 and 5·5 m lengths. A skip transferred concrete from the transit mixer to a hopper fixed at the top of the column or wall formwork. A flexible pipe attached to this extended to the base of the elements. This arrangement was maintained during placing and concrete discharged to the base throughout (i.e. base-to-top pour). The average rate of rise of concrete was 14·0 m/h for the column and 6·0 m/h for the walls.

At site 2, two walls (elements 5 and 6) were tested. These were 5·0 and 5·8 m high, 7·5 and 5·7 m long and 200 mm wide. In both cases, concrete was placed by way of a fixed hopper and flexible pipe in a similar arrangement to site 1. The pipe was removed when concrete was 100–200 mm from the top of the element and the remainder poured through the hopper. The average rates of rise were 2·7 and 7·0 m/h for elements 5 and 6 respectively. At Site 3, a wall (element 7) 2·5 m high, 2·7 m wide and 24·0 m long was examined. In this case, concrete was transported by pump and poured from a height of about 0·5 m above the rising surface. This was carried out at an average rate of rise of about 0·3 m/h.

Table 2 Construction details and fresh properties of concrete (site tests)

Element no. and type	Formwork height: m	Plastic density: kg/m^3	Slump flow: mm	Concrete rate of rise: m/h	Concrete temperature: °C	Pour arrangement
1. Column	4·8	2360	780	14·0	21·0	B to T*
2. Wall	4·8	2360	730	6·0	23·0	B to T
3. Wall	4·8	2410	650	6·0	21·0	B to T
4. Wall	4·8	2405	680	6·0	22·0	B to T
5. Wall	6·0	2340	600	2·7	15·5	B to T
6. Wall	5·8	2335	700	7·0	11·0	B to T
7. Wall	2·5	2430	700	0·3	13·0	From top

* B to T: Base-to-top pour method.

The results from these tests are given in Figure 6 and, as LS is not covered in CIRIA 108, pressures using this method are indicated as a range (with C_2 as 0·3 or 0·6 in Equation 1a, i.e. the lower and upper values in the document), except for element 1 where the estimated pressure was Dh (i.e. $C_1\sqrt{R} > h$). As above, h was the pour height (although, with the fixed hopper arrangement at sites 1 and 2 the height of concrete during pouring would have been greater than this). The column at site 1 (element 1) had a maximum concrete pressure, measured at the bottom load cell of 123·5 kN/m^2. The maximum pore water pressure was about two-thirds of this. For the wall elements (elements 2–4), maximum concrete pressures (recorded at the lowest load cells) were between 55·0 and 80·0 kN/m^2, with maximum pore water pressures below and above these. A comparison of maximum concrete or pore water pressures (where higher), with the calculated CIRIA 108 pressures, indicates that those measured were higher (element 1), within or just below the estimated range (with element 1 exceeding hydrostatic pressure, 106·5 kN/m^2).

The site 2 results (elements 5 and 6), also shown in Figure 6, gave maximum concrete pressures of 127·5 and 142·5 kN/m^2 at the lowest load cells (despite a relatively low rate of rise for the former). The maximum pore water pressures were less than half of these. For both, the maximum concrete pressures were higher than those estimated by CIRIA 108 (and were greater than hydrostatic pressures, 110·0 and 128·5 kN/m^2). The results indicate differences in maximum pressures between the walls of sites 1 and 2 for similar placing arrangements, although the reason for this is unclear. For site 3 (element 7), the slow rate of rise gave a maximum pressure of 6·5 kN/m^2 when the head above the load cell was about 400 mm (giving a vertical pressure of approximately 10 kN/m^2).

Figure 6 Comparison of maximum pressures for site test elements

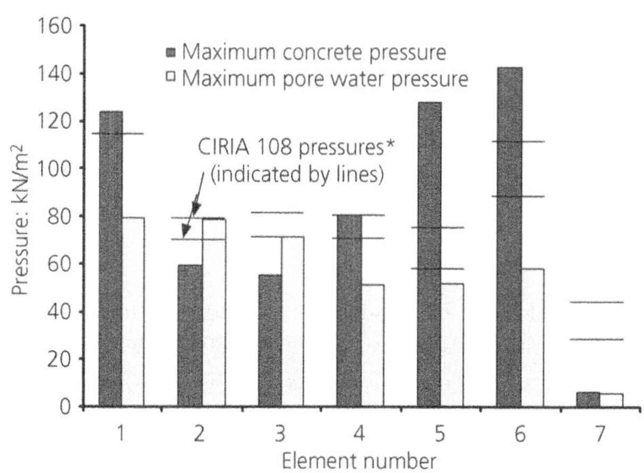

* Element 1: CIRIA 108 pressure calculated by Dh
Elements 2–7 pressure range for C_2 (material coefficient): 0·3 (lower line) and 0·6 (upper line)

The control tests and those at the construction sites indicate some differences between results. Indeed, while the former gave maximum concrete pressures less than those of CIRIA 108 (and hydrostatic), the latter, in some cases, had maximum pressures close to or exceeding this. Given the construction details for these, in other words concrete poured from above to the rising surface in the control tests, compared to the base-to-top method for most of the site elements, it seems that these differences may relate to the placing methods.

Base-to-top pouring may reduce the effects of impact pressure, as concrete entering formwork is absorbed into the body of the material. However, with this arrangement, there is pressure between material in the hopper (at the top of the pour) and that at lower levels, causing it to move upwards and fill the formwork. Furthermore, in this case, concrete near the base, being that most recently introduced to the formwork is likely to retain its fluidity, while that close to the top will stiffen (i.e. the reverse of where concrete is poured from above) and hence placing should be completed before significant hardening occurs. It is possible that such effects may influence concrete pressure on formwork.

6. Comparison with other SCC formwork pressure data

Given the differences noted in concrete pressure between control and site tests, it was decided to examine data from the study with these in the available literature considering this aspect of SCC. A summary of data from selective literature is given in Table 3, while data from the current study are shown in Table 4. Table 3 includes data from laboratory and full-scale site tests, where the information given enabled comparisons. Most concretes contained either FA or LS as additions/fillers, with some including VMA (and air-entraining admixtures). Slump flows were generally in the range 550–800 mm.

Tables 3 and 4 compare maximum pressures recorded and corresponding hydrostatic pressures (from density measurements/mix proportions) for elements at the positions in the formwork where the former were obtained, which were mainly at the lowest measurement points. These indicate that maximum pressures ranged from less than 20% of hydrostatic pressure to values in excess of 100%, see Figure 7. When rates of rise were 3·0 m/h or less, with pouring from above, maximum pressures ranged from 11 to 82% of hydrostatic pressure. More than half of the tests with these relatively low rates of rise gave maximum pressures less than 60% of hydrostatic. Above 3·0 m/h, the corresponding range was 29–100% of hydrostatic pressure, with most greater than 60%, generally increasing with rate of rise.

Where pumping from below, or base-to-top pours were used to place concrete (shaded in Tables 3 and 4), the majority of results were more than 80% of hydrostatic pressure (8 of 11), with most of these between 95 and 116%. Values < 80% were all obtained from site 1 of the current study, although the reason for this is not known. Given the effects noted above with regard to base-to-top pouring, and that when pumping from below, pressures in excess of the concrete head are necessary to move material from the pump, the high pressure levels obtained against formwork are reasonable. Both of these placing arrangements therefore give concrete pressures, which tend to be at the upper end of those measured for SCC.

Table 3 Study details and comparison of maximum pressures with respect to hydrostatic pressures for SCCs (selective literature)

Reference	Element no. and type	Pour height: m*	Additions or fillers in SCC	Slump flow: mm	Concrete rate of rise: m/h	Maximum pressure: kN/m^2	Maximum pressure/ hydrostatic pressure: %
Bernabeu (2000)	8W	2·1–7·5	LS	620–800	6·0–120·0	32·5–111·5	29–100
	5C	2·2–5·0	LS	580–890	10·0–150·0	46·0–114·5	83–100
	3C	5·0	LS	720–890	11·0–60·0	110·5–120·5	96–105
Proske and Graubner (2002)	5C	4·0	FA	550–750	25–60	40·0–90·0	43–97
Billberg (2003)	7W	2·9	LS	650–780	0·8–2·3	12·5–49·5	34–77
Leemann and Hoffmann (2003)	3C	2·6	FA	570–660	8·0	50·5–52·0	87–90
	1W	3·9	FA‡	700	18·8	97·0	110
Vanhove *et al.* (2004)	1C	10·4	LS	700	10·3	174·5	74
	1C	10·4	LS	700	24·0	197·5	80
Fedroff and Frosch (2004)	2W	3·4	FA	660–710	1·2	21·0–26·0	34–37
Tejeda-Dominguez *et al.* (2005)	1W	7·9	FA	600–740	1·2	38·5	24
	1C	2·9	FA	600–740	18·3	54·5	79
Assaad and Khayat (2006)	2C	2·8	FA/SF	540–745	10·0	49·5–58·0	88–94
Cauberg and Desmyter (2007)	3C	3·8	NA	685–810	10·0	69·0–85·0	85–97

* Height from top of element to lowest pressure measurement point; C: column; W: wall (using CIRIA 108 definition).
‡ Involved a retarder.
Shaded areas: Pumping from below, except one case (Bernabeu, 2000) when the discharge pipe was kept under SCC level during the pour.
SF: silica fume; NA: information not available.

Table 4 Comparison of maximum pressures (concrete or pore water) with respect to hydrostatic pressures for SCCs (current study)

Part of study	Element type	Pour height: m*	Additions or fillers in SCC	Slump flow: mm	Concrete rate of rise: m/hr	Maximum pressure: kN/m^2	Maximum pressure/ hydrostatic pressure: %
Control tests	1W	8·0	GGBS	750	3·0	46·0	25
	4C	7·7–7·9	FA, GGBS	640–680	3·0–80·0	52·5–121·0 (162·0)	61–82 (91)
Site tests	1W	2·4	LS	700	0·3	6·5	11
	5W	4·6–5·6	LS	600–730	2·7–7·0	55·5 (71·5)–142·5	51 (66)–116
	1C	4·6	LS	780	14·0	123·5	116

* Height from top of element to lowest measuring point; C: Column, W: Wall (using CIRIA 108 definition).
Shaded areas: Base-to-top pours.
( ) indicates where pore water pressure exceeded concrete pressure during measurements.
Note the higher measured pressures (concrete or pore water) were used for the data shown in Figures 7 and 8.

Figure 7 Comparison of SCC maximum pressures with respect to hydrostatic pressures (current study and selective literature)

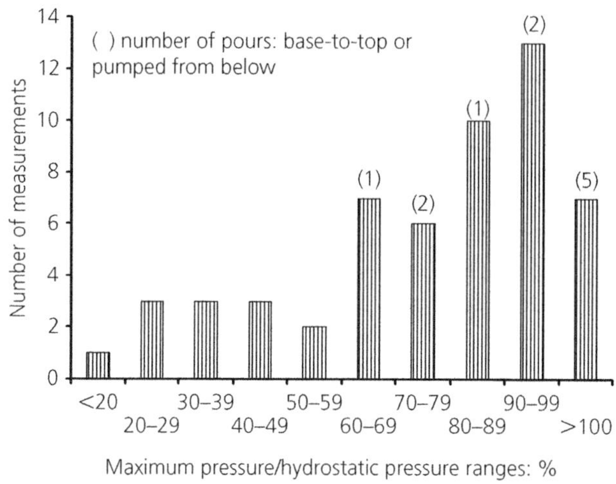

Clearly, concrete pressure on formwork is influenced by several factors (Clear and Harrison, 1985) and the above data correspond to particular details of the individual studies. For example, the following could have influenced the results:

(*a*) the concretes used (e.g. it has been suggested between SCCs, that variations in flow properties (Proske and Graubner, 2002), thixotropy (Assaad *et al.*, 2003; Bartos, 2005), and aggregate friction due to aggregate sizes and contents (Assaad and Khayat, 2004) may influence concrete pressure)

(*b*) the wall/column details, formwork and placing arrangements (e.g. element dimensions and discharge of concrete from different heights)

(*c*) the methods of pressure measurement (including instrumented ties, or sensors in formwork).

However, as indicated, it was possible to identify general behaviour with regard to SCC effects on concrete pressure and the influence of placing method on this.

7. Comparison of measured and CIRIA 108-estimated pressures

The results from the control tests of the current study, as noted above, suggest that CIRIA 108 could be applied to SCC, but data from the site tests gave pressures in excess of those estimated by this method, which may relate to how concrete was placed. In order to examine this further, a comparison between calculated (CIRIA 108) pressures using Equation 1 and maximum pressures recorded in the current study and the literature (given in Table 3) was made. For the calculations, C_1 was based on the definition for walls and columns in CIRIA 108 (i.e. column unless either plan dimension exceeds 2·0 m). C_2 was taken as 0·45 for SCCs (the concretes considered mainly contained, FA, GABS, or LS). The pour height was from the top of the element to the lowest pressure measurement point. Concrete rates of rise and temperatures were as reported (or used in making similar type comparisons).

The data from this comparison are shown in Figure 8. It should be noted that in carrying this out about 60% of the CIRIA 108 pressures were obtained from Dh (given that $C_1\sqrt{R} > h$), with the fast rates of rise of concrete in these cases. The results indicate that most of the maximum concrete pressures with conventional pouring (i.e. from above to the rising concrete surface) were below those calculated by CIRIA 108. The point above the line may have been due to the particular concrete composition and placing arrangements used. Where pumping from below or base-to-top pour methods were used, maximum concrete pressures similar to or above those of CIRIA 108 were obtained (with the latter tending to be the higher). This generally agrees with the existing guidance given by the method with regard to placing, which indicates that it does not apply to concrete being pumped from below, where pressures are likely to be higher. Therefore, given SCC lends itself to these types of placing arrangement, this is likely to be an important issue in the development of guidance for estimating concrete pressures on formwork for the material.

It should also be recognised that SCCs cover a range of compositions and properties, and, as noted above, the data in Figure 8 correspond to the particular materials, procedures and test methods of the various studies and are indicative of general behaviour for these types of concrete in comparison to the estimation method.

Figure 8 Comparison of measured maximum and calculated (CIRIA 108) pressures for SCCs (current study and selective literature)

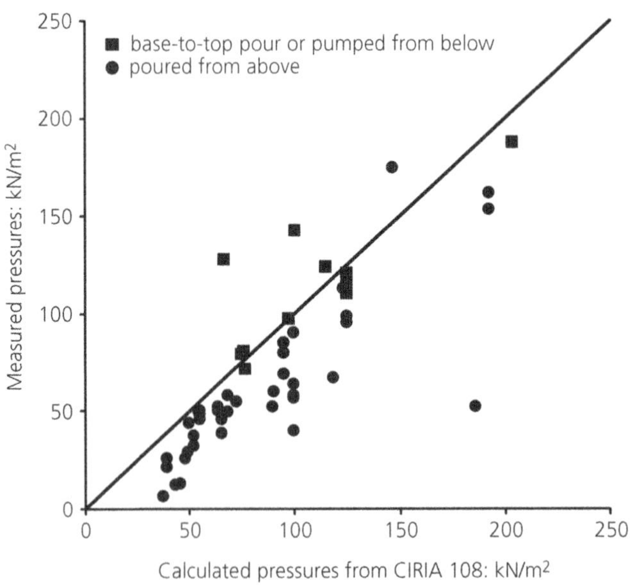

8. Conclusions

8.1 Control and site tests

Based on the concretes, test conditions and methods used in the study, the following conclusions were drawn from the control and site tests.

SCC poured from above through a pipe, kept at around 0·5 m above the rising concrete surface, gave maximum pressures slightly less than those of PC and SP-HC concretes. This was observed in tests for both slow and fast rates of rise of concrete in formwork (3·0 and 80·0 m/h) and with different element types (columns or walls).

For the slow rate of rise, maximum pore water pressure, as a percentage of maximum concrete pressure on formwork, tended to be higher for SCC than the other concretes. However, all values recorded were less than the 90% level noted for this in the literature. At the fast rate of rise SCC and SP-HC gave greater maximum pore water pressures than those measured for concrete pressure, with SCC tending to be slightly higher. Concrete and pore water pressures in the control tests were less than those estimated using CIRIA 108.

Tests at construction sites on actual elements using SCC indicated that base-to-top pouring techniques (i.e. the supply hopper was fixed at the top of the formwork and the discharge pipe kept at its base throughout the pour) were mainly used with the material. The results from these indicated that the pressures of SCC on formwork in several cases were higher than hydrostatic and those estimated using CIRIA 108.

8.2 Comparison with other SCC formwork pressure data

To examine this further, data from the current study and selective literature investigating pressure on formwork from SCC were considered collectively. This indicated that pressures at the lower end of the range for SCC, with respect to hydrostatic, were obtained with concrete poured from above (to the rising concrete surface), generally increasing with rate of rise. In most cases, where base-to-top pouring or pumping from the base of formwork were used, maximum concrete pressures were at the higher end of the range of pressures for SCC, that is between 80 and 116% of hydrostatic pressure.

8.3 Comparison of measured and CIRIA 108-estimated pressures

An attempt was made to compare concrete pressures calculated using CIRIA 108 and maximum pressures measured in the current study and literature for SCC. This indicated that in cases where concrete was poured from above, the measured pressures were generally less than those determined by the method. However, elements produced by base-to-top pouring, or pumping from the base of formwork, had maximum concrete pressures similar to or above those estimated using CIRIA 108. This suggests that for base-to-top pours, alternative methods would be required to evaluate concrete pressure (CIRIA 108 does not apply to pumping of concrete from below).

Acknowledgements

The authors acknowledge the financial support for the project from the Department of Trade and Industry and industrial partners, and the guidance given by their representatives namely

Castle Cement Ltd (P. Livesey), Cementitious Slag Makers Association (D. D. Higgins), CIRIA (A. Pitchford), Degussa Construction Chemicals UK (R. A. Jones), Edmund Nuttall Ltd (J. Payne), Ischebeck Titan Ltd (D. Smith), John Doyle Construction Ltd (P. Goring), Lafarge Aggregates Ltd (P. Barker), Alfred McAlpine Capital Projects (A. G. Peters), AWG Construction Services Ltd (A. MacLennan), Quarry Products Association (T. A. Harrison), Rugby Cement (B. Patel), SGB Formwork (C. J. C. Carey), The Concrete Society (R. Day), WS Atkins Consultants Ltd (R. P. Wolstenholme) and the British Cement Association (C. A. Clear).

REFERENCES

Assaad JJ, Khayat KH and Mesbah H (2003) Variation of formwork pressure with thixotropy of self-consolidating concrete. *ACI Material Journal* **100(1)**: 29–37.

Assaad JJ and Khayat KH (2004) Variations of lateral and pore water pressure of self-consolidating concrete at early age. *ACI Material Journal* **101(4)**: 310–317.

Assaad JJ and Khayat KH (2006) Effect of mixture consistency on formwork pressure exerted by highly flowable concrete. *ASCE Journal of Materials in Civil Engineering* **18(6)**: 786–791.

Banfill PFG, Xu Y and Domone PLJ (1999) Relationship between the rheology of unvibrated fresh concrete and its flow under vibration in a vertical pipe. *Magazine of Concrete Research* **51(3)**: 181–190.

Bartos PJM (2005) *Self-compacting Concrete in Bridge Construction*. The Concrete Society, Concrete Bridge Development Group, Camberley, Surrey, Technical Guide No. 7.

Bernabeu A (2000) *SCC – Form System and Surface Quality*. Brite EuRam Proposal No. BE96-3801 (GTM Construction, Marolles, France).

Billberg P (2003) Form pressure generated by self-compacting concrete. In *Proceedings of the 3rd International RILEM Symposium on Self-Compacting Concrete, Reykjavic, Iceland* (Wallevik O and Nielsson I (eds)). PRO 33, RILEM Publications SARL, Bagneaux, France, pp. 271–280.

Brameshuber W and Uebachs S (2003) Investigations on formwork pressure using self-compacting concrete. In *Proceedings of the 3rd International RILEM Symposium on Self-Compacting Concrete, Reykjavic, Iceland* (Wallevik O and Nielsson I (eds)). PRO 33, RILEM Publications SARL, Bagneaux, France, pp. 281–287.

BSI (British Standards Institution) (1992) BS 6699: Specification for ground granulated blastfurnace slag for use with Portland cement. BSI, London.

BSI (2000) BS EN 197-1: Definitions and composition, Cement: composition, specifications and conformity criteria. BSI, London.

BSI (2001) BS EN 934-2: Admixtures for concrete, mortar and grout. Concrete admixtures – definitions, requirements, conformity, marking and labelling. BSI, London.

BSI (2002) BS EN 12620: Specifications for aggregate for concrete including those for use in roads and pavements. BSI, London.

BSI (2005) BS EN 450-1: Fly ash for concrete – Part 1, Definitions, specifications and conformity criteria. BSI, London.

Cauberg N and Desmyter J (2007) Measuring the formwork pressure of self-compacting concrete. *Proceedings of the 5th International RILEM Symposium on Self-Compacting Concrete, Ghent, Belgium* (De Shutter G and Boel V (eds)). PRO 54, RILEM Publications SARL, Bagneaux, France, pp. 503–508.

Clear CA and Harrison TA (1985) *Concrete Pressure on Formwork*. CIRIA (Construction Industry Research and Information Association), London, Report 108.

Concrete Society (2005) *Self-compacting Concrete – A Review*. The Concrete Society, Camberley, Surrey, Technical Report No. 62.

Dhir RK, McCarthy MJ, Caliskan S and Ashraf MK (2004) *Design Formwork Pressures for the Range of New Cement, Superplasticized and Self-Compaction Concretes*. Department of Trade and Industry Research Contract No. 39/3/739, September 2004 (University of Dundee, Scotland).

Dhir RK, McCarthy MJ, Caliskan S and Ashraf MK (2009) Concrete pressure on formwork: influence of cement combinations and superplasticizing admixtures. *Magazine of Concrete Research* **61(6)**: 407–417.

Edmeades RM and Hewlett PC (1998) Cement admixtures. In *Leas' Chemistry of Cement and Concrete* (Hewlett PC (ed.)). Arnold, London, pp. 901–901.

EFNARC (European Federation for Specialist Construction and Concrete Systems) (2005) *The European Guidelines for Self-Compacting Concrete, Specification, Production and Use*. EFCA/EFNARC.

Fedroff D and Frosch RJ (2004) Formwork for self consolidating concrete. *Concrete International* **26(10)**: 32–37.

Galeota D, Giammatteo MM, Gregori A, Marino R and Shah SP (2007) Formwork pressure of self-compacting concrete. *Proceedings of the 5th International RILEM Symposium on Self-Compacting Concrete, Ghent, Belgium* (De Shutter G and Boel V (eds)). PRO 54, RILEM Publicatons SARL, Bagneux, France, pp. 1059–1064.

Gardner NJ (1986) Concrete pressure on formwork. *Proceedings of the Institution of Civil Engineers, Part 1* **80(1)**: 145–159.

Habgood MG (1982) *Site Measurements of Concrete Pressure Recorded during the Period March 1980 to June 1981*. Cement and Concrete Association, DN/2058.

Harrison TA (1983) *Pressure on Vertical Formwork when Concrete is Placed in Wide Sections*. Cement and Concrete Association, Research Report 22.

Holton I (2004) *Self-compacting Concrete*. Building Research Establishment, Garston, Watford, BRE IP3/04.

Khayat KH, Assaad JJ, Mesbah H and Lessard M (2005) Effect of section width and casting rate on variations of formwork pressure. *Material and Structures* **38(1)**: 73–78.

Leemann A and Hoffmann C (2003) Pressure of self-compacting concrete on the formwork. *Proceedings of the 3rd International RILEM Symposium on Self-Compacting Concrete, Reykjavic, Iceland* (Wallevik O and Nielsson I (eds)). PRO 33, RILEM Publications SARL, Bagneaux, France, pp. 288–295.

Proske T and Graubner CA (2002) Self-compacting concrete – pressure on formwork and ability to de-aerate. *Annual Journal on Concrete and Concrete Structures*, Darmstadt Concrete 17 (Technical University of Darmstadt, Germany).

Tejeda-Dominguez F, Lange DA and D'Ambrosia MD (2005) Formwork pressure of self-consolidating concrete in tall wall field applications. *Transportation Research Record: Journal of the Transportation Research Board* **1914**: 1–7.

Teychenne D, Nicholls JC, Franklin RE and Hobbs DW (1997) *Design of Normal Concrete Mixes*. Building Research Establishment, Garston, Watford, UK.

Vanhove Y, Djelal C and Magnin A (2004) Prediction of the lateral pressure exerted by self-compacting concrete on formwork. *Magazine of Concrete Research* **56(1)**: 55–62.

ICE Publishing

Dhir and Lye
ISBN 978-0-7277-6461-4
https://doi.org/10.1680/icetscc.64614.447

Chapter 20
UK contractors' views on self-compacting concrete in construction

David Rich MEng
Research Engineer, School of Civil and Building Engineering, Loughborough University, UK

Jacqui Glass BA(Hons), PhD, DipArch, CertHE, DipBRS, FHEA
Senior Lecturer in Architecture and Sustainable Construction, School of Civil and Building Engineering, Loughborough University, UK

Alistair G. F. Gibb BSc, PhD, CEng, MICE, MCIOB
Professor of Construction Engineering Management, School of Civil and Building Engineering, Loughborough University, UK

Chris Goodier BEng, PhD, pgCert, MCIOB, MICT, FHEA
Lecturer in Structures and Materials, School of Civil and Building Engineering, Loughborough University, UK

Self-compacting concrete is claimed to offer faster construction, safer sites and more consistent concrete quality, but little corroborative research data exist on performance advantages, particularly in comparison with traditional construction. Industry opinions also appear to be divided. For these reasons, an extensive interview programme was undertaken with UK contractors – from large national concrete frame contractors to small, locally based housebuilders – to assess whether benefits were being achieved and to try to understand the reasons why self-compacting concrete is, or is not, being used. The 48 participants reported that decisions on the suitability of self-compacting concrete were inherently complex and, if selected, there were challenges in understanding 'how' construction should be planned and managed to accommodate the use of self-compacting concrete and to fully utilise its advantages. The findings identify the need for a step change in the industry's perception of self-compacting concrete, such that it should be considered as a construction method, not simply as a material.

1. Introduction

Despite its traditional culture, innovations can be found in the construction industry, with a select few acknowledged as enhancing construction processes. Self-compacting concrete (SCC) is one such innovation due to its effects on the construction process. While still regarded in the industry as a recent innovation, it has been available in the UK for more than 10 years, with the technology being available even before the creation of the term SCC.

Simply explained, SCC is a concrete that requires no external energy input (Concrete Society/BRE, 2005; Damtoft *et al.*, 2008; Holton, 2003) in order to achieve full compaction, which is vital in achieving robust and durable concrete. For a concrete to be considered a true SCC it must possess three distinct properties; resistance to segregation, which is self-explanatory, flowing ability and passing ability. Flowing ability refers to the concrete's ability, under its own self-weight, to flow and completely fill the form into which it is placed. When used in applications consisting of complex shapes or with dense reinforcement there is a need for the concrete to have greater passing ability through and around obstructions without causing blockages, which can result in internal voids (Gaimster and Gibbs, 2001; Goodier, 2003; RILEM Technical Committee, 2006). Together these properties are particularly helpful.

However, since its inception and commercialisation SCC has remained somewhat under-used. Extensive research has been carried out into the material's structural and physical performance criteria (De Schutter *et al.*, 2008; Khayat, 1999; Okamura and Ouchi, 2003; Shobha *et al.*, 2006), but research on the effects of SCC on the construction process has generally been more subjective and indirect (Damtoft *et al.*, 2008; Gaimster and Foord, 2000; Goodier, 2003; Henderson, 2000; Walraven, 2003).

In response to this, a project was established with the objective of identifying the implications that SCC can have on construction, while providing information and tools for exemplar use. The results presented here form an integral part of this research by considering the views held by a range of contractors within the UK construction industry.

Due to the low uptake and lack of information on practical applications, the aim of this research was to clarify the views and perceptions of contractors and to understand the effects of SCC on construction. Research was directed at establishing the reasons and drivers for using SCC and whether these align with views and findings within academic and industry literature. Other aspects considered were the decision-making process surrounding new methods or innovations and planning of the construction phase. Through these results a more fulsome and up-to-date understanding of the industry's views on SCC were obtained; an important research study that has not been replicated previously. This research forms part of a wider programme of research which is focused on assessing the implications of SCC in construction, and will move on to establish direct, quantifiable results linked to its application.

2. Background

Self-compacting concrete is seen as a specialist material (Clear, 2006; Holton, 2003) but one that is gaining more recognition within a wider range of construction applications (Concrete Society/BRE, 2005). Some view SCC (Figure 1) as a material whose use is limited to situations where it can perform as a problem solver (Clear, 2006; Okamura and Ouchi, 2003) or as an architectural tool due to the high quality finishes available (Grimes, 2005). Several factors identified previously as drivers for the uptake of SCC are improved durability, versatility, skilled labour shortages and improvements in performance. Durability and versatility are enhanced by the physical properties of the material; the flowable nature enabling greater confidence in formwork-filling and final quality (Grimes, 2005; Walraven, 2003); these in turn result in more uniform and dense elements (Skarendahl and RILEM, 2003), subsequently improving the resistance to chloride diffusion, sulfate attack and freeze–thaw problems

Figure 1 Self-compacting concrete in use (courtesy of Lafarge)

(De Schutter *et al.*, 2008). The ability of SCC to be placed without compaction has removed the need for skilled labour input and decreased impact on operatives (Damtoft *et al.*, 2008; Concrete Society/BRE, 2005). Financially, the material can be cost effective if an holistic calculation is made, taking into account the ability of SCC to reduce labour, remove plant, reduce remedials and, as a result of improved rates of casting, to reduce project time (Gaimster and Foord, 2000; Goodier 2003).

Although in situ use of SCC is not widespread on site, according to Holton (2004), 60% of the structural precast sector employed SCC in 2004, increasing to over 75% in 2008 (Goodier, 2008). The key to this uptake is the result of all operations being in one place, with the entire batching and casting operation under total control of a single organisation (Skarendahl and RILEM, 2003); as such, any changes are easy to manage and benefits are easier to measure and obtain.

Application has not been replicated to this extent in in situ applications due to the gearing of site practices towards traditional vibrated concrete (Okamura and Ouchi, 2003). If SCC use is to increase, change is required in the early project stages, conceptual and preliminary design and also specification (Concrete Society/BRE, 2005). A number of publications has been made available to address the aforementioned issues, not least *The European Guidelines for Self-Compacting Concrete* by the Self-Compacting Concrete European Project Group (2005) and the joint report (TR62) into SCC (Concrete Society/BRE, 2005).

Sustainability is a major concern within the construction industry and therefore needs to be considered with regard to SCC. The increased cement volumes in SCC suggest an increased environmental impact (Gaimster and Gibbs, 2001) due to the carbon dioxide emissions during production. However, SCC can improve productivity, improve the work environment, reduce repair and replacement, and as such the overall environmental impact of the project is reduced (Damtoft *et al.*, 2008).

However, based on current literature the case for SCC remains unclear. Recent literature on the application of SCC is over 5 years old and no recent work has been undertaken to revisit and re-research the case for SCC in the UK industry. Some of the key literature to date [IP3/04 (BRE/Holton, 2004) and BRE (Concrete Society/BRE, 2005)] that has been used to further the case for SCC can, in part, be said to lack validation; for example canvassing opinion across the industry. It is this lack of wider consultation, together with the age and nature of available information on SCC that have served as key drivers for the research presented here.

3. Methodology

This research aimed to establish current industry perceptions, opinions and ideas on SCC, including the following topics.

(*a*) The perceptions that are held on SCC as a construction option and material.
(*b*) How the decision is made to use SCC and/or conventional concrete and any other construction innovations.
(*c*) How the decision-making process surrounding material and method can be improved.
(*d*) The influence of the timing of construction decisions on the choice of material and method.
(*e*) The rationale for such decisions and the identification of those responsible.

Interviews were chosen as the method of data capture due to their flexibility and their capability to derive a large amount of information, when compared with questionnaires. It was the lack of ability to interrogate and expand on responses combined with an inability to encourage contractors who were less enthusiastic about SCC to participate that supported the selection of interviews. In the process of designing the interview protocol, semi-structured interviews were identified as most appropriate. These provide a basis for transferable questions, while retaining the option to explore responses and redirect questioning (Bryman, 2004).

Distinct approaches were adopted in order to identify potential interviewees. Initial participants were members of the Construct organisation, representing large nationally-operating contractors (Construct is an association of UK organisations looking to improve the efficiency of in situ concrete frame construction: see www.construct.org.uk). This also includes concrete frame contractors operating under the Specialist Concrete Contractor (SpeCC) scheme. The SpeCC scheme was devised to raise standards within the concrete frame industry (Figure 2),

Figure 2 First major UK concrete frame constructed with SCC (courtesy of Lafarge)

and acceptance is dictated by adherence to minimum standards and annual audits to ensure compliance. Initial contact was made via letters to technical directors or their equivalent, explaining the research and giving an indicative set of questions. Subsequent to this, follow-up telephone calls were made to arrange in-depth, face-to-face or telephone interviews.

To obtain a broader sample from the contracting industry, a further group of small locally based UK contractors was also interviewed (Figure 3). These were drawn from the UK-customer database of a global construction materials supplier, through which it was possible to obtain a direct link with the contractors. These firms included general builders, house builders, ground workers, precasters, concrete frame contractors, screeders and pumping contractors (pumping contractors are contractors typically who only provide pumping services but in these cases have expanded their business to include concrete placement). Those interviewed ranged from on-site general operatives to directors and owners of said contactors.

Figure 3 Typical site of a group 2 participant (courtesy of Lafarge)

Construct's aim is to 'improve the efficiency of building in situ concrete frames and associated structures', which may be reflected in the willingness of the SpeCC members to participate. However, this data should be treated with care as, by their nature, such organisations are inclined to be proactive in the development of new products and the transfer of information, which may not be representative of the wider construction industry. The second group of contractors were selected and categorised based on their being

(*a*) regular users of SCC
(*b*) occasional users of SCC
(*c*) former users or non-users of SCC.

This approach provided a range of balanced and representative views. Within these groups a potential for bias exists (particularly group (*a*)), in that interviewees happen to be more interested in SCC and construction innovation than the wider industry. Therefore, steps have been taken to mitigate this through interviews with non-users, as can be seen in Figure 4. Group 1 (Construct members) were selected to provide information on larger-scale projects and group 2 (other contractors) to address the smaller-scale and less complex projects, which currently represent the majority of applications of SCC within the UK.

In group 1, out of 22 companies two Construct members declined to participate and ten did not respond. The remaining ten contractors took part fully, providing a response rate of 45%. In group 2, 38 participants were interviewed, out of 60 approached (63% of the sample).

The combination of participants provided an overall response rate of 59%; namely 48 participants, representing a range of contractors and specialist firms, the breakdown of contractors is shown in Figure 5.

Figure 4 Breakdown of group 2 user experience

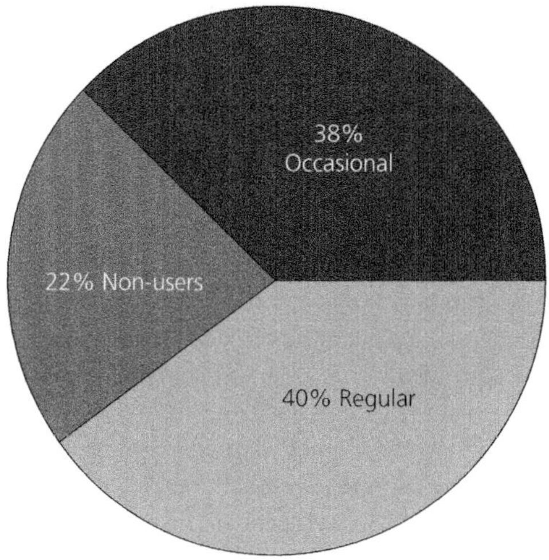

Figure 5 Breakdown of interviewees by activity

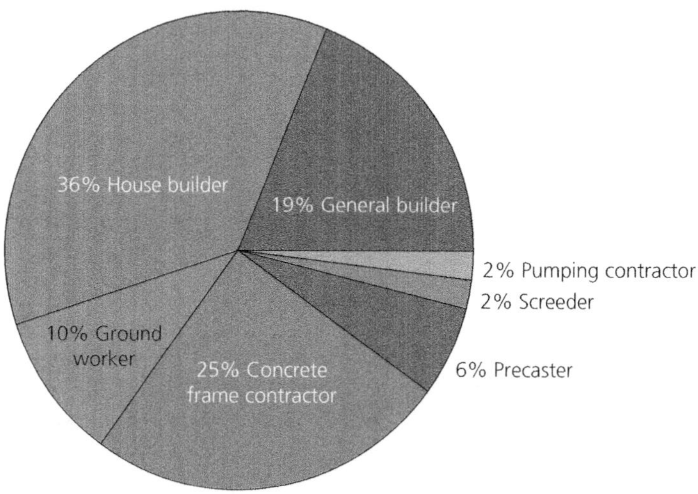

Most of the respondents were located in England, with three contractors each in Scotland and Wales (in the areas surrounding Glasgow and Cardiff). Within England a significant proportion of participants was based in the north east (27%), south and east Midlands (23%) and London and the south east (10%). The Construct participants (20%), although having headquarters mainly in the London and the south east, typically operate nationally and so are not limited to a particular geographical area.

4. Results and data analysis

This section presents an overview of the results of the interview programme and includes verbatim quotes where appropriate, with the respondent's role indicated in brackets after each quote.

4.1 Reasons for using SCC

There seems to be a clear distinction between knowledge and experience of SCC – experience being based on practical use and knowledge based on one's impression of the material. Although most study participants (83%) had some previous experience of SCCs, the range of applications was limited. This limited use suggests either a lack of specific or universal applications for SCC or a lack in understanding of its potential. Of the drivers and applications cited for use, two were most prevalent – as a 'problem solver' and for housing slabs. SCC was said to be able to 'resolve and remove problems' and enable 'risk reduction' (concrete frame contractor) according to 23% of participants. The majority of use by concrete frame contractors can be described as reactionary; that is, when conventional concrete could not achieve the desired results, typically where there is congested reinforcement, poor access, site restrictions or a need for a high-quality finish. Slab applications (Figure 6) accounted for 53% of previous use; principally with house builders, general builders and groundworkers. Other applications are presented in Figures 7 to 9.

There was no coherent overarching view of SCC, with participants stating 'it is difficult to see where you can actually make savings' (housebuilder) or that cost differences 'can be returned through time saved, reduced labour and removal of powerfloating' (housebuilder). No single ideal opportunity for the material was presented.

Figure 6 The most popular application of SCC was for slabs (courtesy of Lafarge)

Figure 7 Simplified placing, dappling to a finish, reduces the impact on operatives (courtesy of Lafarge)

Self-compacting concrete was generally viewed as a positive option but contractors were discouraged by certain problems. It is in response to this that some said the material could only be used if specified – in other words, many contractors did not want to take the responsibility for its selection.

It is interesting to establish the perceptions of SCC, both positive and negative. Participants identified that SCC could reduce 'effort levels in placement' (general builder) and enable 'faster and more accurate' (general builder) construction while mitigating 'workmanship issues' (concrete frame contractor). These comments corroborate existing literature (Damtoft *et al.*, 2008; Gaimster and Foord, 2000; Goodier 2003; Henderson, 2000). Following these positive statements, reflection of knowledge in literature can also be said to be present regarding weaknesses. However, in certain cases these have been contradicted, one such example is that 'labour skill changes are not correct' and that a 'traditional concrete gang'

Figure 8 Placement of SCC into a confined space removes the difficult task of compacting (courtesy of Lafarge)

(concrete frame contractor) would still need to utilised, where literature states skill levels can be reduced (Goodier *et al.*, 2002).

This however must be put into context; if the material is only to be used on a single application then this action is understandable. However, the issue in question is the use of the material as part of a larger programme of works; in other words, is a significant reduction in labour only viable once SCC is used to a large extent across a whole project?

Nearly 40% of participants maintained that 'cost is prohibitive' to the use of SCC and is the 'main problem with the material' (house builder). 'Cost' was often used to describe the first cost or tender price; this interpretation fails to identify savings in other parts in the construction process, which result from using SCC. Typical project costs were said to be approximately '15% concrete, 15% steel, 33% labour and 33% overheads' (concrete frame contractor) which, if no clear value can be attached to SCC, presents a significant barrier.

Figure 9 Reflection of timber formwork material on final finished face of an SCC wall (courtesy of Lafarge)

So, to negate a focus on cost, the concept of value was put forward to interviewees, where value was described as the impact of SCC on a whole project. Twenty-one percent responded that they could see no value in SCC in construction, suggesting they saw SCC as a material rather than a method (necessitating a different approach to planning and implementation). Further to this its inclusion was said to be detrimental to the acquisition of work, due to increased tender prices. While the concept of value was clearly prevalent in the industry, it has yet to be integrated into projects with 'more talking about value than actually considering it' (concrete frame contractor). In precast applications it was said that companies 'cannot justify savings on a balance sheet, but they do exist'. However, overall within the industry 'cost is king' (concrete frame contractor) and in this respect it was said the cheapest option will always be chosen, regardless of market buoyancy or economic downturn.

4.2 Decision making

The decision to employ SCC in a project appears to originate from three circumstances, with the first two being most prominent.

(*a*) A strategic change from conventional methods as part of a balanced assessment of the material and its effects on construction.
(*b*) Reactionary, in order to address a specific issue or problem.
(*c*) Specification of the material or being taken on board as a preconceived construction option.

The strategic decision to use SCC was referred to by 14 contractors, of which five found that, on balance, SCC in a specific application could add value. It is clear that use and value need to be judged on an application by application basis. Ten stated that SCC was used as a reaction to an issue or problem, with one citing that the only viable solution on these occasions was SCC. Only two participants had experience of being required to use SCC, with seven saying specification was the only route to application.

> The material is rarely in a specification, (there have only been only a) few cases of specification. Used only on jobs when a problem occurs, application-led rather than a conscious decision. Need to balance risk versus reward, rework potential associated with conventional concrete in an application; strike a balance (between materials). (concrete frame contractor)

Whether SCC is chosen as a reactionary solution, a strategic change or a preconceived option, the main decision-makers were reported to be the client, architect and engineer and contractor – so it is they who need to be convinced about its adoption. Without SCC being specified, introduction occurs at site level, where approval is then sought from senior personnel in the project management structure. This can present problems with 'educating the client or engineer on the material, the role it can have in construction and the reasons for inclusion' (concrete frame contractor). The rationale that SCC was seen as a problem solver is more straightforward than in any other application. 'Narrow column design with high levels of reinforcement raised the potential problem of limited poker access' (house builder), thus SCC is chosen when or where a conventional concrete cannot be used.

Selection based on a considered change from conventional concrete appeared to be grounded in decisions driven by a balanced assessment of construction effects, considering not only costs, but also changes to methods and practices.

> (The) material was selected to speed up construction times; (we) undertook a cost comparison with conventional, combination of labour and material cost balanced against SCC. (Its) selection was based on time, effort, labour and finish quality. (groundworker).

The selection of SCC in these cases is on a job-by-job basis showing that the material is not a direct replacement for conventional concrete. Selection in this manner requires an understanding of the design process. Indeed, it was clearly stated by one concrete frame contractor that material choice is second to construction method when designing or developing a project, with the best construction option selected first.

Construction teams (who typically retain most knowledge of SCC and methods of innovative construction) are involved once a design is completed. It is at this point where they can 'make suggestions on materials', but generally can only provide 'a best price and advice' (concrete frame contractor) on construction, creating an inbuilt barrier to innovation and SCC use.

4.3 Use of SCC in precast

A general indication of views from the precast industry was provided by three precast plant operators. The perception and use of SCC has changed significantly over the last 10 years when it was 'not possible to achieve prescribed results' and the 'additional cost made the material unviable' (precaster). All three manufacturers responded positively to SCC, stating it was now possible to realise 'savings in labour, time and plant' (precaster), with another reporting it is 'possible to reduce placing time from 3 h to 45 min or 1 h' (precaster). However, it was said that these benefits were difficult to quantify and reflect financially and in older factories a significant plant overhaul is required to improve standards to accept SCC.

4.4 Sustainability

Questions on sustainability were directed principally to the group 1 (Construct) participants, based on perceptions regarding membership and a desire to improve construction. Only two of the ten respondents were actively pursuing sustainability improvements. Other contributors identified typical industry characteristics as barriers, such as a resistance to change and the desire to reduce costs, based on the assumption that sustainable approaches were inherently more expensive.

With the industry requiring 'work to be carried out on a lowest cost basis' (concrete frame contractor), integration of 'green' initiatives were thought to increase tender prices and reduce work. As a result four of the ten participants stated that the client must drive sustainability agendas.

Throughout the Construct interviews, a range of views was presented. SCC was seen as more sustainable because the 'need for additional finishes' (concrete frame contractor) had been removed, however it was also said to be worse 'as more carbon dioxide is generated' (concrete frame contractor) due to increased cement content. The most viable response is that 'not enough evidence or detail is available' and one 'still has the same concerns as with conventional concrete' (concrete frame contractor).

As none could offer a coherent account of the sustainability credentials of SCC, it is clear that either there is a problem with knowledge transfer or a lack of research into the subject.

4.5 Implementation

Time was identified as the overriding factor in the implementation of new materials or methods. It was said that the earlier a change is introduced the easier it is to assess its viability, where viability can be judged to be a positive effect on a project. In conjunction with time, project flexibility was cited to be essential in enabling design or construction methods to be altered.

Interviewees said that SCC had 'made it possible to reduce both time and manpower' (general builder), remove construction activities, but also 'needs to be judged on its effects on the critical path' (concrete frame contractor). Improvements to the critical path presented the opportunity to make dramatic savings in project duration and, in turn, overheads.

Change, on a large scale, required approval by the client, architect or engineer. SCC use was thought to be driven by contractors typically, who needed to influence and educate those higher in the project hierarchy, the engineer, architect and client. A lead time of '2–3 months rather than 4 weeks' (concrete frame contractor) was required as late involvement would result in an inability to develop and introduce change. Without early consideration of any new innovation or material the probability of inclusion is slim, unless it is used to address a specific issue.

The lack of an upfront opportunity could be counteracted by review processes and post-project appraisals. However, when participants were challenged on appraisals, there was a mixed response, with appraisals carried out in an ad hoc manner. On the majority of occasions when they were carried out 'there have been problems' and are 'typically focused on methods' (concrete frame contractor) which can leave SCC unconsidered.

5. Discussion

The results from the interviews present complementary and contrasting views on SCC when compared with the existing literature. It remains clear that there is a lack of quantifiable information on the use and effect of SCC on construction. The literature has highlighted two distinct circumstances for the application of SCC. These were as a problem solver (Clear, 2006; Concrete Society/BRE, 2005; Okamura and Ouchi, 2003) or as an architectural tool (Grimes, 2005), but through this research it has been possible to clarify three circumstances for use, as stated within the results. SCC is still widely used as a problem solver which signifies that perceptions have not dramatically changed.

Strengths and weaknesses identified in the literature relating to labour, quality and workmanship (Damtoft *et al.*, 2008; De Schutter *et al.*, 2008) have, to an extent, been confirmed by responses. There remains a contradiction on cost and the impact of SCC on construction costs, although it is stated in the literature that SCC can reduce total project costs (Gaimster and Foord, 2000; Goodier, 2003). This is contradicted in part by the present findings as its price has been cited as prohibitive; nevertheless a proportion still used the material after a balanced assessment, thus viewing SCC as just a material.

As a method, SCC takes into account wider implications and in this respect value can be considered rather than cost. For example it is useful to borrow a concept from preconstruction planning regarding site and ground investigations, where an initial capital outlay can have a dramatic effect on reducing unforeseen problems and in turn unforeseen expenditure. It is conceivable to apply this concept to above-ground works to develop construction processes which can be adapted for SCCs and new innovations with the long-term result being a net reduction in costs.

Precast concrete, according to the literature, has seen a considerable increase in use of SCC that has not been reflected in in situ applications. Although from a very small sample, it did seem that there is an increased willingness to use SCC, with all precast participants having previously used or be currently using SCC and looking to convert some or all of their facilities. All interviewees noted that SCC has improved, or could improve, their processes but they have not been able to identify the exact 'value', monetary or otherwise, to their business. The inability to quantify this directly is replicated in the in situ industry where there is little robust information available.

On sustainability there is no clear position in either the literature or the industry; several conflicting and contrasting views have been put forward.

Construction is geared towards traditional concretes and traditional methods. Early consideration was required to enable uptake of SCC, however this requires approval by the client, architect, engineer and higher project teams. So, education was stated to be key in changing the approaches taken by companies and to overcome conservatism (perceived to exist at higher levels in project teams). Conservatism could be interpreted as site teams not being willing to fully understand SCC themselves, pushing decisions upwards and removing their risk. Where change was embraced, SCC had been perceived as a 'method', not just as a material. To ease this situation the industry 'needs to increase the knowledge of (SCC), how it works and how it can be designed into construction projects' (groundworker). For example, a balanced assessment or study of site-based costs, technical information on the application of SCC, and guidance on how to adapt construction. A curious example was also provided in respect of current publicised guidelines and aids [e.g. *The European Guidelines for Self-Compacting Concrete* (Self–Compacting Concrete European Project Group, 2005) and TR62 (Concrete Society/BRE, 2005]). The interviewees displayed an apparent lack of awareness of these documents, perhaps suggesting that they are either not relevant or unknown, but on the other hand the documents may be so well-known that they are a given and so remain unmentioned. This may merit further study.

6. Conclusion

In the 5 or 6 years since the last significant work into SCC, regarding its application in the industry, it is clear that there has been little progress and there remains a lack of unanimous or general consensus on SCC and its role within construction processes. There has been little research on its effects on mainstream construction. SCC has been described as a viable material, that offers distinct benefits to construction projects, or hinders operation in a competitive market place but this is dictated by scale. Its position still therefore remains unclear and requires further research.

This research has gone part way to address the research questions of how, when and where to incorporate SCC into a project but a major research question, with respect to decision rationale, process planning and timing of construction remains. That said the historical structure and organisation of the construction industry were cited by industry as the basis for current management structures, project control and project implementation. All of these aspects influence the use (or not) of SCC, particularly the time at which contractors become involved in projects, by those who are responsible for decisions and the fiscal arrangement of project procurement.

Processes surrounding construction decisions (i.e. how a project is constructed) are focused (and correctly so) on selecting the 'best' method, with material choice usually a secondary consideration. SCC is currently considered as a material, which does not encourage the contractor to consider its wider effects and benefits. If SCC is considered as a method (a distinctive step forward from previous studies), there is recognition that SCC needs and requires greater planning and understanding, in order for the material and its associated benefits to attain their full potential. Identification as a method requires that the complete construction phase is geared towards SCC and adapted to suit its distinctive properties; it is a development from previous concepts that SCC is no longer part of the process (IP3/04: BRE/Holton, 2004) but is the process itself. For its uptake to grow, it is absolutely essential that the material is viewed and considered in this regard, rather than on a simple like-for-like basis with other materials.

7. Recommendations

The following list presents industry recommendations.

(*a*) Consider SCC as a construction method rather than as a material and interpret its implications on the wider construction process.
(*b*) Introduce contractors into the early (design) stages of projects to increase collaboration efforts and enable the uptake of new construction methods and innovations such as SCC.
(*c*) Increase upfront investment in projects, at the preliminary/conceptual stage to enable additional value to be sought through the assessment of innovations and new methods of construction, prior to the commencement of works.

The following list presents research recommendations.

(*a*) Develop and establish guidelines on how SCC, as a 'construction option and/or method', can be integrated into construction projects and determine the changes or adaptations needed in current construction practice.
(*b*) Ascertain the effect that SCC can have on the construction process by quantifying benefits and savings.
(*c*) Interpret and understand the roles and requirements of key decision-makers within the construction chain. Establish industry views on risk to develop a strategy for contractors for implementing SCC into construction projects.

Achievement and adherence to these recommendations would enable the integration of SCC more widely into mainstream construction, and have broader potential to ease the development and inclusion of other innovative construction methodologies. The process of addressing these recommendations would help overcome several of the mainstream issues and barriers to SCC and support the further development of SCC in construction.

Acknowledgements

The authors would like to acknowledge the contribution of Lafarge Aggregates Ltd, whose assistance eased the acquisition of data, along with all the participants without whom the research would not have been possible. This work was carried out as part of an Engineering Doctorate, funded by the EPSRC, via the Centre for Innovative and Collaborative Engineering.

REFERENCES

BRE (Building Research Establishment)/Holton I (2004) Self-Compacting Concrete. IHS BRE, IP3/04. BRE/Holton, Watford, UK.

Bryman A (2004) *Social Research Methods*, 2nd edn. Oxford University Press, Oxford, UK.

Clear CA (2006) *Fact Sheet 5: Self-Compacting Concrete (SCC)*. British Cement Association, Camberley, UK.

Concrete Society/Building Research Establishment (BRE) (2005) *Self-compacting Concrete: a Review,* Report of a Joint Working Group. The Concrete Society and BRE, Camberley, Surrey, UK, Technical Report No. 62.

Damtoft JS, Lukasik J, Herfort D, Sorrentino D and Gartner EM (2008) Sustainable development and climate change initiatives. *Cement and Concrete Research* **38(2)**: 115–127.

De Schutter G, Bartos P, Domone P and Gibbs J (2008) *Self-Compacting Concrete.* Taylor and Francis, Dunbeath, UK.

Gaimster R and Foord C (2000) Self-compacting concrete. *Concrete* **34(4)**: 23–25.

Gaimster R and Gibbs J (2001) Current practice sheet 123: Self-compacting concrete. Part 1: The material and its properties. *Concrete* **35(7)**: 32–34.

Goodier CI (2003) Development of self-compacting concrete. *Proceedings of the Institution of Civil Engineers – Structures and Buildings* **156(4)**: 405.

Goodier CI (2008) Skills and training in the UK precast concrete manufacturing sector. *Construction Information Quarterly* **10(1)**: 405–414.

Goodier C, Holton I and Sparksman G (2002) *Self-Compacting Concrete: Interim Report Detailing Industry Perceptions of the Advantages and Disadvantages of SCC.* Department of Trade and Industry and BRE, Watford, UK.

Grimes P (2005) A vision for Lincoln. *Concrete* **39(1)**: 10–12.

Henderson N (2000) Self-compacting concrete at Millennium Point. *Concrete* **34(4)**: 26–27.

Holton I (2003) *Interim Report Detailing the Effects of SCC on the Construction Process.* Department of Trade and Industry and BRE, Watford, UK.

Holton I (2004) *Update on Research and Development taking place in Self-compacting Concrete.* Department of Trade and Industry and BRE, Watford, UK.

Khayat KH (1999) Workability, testing, and performance of self-consolidating concrete. *ACI Materials Journal* **96(3)**: 346–353.

Okamura H and Ouchi M (2003) Self-compacting concrete. *Journal of Advanced Concrete Technology* **1(1)**: 5–15.

RILEM Technical Committee (2006) *188-CSC: Casting of Self Compacting Concrete.* RILEM Publications, Bagneux, France.

Self-Compacting Concrete European Project Group (2005) *The European Guidelines for Self-Compacting Concrete.* The European Project Group (BIBM, CEMBUREAU, ERMCO, EFCA, EFNARC). See www.efnarc.org/pdf/SCCguidelinesmay2005.pdf (accessed 06/10/2011).

Shobha M, Harish Mohan D and Raju PSN (2006) Aggregate size and behaviours of self-compacting concrete. *Proceedings of the Institution of Civil Engineers – Construction Materials* **159(2)**: 147–152.

Skarendahl A RILEM (2003) The present – the future. *Rilem Proceedings Pro* **33**: 6–14.

Walraven J (2003) Structural aspects of self-compacting concrete. In *Proceeding of the 3rd International RILEM Symposium on Self-Compacting Concrete, Reykjavik, Iceland, 17–20 August 2003.* (Wallevik O and Nielsson I (eds)). PRO 33, RILEM Publication, Bagneux, France, pp. 15–22.

ice Publishing

Dhir and Lye
ISBN 978-0-7277-6461-4
https://doi.org/10.1680/icetscc.64614.465
ICE Publishing: All rights reserved

Index

Page numbers followed by f and t indicate figures and tables, respectively.